Lecture Notes in Artificial Intelligence   8776

Subseries of Lecture Notes in Computer Science

LNAI Series Editors

Randy Goebel
  University of Alberta, Edmonton, Canada
Yuzuru Tanaka
  Hokkaido University, Sapporo, Japan
Wolfgang Wahlster
  DFKI and Saarland University, Saarbrücken, Germany

LNAI Founding Series Editor

Joerg Siekmann
  DFKI and Saarland University, Saarbrücken, Germany

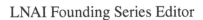

T0236188

Peter Auer   Alexander Clark
Thomas Zeugmann   Sandra Zilles (Eds.)

Algorithmic Learning Theory

25th International Conference, ALT 2014
Bled, Slovenia, October 8-10, 2014
Proceedings

 Springer

Volume Editors

Peter Auer
Montanuniversität Leoben, Austria
E-mail: auer@unileoben.ac.at

Alexander Clark
King's College London, UK
Department of Philosophy
E-mail: alexander.clark@kcl.ac.uk

Thomas Zeugmann
Hokkaido University
Division of Computer Science
Sapporo, Japan
E-mail: thomas@ist.hokudai.ac.jp

Sandra Zilles
University of Regina
Department of Computer Science
Regina, SK, Canada
E-mail: zilles@cs.uregina.ca

ISSN 0302-9743   e-ISSN 1611-3349
ISBN 978-3-319-11661-7   e-ISBN 978-3-319-11662-4
DOI 10.1007/978-3-319-11662-4
Springer Cham Heidelberg New York Dordrecht London

Library of Congress Control Number: 2014948640

LNCS Sublibrary: SL 7 – Artificial Intelligence

Typesetting: Camera-ready by author, data conversion by Scientific Publishing Services, Chennai, India

Printed on acid-free paper

Springer is part of Springer Science+Business Media (www.springer.com)

Preface

This volume contains the papers presented at the 25th International Conference on Algorithmic Learning Theory (ALT 2014), which was held in Bled, Slovenia, during October 8–10, 2014. ALT 2014 was co-located with the 17th International Conference on Discovery Science (DS 2014). The technical program of ALT 2014 had 4 invited talks (presented jointly to both ALT 2014 and DS 2014) and 21 papers selected from 50 submissions by the ALT Program Committee.

ALT 2014 took place in the hotel Golf in a beautiful park full of old trees in the very heart of Bled. It provided a stimulating interdisciplinary forum to discuss the theoretical foundations of machine learning as well as their relevance to practical applications.

ALT is dedicated to the theoretical foundations of machine learning and provides a forum for high-quality talks and scientific interaction in areas such as reinforcement learning, inductive inference and grammatical inference, learning from queries, active learning, probably approximate correct learning, online learning, bandit theory, statistical learning theory, Bayesian and stochastic learning, un-supervised or semi-supervised learning, clustering, universal prediction, stochastic optimization, high dimensional and non-parametric inference, information-based methods, decision tree methods, kernel-based methods, graph methods and/or manifold-based methods, sample complexity, complexity of learning, privacy preserving learning, learning based on Kolmogorov complexity, new learning models, and applications of algorithmic learning theory.

The present volume of LNAI contains the text of the 21 papers presented at ALT 2014, as well as the texts/abstracts of the invited talks:

- Zoubin Ghahramani (University of Cambridge, Cambridge, UK), "Building an Automated Statistician" (joint invited speaker for ALT 2014 and DS 2014)
- Luc Devroye (McGill University, Montreal, Canada), "Cellular Tree Classifiers" (invited speaker for ALT 2014),
- Eyke Hüllermeier (Universität Paderborn, Germany), "A Survey of Preference-Based Online Learning with Bandit Algorithms" (tutorial speaker for ALT 2014),
- Anuška Ferligoj (University of Ljubljana, Slovenia). "Social Network Analysis" (tutorial speaker for DS 2014)

Since 1999, ALT has been awarding the E. M. Gold Award for the most outstanding student contribution. This year, the award was given to Hasan Abasi and Ali Z. Abdi for their paper "Learning Boolean Halfspaces with Small Weights from Membership Queries" co-authored by Nader H. Bshouty.

ALT 2014 was the 25th meeting in the ALT conference series, established in Japan in 1990. The ALT series is supervised by its Steering Committee: Peter

Auer (University of Leoben, Austria), Shai Ben-David (University of Waterloo, Canada), Nader H. Bshouty (Technion - Israel Institute of Technology, Israel), Alexander Clark (King's College London, UK), Marcus Hutter (Australian National University, Canberra, Australia), Jyrki Kivinen (University of Helsinki, Finland), Frank Stephan (National University of Singapore, Republic of Singapore), Gilles Stoltz (Ecole normale supérieure, Paris, France), Csaba Szepesvári (University of Alberta, Edmonton, Canada), Eiji Takimoto (Kyushu University, Fukuoka, Japan), György Turán (University of Illinois at Chicago, USA, and University of Szeged, Hungary), Akihiro Yamamoto (Kyoto University, Japan), Thomas Zeugmann (Chair, Hokkaido University, Sapporo, Japan), and Sandra Zilles (Co-chair, University of Regina, Saskatchewan, Canada).

We thank various people and institutions who contributed to the success of the conference. Most importantly, we would like to thank the authors for contributing and presenting their work at the conference. Without their contribution this conference would not have been possible. We would like to thank the Office of Naval Research Global for the generous financial support for the conference ALT 2014 provided under ONRG GRANT N62909-14-1-C195.

ALT 2014 and DS 2014 were organized by the Jožef Stefan Institute (JSI) and the University of Ljubljana. We are very grateful to the Department of Knowledge Technologies (and the project MAESTRA) at JSI for sponsoring the conferences and providing administrative support. In particular, we thank the local arrangement chair, Mili Bauer, and her team, Tina Anžič, Nikola Simidjievski, and Jurica Levatić from JSI for their efforts in organizing the two conferences.

We are grateful for the collaboration with the conference series Discovery Science. In particular we would like to thank the general chair of DS 2014 and ALT 2014 Ljupčo Todorovski and the DS 2014 Program Committee chairs Sašo Džeroski, Dragi Kocev, and Panče Panov.

We are also grateful to EasyChair, the excellent conference management system, which was used for putting together the program for ALT 2014. EasyChair was developed mainly by Andrei Voronkov and is hosted at the University of Manchester. The system is cost-free.

We are grateful to the members of the Program Committee for ALT 2014 and the subreferees for their hard work in selecting a good program for ALT 2014. Last but not the least, we thank Springer for their support in preparing and publishing this volume in the Lecture Notes in Artificial Intelligence series.

August 2014   Peter Auer
                                                   Alexander Clark
                                                   Thomas Zeugmann
                                                    Sandra Zilles

Organization

General Chair for ALT 2014 and DS 2014

Ljupčo Todorovski   University of Ljubljana, Slovenia

Program Committee

Nir Ailon	Technion, Israel
András Antos	SZIT, Hungary
Peter Auer (Chair)	Montanuniversität Leoben, Austria
Shai Ben-David	University of Waterloo, Canada
Sébastien Bubeck	Princeton University, USA
Alexander Clark (Chair)	King's College London, UK
Corinna Cortes	Google, USA
Vitaly Feldman	IBM Research, USA
Claudio Gentile	Università degli Studi dell'Insubria, Italy
Steve Hanneke	Carnegie Mellon University, USA
Kohei Hatano	Kyushu University, Japan
Sanjay Jain	National University of Singapore, Singapore
Timo Kötzing	Friedrich-Schiller-Universität, Germany
Eric Martin	University of New South Wales, Australia
Mehryar Mohri	Courant Institute of Mathematical Sciences, USA
Rémi Munos	Inria, France
Ronald Ortner	Montanuniversität Leoben, Austria
Lev Reyzin	University of Illinois at Chicago, USA
Daniil, Ryabko	Inria, France
Sivan Sabato	Microsoft Research New England, USA
Masashi Sugiyama	Tokyo Institute of Technology, Japan
Csaba Szepesvári	University of Alberta, Canada
John Shawe-Taylor	University College London, UK
Vladimir Vovk	Royal Holloway, University of London, UK
Sandra Zilles	University of Regina, Canada

Local Arrangements Chair

Mili Bauer   Jožef Stefan Institute, Ljubljana

Subreferees

Abbasi-Yadkori, Yasin
Allauzen, Cyril
Amin, Kareem
Ávila Pires, Bernardo
Cesa-Bianchi, Nicolò
Chernov, Alexey
Ge, Rong
Gravin, Nick
Kameoka, Hirokazu
Kanade, Varun
Kanamori, Takafumi
Kocák, Tomáš
Kuznetsov, Vitaly
Lazaric, Alessandro
Lever, Guy
London, Ben
Long, Phil
Ma, Yao
Maillard, Odalric-Ambrym

Mens, Irini-Eleftheria
Morimura, Tetsuro
Munoz, Andres
Nakajima, Shinichi
Neu, Gergely
Procopiuc, Cecilia
Russo, Daniel
Sakuma, Jun
Semukhin, Pavel
Shamir, Ohad
Slivkins, Aleksandrs
Smith, Adam
Syed, Umar
Takimoto, Eiji
Telgarsky, Matus
Wen, Zheng
Yamada, Makoto
Yaroslavtsev, Grigory
Zolotykh, Nikolai

Sponsoring Institutions

Office of Naval Research Global, ONRG GRANT N62909-14-1-C195
Jožef Stefan Institute, Ljubljana
University of Ljubljana

Invited Abstracts

A Survey of Preference-Based Online Learning with Bandit Algorithms

Róbert Busa-Fekete and Eyke Hüllermeier

Department of Computer Science
University of Paderborn, Germany
{busarobi,eyke}@upb.de

Abstract. In machine learning, the notion of *multi-armed bandits* refers to a class of online learning problems, in which an agent is supposed to simultaneously explore and exploit a given set of choice alternatives in the course of a sequential decision process. In the standard setting, the agent learns from stochastic feedback in the form of real-valued rewards. In many applications, however, numerical reward signals are not readily available—instead, only weaker information is provided, in particular relative preferences in the form of qualitative comparisons between pairs of alternatives. This observation has motivated the study of variants of the multi-armed bandit problem, in which more general representations are used both for the type of feedback to learn from and the target of prediction. The aim of this paper is to provide a survey of the state-of-the-art in this field, that we refer to as *preference-based multi-armed bandits*. To this end, we provide an overview of problems that have been considered in the literature as well as methods for tackling them. Our systematization is mainly based on the assumptions made by these methods about the data-generating process and, related to this, the properties of the preference-based feedback.

Keywords: Multi-armed bandits, online learning, preference learning, ranking, top-k selection, exploration/exploitation, cumulative regret, sample complexity, PAC learning.

Cellular Tree Classifiers

Gérard Biau[1,2] and Luc Devroye[3]

[1] Sorbonne Universités, UPMC Univ Paris 06, France
[2] Institut universitaire de France
[3] McGill University, Canada

Abstract. Suppose that binary classification is done by a tree method in which the leaves of a tree correspond to a partition of d-space. Within a partition, a majority vote is used. Suppose furthermore that this tree must be constructed recursively by implementing just two functions, so that the construction can be carried out in parallel by using "cells": first of all, given input data, a cell must decide whether it will become a leaf or an internal node in the tree. Secondly, if it decides on an internal node, it must decide how to partition the space linearly. Data are then split into two parts and sent downstream to two new independent cells. We discuss the design and properties of such classifiers.

Social Network Analysis

Anuška Ferligoj

Faculty of Social Sciences,
University of Ljubljana
anuska.ferligoj@fdv.uni-lj.si

Abstract. Social network analysis has attracted considerable interest from social and behavioral science community in recent decades. Much of this interest can be attributed to the focus of social network analysis on relationship among units, and on the patterns of these relationships. Social network analysis is a rapidly expanding and changing field with broad range of approaches, methods, models and substantive applications. In the talk special attention will be given to:

1. General introduction to social network analysis:
   - What are social networks?
   - Data collection issues.
   - Basic network concepts: network representation; types of networks; size and density.
   - Walks and paths in networks: length and value of path; the shortest path, k-neighbours; acyclic networks.
   - Connectivity: weakly, strongly and bi-connected components; contraction; extraction.
2. Overview of tasks and corresponding methods:
   - Network/node properties: centrality (degree, closeness, betweenness); hubs and authorities.
   - Cohesion: triads, cliques, cores, islands.
   - Partitioning: blockmodeling (direct and indirect approaches; structural, regular equivalence; generalised blockmodeling); clustering.
   - Statistical models.
3. Software for social network analysis (UCINET, PAJEK, ...)

Building an Automated Statistician

Zoubin Ghahramani

Department of Engineering,
University of Cambridge,
Trumpington Street
Cambridge CB2 1PZ, UK
zoubin@eng.cam.ac.uk

Abstract. We will live an era of abundant data and there is an increasing need for methods to automate data analysis and statistics. I will describe the "Automated Statistician", a project which aims to automate the exploratory analysis and modelling of data. Our approach starts by defining a large space of related probabilistic models via a grammar over models, and then uses Bayesian marginal likelihood computations to search over this space for one or a few good models of the data. The aim is to find models which have both good predictive performance, and are somewhat interpretable. Our initial work has focused on the learning of unknown nonparametric regression functions, and on learning models of time series data, both using Gaussian processes. Once a good model has been found, the Automated Statistician generates a natural language summary of the analysis, producing a 10-15 page report with plots and tables describing the analysis. I will discuss challenges such as: how to trade off predictive performance and interpretability, how to translate complex statistical concepts into natural language text that is understandable by a numerate non-statistician, and how to integrate model checking. This is joint work with James Lloyd and David Duvenaud (Cambridge) and Roger Grosse and Josh Tenenbaum (MIT).

Table of Contents

Reinforcement Learning

Online Learning and Learning with Bandit Information

Statistical Learning Theory

Privacy, Clustering, MDL, and Kolmogorov Complexity

Editors' Introduction

Peter Auer, Alexander Clark, Thomas Zeugmann, and Sandra Zilles

The aim of the series of conferences on Algorithmic Learning Theory (ALT) is to look at learning from an algorithmic and mathematical perspective. Over time several models of learning have been developed which study different aspects of learning. In the following we describe in brief the invited talks and the contributed papers for ALT 2014 held in Bled, Slovenia..

Invited Talks. Following the tradition of the co-located conferences ALT and DS all invited lectures are shared by the two conferences. The invited speakers are eminent researchers in their fields and present either their specific research area or lecture about a topic of broader interest.

This year's joint invited speaker for ALT 2014 and DS 2014 is Zoubin Ghahramani, who is Professor of Information Engineering at the University of Cambridge, UK, where he leads a group of about 30 researchers. He studied computer science and cognitive science at the University of Pennsylvania, obtained his PhD from MIT in 1995 under the supervision of Michael Jordan, and was a postdoctoral fellow at the University of Toronto with Geoffrey Hinton. His academic career includes concurrent appointments as one of the founding members of the Gatsby Computational Neuroscience Unit in London, and as a faculty member of CMU's Machine Learning Department for over 10 years. His current research focuses on nonparametric Bayesian modeling and statistical machine learning. He has also worked on applications to bioinformatics, econometrics, and a variety of large-scale data modeling problems. He has published over 200 papers, receiving 25,000 citations (an h-index of 68). His work has been funded by grants and donations from EPSRC, DARPA, Microsoft, Google, Infosys, Facebook, Amazon, FX Concepts and a number of other industrial partners. In 2013, he received a $750,000 Google Award for research on building the Automatic Statistician. In his invited talk *Building an Automated Statistician* (joint work with James Lloyd, David Duvenaud, Roger Grosse, and Josh Tenenbaum) Zoubin Ghahramani addresses the problem of abundant data and the increasing need for methods to automate data analysis and statistics. The Automated Statistician project aims to automate the exploratory analysis and modeling of data. The approach uses Bayesian marginal likelihood computations to search over a large space of related probabilistic models. Once a good model has been found, the Automated Statistician generates a natural language summary of the analysis, producing a 10-15 page report with plots and tables describing the analysis. Zoubin Ghahramani discusses challenges such as: how to trade off predictive performance and interpretability, how to translate complex statistical concepts into natural language text that is understandable by a numerate non-statistician, and how to integrate model checking.

The invited speaker for ALT 2014 is Luc Devroye, who is a James McGill Professor in the School of Computer Science of McGill University in Montreal. He

P. Auer et al. (Eds.): ALT 2014, LNAI 8776, pp. 1–7, 2014.

studied at Katholieke Universiteit Leuven and subsequently at Osaka University and in 1976 received his PhD from University of Texas at Austin under the supervision of Terry Wagner. Luc Devroye specializes in the probabilistic analysis of algorithms, random number generation and enjoys typography. Since joining the McGill faculty in 1977 he has won numerous awards, including an E.W.R. Steacie Memorial Fellowship (1987), a Humboldt Research Award (2004), the Killam Prize (2005) and the Statistical Society of Canada gold medal (2008). He received an honorary doctorate from the Université catholique de Louvain in 2002, and an honorary doctorate from Universiteit Antwerpen in 2012. The invited paper *Cellular Tree Classifiers* (joint work with Gerard Biau) deals with classification by decision trees, where the decision trees are constructed recursively by using only two local rules: (1) given the input data to a node, it must decide whether it will become a leaf or not, and (2) a non-leaf node needs to decide how to split the data for sending them downstream. The important point is that each node can make these decisions based only on its local data, such that the decision tree construction can be carried out in parallel. Somewhat surprisingly there are such local rules that guarantee convergence of the decision tree error to the Bayes optimal error. Luc Devroye discusses the design and properties of such classifiers.

The ALT 2014 tutorial speaker is Eyke Hüllermeier, who is professor and head of the Intelligent Systems Group at the Department of Computer Science of the University of Paderborn. He received his PhD in Computer Science from the University of Paderborn in 1997 and he also holds an MSc degree in business informatics. He was a researcher in artificial intelligence, knowledge-based systems, and statistics at the University of Paderborn and the University of Dortmund and a Marie Curie fellow at the Institut de Recherche en Informatique de Toulouse. He has held already a full professorship in the Department of Mathematics and Computer Science at Marburg University before rejoining the University of Paderborn. In his tutorial *A Survey of Preference-based Online Learning with Bandit Algorithms* (joint work with Róbert Busa-Fekete) Eyke Hüllermeier reports on learning with bandit feedback that is weaker than the usual real-value reward. When learning with bandit feedback the learning algorithm receives feedback only from the decisions it makes, but no information from other alternatives. Thus the learning algorithm needs to simultaneously explore and exploit a given set of alternatives in the course of a sequential decision process. In many applications the feedback is not a numerical reward signal but some weaker information, in particular relative preferences in the form of qualitative comparisons between pairs of alternatives. This observation has motivated the study of variants of the multi-armed bandit problem, in which more general representations are used both for the type of feedback to learn from and the target of prediction. The aim of the tutorial is to provide a survey of the state-of-the-art in this area which is referred to as preference-based multi-armed bandits. To this end, Eyke Hüllermeier provides an overview of problems that have been considered in the literature as well as methods for tackling them. His systematization is mainly based on the assumptions made by these methods about the data-generating process and, related to this, the properties of the preference-based feedback.

The DS 2014 tutorial speaker is Anuška Ferligoj, who is professor of Multivariate Statistical Methods at the University of Ljubljana. She is a Slovenian mathematician who earned international recognition by her research work on network analysis. Her interests include multivariate analysis (constrained and multicriteria clustering), social networks (measurement quality and blockmodeling), and survey methodology (reliability and validity of measurement). She is a fellow of the European Academy of Sociology. She has also been an editor of the journal Advances in Methodology and Statistics (Metodoloski zvezki) since 2004 and is a member of the editorial boards of the Journal of Mathematical Sociology, Journal of Classification, Social Networks, Statistic in Transition, Methodology, Structure and Dynamics: eJournal of Anthropology and Related Sciences. She was a Fulbright scholar in 1990 and visiting professor at the University of Pittsburgh. She was awarded the title of Ambassador of Science of the Republic of Slovenia in 1997. Social network analysis has attracted considerable interest from the social and behavioral science community in recent decades. Much of this interest can be attributed to the focus of social network analysis on relationship among units, and on the patterns of these relationships. Social network analysis is a rapidly expanding and changing field with broad range of approaches, methods, models and substantive applications. In her tutorial *Social Network Analysis* Anuška Ferligoj gives a general introduction to social network analysis and an overview of tasks and corresponding methods, accompanied by pointers to software for social network analysis.

Inductive Inference. There are a number of papers in the field of inductive inference, the most classical branch of algorithmic learning theory. First, *A Map of Update Constraints in Inductive Inference* by Timo Kötzing and Raphaela Palenta provides a systematic overview of various constraints on learners in inductive inference problems. They focus on the question of which constraints and combinations of constraints reduce the learning power, meaning the class of languages that are learnable with respect to certain criteria.

On a related theme, the paper *On the Role of Update Constraints and Text-Types in Iterative Learning* by Sanjay Jain, Timo Kötzing, Junqi Ma, and Frank Stephan looks more specifically at the case where the learner has no memory beyond the current hypothesis. In this situation the paper is able to completely characterize the relations between the various constraints.

The paper *Parallel Learning of Automatic Classes of Languages* by Sanjay Jain and Efim Kinber continues the line of research on learning automatic classes of languages initiated by Jain, Luo and Stephan in 2012, in this case by considering the problem of learning multiple distinct languages at the same time.

Laurent Bienvenu, Benoît Monin and Alexander Shen present a negative result in their paper *Algorithmic Identification of Probabilities is Hard*. They show that it is impossible to identify in the limit the exact parameter—in the sense of the Turing code for a computable real number—of a Bernoulli distribution, though it is of course easy to approximate it.

Exact Learning from Queries. In cases where the instance space is discrete, it is reasonable to aim at exact learning algorithms where the learner is required to produce a hypothesis that is exactly correct.

The paper winning the E.M. Gold Award, *Learning Boolean Halfspaces with Small Weights from Membership Queries* by the student authors Hasan Abasi and Ali Z. Abdi and co-authored by Nader H. Bshouty, presents a significantly improved algorithm for learning Boolean Halfspaces in $\{0, 1\}^n$ with integer weights $\{0, \ldots, t\}$ from membership queries only. It is shown that this algorithm needs only $n^{O(t)}$ membership queries, which improves over previous algorithms with $n^{O(t^5)}$ queries and closes the gap to the known lower bound n^t.

The paper by Hasan Abasi, Nader H. Bshouty and Hanna Mazzawi *On Exact Learning Monotone DNF from Membership Queries* presents learning results on learnability by membership queries of monotone DNF (disjunctive normal forms) with a bounded number of terms and a bounded number of variables per term.

Dana Angluin and Dana Fisman look at exact learning using membership queries and equivalence queries in their paper *Learning Regular Omega Languages*. Here the class concerned is that of regular languages over infinite words; the authors consider three different representations which vary in their succinctness. This problem has applications in verification and synthesis of reactive systems.

Reinforcement Learning. Reinforcement learning continues to be a centrally important area of learning theory, and this conference contains a number of contributions in this field. Ronald Ortner, Odalric-Ambrym Maillard and Daniil Ryabko present a paper *Selecting Near-Optimal Approximate State Representations in Reinforcement Learning*, which looks at the problem where the learner does not have direct information about the states in the underlying Markov Decision Process (MDP); in contrast to Partially Observable MDPs, here the information is via various models that map the histories to states.

L.A. Prashanth considers risk constrained reinforcement learning in his paper *Policy Gradients for CVaR-Constrained MDPs*, focusing on the stochastic shortest path problem. For a risk constrained problem not only the expected sum of costs per step $\mathbb{E}[\sum_m g(s_m, a_m)]$ is to be minimized, but also the sum of an additional cost measure $C = \sum_m c(s_m, a_m)$ needs to be bounded from above. Usually the Value at Risk, $\mathrm{VaR}_\alpha = \inf\{\xi | \mathbb{P}(C \leq \xi) \geq \alpha\}$, is constrained, but such constrained problems are hard to optimize. Instead, the paper proposes to constrain the Conditional Value at Risk, $\mathrm{CVaR}_\alpha = \mathbb{E}[C | C \geq \mathrm{VaR}_\alpha]$, which allows to apply standard optimization techniques. Two algorithms are presented that converge to a locally risk-optimal policy using stochastic approximation, mini batches, policy gradients, and importance sampling.

In contrast to the usual MDP setting for reinforcement learning, the two following papers consider more general reinforcement learning. *Bayesian Reinforcement Learning with Exploration* by Tor Lattimore and Marcus Hutter improves some of their earlier work on general reinforcement learning. Here the true environment does not need to be Markovian, but it is known to be drawn at random from a finite class of possible environments. An algorithm is presented that alternates between periods of playing the Bayes optimal policy and periods of forced experimentation. Upper bounds on the sample complexity are established, and it is shown that for some classes of environments this bound cannot be improved by more than a logarithmic factor.

Marcus Hutter's paper *Extreme State Aggregation beyond MDPs* considers how an arbitrary (non-Markov) decision process with a finite number of actions can be approximated by a finite-state MDP. For a given feature function $\phi :$ $H \to S$ mapping histories h of the general process to some finite state space S, the transition probabilities of the MDP can be defined appropriately. It is shown that the MDP approximates the general process well if the optimal policy for the general process is consistent with the feature function, $\pi^*(h_1) = \pi^*(h_2)$ for $\phi(h_1) = \phi(h_2)$, or if the optimal Q-value function is consistent with the feature function, $|Q^*(h_1, a) - Q^*(h_2, a)| < \varepsilon$ for $\phi(h_1) = \phi(h_2)$ and all a. It is also shown that such a feature function always exists.

Online Learning and Learning with Bandit Information. The paper *On Learning the Optimal Waiting Time* by Tor Lattimore, András György, and Csaba Szepesvári, addresses the problem of how long to wait for an event with independent and identically distributed (i.i.d.) arrival times from an unknown distribution. If the event occurs during the waiting time, then the cost is the time until arrival. If the event occurs after the waiting time, then the cost is the waiting time plus a fixed and known amount. Algorithms for the full information setting and for bandit information are presented that sequentially choose waiting times over several rounds in order to minimize the regret in respect to an optimal waiting time. For bandit information the arrival time is only revealed if it is smaller than the waiting time, and in the full information setting it is revealed always. The performance of the algorithms nearly matches the minimax lower bound on the regret.

In many application areas, e.g. recommendation systems, the learning algorithm should return a ranking: a permutation of some finite set of elements. This problem is studied in the paper by Nir Ailon, Kohei Hatano, and Eiji Takimoto titled *Bandit Online Optimization Over the Permutahedron* when the cost of a rankings is calculated as $\sum_{i=1}^{n} \pi(i)s(i)$, where $\pi(i)$ is the rank of item i and $s(i)$ is its cost. In the bandit setting in each iteration an unknown cost vector $\mathbf{s}_t$ is chosen, and the goal of the algorithm is to minimize the regret in respect to the best fixed ranking of the items.

Marcus Hutter's paper *Offline to Online Conversion* introduces the problem of turning a sequence of distributions q_n on strings in X^n, $n = 1, \ldots, n$, into a stochastic online predictor for the next symbol $\tilde{q}(x_n|x_1, \ldots, x_{n-1})$, such that the induced probabilities $\tilde{q}(x_1, \ldots, x_n)$ are close to $q_n(x_1, \ldots, x_n)$ for all sequences $x_1, x_2, \ldots$ The paper considers four strategies for doing such a conversion, showing that naïve approaches might not be satisfactory but that a good predictor can always be constructed, at the cost of possible computational inefficiency. One examples of such a conversion gives a simple combinatorial derivation of the Good-Turing estimator.

Statistical Learning Theory. Andreas Maurer's paper *A Chain Rule for the Expected Suprema of Gaussian Processes* investigates the problem of assessing generalization of a learner who is adapting a feature space while also learning the target function. The approach taken is to consider extensions of bounds on Gaussian averages to the case where there is a class of functions that create features and a class of mappings from those features to outputs. In the applications considered in the paper this corresponds to a two layer kernel machine, multitask learning, and through an iteration of the application of the bound to multilayer networks and deep learners.

A standard assumption in statistical learning theory is that the data are generated independently and identically distributed from some fixed distribution; in practice, this assumption is often violated and a more realistic assumption is that the data are generated by a process which is only sufficiently fast mixing, and maybe even non-stationary. Vitaly Kuznetsov and Mehryar Mohri in their paper *Generalization Bounds for Time Series Prediction with Non-stationary Processes* consider this case and are able to prove new generalization bounds that depend on the mixing coefficients and the shift of the distribution.

Rahim Samei, Boting Yang, and Sandra Zilles in their paper *Generalizing Labeled and Unlabeled Sample Compression to Multi-label Concept Classes* consider generalizations of the binary VC-dimension to multi-label classification, such that maximum classes of dimension d allow a tight compression scheme of size d. Sufficient conditions for notions of dimensions with this property are derived, and it is shown that some multi-label generalizations of the VC-dimension allow tight compression schemes, while other generalizations do not.

Privacy, Clustering, MDL, and Kolmogorov Complexity. Christos Dimitrakakis, Blaine Nelson, Aikaterini Mitrokotsa, and Benjamin I.P. Rubinstein present the paper *Robust and Private Bayesian Inference*. This paper looks at the problem of privacy in machine learning, where an agent, a statistician for example, might want to reveal information derived from a data set, but without revealing information about the particular data points in the set, which might contain confidential information. The authors show that it is possible to do Bayesian inference in this setting, satisfying differential privacy, provided that the likelihoods and conjugate priors satisfy some properties.

Behnam Neyshabur, Yury Makarychev, and Nathan Srebro in their paper *Clustering, Hamming Embedding, Generalized LSH and the Max Norm* look at asymmetric locality sensitive hashing (LSH) which is useful in many types of machine learning applications. Locality sensitive hashing, which is closely related to the problem of clustering, is a method of probabilistically reducing the dimension of high dimensional data sets; assigning each data point a hash such that similar data points will be mapped to the same hash. The paper shows that by shifting to co-clustering and asymmetric LSH the problem admits a tractable relaxation.

Jan Leike and Marcus Hutter look at martingale theory in *Indefinitely Oscillating Martingales*; as a consequence of their analysis they show a negative result in the theory of Minimum Description Length (MDL) learning, namely that the MDL estimator is in general inductively inconsistent: it will not necessarily converge. The MDL estimator gives the regularized code length, $\mathrm{MDL}(u) = \min_Q\{Q(u) + K(Q)\}$, where Q is a coding function, $K(Q)$ its complexity, and $Q(u)$ the code length for the string u. It is shown that the family of coding functions Q can be constructed such that $\lim_{n\to\infty} \mathrm{MDL}(u_{1:n})$ does not converge for most infinite words u.

As is well-known, the Kolmogorov complexity is not computable. Peter Bloem, Francisco Mota, Steven de Rooij, Luís Antunes, and Pieter Adriaans in their paper *A Safe Approximation for Kolmogorov Complexity* study the problem of approximating this quantity using a restriction to a particular class of models, and a probabilistic bound on the approximation error.

Cellular Tree Classifiers

Gérard Biau[1,2] and Luc Devroye[3]

[1] Sorbonne Universités, UPMC Univ Paris 06, France
[2] Institut universitaire de France
[3] McGill University, Canada

Abstract. Suppose that binary classification is done by a tree method in which the leaves of a tree correspond to a partition of d-space. Within a partition, a majority vote is used. Suppose furthermore that this tree must be constructed recursively by implementing just two functions, so that the construction can be carried out in parallel by using "cells": first of all, given input data, a cell must decide whether it will become a leaf or an internal node in the tree. Secondly, if it decides on an internal node, it must decide how to partition the space linearly. Data are then split into two parts and sent downstream to two new independent cells. We discuss the design and properties of such classifiers.

1 Introduction

We explore in this note a new way of dealing with the supervised classification problem, inspired by greedy approaches and the divide-and-conquer philosophy. Our point of view is novel, but has a wide reach in a world in which parallel and distributed computation are important. In the short term, parallelism will take hold in massive data sets and complex systems and, as such, is one of the exciting questions that will be asked to the statistics and machine learning fields.

The general context is that of classification trees, which make decisions by recursively partitioning $\mathbb{R}^d$ into regions, sometimes called cells. In the model we promote, a basic computational unit in classification, a cell, takes as input training data, and makes a decision whether a majority rule should be locally applied. In the negative, the data should be split and each part of the partition should be transmitted to another cell. What is original in our approach is that all cells must use **exactly** the same protocol to make their decision—their function is not altered by external inputs or global parameters. In other words, the decision to split depends only upon the data presented to the cell, independently of the overall edifice. Classifiers designed according to this autonomous principle will be called cellular tree classifiers, or simply cellular classifiers.

Decision tree learning is a method commonly used in data mining (see, e.g., [27]). For example, in CART (Classification and Regression Trees, [5]), splits are made perpendicular to the axes based on the notion of Gini impurity. Splits are performed until all data are isolated. In a second phase, nodes are recombined from the bottom-up in a process called pruning. It is this second process that makes the CART trees non-cellular, as global information is shared to manage

P. Auer et al. (Eds.): ALT 2014, LNAI 8776, pp. 8–17, 2014.

the recombination process. Quinlan's C4.5 [26] also prunes. Others split until all nodes or cells are homogeneous (i.e., have the same class)—the prime example is Quinlan's ID3 [25]. This strategy, while compliant with the cellular framework, leads to non-consistent rules, as we point out in the present paper. In fact, the choice of a good stopping rule for decision trees is very hard—we were not able to find any in the literature that guarantee convergence to the Bayes error.

2   Tree Classifiers

In the design of classifiers, we have an unknown distribution of a random prototype pair $(\mathbf{X}, Y)$, where $\mathbf{X}$ takes values in $\mathbb{R}^d$ and Y takes only finitely many values, say 0 or 1 for simplicity. Classical pattern recognition deals with predicting the unknown nature Y of the observation $\mathbf{X}$ via a measurable classifier $g : \mathbb{R}^d \to \{0, 1\}$. We make a mistake if $g(\mathbf{X})$ differs from Y, and the probability of error for a particular decision rule g is $L(g) = \mathbb{P}\{g(\mathbf{X}) \neq Y\}$. The Bayes classifier

$$g^\star(\mathbf{x}) = \begin{cases} 1 & \text{if } \mathbb{P}\{Y = 1 | \mathbf{X} = \mathbf{x}\} > \mathbb{P}\{Y = 0 | \mathbf{X} = \mathbf{x}\} \\ 0 & \text{otherwise} \end{cases}$$

has the smallest probability of error, that is

$$L^\star = L(g^\star) = \inf_{g:\mathbb{R}^d \to \{0,1\}} \mathbb{P}\{g(\mathbf{X}) \neq Y\}$$

(see, for instance, Theorem 2.1 in [7]). However, most of the time, the distribution of $(\mathbf{X}, Y)$ is unknown, so that the optimal decision $g^\star$ is unknown too. We do not consult an expert to try to reconstruct $g^\star$, but have access to a database $\mathscr{D}_n = (\mathbf{X}_1, Y_1), \ldots, (\mathbf{X}_n, Y_n)$ of i.i.d. copies of $(\mathbf{X}, Y)$, observed in the past. We assume that $\mathscr{D}_n$ and $(\mathbf{X}, Y)$ are independent. In this context, a classification rule $g_n(\mathbf{x}; \mathscr{D}_n)$ is a Borel measurable function of $\mathbf{x}$ and $\mathscr{D}_n$, and it attempts to estimate Y from $\mathbf{x}$ and $\mathscr{D}_n$. For simplicity, we suppress $\mathscr{D}_n$ in the notation and write $g_n(\mathbf{x})$ instead of $g_n(\mathbf{x}; \mathscr{D}_n)$.

The probability of error of a given classifier g_n is the random variable

$$L(g_n) = \mathbb{P}\{g_n(\mathbf{X}) \neq Y | \mathscr{D}_n\},$$

and the rule is consistent if

$$\lim_{n \to \infty} \mathbb{E}L(g_n) = L^\star.$$

It is universally consistent if it is consistent for all possible distributions of $(\mathbf{X}, Y)$. Many popular classifiers are universally consistent. These include several brands of histogram rules, k-nearest neighbor rules, kernel rules, neural networks, and tree classifiers. There are too many references to be cited here, but the monographs by [7] and [15] will provide the reader with a comprehensive introduction to the domain and a literature review.

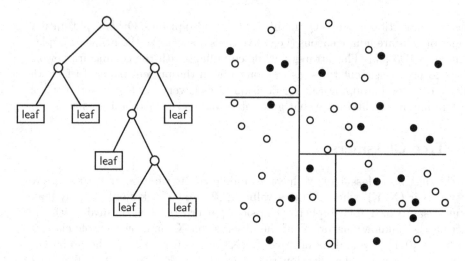

Fig. 1. A binary tree (left) and the corresponding partition (right)

Trees have been suggested as tools for classification for more than thirty years. We mention in particular the early work of Fu [36, 1, 21, 18, 24]. Other references from the 1970s include [20, 3, 23, 30, 34, 12, 8]. Most influential in the classification tree literature was the CART proposal by [5]. While CART proposes partitions by hyperrectangles, linear hyperplanes in general position have also gained in popularity—the early work on that topic is by [19], and [22]. Additional references on tree classification include [14, 2, 16, 17, 35, 33, 31, 6, 9, 10, 32, 13].

3 Cellular Trees

In general, classification trees partition $\mathbb{R}^d$ into regions, often hyperrectangles parallel to the axes (an example is depicted in Figure 1). Of interest in this article are binary trees, where each node has exactly 0 or 2 children. If a node u represents the set A and its children u_1, u_2 represent A_1, A_2, then it is required that $A = A_1 \cup A_2$ and $A_1 \cap A_2 = \emptyset$. The root of the tree represents $\mathbb{R}^d$, and the terminal nodes (or leaves), taken together, form a partition of $\mathbb{R}^d$. If a leaf represents region A, then the tree classifier takes the simple form

$$g_n(\mathbf{x}) = \begin{cases} 1 & \text{if } \sum_{i=1}^n \mathbb{1}_{[\mathbf{X}_i \in A, Y_i=1]} > \sum_{i=1}^n \mathbb{1}_{[\mathbf{X}_i \in A, Y_i=0]}, \quad \mathbf{x} \in A \\ 0 & \text{otherwise.} \end{cases}$$

That is, in every leaf region, a majority vote is taken over all $(\mathbf{X}_i, Y_i)$'s with $\mathbf{X}_i$'s in the same region. Ties are broken, by convention, in favor of class 0.

The tree structure is usually data-dependent, and indeed, it is in the construction itself where different trees differ. Thus, there are virtually infinitely many possible strategies to build classification trees. Nevertheless, despite this great diversity, all tree species end up with two fundamental questions at each node:

① Should the node be split?

② In the affirmative, what are its children?

These two questions are typically answered using **global** information on the tree, such as, for example, a function of the data $\mathscr{D}_n$, the level of the node within the tree, the size of the data set and, more generally, any parameter connected with the structure of the tree. This parameter could be, for example, the total number k of cells in a k-partition tree or the penalty term in the pruning of the CART algorithm (e.g., [5] and [11]).

Our cellular trees proceed from a different philosophy. In short, a cellular tree should, at each node, be able to answer questions ① and ② using **local** information only, without any help from the other nodes. In other words, each cell can perform as many operations as it wishes, provided it uses only the data that are transmitted to it, regardless of the general structure of the tree. Just imagine that the calculations to be carried out at the nodes are sent to different computers, eventually asynchronously, and that the system architecture is so complex that computers do not communicate. Thus, once a computer receives its data, it has to make its own decisions on ① and ② based on this data subset only, independently of the others and without knowing anything of the overall edifice. Once a data set is split, it can be given to another computer for further splitting, since the remaining data points have no influence.

Formally, a cellular binary classification tree is a machine that partitions the space recursively in the following manner. With each node we associate a subset of $\mathbb{R}^d$, starting with $\mathbb{R}^d$ for the root node. We consider binary tree classifiers based on a class $\mathcal{C}$ of possible Borel subsets of $\mathbb{R}^d$ that can be used for splits. A typical example of such a class is the family of all hyperplanes, or the class of all hyperplanes that are perpendicular to one of the axes. Higher order polynomial splitting surfaces can be imagined as well. The class is parametrized by a vector $\sigma \in \mathbb{R}^p$. There is a splitting function $f(\mathbf{x}, \sigma)$, $\mathbf{x} \in \mathbb{R}^d, \sigma \in \mathbb{R}^p$, such that $\mathbb{R}^d$ is partitioned into $A = \{\mathbf{x} \in \mathbb{R}^d : f(\mathbf{x}, \sigma) \geq 0\}$ and $B = \{\mathbf{x} \in \mathbb{R}^d : f(\mathbf{x}, \sigma) < 0\}$. Formally, a cellular split can be viewed as a family of measurable mappings $(\sigma_m)_m$ from $(\mathbb{R}^d \times \{0, 1\})^m$ to $\mathbb{R}^p$. In this model, m is the size of the data set transmitted to the cell. Thus, for each possible input size m, we have a map. In addition, there is a family of measurable mappings $(\theta_m)_m$ from $(\mathbb{R}^d \times \{0, 1\})^m$ to $\{0, 1\}$ that indicate decisions: $\theta_m = 1$ indicates that a split should be applied, while $\theta_m = 0$ corresponds to a decision not to split. In that case, the cell acts as a leaf node in the tree. We note that $(\theta_m)_m$ and $(\sigma_m)_m$ correspond to the decisions given in ① and ②.

Let the set data set be $\mathscr{D}_n$. If $\theta(\mathscr{D}_n) = 0$, the root cell is final, and the space is not split. Otherwise, $\mathbb{R}^d$ is split into

$$A = \left\{\mathbf{x} \in \mathbb{R}^d : f\left(\mathbf{x}, \sigma(\mathscr{D}_n)\right) \geq 0\right\} \quad \text{and} \quad B = \left\{\mathbf{x} \in \mathbb{R}^d : f\left(\mathbf{x}, \sigma(\mathscr{D}_n)\right) < 0\right\}.$$

The data $\mathscr{D}_n$ are partitioned into two groups–the first group contains all $(\mathbf{X}_i, Y_i)$, $i = 1, \ldots, n$, for which $\mathbf{X}_i \in A$, and the second group all others. The groups are sent to child cells, and the process is repeated. When $\mathbf{x} \in \mathbb{R}^d$ needs to be classified, we first determine the unique leaf set $A(\mathbf{x})$ to which $\mathbf{x}$ belongs, and then take votes among the $\{Y_i : \mathbf{X}_i \in A(\mathbf{x}), i = 1, \ldots, n\}$. Classification proceeds by a majority vote, with the majority deciding the estimate $g_n(\mathbf{x})$. In case of a tie, we set $g_n(\mathbf{x}) = 0$.

A cellular binary tree classifier is said to be randomized if each node in the tree has an independent copy of a uniform $[0, 1]$ random variable associated with it, and θ and σ are mappings that have one extra real-valued component in the input. For example, we could flip an unbiased coin at each node to decide whether $\theta_m = 0$ or $\theta_m = 1$.

4   A Consistent Cellular Tree Classifier

At first sight, it appears that there are no universally consistent cellular tree classifiers. Consider for example complete binary trees with k full levels, i.e., there are 2^k leaf regions. We can have consistency when k is allowed to depend upon n. An example is the median tree (see Section 20.3 in [7]). When $d = 1$, split by finding the median element among the $\mathbf{X}_i$'s, so that the child sets have cardinality given by $\lfloor (n-1)/2 \rfloor$ and $\lceil (n-1)/2 \rceil$, where $\lfloor . \rfloor$ and $\lceil . \rceil$ are the floor and ceiling functions. The median itself does stay behind and is not sent down to the subtrees, with an appropriate convention for breaking cell boundaries as well as empty cells. Keep doing this for k rounds—in d dimensions, one can either rotate through the coordinates for median splitting, or randomize by selecting uniformly at random a coordinate to split orthogonally.

This rule is known to be consistent as soon as the marginal distributions of $\mathbf{X}$ are nonatomic, provided $k \to \infty$ and $k2^k/n \to 0$. However, this is not a cellular tree classifier. While we can indeed specify σ_m, it is impossible to define θ_m because θ_m cannot be a function of the global value of n. In other words, if we were to apply median splitting and decide to split for a fixed k, then the leaf nodes would all correspond to a fixed proportion of the data points. It is clear that the decisions in the leaves are off with a fair probability if we have, for example, Y independent of $\mathbf{X}$ and $\mathbb{P}\{Y = 1\} = 1/2$. Thus, we cannot create a cellular tree classifier in this manner.

In view of the preceding discussion, it seems paradoxical that there indeed exist universally consistent cellular tree classifiers. (We note here that we abuse the word "universal"—we will assume throughout, to keep the discussion at a manageable level, that the marginal distributions of $\mathbf{X}$ are nonatomic. But no other conditions on the joint distribution of $(\mathbf{X}, Y)$ are imposed.) Our construction follows the median tree principle and uses randomization. The original work on the solution appears in [4].

From now on, to keep things simple, it is assumed that the marginal distributions of $\mathbf{X}$ are nonatomic. The cellular splitting method σ_m described in this section mimics the median tree classifier discussed above. We first choose a dimension to cut, uniformly at random from the d dimensions, as rotating through the dimensions by level number would violate the cellular condition. The selected dimension is then split at the data median, just as in the classical median tree. Repeating this for k levels of nodes leads to 2^k leaf regions. On any path of length k to one of the 2^k leaves, we have a deterministic sequence of cardinalities $n_0 = n(\text{root}), n_1, n_2, \ldots, n_k$. We always have $n_i/2 - 1 \le n_{i+1} \le n_i/2$. Thus, by induction, one easily shows that, for all i,

$$\frac{n}{2^i} - 2 \le n_i \le \frac{n}{2^i}.$$

In particular, each leaf has at least $\max(n/2^k - 2, 0)$ points and at most $n/2^k$. The novelty is in the choice of the decision function. This function ignores the data altogether and uses a randomized decision that is based on the size of the input. More precisely, consider a nonincreasing function $\varphi : \mathbb{N} \to (0, 1]$ with $\varphi(0) = \varphi(1) = 1$. Cells correspond in a natural way to sets of $\mathbb{R}^d$. So, we can and will speak of a cell A, where $A \subset \mathbb{R}^d$. The number of data points in A is denoted by $N(A)$:

$$N(A) = \sum_{i=1}^{n} \mathbb{1}_{[\mathbf{X}_i \in A]}.$$

Then, if U is the uniform $[0, 1]$ random variable associated with the cell A and the input to the cell is $N(A)$, the stopping rule ① takes the form:

① Put $\theta = 0$ if

$$U \le \varphi\left(N(A)\right).$$

In this manner, we obtain a possibly infinite randomized binary tree classifier. Splitting occurs with probability $1 - \varphi(m)$ on inputs of size m. Note that no attempt is made to split empty sets or singleton sets. For consistency, we need to look at the random leaf region to which $\mathbf{X}$ belongs. This is roughly equivalent to studying the distance from that cell to the root of the tree.

In the sequel, the notation $u_n = o(v_n)$ (respectively, $u_n = \omega(v_n)$ and $u_n = O(v_n)$) means that $u_n/v_n \to 0$ (respectively, $v_n/u_n \to 0$ and $u_n \le Cv_n$ for some constant C) as $n \to \infty$. Many choices $\varphi(m) = o(1)$, but not all, will do for us. The next lemma makes things more precise.

Lemma 1. *Let $\beta \in (0, 1)$. Define*

$$\varphi(m) = \begin{cases} 1 & \text{if } m < 3 \\ 1/\log^\beta m & \text{if } m \ge 3. \end{cases}$$

Let $K(\mathbf{X})$ denote the random path distance between the cell of $\mathbf{X}$ and the root of the tree. Then

$$\lim_{n \to \infty} \mathbb{P}\{K(\mathbf{X}) \ge k_n\} = \begin{cases} 0 & \text{if } k_n = \omega(\log^\beta n) \\ 1 & \text{if } k_n = o(\log^\beta n). \end{cases}$$

Proof. Let us recall that, at level k, each cell of the underlying median tree contains at least $\max(n/2^k - 2, 0)$ points and at most $n/2^k$. Since the function $\varphi(.)$ is nonincreasing, the first result follows from this:

$$\mathbb{P}\left\{K(\mathbf{X}) \geq k_n\right\} \leq \prod_{i=0}^{k_n-1}\left(1 - \varphi\left(\lfloor n/2^i \rfloor\right)\right)$$

$$\leq \exp\left(-\sum_{i=0}^{k_n-1} \varphi\left(\lfloor n/2^i \rfloor\right)\right)$$

$$\leq \exp\left(-k_n\varphi(n)\right).$$

The second statement follows from

$$\mathbb{P}\left\{K(\mathbf{X}) < k_n\right\} \leq \sum_{i=0}^{k_n-1} \varphi\left(\lceil n/2^i - 2 \rceil\right) \leq k_n\varphi\left(\lceil n/2^{k_n} \rceil\right),$$

valid for all n large enough since $n/2^{k_n} \to \infty$ as $n \to \infty$.   □

Lemma 1, combined with the median tree consistency result of [7], suffices to establish consistency of the randomized cellular tree classifier.

Theorem 1. *Let β be a real number in $(0,1)$. Define*

$$\varphi(m) = \begin{cases} 1 & \text{if } m < 3 \\ 1/\log^\beta m & \text{if } m \geq 3. \end{cases}$$

Let g_n be the associated randomized cellular binary tree classifier. Assume that the marginal distributions of $\mathbf{X}$ are nonatomic. Then the classification rule g_n is consistent:

$$\lim_{n\to\infty} \mathbb{E}L(g_n) = L^\star.$$

Proof. By diam(A) we mean the diameter of the cell A, i.e., the maximal distance between two points of A. We recall a general consistency theorem for partitioning classifiers whose cell design depends on the $\mathbf{X}_i$'s only (see Theorem 6.1 in [7]). According to this theorem, such a classifier is consistent if both

1. diam$(A(\mathbf{X})) \to 0$ in probability as $n \to \infty$, and
2. $N(A(\mathbf{X})) \to \infty$ in probability as $n \to \infty$,

where $A(\mathbf{X})$ is the cell of the random partition containing $\mathbf{X}$.

Condition 2. is proved in Lemma 1. Notice that

$$N(A(\mathbf{X})) \geq \frac{n}{2^{K(\mathbf{X})}} - 2$$

$$\geq \mathbb{1}_{[K(\mathbf{X})<\log^{(\beta+1)/2} n]}\left(\frac{n}{2^{\log^{(\beta+1)/2} n}} - 2\right)$$

$$= \omega(1)\mathbb{1}_{[K(\mathbf{X})<\log^{(\beta+1)/2} n]}.$$

Therefore, by Lemma 1, $N(A(\mathbf{X})) \to \infty$ in probability as $n \to \infty$.

To show that diam$(A(\mathbf{X})) \to 0$ in probability, observe that on a path of length $K(\mathbf{X})$, the number of times the first dimension is cut is binomial $(K(\mathbf{X}), 1/d)$. This tends to infinity in probability. Following the proof of Theorem 20.2 in [7], the diameter of the cell of $\mathbf{X}$ tends to 0 in probability with n. Details are left to the reader.   □

Let us finally take care of the randomization. Can one do without randomization? The hint to the solution of that enigma is in the hypothesis that the data elements in $\mathscr{D}_n$ are i.i.d. The median classifier does not use the ordering in the data. Thus, one can use the randomness present in the permutation of the observations, e.g., the ℓ-th components of the $\mathbf{X}_i$'s can form $n!$ permutations if ties do not occur. This corresponds to $(1 + o(1))n \log_2 n$ independent fair coin flips, which are at our disposal. Each decision to split requires on average at most 2 independent bits. The selection of a random direction to cut requires no more than $1 + \log_2 d$ independent bits. Since the total tree size is, with probability tending to 1, $O(2^{\log^{\beta+\varepsilon} n})$ for any $\varepsilon > 0$, a fact that follows with a bit of work from summing the expected number of nodes at each level, the total number of bits required to carry out all computations is

$$O\left((3 + \log_2 d)2^{\log^{\beta+\varepsilon} n}\right),$$

which is orders of magnitude smaller than n provided that $\beta + \varepsilon < 1$. Thus, there is sufficient randomness at hand to do the job. How it is actually implemented is another matter, as there is some inevitable dependence between the data sets that correspond to cells and the data sets that correspond to their children. We will not worry about the finer details of this in the present paper.

References

[1] Anderson, A.C., Fu, K.S.: Design and development of a linear binary tree classifier for leukocytes. Technical Report TR-EE-79-31, Purdue University (1979)

[2] Argentiero, P., Chin, R., Beaudet, P.: An automated approach to the design of decision tree classifiers. IEEE Transactions on Pattern Analysis and Machine Intelligence 4, 51–57 (1982)

[3] Bartolucci, L.A., Swain, P.H., Wu, C.: Selective radiant temperature mapping using a layered classifier. IEEE Transactions on Geosciences and Electronics 14, 101–106 (1976)

[4] Biau, G., Devroye, L.: Cellular tree classifiers. Electronic Journal of Statistics 7, 1875–1912 (2013)

[5] Breiman, L., Friedman, J.H., Olshen, R.A., Stone, C.J.: Classification and Regression Trees. Chapman & Hall, New York (1984)

[6] Chou, P.A.: Optimal partitioning for classification and regression trees. IEEE Transactions on Pattern Analysis and Machine Intelligence 13, 340–354 (1991)

[7] Devroye, L., Györfi, L., Lugosi, G.: A Probabilistic Theory of Pattern Recognition. Springer, New York (1996)

[8] Friedman, J.H.: A tree-structured approach to nonparametric multiple regression. In: Gasser, T., Rosenblatt, M. (eds.) Smoothing Techniques for Curve Estimation. Lecture Notes in Mathematics, vol. 757, pp. 5–22. Springer, Heidelberg (1979)

[9] Gelfand, S.B., Delp, E.J.: On tree structured classifiers. In: Sethi, I.K., Jain, A.K. (eds.) Artificial Neural Networks and Statistical Pattern Recognition, Old and New Connections, pp. 71–88. Elsevier Science Publishers, Amsterdam (1991)

[10] Gelfand, S.B., Ravishankar, C.S., Delp, E.J.: An iterative growing and pruning algorithm for classification tree design. IEEE Transactions on Pattern Analysis and Machine Intelligence 13, 163–174 (1991)

[11] Gey, S., Nédélec, E.: Model selection for CART regression trees. IEEE Transactions on Information Theory 51, 658–670 (2005)

[12] Gordon, L., Olshen, R.A.: Asymptotically efficient solutions to the classification problem. The Annals of Statistics 6, 515–533 (1978)

[13] Guo, H., Gelfand, S.B.: Classification trees with neural network feature extraction. IEEE Transactions on Neural Networks 3, 923–933 (1992)

[14] Gustafson, D.E., Gelfand, S., Mitter, S.K.: A nonparametric multiclass partitioning method for classification. In: Proceedings of the Fifth International Conference on Pattern Recognition, pp. 654–659 (1980)

[15] Györfi, L., Kohler, M., Krzyżak, A., Walk, H.: A Distribution-Free Theory of Nonparametric Regression. Springer, New York (2002)

[16] Hartmann, C.R.P., Varshney, P.K., Mehrotra, K.G., Gerberich, C.L.: Application of information theory to the construction of efficient decision trees. IEEE Transactions on Information Theory 28, 565–577 (1982)

[17] Kurzynski, M.W.: The optimal strategy of a tree classifier. Pattern Recognition 16, 81–87 (1983)

[18] Lin, Y.K., Fu, K.S.: Automatic classification of cervical cells using a binary tree classifier. Pattern Recognition 16, 69–80 (1983)

[19] Loh, W.Y., Vanichsetakul, N.: Tree-structured classification via generalized discriminant analysis. Journal of the American Statistical Association 83, 715–728 (1988)

[20] Meisel, W.S., Michalopoulos, D.A.: A partitioning algorithm with application in pattern classification and the optimization of decision trees. IEEE Transactions on Computers 22, 93–103 (1973)

[21] Mui, J.K., Fu, K.S.: Automated classification of nucleated blood cells using a binary tree classifier. IEEE Transactions on Pattern Analysis and Machine Intelligence 2, 429–443 (1980)

[22] Park, Y., Sklansky, J.: Automated design of linear tree classifiers. Pattern Recognition 23, 1393–1412 (1990)

[23] Payne, H.J., Meisel, W.S.: An algorithm for constructing optimal binary decision trees. IEEE Transactions on Computers 26, 905–916 (1977)

[24] Qing-Yun, S., Fu, K.S.: A method for the design of binary tree classifiers. Pattern Recognition 16, 593–603 (1983)

[25] Quinlan, J.R.: Induction of decision trees. Machine Learning 1, 81–106 (1986)

[26] Quinlan, J.R.: C4.5: Programs for Machine Learning. Machine Learning. Morgan Kaufmann Publishers, San Mateo (1993)

[27] Rokach, L., Maimon, O.: Data Mining with Decision Trees: Theory and Applications. World Scientific, Singapore (2008)

[28] Samet, H.: The quadtree and related hierarchical data structures. Computing Surveys 16, 187–260 (1984)

[29] Samet, H.: The Design and Analysis of Spatial Data Structures. Addison-Wesley, Reading (1990)

[30] Sethi, I.K., Chatterjee, B.: Efficient decision tree design for discrete variable pattern recognition problems. Pattern Recognition 9, 197–206 (1977)

[31] Shlien, S.: Multiple binary decision tree classifiers. Pattern Recognition 23, 757–763 (1990)

[32] Simon, H.U.: The Vapnik-Chervonenkis dimension of decision trees with bounded rank. Information Processing Letters 39, 137–141 (1991)

[33] Suen, C.Y., Wang, Q.R.: Large tree classifier with heuristic search and global training. IEEE Transactions on Pattern Analysis and Machine Intelligence 9, 91–101 (1987)

[34] Swain, P.H., Hauska, H.: The decision tree classifier: Design and potential. IEEE Transactions on Geosciences and Electronics 15, 142–147 (1977)

[35] Wang, Q.R., Suen, C.Y.: Analysis and design of a decision tree based on entropy reduction and its application to large character set recognition. IEEE Transactions on Pattern Analysis and Machine Intelligence 6, 406–417 (1984)

[36] You, K.C., Fu, K.S.: An approach to the design of a linear binary tree classifier. In: Proceedings of the Symposium of Machine Processing of Remotely Sensed Data, West Lafayette, pp. 3A-1–3A-10. Purdue University (1976)

A Survey of Preference-Based Online Learning with Bandit Algorithms

Róbert Busa-Fekete and Eyke Hüllermeier

Department of Computer Science
University of Paderborn, Germany
{busarobi,eyke}@upb.de

Abstract. In machine learning, the notion of *multi-armed bandits* refers to a class of online learning problems, in which an agent is supposed to simultaneously explore and exploit a given set of choice alternatives in the course of a sequential decision process. In the standard setting, the agent learns from stochastic feedback in the form of real-valued rewards. In many applications, however, numerical reward signals are not readily available—instead, only weaker information is provided, in particular relative preferences in the form of qualitative comparisons between pairs of alternatives. This observation has motivated the study of variants of the multi-armed bandit problem, in which more general representations are used both for the type of feedback to learn from and the target of prediction. The aim of this paper is to provide a survey of the state-of-the-art in this field, that we refer to as *preference-based multi-armed bandits*. To this end, we provide an overview of problems that have been considered in the literature as well as methods for tackling them. Our systematization is mainly based on the assumptions made by these methods about the data-generating process and, related to this, the properties of the preference-based feedback.

Keywords: Multi-armed bandits, online learning, preference learning, ranking, top-k selection, exploration/exploitation, cumulative regret, sample complexity, PAC learning.

1 Introduction

Multi-armed bandit (MAB) algorithms have received considerable attention and have been studied quite intensely in machine learning in the recent past. The great interest in this topic is hardly surprising, given that the MAB setting is not only theoretically challenging but also practically useful, as can be seen from their use in a wide range of applications. For example, MAB algorithms turned out to offer effective solutions for problems in medical treatment design [35, 34], online advertisement [16], and recommendation systems [33], just to mention a few.

The multi-armed bandit problem, or bandit problem for short, is one of the simplest instances of the sequential decision making problem, in which a *learner* (also called decision maker or agent) needs to select *options* from a given set of alternatives repeatedly in an online manner—referring to the metaphor of the

P. Auer et al. (Eds.): ALT 2014, LNAI 8776, pp. 18–39, 2014.

eponymous gambling machine in casinos, these options are also associated with "arms" that can be "pulled". More specifically, the agent selects one option at a time and observes a numerical (and typically noisy) *reward* signal providing information on the quality of that option. The goal of the learner is to optimize an evaluation criterion such as the *error rate* (the expected percentage of playing a suboptimal arm) or the *cumulative regret* (the expected difference between the sum of the rewards actually obtained and the sum of rewards that could have been obtained by playing the best arm in each round). To achieve the desired goal, the online learner has to cope with the famous exploration/exploitation dilemma [5, 14, 35]: It has to find a reasonable compromise between playing the arms that produced high rewards in the past (exploitation) and trying other, possibly even better arms the (expected) reward of which is not precisely known so far (exploration).

The assumption of a numerical reward signal is a potential limitation of the MAB setting. In fact, there are many practical applications in which it is hard or even impossible to quantify the quality of an option on a numerical scale. More generally, the lack of precise feedback or exact supervision has been observed in other branches of machine learning, too, and has led to the emergence of fields such as *weakly supervised learning* and *preference learning* [25]. In the latter, feedback is typically represented in a purely qualitative way, namely in terms of pairwise comparisons or rankings. Feedback of this kind can be useful in online learning, too, as has been shown in online information retrieval [28, 42].

As another example, think of crowd-sourcing services like the Amazon Mechanical Turk, where simple questions such as pairwise comparisons between decision alternatives are asked to a group of annotators. The task is to approximate an underlying target ranking on the basis of these pairwise comparisons, which are possibly noisy and partially inconsistent [17]. Another application worth mentioning is the ranking of XBox gamers based on their pairwise online duels; the ranking system of XBox is called TrueSkill™ [26].

Extending the multi-armed bandit setting to the case of preference-based feedback, i.e., the case in which the online learner is allowed to compare arms in a qualitative way, is therefore a promising idea. And indeed, extensions of that kind have received increasing attention in the recent years. The aim of this paper is to provide a survey of the state-of-the-art in the field of preference-based multi-armed bandits (PB-MAB). After recalling the basic setting of the problem in Section 2, we provide an overview of methods that have been proposed to tackle PB-MAB problems in Sections 3 and 4. Our main criterion for systematization is the assumptions made by these methods about the data-generating process or, more specifically, the properties of the pairwise comparisons between arms. Our survey is focused on the *stochastic* MAB setup, in which feedback is generated according to an underlying (unknown but stationary) probabilistic process; we do not cover the case of an *adversarial* data-generating processes, although this setting has recently received a lot of attention, too [1, 15, 14].

2   The Preference-Based Bandit Problem

The stochastic MAB problem with pairwise comparisons as actions has been studied under the notion of "dueling bandits" in several papers [45, 44]. However, since this term is associated with specific modeling assumptions, we shall use the more general term "preference-based bandits" throughout this paper.

Consider a fixed set of arms (options) $\mathcal{A} = \{a_1, \ldots, a_K\}$. As actions, the learning algorithm (or simply the learner or agent) can perform a comparison between any pair of arms a_i and a_j, i.e., the action space can be identified with the set of index pairs (i, j) such that $1 \leq i \leq j \leq K$. We assume the feedback observable by the learner to be generated by an underlying (unknown) probabilistic process characterized by a *preference relation*

$$\mathbf{Q} = [q_{i,j}]_{1 \leq i,j \leq K} \in [0, 1]^{K \times K} .$$

More specifically, for each pair of actions (a_i, a_j), this relation specifies the probability

$$\mathbf{P}(a_i \succ a_j) = q_{i,j} \tag{1}$$

of observing a preference for a_i in a direct comparison with a_j. Thus, each $q_{i,j}$ specifies a Bernoulli distribution. These distributions are assumed to be stationary and independent, both across actions and iterations. Thus, whenever the learner takes action (i, j), the outcome is distributed according to (1), regardless of the outcomes in previous iterations.

The relation $\mathbf{Q}$ is reciprocal in the sense that $q_{i,j} = 1 - q_{j,i}$ for all $i, j \in [K] = \{1, \ldots, K\}$. We note that, instead of only observing strict preferences, one may also allow a comparison to result in a *tie* or an *indifference*. In that case, the outcome is a trinomial instead of a binomial event. Since this generalization makes the problem technically more complicated, though without changing it conceptually, we shall not consider it further. In [12, 11], indifference was handled by giving "half a point" to both arms, which, in expectation, is equivalent to deciding the winner by flipping a coin. Thus, the problem is essentially reduced to the case of binomial outcomes.

We say arm a_i beats arm a_j if $q_{i,j} > 1/2$, i.e., if the probability of winning in a pairwise comparison is larger for a_i than it is for a_j. Clearly, the closer $q_{i,j}$ is to $1/2$, the harder it becomes to distinguish the arms a_i and a_j based on a finite sample set from $\mathbf{P}(a_i \succ a_j)$. In the worst case, when $q_{i,j} = 1/2$, one cannot decide which arm is better based on a finite number of pairwise comparisons. Therefore,

$$\Delta_{i,j} = q_{i,j} - \frac{1}{2}$$

appears to be a reasonable quantity to characterize the hardness of a PB-MAB task (whatever goal the learner wants to achieve). Note that $\Delta_{i,j}$ can also be negative (unlike the value-based setting, in which the quantity used for characterizing the complexity of a multi-armed bandit task is always positive and depends on the gap between the means of the best arm and the suboptimal arms).

2.1 Pairwise Probability Estimation

The decision making process iterates in discrete steps, either through a finite time horizon $\mathbb{T} = [T]$ or an infinite horizon $\mathbb{T} = \mathbb{N}$. As mentioned above, the learner is allowed to compare two actions in each iteration $t \in \mathbb{T}$. Thus, in each iteration t, it selects an index pair $1 \leq i(t) \leq j(t) \leq K$ and observes

$$\begin{cases} a_{i(t)} \succ a_{j(t)} & \text{with probability } q_{i(t),j(t)} \\ a_{j(t)} \succ a_{i(t)} & \text{with probability } q_{j(t),i(t)} \end{cases}$$

The pairwise probabilities $q_{i,j}$ can be estimated on the basis of finite sample sets. Consider the set of time steps among the first t iterations, in which the learner decides to compare arms a_i and a_j, and denote the size of this set by $n_{i,j}^t$. Moreover, denoting by $w_{i,j}^t$ and $w_{j,i}^t$ the frequency of "wins" of a_i and a_j, respectively, the proportion of wins of a_i against a_j up to iteration t is then given by

$$\widehat{q}_{i,j}^t = \frac{w_{i,j}^t}{n_{i,j}^t} = \frac{w_{i,j}^t}{w_{i,j}^t + w_{j,i}^t} .$$

Since our samples are assumed to be independent and identically distributed (i.i.d.), $\widehat{q}_{i,j}^t$ is a plausible estimate of the pairwise probability (1). Yet, this estimate might be biased, since $n_{i,j}^t$ depends on the choice of the learner, which in turn depends on the data; therefore, $n_{i,j}^t$ itself is a random quantity. A high probability confidence interval for $q_{i,j}$ can be obtained based on the Hoeffding bound [27], which is commonly used in the bandit literature. Although the specific computation of the confidence intervals may differ from case to case, they are generally of the form $[\widehat{q}_{i,j}^t \pm c_{i,j}^t]$. Accordingly, if $\widehat{q}_{i,j}^t - c_{i,j}^t > 1/2$, arm a_i beats arm a_j with high probability; analogously, a_i is beaten by arm a_j with high probability, if $\widehat{q}_{j,i}^t + c_{j,i}^t < 1/2$.

2.2 Evaluation Criteria

The goal of the online learner is usually stated as minimizing some kind of cumulative regret. Alternatively, in the "pure exploration" scenario, the goal is to identify the best arm (or the best k arms, or a ranking of all arms) both quickly and reliably. As an important difference between these two types of targets, note that the regret of a comparison of arms depends on the concrete arms being chosen, whereas the sample complexity penalizes each comparison equally.

It is also worth mentioning that the notion of optimality of an arm is far less obvious in the preference-based setting than it is in the value-based (numerical) setting. In the latter, the optimal arm is simply the one with the highest expected reward—more generally, the expected reward induces a natural total order on the set of actions $\mathcal{A}$. In the preference-based case, the connection between the pairwise preferences $\mathbf{Q}$ and the order induced by this relation on $\mathcal{A}$ is less trivial; in particular, the latter may contain preferential cycles. We shall postpone a more detailed discussion of these issues to subsequent sections, and for the time being simply assume the existence of an arm a_{i*} that is considered optimal.

2.3   Cumulative Regret

In a preference-based setting, defining a reasonable regret is not as straightforward as in the value-based setting, where the sub-optimality of an action can be expressed easily on a numerical scale. In particular, since the learner selects two arms to be compared in an iteration, the sub-optimality of both of these arms should be taken into account. A commonly used definition of regret is the following [46, 43, 47, 45]: Suppose the learner selects arms $a_{i(t)}$ and $a_{j(t)}$ in time step t. Then, the *cumulative regret* incurred by the learner A up to time T is

$$R_A^T = \sum_{t=1}^{T} r^t = \sum_{t=1}^{T} \frac{\Delta_{i^*, i(t)} + \Delta_{i^*, j(t)}}{2} . \tag{2}$$

This regret takes into account the optimality of both arms, meaning that the learner has to select two nearly optimal arms to incur small regret. Note that this regret is zero if the optimal arm a_{i^*} is compared to itself, i.e., if the learner effectively abstains from gathering further information and instead fully commits to the arm a_{i^*}.

2.4   Regret Bounds

In a theoretical analysis of a MAB algorithm, one is typically interested in providing a bound on the (cumulative) regret produced by that algorithm. We are going to distinguish two types of regret bound. The first one is the *expected regret bound*, which is of the form

$$\mathbf{E}\left[R^T\right] \le B(\mathbf{Q}, K, T) , \tag{3}$$

where $\mathbf{E}\left[\cdot\right]$ is the expected value operator, R^T is the regret accumulated till time step T, and $B(\cdot)$ is a positive real-valued function with the following arguments: the pairwise probabilities $\mathbf{Q}$, the number of arms K, and the iteration number T. This function may additionally depend on parameters of the learner, however, we neglect this dependence here. The expectation is taken with respect to the stochastic nature of the data-generating process and the (possible) internal randomization of the online learner. The regret bound (3) is technically akin to the expected regret bound of value-based multi-armed bandit algorithms like the one that is calculated for UCB [5], although the parameters used for characterizing the complexity of the learning task are different.

The bound in (3) does not inform about how the regret achieved by the learner is concentrated around its expectation. Therefore, we consider a second type of regret bound, namely one that holds with high probability. This bound can be written in the form

$$\mathbf{P}\left(R^T < B(\mathbf{Q}, K, T, \delta) \right) \ge 1 - \delta .$$

For simplicity, we also say that the regret achieved by the online learner is $\mathcal{O}(B(\mathbf{Q}, K, T, \delta))$ with high probability.

2.5   Sample Complexity

The sample complexity analysis is considered in a "pure exploration" setup where the learner, in each iteration, must either select a pair of arms to be compared or terminate and return its recommendation. The *sample complexity of the learner* is then the number of pairwise comparisons it queries prior to termination, and the corresponding bound is denoted $B(\mathbf{Q}, K, \delta)$. Here, $1 - \delta$ specifies a lower bound on the probability that the learner terminates and returns the correct solution.[1] Note that only the number of the pairwise comparisons is taken into account, which means that pairwise comparisons are equally penalized, independently of the suboptimality of the arms chosen.

The recommendation of the learner depends on the task to be solved. In the simplest case, it consists of the best arm. However, as will be discussed in Section 4, more complex predictions are conceivable, such as a complete ranking of all arms.

The above sample complexity bound is valid most of the time (more than $1 - \delta$ of the runs). However, in case an error occurs and the correct recommendation is not found by the algorithm, the bound does not guarantee anything. Therefore, it cannot be directly linked to the expected sample complexity. In order to define the expected sample complexity, the learning algorithm needs to terminate in a finite number of steps with probability 1. Under this condition, running a learning algorithm on the same bandit instance results in a finite sample complexity, which is a random number distributed according to an unknown law $\mathbf{P} : \mathbb{N} \to [0, 1]$. The distribution $\mathbf{P}$ has finite support, since the algorithm terminates in a finite number of steps in every case. By definition, the *expected sample complexity* of the learning algorithm is the finite mean of the distribution $\mathbf{P}$. Moreover, the *worst case sample complexity* is the upper bound of the support of $\mathbf{P}$.

2.6   PAC Algorithms

In many applications, one might be interested in gaining efficiency at the cost of optimality: The algorithm is allowed to return a solution that is only approximately optimal, though it is supposed to do so more quickly. For standard bandit problems, for example, this could mean returning an arm the expected reward of which deviates by at most some ϵ from the expected reward of the optimal arm.

In the preference-based setup, approximation errors are less straightforward to define. Nevertheless, the sample complexity can also be analyzed in a PAC-framework as originally introduced by Even-Dar *et al.* [20] for value-based MABs. A preference-based MAB algorithm is called (ϵ, δ)-PAC preference-based MAB algorithm with a *sample complexity* $B(\mathbf{Q}, K, \epsilon, \delta)$, if it terminates and returns an ϵ-optimal arm with probability at least $1 - \delta$, and the number of comparisons

---

[1] Here, we consider the pure exploration setup with fixed confidence. Alternatively, one can fix the horizon and control the error of the recommendation [4, 8, 9].

taken by the algorithm is at most $B(\mathbf{Q}, K, \epsilon, \delta)$. If the problem is to select a single arm, ϵ-optimality could mean, for example, that $\Delta_{i^*,j} < \epsilon$, although other notions of approximation can be used as well.

2.7   Explore-then-Exploit Algorithms

Most PB-MAB algorithms for optimizing regret are based on the idea of decoupling the exploration and exploitation phases: First, the algorithm tries to identify the best arm with high probability, and then fully commits to the arm found to be best for the rest of the time (i.e., repeatedly compares this arm to itself). Algorithms implementing this principle are called "explore-then-exploit" algorithms.

Such algorithms need to know the time horizon T in advance, since being aware of the horizon, the learning algorithm is able to control the regret incurred in case it fails to identify the best arm. More specifically, assume a so-called exploratory algorithm A to be given, which is able to identify the best arm a_{i^*} with probability at least $1 - \delta$. By setting δ to $1/T$, algorithm A guarantees that $\mathbf{P}(\widehat{i}^* = i^*) > 1 - 1/T$ if it terminates before iteration step T, where $\widehat{i}^*$ is the arm index returned by A. Thus, if A terminates and commits a mistake, i.e., $\widehat{i}^* \neq i^*$, then the expected regret incurred in the exploitation phase is $1/T \cdot \mathcal{O}(T) = \mathcal{O}(1)$, since the per-round regret is upper-bounded by one and the exploitation phase consists of at most T steps. Consequently, the expected regret of an explore-then-exploit algorithm is

$$\mathbf{E}[R^T] \leq (1 - 1/T)\,\mathbf{E}[R_A^T] + (1/T)\,\mathcal{O}(T) = \mathcal{O}\left(\mathbf{E}[R_A^T] + 1\right) \ .$$

Note that the inequality is trivially valid if A does not terminate before T.

The same argument as given above for the case of expected regret also holds for high probability regret bounds in the explore-then-exploit framework. In summary, the performance of an explore-then-exploit algorithm is bounded by the performance of the exploration algorithm. More importantly, since the per round regret is at most one, the sample complexity of the exploration algorithm readily upper-bounds the expected regret; this fact was pointed out in [46, 44]. Therefore, like in the case of value-based MABs, explore-then-exploit algorithms somehow blur the distinction between the "pure exploration" and regret optimization setting.

However, in a recent study [47], a novel preference-based MAB algorithm is proposed that optimizes the cumulative regret without decoupling the exploration from the exploitation phase (for more details see Section 3.1). Without decoupling, there is no need to know the horizon in advance, which allows one to provide a *horizonless* regret bound that holds for any time step T.

The regret defined in (2) reflects the average quality of the decision made by the learner. Obviously, one can define a more strict or less strict regret by taking the maximum or minimum, respectively, instead of the average. Formally, the strong and weak regret in time step t are defined, respectively, as

$$r^t_{\max} = \max\left\{\Delta_{i^*,i(t)}, \Delta_{i^*,j(t)}\right\} \ ,$$
$$r^t_{\min} = \min\left\{\Delta_{i^*,i(t)}, \Delta_{i^*,j(t)}\right\} \ .$$

From a theoretical point of view, when the number of pairwise comparisons is bounded by a known horizon, these regret definitions do not lead to a fundamentally different problem. Roughly speaking, this is because most of the methods designed for optimizing regret seek to identify the best arm with high probability in the exploration phase, based on as few sample as possible.

3   Learning from Consistent Pairwise Comparisons

As explained in Section 2.1, learning in the preference-based MAB setting essentially means estimating the pairwise preference matrix $\mathbf{Q}$, i.e., the pairwise probabilities $q_{i,j}$. The target of the agent's prediction, however, is not the relation $\mathbf{Q}$ itself, but the best arm or, more generally, a ranking $\succ$ of all arms $\mathcal{A}$. Consequently, the least assumption to be made is a connection between $\mathbf{Q}$ and $\succ$, so that information about the former is indicative of the latter. Or, stated differently, the pairwise probabilities $q_{i,j}$ should be sufficiently consistent, so as to allow the learner to approximate and eventually identify the target (at least in the limit when the sample size grows to infinity). For example, if the target is a ranking $\succ$ on $\mathcal{A}$, then the $q_{i,j}$ should be somehow consistent with that ranking, e.g., in the sense that $a_i \succ a_j$ implies $q_{i,j} > 1/2$.

While this is only an example of a consistency property that might be required, different consistency or regularity assumptions on the pairwise probabilities $\mathbf{Q}$ have been proposed in the literature—needless to say, these assumptions have a major impact on how PB-MAB problems are tackled algorithmically. In this section and the next one, we provide an overview of approaches to such problems, categorized according to these assumptions (see Figure 1).

3.1   Axiomatic Approaches

The seminal work of Yue et al. [44] relies on three regularity properties on the set of arms and their pairwise probabilities:

- *Total order over arms*: there exists a total order $\succ$ on $\mathcal{A}$, such that $a_i \succ a_j$ implies $\Delta_{i,j} > 0$.
- *Strong stochastic transitivity*: for any triplet of arms such that $a_i \succ a_j \succ a_k$, the pairwise probabilities satisfy $\Delta_{i,k} \geq \max\left(\Delta_{i,j}, \Delta_{j,k}\right)$.
- *Stochastic triangle inequality*: for any triplet of arms such that $a_i \succ a_j \succ a_k$, the pairwise probabilities satisfy $\Delta_{i,k} \leq \Delta_{i,j} + \Delta_{j,k}$.

The first assumption of a total order with arms separated by positive margins ensures the existence of a unique best arm, which in this case coincides with the Condorcet winner.[2] The second and third assumptions induce a strong structure on the pairwise preferences, which allows one to devise efficient algorithms.

---

[2] In voting and choice theory, an option is a Condorcet winner if it beats all other options in a pairwise comparison. In our context, this means an arm a_i is considered a Condorcet winner if $\Delta_{i,j} > 1/2$ for all $j \in [K]$.

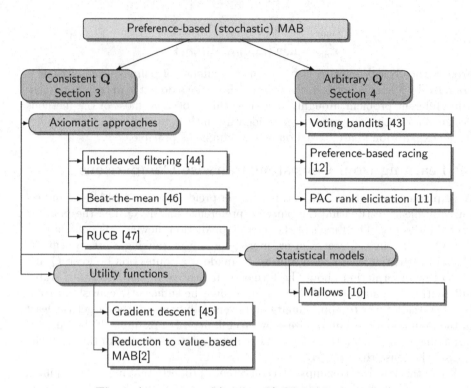

Fig. 1. A taxonomy of (stochastic) PB-MAB algorithms

Interleaved Filtering. Yue *et al.* [44] propose an explore-then-exploit algorithm. The exploration step consists of a simple sequential elimination strategy, called INTERLEAVED FILTERING (IF), which identifies the best arm with probability at least $1-\delta$. The IF algorithm successively selects an arm which is compared to other arms in a one-versus-all manner. More specifically, the currently selected arm a_i is compared to the rest of the active (not yet eliminated) arms. If an arm a_j beats a_i, that is, $\widehat{q}_{i,j} + c_{i,j} < 1/2$, then a_i is eliminated, and a_j is compared to the rest of the (active) arms, again in a one-versus-all manner. In addition, a simple pruning technique can be applied: if $\widehat{q}_{i,j} - c_{i,j} > 1/2$ for an arm a_j at any time, then a_j can be eliminated, as it cannot be the best arm anymore (with high probability). After the exploration step, the exploitation step simply takes the best arm $a_{\widehat{i}_*}$ found by IF and repeatedly compares $a_{\widehat{i}_*}$ to itself.

The authors analyze the expected regret achieved by IF. Assuming the horizon T to be finite and known in advance, they show that IF incurs an expected regret

$$\mathbf{E}\left[R_{\mathrm{IF}}^T\right] = \mathcal{O}\left(\frac{K}{\min_{j \neq i^*} \Delta_{i^*,j}} \log T\right) .$$

Beat the Mean. In a subsequent work, Yue and Joachims [46] relax the strong stochastic transitivity property and only require a so-called *relaxed stochastic* *transitivity* for the pairwise probabilities: There is a $\gamma \geq 1$ such that, for any triplet of arms such that $a_{i*} \succ a_i \succ a_j$ with respect to the total order $\succ$,

$$\gamma \Delta_{i*,j} \geq \max\{\Delta_{i*,i}, \Delta_{i,j}\} \ .$$

Obviously, the strong stochastic transitivity is recovered for $\gamma = 1$, albeit it is still restricted to triplets involving the best arm a_{i*}. The stochastic triangle inequality is relaxed in a similar way, and again, it is required to hold only relative to the best arm.

With these relaxed properties, Yue and Joachims [46] propose a preference-based online learning algorithm called BEAT-THE-MEAN (BTM), which is an elimination strategy resembling IF. However, while IF compares a single arm to the rest of the (active) arms in a one-versus-all manner, BTM selects an arm with the fewest comparisons so far and pairs it with a randomly chosen arm from the set of active arms (using the uniform distribution). Based on the outcomes of the pairwise comparisons, a score b_i is assigned to each active arm a_i, which is an empirical estimate of the probability that a_i is winning in a pairwise comparison (not taking into account which arm it was compared to). The idea is that comparing an arm a_i to the "mean" arm, which beats half of the arms, is equivalent to comparing a_i to an arm randomly selected from the active set. One can deduce a confidence interval for the b_i scores, which allows for deciding whether the scores for two arms are significantly different. An arm is then eliminated as soon as there is another arm with a significantly higher score.

In the regret analysis of BTM, a high probability bound is provided for a finite time horizon. More precisely, the regret accumulated by BTM is

$$\mathcal{O}\left(\frac{\gamma^7 K}{\min_{j \neq i*} \Delta_{i*,j}} \log T\right)$$

with high probability. This result is stronger than the one proven for IF, in which only the expected regret is upper bounded. Moreover, this high probability regret bound matches with the expected regret bound in the case $\gamma = 1$ (strong stochastic transitivity). The authors also analyze the BTM algorithm in a PAC setting, and find that BTM is an (ϵ, δ)-PAC preference-based learner (by setting its input parameters appropriately) with a sample complexity of $\mathcal{O}(\frac{\gamma^6 K}{\epsilon^2} \log \frac{KN}{\delta})$ if N is large enough, that is, N is the smallest positive integer for which $N = \left\lceil \frac{36\gamma^6}{\epsilon^2} \log \frac{K^3 N}{\delta} \right\rceil$. One may simplify this bound by noting that $N < N' = \left\lceil \frac{864\gamma^6}{\epsilon^2} \log \frac{K}{\delta} \right\rceil$. Therefore, the sample complexity is

$$\mathcal{O}\left(\frac{\gamma^6 K}{\epsilon^2} \log \frac{K\gamma \log(K/\delta)}{\delta\epsilon}\right) \ .$$

Preference-based UCB. In a very recent work by Zoghi *et al.* [47], the well-known UCB [5] algorithm is adapted from the value-based to the preference-based MAP setup. One of the main advantages of the proposed algorithm, called RUCB (for Relative UCB), is that only the existence of a Condorcet winner is required. Consequently, it is more broadly applicable. The RUCB algorithm is based on the "optimism in the face of uncertainty" principle, which means that the arms to be compared next are selected based on the optimistic estimates of the pairwise probabilities, that is, based on the upper boundaries $\widehat{q}_{i,j} + c_{i,j}$ of the confidence intervals. In an iteration step, RUCB selects the set of potential Condorcet winners for which all $\widehat{q}_{i,j} + c_{i,j}$ values are above $1/2$, and then selects an arm a_i from this set uniformly at random. Finally, a_i is compared to the arm a_j, $j = \mathrm{argmax}_{\ell \neq i} \widehat{q}_{i,\ell} + c_{i,\ell}$, that may lead to the smallest regret, taking into account the optimistic estimates.

In the analysis of the RUCB algorithm, horizonless regret bounds are provided, both for the expected regret and high probability bound. Thus, unlike the bounds for IF and BTM, these bounds are valid for each time step. Both the expected regret bound and high probability bound of RUCB are $\mathcal{O}(K \log T)$. However, while the regret bounds of IF and BTM only depend on $\min_{j \neq i^*} \Delta_{i^*,j}$, the constants are now of different nature, despite being still calculated based on the $\Delta_{i,j}$ values. Therefore, the regret bounds for RUCB are not directly comparable with those given for IF and BTM.

3.2 Regularity through Latent Utility Functions

The representation of preferences in terms of utility functions has a long history in decision theory [22]. The idea is that the absolute preference for each choice alternative can be reflected by a real-valued utility degree. Obviously, such degrees immediately impose a total order on the set of alternatives. Typically, however, the utility degrees are assumed to be latent and not directly observable.

In [45], a preference-based stochastic MAB setting is introduced in which the pairwise probabilities are directly derived from the (latent) utilities of the arms. More specifically, the authors assume a space $\mathcal{S}$ of arms, which is not necessarily finite.[3] The probability of an arm $a \in \mathcal{S}$ beating arm $a' \in \mathcal{S}$ is given by

$$\mathbf{P}(a \succ a') = \frac{1}{2} + \delta(a, a')$$

where $\delta : \mathcal{S} \times \mathcal{S} \to [-1/2, 1/2]$. Obviously, the closer the value of the function δ is to 0, the harder it becomes to compare the corresponding pair of arms. The authors furthermore assume the pairwise δ-values to be connected to an underlying (differentiable and strictly concave) utility function $u : \mathcal{S} \to \mathcal{R}$:

$$\frac{1}{2} + \delta(a, a') = \sigma\big(u(a) - u(a')\big) \ ,$$

where $\sigma : \mathbb{R} \to [0, 1]$ is called *link function*, as it establishes a connection between the pairwise probabilities and utilities. This function is assumed to satisfy the following conditions: $\lim_{x \to \infty} \sigma(x) = 1$ and $\lim_{x \to -\infty} \sigma(x) = 0$, $\sigma(x) = 1 - \sigma(x)$,

---

[3] This space corresponds to our set of arms $\mathcal{A}$. However, as we assume $\mathcal{A}$ to be finite, we use another notation here.

$\sigma(0) = 1/2$. An example of such a function is the logistic function, which was used in [45].

The problem of finding the optimal arm can be viewed as a noisy optimization task [21]. The underlying search space is $\mathcal{S}$, and the function values cannot be observed directly; instead, only noisy pairwise comparisons of function values (utilities) are available. In this framework, it is hard to have a reasonable estimate for the gradient, therefore the authors opted for applying an online convex optimization method [23], which does not require the gradient to be calculated explicitly.

In the theoretical analysis of the proposed method, the regret definition is similar to the one in (2), and can be written as

$$R^T = \sum_{t=1}^{T} \delta(a_*, a_t) + \delta(a_*, a_t') \ .$$

Here, however, the reference arm a_* is the best one known only in hindsight. In other words, a_* is the best arm among those evaluated during the search process.

Under a strong convexity assumption on ϵ, an expected regret bound for the proposed algorithm is computed as follows. Assuming the search space $\mathcal{S}$ to be given by the d-dimensional ball of radius R, the expected regret is

$$\mathbf{E}[R^T] \leq 2T^{3/4}\sqrt{10RdL} \ .$$

Ailon *et al.* [47] propose various methodologies to reduce the utility-based PB-MAB problem to the standard value-based MAB problem. In their setup, the utility of an arm is assumed to be in $[0, 1]$. Formally, $u : \mathcal{S} \to [0, 1]$, and the link function is a linear function $\sigma_{lin}(x) = \frac{1}{2}x$. Therefore, the probability of an arm $a \in \mathcal{S}$ beating another arm $a' \in \mathcal{S}$ is

$$\mathbf{P}(a \succ a') = \frac{1 + u(a) - u(a')}{2} \ ,$$

which is again in $[0, 1]$. The regret considered is the one defined in (2), where the reference arm a_{i^*} is the globally best arm with maximal utility.

In [47], two reduction techniques are proposed for a finite and an infinite set of arms. In both techniques, value-based MAB algorithms such as UCB [5] are used as a black box for driving the search in the space of arms. For a finite number of arms, value-based bandit instances are assigned to each arm, and these bandit algorithms are run in parallel. More specifically, assume that an arm $i(t)$ is selected in iteration t (to be explained in more detail shortly). Then, the bandit instance that belongs to arm $i(t)$ suggests another arm $j(t)$. These two arms are then compared in iteration t, and the reward, which is 0 or 1, is assigned to the bandit algorithm that belongs to $i(t)$. In iteration $t + 1$, the arm $j(t)$ suggested by the bandit algorithm is compared, that is, $i(t+1) = j(t)$. What is nice about this reduction technique is that, under some mild conditions on the performance of the bandit algorithm, the preference-based expected regret defined in (2) is asymptotically identical to the one achieved by the value-based algorithm for the standard value-based MAB task.

For infinitely many arms, the reduction technique can be viewed as a two player game. A run is divided into epochs: the ℓ-th epoch starts in round $t = 2^{\ell}$ and ends in round $t = 2^{\ell+1} - 1$, and in each epoch the players start a new game. During the ℓth epoch, the second player plays adaptively according to a strategy provided by the value-based bandit instance, which is able to handle infinitely many arms, such as the ConfidenceBall algorithm by Dani *et al.* [19]. The first player obeys some stochastic strategy, which is based on the strategy of the second player from the previous epoch. That is, the first player always draws a random arm from the multi-set of arms that contains the arms selected by the second player in the previous epoch. This reduction technique incurs an extra $\log T$ factor to the expected regret of the value-based bandit instance.

3.3 Regularity through Statistical Models

Since the most general task in the realm of preference-based bandits is to elicit a ranking of the complete set of arms based on noisy (probabilistic) feedback, it is quite natural to establish a connection to statistical models of rank data [37]. This idea was recently put forward by Busa-Fekete *et al.* [10], who assume the underlying data-generating process to be given in the form of a probability distribution $\mathbf{P} : \mathbb{S}_K \to [0, 1]$. Here, $\mathbb{S}_K$ is the set of all permutations of $[K]$ (the symmetric group of order K) or, via a natural bijection, the set of all rankings (total orders) of the K arms.

The probabilities for pairwise comparisons are then obtained as marginals of $\mathbf{P}$. More specifically, with $\mathbf{P}(\mathbf{r})$ the probability of observing the ranking $\mathbf{r}$, the probability $q_{i,j}$ that a_i is preferred to a_j is obtained by summing over all rankings $\mathbf{r}$ in which a_i precedes a_j:

$$q_{i,j} = \mathbf{P}(a_i \succ a_j) = \sum_{\mathbf{r} \in \mathcal{L}(r_j > r_i)} \mathbf{P}(\mathbf{r}) \qquad (4)$$

where $\mathcal{L}(r_j > r_i) = \{\mathbf{r} \in \mathbb{S}_K \,|\, r_j > r_i\}$ denotes the subset of permutations for which the rank r_j of a_j is higher than the rank r_i of a_i (smaller ranks indicate higher preference).

In this setting, the learning problem essentially comes down to making inference about $\mathbf{P}$ based on samples in the form of pairwise comparisons. Concretely, three different goals of the learner are considered, depending on whether the application calls for the prediction of a single arm, a full ranking of all arms, or the entire probability distribution:

- The **MPI** problem consists of finding the most preferred item i^*, namely the item whose probability of being top-ranked is maximal:

$$i^* = \operatorname*{argmax}_{1 \leq i \leq K} \mathbf{E}_{\mathbf{r} \sim \mathbf{P}} \, \mathbb{I}\{r_i = 1\} = \operatorname*{argmax}_{1 \leq i \leq K} \sum_{\mathbf{r} \in \mathcal{L}(r_i = 1)} \mathbf{P}(\mathbf{r}) \ ,$$

where $\mathbb{I}\{\cdot\}$ denotes the indicator function.

- The **MPR** problem consists of finding the most probable ranking $\mathbf{r}^*$:

$$\mathbf{r}^* = \underset{\mathbf{r} \in \mathbb{S}_K}{\operatorname{argmax}} \, \mathbf{P}(\mathbf{r})$$

- The **KLD** problem calls for producing a good estimate $\widehat{\mathbf{P}}$ of the distribution $\mathbf{P}$, that is, an estimate with small KL divergence:

$$\mathrm{KL}\left(\mathbf{P}, \widehat{\mathbf{P}}\right) = \sum_{\mathbf{r} \in \mathbb{S}_K} \mathbf{P}(\mathbf{r}) \log \frac{\mathbf{P}(\mathbf{r})}{\widehat{\mathbf{P}}(\mathbf{r})} < \epsilon$$

All three goals are meant to be achieved with probability at least $1 - \delta$.

Busa-Fekete *et al.* [10] assume the underlying probability distribution $\mathbf{P}$ to be a Mallows model [36], one of the most well-known and widely used statistical models of rank data [37]. The Mallows model or, more specifically, Mallows ϕ-distribution is a parameterized, distance-based probability distribution that belongs to the family of exponential distributions:

$$\mathbf{P}(\mathbf{r} \mid \theta, \widetilde{\mathbf{r}}) = \frac{1}{Z(\phi)} \phi^{d(\mathbf{r}, \widetilde{\mathbf{r}})} \tag{5}$$

where ϕ and $\widetilde{\mathbf{r}}$ are the parameters of the model: $\widetilde{\mathbf{r}} = (\tilde{r}_1, \ldots, \tilde{r}_K) \in \mathbb{S}_K$ is the location parameter (center ranking) and $\phi \in (0, 1]$ the spread parameter. Moreover, $d(\cdot, \cdot)$ is the Kendall distance on rankings, that is, the number of discordant pairs:

$$d(\mathbf{r}, \widetilde{\mathbf{r}}) = \sum_{1 \le i < j \le K} \mathbb{I}\{ (r_i - r_j)(\tilde{r}_i - \tilde{r}_j) < 0 \} \; .$$

The normalization factor in (5) can be written as

$$Z(\phi) = \sum_{\mathbf{r} \in \mathbb{S}_K} \mathbf{P}(\mathbf{r} \mid \theta, \widetilde{\mathbf{r}}) = \prod_{i=1}^{K-1} \sum_{j=0}^{i} \phi^j$$

and thus only depends on the spread [24]. Note that, since $d(\mathbf{r}, \widetilde{\mathbf{r}}) = 0$ is equivalent to $\mathbf{r} = \widetilde{\mathbf{r}}$, the center ranking $\widetilde{\mathbf{r}}$ is the mode of $\mathbf{P}(\cdot \mid \theta, \widetilde{\mathbf{r}})$, that is, the most probable ranking according to the Mallows model.

In the case of Mallows, it is easy to see that $\tilde{r}_i < \tilde{r}_j$ implies $q_{i,j} > 1/2$ for any pair of items a_i and a_j. That is, the center ranking defines a total order on the set of arms: If an arm a_i precedes another arm a_j in the (center) ranking, then a_i beats a_j in a pairwise comparison.[4] Moreover, as shown by Mallows [36], the pairwise probabilities can be calculated analytically as functions of the model parameters ϕ and $\widetilde{\mathbf{r}}$ as follows: Assume the Mallows model with parameters ϕ and $\widetilde{\mathbf{r}}$. Then, for any pair of items i and j such that $\tilde{r}_i < \tilde{r}_j$, the pairwise probability is given by $q_{i,j} = g(\tilde{r}_i, \tilde{r}_j, \phi)$, where

$$g(i, j, \phi) = h(j - i + 1, \phi) - h(j - i, \phi)$$

---

[4] Recall that this property is an axiomatic assumption underlying the IF and BTM algorithms. Interestingly, the stochastic triangle inequality, which is also assumed by Yue *et al.* [44], is not satisfied for Mallows ϕ-model [36].

with $h(k, \phi) = k/(1 - \phi^k)$. Based on this result, one can show that the "margin"

$$\min_{i \neq j} |1/2 - q_{i,j}|$$

around $1/2$ is relatively wide; more specifically, there is no $q_{i,j} \in (\frac{\phi}{1+\phi}, \frac{1}{1+\phi})$. Moreover, the result also implies that $q_{i,j} - q_{i,k} = O(\ell\phi^\ell)$ for arms a_i, a_j, a_k satisfying $\tilde{r}_i = \tilde{r}_j - \ell = \tilde{r}_k - \ell - 1$ with $1 < \ell$, and $q_{i,k} - q_{i,j} = O(\ell\phi^\ell)$ for arms a_i, a_j, a_k satisfying $\tilde{r}_i = \tilde{r}_j + \ell = \tilde{r}_k + \ell + 1$ with $1 < \ell$. Therefore, deciding whether an arm a_j has higher or lower rank than a_i (with respect to $\tilde{r}$) is easier than selecting the preferred option from two candidates a_j and a_k for which $j, k \neq i$.

Based on these observations, one can devise an efficient algorithm for identifying the most preferred arm when the underlying distribution is Mallows. The algorithm proposed in [10] for the **MPI** problem, called MALLOWSMPI, is similar to the one used for finding the largest element in an array. However, since a stochastic environment is assumed in which the outcomes of pairwise comparisons are random variables, a single comparison of two arms a_i and a_j is not enough; instead, they are compared until

$$1/2 \notin \left[\widehat{q}_{i,j} - c_{i,j}, \widehat{q}_{i,j} + c_{i,j} \right] . \tag{6}$$

This simple strategy finds the most preferred arm with probability at least $1 - \delta$ for a sample complexity that is of the form $\mathcal{O}\left(\frac{K}{\rho^2} \log \frac{K}{\delta\rho} \right)$, where $\rho = \frac{1-\phi}{1+\phi}$.

For the **MPR** problem, a sampling strategy called MALLOWSMERGE is proposed, which is based on the merge sort algorithm for selecting the arms to be compared. However, as in the case of **MPI**, two arms a_i and a_j are not only compared once but until condition (6) holds. The MALLOWSMERGE algorithm finds the most probable ranking, which coincides with the center ranking of the Mallows model, with a sample complexity of

$$\mathcal{O}\left(\frac{K \log_2 K}{\rho^2} \log \frac{K \log_2 K}{\delta\rho} \right) ,$$

where $\rho = \frac{1-\phi}{1+\phi}$. The leading factor of the sample complexity of MALLOWSMERGE differs from the one of MALLOWSMPI by a logarithmic factor. This was to be expected, and simply reflects the difference in the worst case complexity for finding the largest element in an array and sorting an array using the merge sort strategy.

The **KLD** problem turns out to be very hard for the case of Mallows, and even for small K, the sample complexity required for a good approximation of the underlying Mallows model is extremely high with respect to ϵ. In [10], the existence of a polynomial algorithm for this problem (under the assumption of the Mallows model) was left as an open question.

4   Learning from Inconsistent Pairwise Comparisons

The methods presented in the previous section essentially proceed from a given target, for example a ranking $\succ$ of all arms, which is considered as a "ground truth".

The preference feedback in the form of (stochastic) pairwise comparisons provide information about this target and, consequently, should obey certain consistency or regularity properties. This is perhaps most explicitly expressed in Section 3.3, in which the $q_{i,j}$ are derived as marginals of a probability distribution on the set of all rankings, which can be seen as modeling a noisy observation of the ground truth given in the form of the center ranking.

Another way to look at the problem is to start from the pairwise preferences $\mathbf{Q}$ themselves, that is to say, to consider the pairwise probabilities $q_{i,j}$ as the ground truth. In tournaments in sports, for example, the $q_{i,j}$ may express the probabilities of one team a_i beating another one a_j. In this case, there is no underlying ground truth ranking from which these probabilities are derived. Instead, it is just the other way around: A ranking is derived from the pairwise comparisons. Moreover, there is no reason for why the $q_{i,j}$ should be consistent in a specific sense. In particular, preferential cyclic and violations of transitivity are commonly observed in many applications.

This is exactly the challenge faced by *ranking procedures*, which have been studied quite intensely in operations research and decision theory [40, 18]. A ranking procedure $\mathcal{R}$ turns $\mathbf{Q}$ into a complete preorder relation $\succ^{\mathcal{R}}$ of the alternatives under consideration. Thus, another way to pose the preference-based MAB problem is to instantiate $\succ$ with $\succ^{\mathcal{R}}$ as the target for prediction—the connection between $\mathbf{Q}$ and $\succ$ is then established by the ranking procedure $\mathcal{R}$, which of course needs to be given as part of the problem specification.

Formally, a ranking procedure $\mathcal{R}$ is a map $[0,1]^{K \times K} \to \mathcal{C}_K$, where $\mathcal{C}_K$ denotes the set of complete preorders on the set of alternatives. We denote the complete preorder produced by the ranking procedure $\mathcal{R}$ on the basis of $\mathbf{Q}$ by $\succ^{\mathcal{R}}_{\mathbf{Q}}$, or simply by $\succ^{\mathcal{R}}$ if $\mathbf{Q}$ is clear from the context. In [12], three instantiations of the ranking procedure $\mathcal{R}$ are considered:

- Copeland's ranking (CO) is defined as follows [40]: $a_i \succ^{CO} a_j$ if and only if $d_i > d_j$, where $d_i = \#\{k \in [K] \mid 1/2 < q_{i,k}\}$. The interpretation of this relation is very simple: An option a_i is preferred to a_j whenever a_i "beats" more options than a_j does.
- The sum of expectations (SE) (or Borda) ranking is a "soft" version of CO: $a_i \succ^{SE} a_j$ if and only if

$$q_i = \frac{1}{K-1} \sum_{k \neq i} q_{i,k} > \frac{1}{K-1} \sum_{k \neq j} q_{j,k} = q_j \ . \tag{7}$$

- The idea of the random walk (RW) ranking is to handle the matrix $\mathbf{Q}$ as a transition matrix of a Markov chain and order the options based on its stationary distribution. More precisely, RW first transforms $\mathbf{Q}$ into the stochastic matrix $\mathbf{S} = [s_{i,j}]_{K \times K}$ where $s_{i,j} = q_{i,j} / \sum_{\ell=1}^{K} q_{i,\ell}$. Then, it determines the stationary distribution $(v_1, \ldots, v_K)$ for this matrix (i.e., the eigenvector corresponding to the largest eigenvalue 1). Finally, the options are sorted according to these probabilities: $a_i \succ^{RW} a_j$ iff $v_i > v_j$. The RW ranking is directly motivated by the PageRank algorithm [7], which has been well studied in social choice theory [3, 6] and rank aggregation [41], and which is widely used in many application fields [7, 32].

Top-k Selection. The learning problem considered in [12] is to find, for some $k < K$, the top-k arms with respect to the above ranking procedures with high probability. To this end, three different learning algorithms are proposed in the finite horizon case, with the horizon given in advance. In principle, these learning problems are very similar to the value-based racing task [38, 39], where the goal is to select the k arms with the highest means. However, in the preference-based case, the ranking over the arms is determined by the ranking procedure instead of the means. Accordingly, the algorithms proposed in [12] consist of a successive selection and rejection strategy. The sample complexity bounds of all algorithms are of the form $\mathcal{O}(K^2 \log T)$. Thus, they are not as tight in the number of arms as those considered in Section 3. This is mainly due to the lack of any assumptions on the structure of $\mathbf{Q}$. Since there are no regularities, and hence no redundancies in $\mathbf{Q}$ that could be exploited, a sufficiently good estimation of the entire relation is needed to guarantee a good approximation of the target ranking in the worst case.

PAC Rank Elicitation. In a subsequent work by Busa-Fekete *et al.* [11], an extended version of the top-k selection problem is considered. In the PAC *rank elicitation problem*, the goal is to find a ranking that is "close" to the ranking produced by the ranking procedure with high probability. To make this problem feasible, more practical ranking procedures are considered. In fact, the problem of ranking procedures like Copeland is that a minimal change of a value $q_{i,j} \approx \frac{1}{2}$ may strongly influence the induced order relation $\succ^{\mathrm{CO}}$. Consequently, the number of samples needed to assure (with high probability) a certain approximation quality may become arbitrarily large. A similar problem arises for $\succ^{\mathrm{SE}}$ as a target order if some of the individual scores q_i are very close or equal to each other.

As a practical (yet meaningful) solution to this problem, the relations $\succ^{\mathrm{CO}}$ and $\succ^{\mathrm{SE}}$ are made a bit more "partial" by imposing stronger requirements on the order. To this end, let $d_i^* = \#\{k \mid 1/2 + \epsilon < q_{i,k}, i \neq k\}$ denote the number of options that are beaten by a_i with a margin $\epsilon > 0$, and let

$$s_i^* = \#\{k : |1/2 - q_{i,k}| \leq \epsilon, i \neq k\} \ .$$

Then, the ϵ-insensitive Copeland relation is defined as follows: $a_i \succ^{\mathrm{CO}_\epsilon} a_j$ if and only if $d_i^* + s_i^* > d_j^*$. Likewise, in the case of $\succ^{\mathrm{SE}}$, small differences of the q_i are neglected the ϵ-insensitive sum of expectations relation is defined as follows: $a_i \succ^{\mathrm{SE}_\epsilon} a_j$ if and only if $q_i + \epsilon > q_j$.

These ϵ-insensitive extensions are interval (and hence partial) orders, that is, they are obtained by characterizing each option a_i by the interval $[d_i^*, d_i^* + s_i^*]$ and sorting intervals according to $[a, b] \succ [a', b']$ iff $b > a'$. It is readily shown that $\succ^{\mathrm{CO}_\epsilon} \subseteq \succ^{\mathrm{CO}_{\epsilon'}} \subseteq \succ^{\mathrm{CO}}$ for $\epsilon > \epsilon'$, with equality $\succ^{\mathrm{CO}_0} \equiv \succ^{\mathrm{CO}}$ if $q_{i,j} \neq 1/2$ for all $i \neq j \in [K]$ (and similarly for SE). The parameter ϵ controls the strictness of the order relations, and thereby the difficulty of the rank elicitation task.

As mentioned above, the task in PAC rank elicitation is to approximate $\succ^{\mathcal{R}}$ without knowing the $q_{i,j}$. Instead, relevant information can only be obtained through sampling pairwise comparisons from the underlying distribution. Thus, the options can be compared in a pairwise manner, and a single sample essentially informs about a pairwise preference between two options a_i and a_j. The goal is to devise a *sampling strategy* that keeps the size of the sample (the sample complexity) as small as possible while producing an estimation $\succ$ that is "good" in a PAC sense: $\succ$ is supposed to be sufficiently "close" to $\succ^{\mathcal{R}}$ with high probability. Actually, the algorithms in [11] even produce a total order as a prediction, i.e., $\succ$ is a ranking that can be represented by a permutation τ of order K, where τ_i denotes the rank of option a_i in the order.

To formalize the notion of "closeness", appropriate distance measures are applied that compare a (predicted) permutation τ with a (target) order $\succ$. In [11], the following two measures are used: The *number of discordant pairs* (NDP), which is closely connected to Kendall's rank correlation [31], and can be expressed as follows:

$$d_K(\tau, \succ) = \sum_{i=1}^{K} \sum_{j \neq i} \mathbb{I}\{\tau_j < \tau_i\} \mathbb{I}\{a_i \succ a_j\}.$$

The *maximum rank difference* (MRD) is defined as the maximum difference between the rank of an object a_i according to τ and $\succ$, respectively. More specifically, since $\succ$ is a partial but not necessarily total order, τ is compared to the set $\mathcal{L}^{\succ}$ of its linear extensions:[5]

$$d_{\mathcal{M}}(\tau, \succ) = \min_{\tau' \in \mathcal{L}^{\succ}} \max_{1 \leq i \leq K} |\tau_i - \tau_i'|.$$

In [11], the authors propose four different methods for the two ϵ-sensitive ranking procedures, along with the two distance measures described above. Each algorithm calculates a surrogate ranking based on the empirical estimate of the preference matrix whose distance can be upper-bounded again based on some statistics of the empirical estimates of preference. The sampling is carried out in a greedy manner in every case, in the sense that those arms are compared which are supposed to result in a maximum decrease of the upper bound calculated for the surrogate ranking.

An expected sample complexity bound is calculated for the ϵ-sensitive Copeland ranking procedure along with the MRD distance in a similar way like in [29, 30]. The bound is of the form $\mathcal{O}\left(R_1 \log\left(\frac{R_1}{\delta}\right)\right)$, where R_1 is a task dependent constant. More specifically, R_1 depends on the $\Delta_{i,j}$ values, and on the robustness of the ranking procedure to small changes in the preference matrix (i.e., on how much the ranking produced by the ranking procedure might be changed in terms of the MRD distance if the preference matrix is slightly altered). Interestingly, an expected sample complexity can also be calculated for the ϵ-insensitive sum of expectations ranking procedure along with the MRD distance with a similar flavor like for the ϵ-sensitive Copeland ranking procedure. The analysis of the NDP distance is more difficult, since small changes in the preference matrix

---

[5] $\tau \in \mathcal{L}^{\succ}$ iff $\forall i, j \in [K] : (a_i \succ a_j) \Rightarrow (\tau_i < \tau_j)$.

may strongly change the ranking in terms of the NDP distance. The sample complexity analysis for this distance has therefore been left as an open question.

Urvoy *et al.* [43] consider a setup similar to the one in [11]. Again, a ranking procedure is assumed that produces a ranking over the arms, and the goal of the learner is to find a maximal element according to this ranking (instead of the top-k). Note that a ranking procedure only defines a complete preorder, which means there can be more than one "best" element. The authors propose an algorithm called SAVAGE as a general solution to this problem, which can be adapted to various ranking procedure. Concretely, the Copeland and the sum of expectations (or Borda counts) procedure are used in their study. Moreover, they also devise a method to find the Condorcet winner, assuming it exists—a problem that is akin to the axiomatic approaches described in Subsection 3.1.

The sample complexity of the implementations in [43] are of order K^2 in general. Just like in [11], this is the price to pay for a "model-free" learning procedure that does not make any assumptions on the structure of the preference matrix. The analysis of the authors is more general, because they also investigate the infinite horizon case, where a time limit is not given in advance.

5 Summary and Perspectives

This paper provides a survey of the state-of-the-art in preference-based online learning with bandit algorithms, an emerging research field that we referred to as preference-based multi-armed bandits (PB-MAB). In contrast to standard MAB problems, where bandit information is understood as (stochastic) real-valued rewards produced by the arms the learner decided to explore (or exploit), feedback is assumed to be of a more indirect and qualitative nature in the PB-MAB setting. In particular, the work so far has focused on preference information in the form of comparisons between pairs of arms. We have given an overview of instances of the PB-MAP problem that have been studied in the literature, algorithms for tackling them and criteria for evaluating such algorithms.

Needless to say, the field is still in its beginning and far from being mature. The contributions so far are highly interesting, and some of them have already been used in concrete applications, such as preference-based reinforcement learning [13]. Yet, they are still somewhat fragmentary, and a complete and coherent theoretical framework is still to be developed. With this survey, we hope to contribute to the popularization, development and shaping of the field.

We conclude the paper with a short (and certainly non-exhaustive) list of open problems that we consider particularly interesting for future work:

- As we have seen, the difficulty of PB-MAB learning strongly depends on the assumptions on properties of the preference relation **Q**: The more restrictive these assumptions are, the easier the learning task becomes. An interesting question in this regard concerns the "weakest" assumptions one could make while still guaranteeing the existence of an algorithm that scales linearly in the number of arms.
- A similar question can be asked for the regret. The RUCB algorithm achieves a high probability regret bound of order $K \log T$ by merely assuming the

existence of a Condorcet winner. Yet, this assumption is arguable and certainly not always valid.

- For most of the settings discussed in the paper, such as those based on statistical models like Mallows, a lower bound on the sample complexity is not known. Thus, it is difficult to say whether an algorithm is optimal or not. There are a few exceptions, however. For dueling bandits, it is known that, for any algorithm A, there is a bandit problem such that the regret of A is $\Omega(K \log T)$. Obviously, this is also a lower bound for all settings starting from weaker assumptions than dueling bandits, including RUCB.
- Another important problem concerns the development of (statistical) tests for verifying the assumptions made by the different approaches in a real application. In the case of the statistical approach based on the Mallows distribution, for example, the problem would be to decide, based on data in the form of pairwise comparisons, whether the underlying distribution could indeed be Mallows. Similarly, one could ask for methods to test the validity of strong stochastic transitivity and stochastic triangle inequality as required by methods such as IF and BTM.
- Last but not least, it would be important to test the algorithms in real applications—crowd-sourcing platforms appear to provide an interesting testbed in this regard.

Acknowledgments. The authors are grateful for financial support by the German Research Foundation (DFG).

References

[1] Ailon, N., Hatano, K., Takimoto, E.: Bandit online optimization over the permutahedron. CoRR, abs/1312.1530 (2014)

[2] Ailon, N., Karnin, Z., Joachims, T.: Reducing dueling bandits to cardinal bandits. In: Proceedings of the International Conference on Machine Learning (ICML), JMLR W&CP, vol. 32(1), pp. 856–864 (2014)

[3] Altman, A., Tennenholtz, M.: Axiomatic foundations for ranking systems. Journal of Artificial Intelligence Research 31(1), 473–495 (2008)

[4] Audibert, J.Y., Bubeck, S., Munos, R.: Best arm identification in multi-armed bandits. In: Proceedings of the Twenty-third Conference on Learning Theory (COLT), pp. 41–53 (2010)

[5] Auer, P., Cesa-Bianchi, N., Fischer, P.: Finite-time analysis of the multiarmed bandit problem. Machine Learning 47, 235–256 (2002)

[6] Brandt, F., Fischer, F.: PageRank as a weak tournament solution. In: Deng, X., Graham, F.C. (eds.) WINE 2007. LNCS, vol. 4858, pp. 300–305. Springer, Heidelberg (2007)

[7] Brin, S., Page, L.: The anatomy of a large-scale hypertextual web search engine. Computer Networks 30(1-7), 107–117 (1998)

[8] Bubeck, S., Munos, R., Stoltz, G.: Pure exploration in finitely-armed and continuous-armed bandits. Theoretical Computer Science 412, 1832–1852 (2011)

[9] Bubeck, S., Wang, T., Viswanathan, N.: Multiple identifications in multi-armed bandits. In: Proceedings of the International Conference on Machine Learning (ICML), JMLR W&CP, vol. 28(1), pp. 258–265 (2013)

[10] Busa-Fekete, R., Hüllermeier, E., Szörényi, B.: Preference-based rank elicitation using statistical models: The case of Mallows. In: Proceedings of the International
Conference on Machine Learning (ICML), JMLR W&CP, vol. 32(2), pp. 1071–1079
(2014)

[11] Busa-Fekete, R., Szörényi, B., Hüllermeier, E.: PAC rank elicitation through adaptive sampling of stochastic pairwise preferences. In: Proceedings of the Twenty-
Eighth AAAI Conference on Artificial Intelligence, AAAI 2014 (2014)

[12] Busa-Fekete, R., Szörényi, B., Weng, P., Cheng, W., Hüllermeier, E.: Top-k selection based on adaptive sampling of noisy preferences. In: Proceedings of the
International Conference on Machine Learning (ICML), JMLR W&CP, vol. 28(3),
pp. 1094–1102 (2013)

[13] Busa-Fekete, R., Szörényi, B., Weng, P., Cheng, W., Hüllermeier, E.: Preference-
based reinforcement learning: Evolutionary direct policy search using a preference-
based racing algorithm. Machine Learning (page accepted, 2014)

[14] Cesa-Bianchi, N., Lugosi, G.: Prediction, Learning, and Games. Cambridge
University Press, NY (2006)

[15] Cesa-Bianchi, N., Lugosi, G.: Combinatorial bandits. In: Proceedings of the
Twenty-second Conference on Learning Theory (COLT), pp. 237–246 (2009)

[16] Chakrabarti, D., Kumar, R., Radlinski, F., Upfal, E.: Mortal Multi-Armed Bandits. In: Neural Information Processing Systems (NIPS), pp. 273–280. MIT Press
(2008)

[17] Chen, X., Bennett, P.N., Collins-Thompson, K., Horvitz, E.: Pairwise ranking
aggregation in a crowdsourced setting. In: Proceedings of the Sixth ACM International Conference on Web Search and Data Mining, pp. 193–202 (2013)

[18] Chevaleyre, Y., Endriss, U., Lang, J., Maudet, N.: A short introduction to com-
putational social choice. In: van Leeuwen, J., Italiano, G.F., van der Hoek, W.,
Meinel, C., Sack, H., Plášil, F. (eds.) SOFSEM 2007. LNCS, vol. 4362, pp. 51–69.
Springer, Heidelberg (2007)

[19] Dani, V., Hayes, T.P., Kakade, S.M.: Stochastic linear optimization under bandit
feedback. In: Proceedings of the Twenty-first Conference on Learning Theory
(COLT), pp. 355–366 (2008)

[20] Even-Dar, E., Mannor, S., Mansour, Y.: PAC bounds for multi-armed bandit and
markov decision processes. In: Kivinen, J., Sloan, R.H. (eds.) COLT 2002. LNCS
(LNAI), vol. 2375, pp. 255–270. Springer, Heidelberg (2002)

[21] Finck, S., Beyer, H., Melkozerov, A.: Noisy optimization: a theoretical strategy
comparison of ES, EGS, SPSA & IF on the noisy sphere. In: Proceedings of the
13th Annual Conference on Genetic and Evolutionary Computation (GECCO),
pp. 813–820. ACM (2011)

[22] Fishburn, P.C.: Utility theory for decision making. John Wiley and Sons, New
York (1970)

[23] Flaxman, A., Kalai, A.T., McMahan, B.H.: Online convex optimization in the bandit setting: gradient descent without a gradient. In: Proceedings of the Sixteenth
Annual ACM-SIAM Symposium on Discrete Algorithms (SODA), pp. 385–394
(2005)

[24] Fligner, M.A., Verducci, J.S.: Distance based ranking models. Journal of the Royal
Statistical Society. Series B (Methodological) 48(3), 359–369 (1986)

[25] Fürnkranz, J., Hüllermeier, E. (eds.): Preference Learning. Springer (2011)

[26] Guo, S., Sanner, S., Graepel, T., Buntine, W.: Score-based bayesian skill learning.
In: Flach, P.A., De Bie, T., Cristianini, N. (eds.) ECML PKDD 2012, Part I.
LNCS, vol. 7523, pp. 106–121. Springer, Heidelberg (2012)

[27] Hoeffding, W.: Probability inequalities for sums of bounded random variables. Journal of the American Statistical Association 58, 13–30 (1963)

[28] Hofmann, K.: Fast and Reliably Online Learning to Rank for Information Retrieval. PhD thesis, Dutch Research School for Information and Knowledge Systems, Off Page, Amsterdam (2013)

[29] Kalyanakrishnan, S.: Learning Methods for Sequential Decision Making with Imperfect Representations. PhD thesis, University of Texas at Austin (2011)

[30] Kalyanakrishnan, S., Tewari, A., Auer, P., Stone, P.: Pac subset selection in stochastic multi-armed bandits. In: Proceedings of the Twenty-ninth International Conference on Machine Learning (ICML 2012), pp. 655–662 (2012)

[31] Kendall, M.G.: Rank correlation methods. Charles Griffin, London (1955)

[32] Kocsor, A., Busa-Fekete, R., Pongor, S.: Protein classification based on propagation on unrooted binary trees. Protein and Peptide Letters 15(5), 428–434 (2008)

[33] Kohli, P., Salek, M., Stoddard, G.: A fast bandit algorithm for recommendation to users with heterogenous tastes. In: Proceedings of the Twenty-Seventh AAAI Conference on Artificial Intelligence (AAAI 2013) (2013)

[34] Kuleshov, V., Precup, D.: Algorithms for multi-armed bandit problems. CoRR, abs/1402.6028 (2014)

[35] Lai, T.L., Robbins, H.: Asymptotically efficient allocation rules. Advances in Applied Mathematics 6(1), 4–22 (1985)

[36] Mallows, C.: Non-null ranking models. Biometrika 44(1), 114–130 (1957)

[37] Marden, J.I.: Analyzing and Modeling Rank Data. Chapman & Hall (1995)

[38] Maron, O., Moore, A.W.: Hoeffding races: accelerating model selection search for classification and function approximation. In: Proceedings of the Advances in Neural Information Processing Systems, pp. 59–66 (1994)

[39] Maron, O., Moore, A.W.: The racing algorithm: Model selection for lazy learners. Artificial Intelligence Review 5(1), 193–225 (1997)

[40] Moulin, H.: Axioms of cooperative decision making. Cambridge University Press (1988)

[41] Negahban, S., Oh, S., Shah, D.: Iterative ranking from pairwise comparisons. In: Proceedings of the Advances in Neural Information Processing Systems (NIPS), pp. 2483–2491 (2012)

[42] Radlinski, F., Kurup, M., Joachims, T.: How does clickthrough data reflect retrieval quality? In: Proceedings of the 17th ACM Conference on Information and Knowledge Management (CIKM), pp. 43–52 (2008)

[43] Urvoy, T., Clerot, F., Féraud, R., Naamane, S.: Generic exploration and k-armed voting bandits. In: Proceedings of the 30th International Conference on Machine Learning (ICML), JMLR W&CP, vol. 28, pp. 91–99 (2013)

[44] Yue, Y., Broder, J., Kleinberg, R., Joachims, T.: The K-armed dueling bandits problem. Journal of Computer and System Sciences 78(5), 1538–1556 (2012)

[45] Yue, Y., Joachims, T.: Interactively optimizing information retrieval systems as a dueling bandits problem. In: Proceedings of the 26th International Conference on Machine Learning (ICML), pp. 1201–1208 (2009)

[46] Yue, Y., Joachims, T.: Beat the mean bandit. In: Proceedings of the International Conference on Machine Learning (ICML), pp. 241–248 (2011)

[47] Zoghi, M., Whiteson, S., Munos, R., de Rijke, M.: Relative upper confidence bound for the k-armed dueling bandit problem. In: Proceedings of the International Conference on Machine Learning (ICML), JMLR W&CP, vol. 32(1), pp. 10–18 (2014)

A Map of Update Constraints
in Inductive Inference

Timo Kötzing and Raphaela Palenta[*]

Friedrich-Schiller-Universität Jena, Germany
{timo.koetzing,raphaela-julia.palenta}@uni-jena.de

Abstract. We investigate how different learning restrictions reduce learning power and how the different restrictions relate to one another. We give a complete map for nine different restrictions both for the cases of complete information learning and set-driven learning. This completes the picture for these well-studied *delayable* learning restrictions. A further insight is gained by different characterizations of *conservative* learning in terms of variants of *cautious* learning.

Our analyses greatly benefit from general theorems we give, for example showing that learners with exclusively delayable restrictions can always be assumed total.

1 Introduction

This paper is set in the framework of *inductive inference*, a branch of (algorithmic) learning theory. This branch analyzes the problem of algorithmically learning a description for a formal language (a computably enumerable subset of the set of natural numbers) when presented successively all and only the elements of that language. For example, a learner h might be presented more and more even numbers. After each new number, h outputs a description for a language as its conjecture. The learner h might decide to output a program for the set of all multiples of 4, as long as all numbers presented are divisible by 4. Later, when h sees an even number not divisible by 4, it might change this guess to a program for the set of all multiples of 2.

Many criteria for deciding whether a learner h is *successful* on a language L have been proposed in the literature. Gold, in his seminal paper [Gol67], gave a first, simple learning criterion, **TxtGEx**-*learning*[1], where a learner is *successful* iff, on every *text* for L (listing of all and only the elements of L) it eventually stops changing its conjectures, and its final conjecture is a correct description for the input sequence. Trivially, each single, describable language L has a suitable constant function as a **TxtGEx**-learner (this learner constantly outputs a description for L). Thus, we are interested in analyzing for which *classes of languages* $\mathcal{L}$ there is a *single learner* h learning *each* member of $\mathcal{L}$. This framework

---

[*] We would like to thank the reviewers for their very helpful comments.
[1] **Txt** stands for learning from a *text* of positive examples; **G** stands for Gold, who introduced this model, and is used to to indicate full-information learning; **Ex** stands for *explanatory*.

P. Auer et al. (Eds.): ALT 2014, LNAI 8776, pp. 40–54, 2014.
© Springer International Publishing Switzerland 2014

is also sometimes known as *language learning in the limit* and has been studied extensively, using a wide range of learning criteria similar to **TxtGEx**-learning (see, for example, the textbook [JORS99]).

A wealth of learning criteria can be derived from **TxtGEx**-learning by adding restrictions on the intermediate conjectures and how they should relate to each other and the data. For example, one could require that a conjecture which is consistent with the data must not be changed; this is known as *conservative* learning and known to restrict what classes of languages can be learned ([Ang80], we use **Conv** to denote the restriction of conservative learning). Additionally to conservative learning, the following learning restrictions are considered in this paper (see Section 2.1 for a formal definition of learning criteria including these learning restrictions).

In *cautious* learning (**Caut**, [OSW82]) the learner is not allowed to ever give a conjecture for a strict subset of a previously conjectured set. In *non-U-shaped* learning (**NU**, [BCM+08]) a learner may never *semantically* abandon a correct conjecture; in *strongly non-U-shaped* learning (**SNU**, [CM11]) not even syntactic changes are allowed after giving a correct conjecture.

In *decisive* learning (**Dec**, [OSW82]), a learner may never (semantically) return to a *semantically* abandoned conjecture; in *strongly decisive* learning (**SDec**, [Köt14]) the learner may not even (semantically) return to *syntactically* abandoned conjectures. Finally, a number of monotonicity requirements are studied ([Jan91, Wie91, LZ93]): in *strongly monotone* learning (**SMon**) the conjectured sets may only grow; in *monotone* learning (**Mon**) only incorrect data may be removed; and in *weakly monotone* learning (**WMon**) the conjectured set may only grow while it is consistent.

The main question is now whether and how these different restrictions reduce learning power. For example, non-U-shaped learning is known not to restrict the learning power [BCM+08], and the same for strongly non-U-shaped learning [CM11]; on the other hand, decisive learning *is* restrictive [BCM+08]. The relations of the different monotone learning restriction were given in [LZ93]. Conservativeness is long known to restrict learning power [Ang80], but also known to be equivalent to weakly monotone learning [KS95, JS98].

Cautious learning was shown to be a restriction but not when added to conservativeness in [OSW82, OSW86], similarly the relationship between decisive and conservative learning was given. In Exercise 4.5.4B of [OSW86] it is claimed (without proof) that cautious learners cannot be made conservative; we claim the opposite in Theorem 13.

This list of previously known results leaves a number of relations between the learning criteria open, even when adding trivial inclusion results (we call an inclusion trivial iff it follows straight from the definition of the restriction without considering the learning model, for example strongly decisive learning is included in decisive learning; formally, trivial inclusion is inclusion on the level of learning restrictions as predicates, see Section 2.1). With this paper we now give the complete picture of these learning restrictions. The result is shown as a map in Figure 1. A solid black line indicates a trivial inclusion (the lower

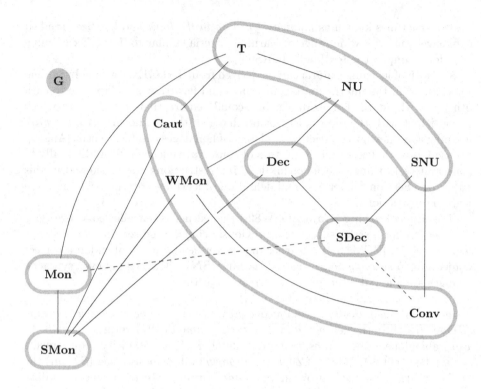

Fig. 1. Relation of criteria

criterion is included in the higher); a dashed black line indicates inclusion (which is not trivial). A gray box around criteria indicates equality of (learning of) these criteria.

A different way of depicting the same results is given in Figure 2 (where solid lines indicate any kind of inclusion). Results involving monotone learning can be found in Section 6, all others in Section 4.

For the important restriction of conservative learning we give the characterization of being equivalent to cautious learning. Furthermore, we show that even two weak versions of cautiousness are equivalent to conservative learning. Recall that cautiousness forbids to return to a strict subset of a previously conjectured set. If we now weaken this restriction to forbid to return to *finite* subsets of a previously conjectured set we get a restriction still equivalent to conservative learning. If we forbid to go down to a correct conjecture, effectively forbidding to ever conjecture a superset of the target language, we also obtain a restriction equivalent to conservative learning. On the other hand, if we weaken it so as to only forbid going to *infinite* subsets of previously conjectured sets, we obtain a restriction equivalent to no restriction. These results can be found in Section 4.

In *set-driven* learning [WC80] the learner does not get the full information about what data has been presented in what order and multiplicity; instead, the learner only gets the set of data presented so far. For this learning model it is

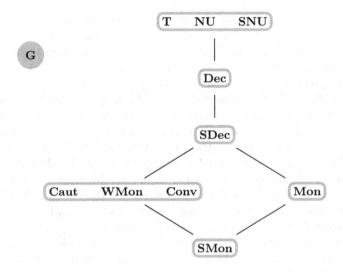

Fig. 2. Partial order of delayable learning restrictions in Gold-style learning

known that, surprisingly, conservative learning is no restriction [KS95]! We complete the picture for set driven learning by showing that set-driven learners can always be assumed conservative, strongly decisive and cautious, and by showing that the hierarchy of monotone and strongly monotone learning also holds for set-driven learning. The situation is depicted in Figure 3. These results can be found in Section 5.

1.1   Techniques

A major emphasis of this paper is on the techniques used to get our results. These techniques include specific techniques for specific problems, as well as general theorems which are applicable in many different settings. The general

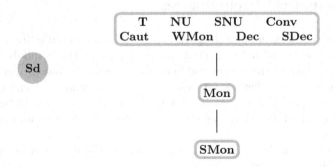

Fig. 3. Hierarchy of delayable learning restrictions in set-driven learning

techniques are given in Section 3, one main general result is as follows. It is well-known that any **TxtGEx**-learner h learning a language L has a *locking sequence*, a sequence σ of data from L such that, for any further data from L, the conjecture does not change and is correct. However, there might be texts such that no initial sequence of the text is a locking sequence. We call a learner such that any text for a target language contains a locking sequence *strongly locking*, a property which is very handy to have in many proofs. Fulk [Ful90] showed that, without loss of generality, a **TxtGEx**-learner can be assumed strongly locking, as well as having many other useful properties (we call this the *Fulk normal form*, see Definition 8). For many learning criteria considered in this paper it might be too much to hope for that all of them allow for learning by a learner in Fulk normal form. However, we show in Corollary 7 that we can always assume our learners to be strongly locking, total, and what we call *syntactically decisive*, never *syntactically* returning to syntactically abandoned hypotheses.

The main technique we use to show that something is decisively learnable, for example in Theorem 22, is what we call *poisoning* of conjectures. In the proof of Theorem 22 we show that a class of languages is decisively learnable by simulating a given monotone learner h, but changing conjectures as follows. Given a conjecture e made by h, if there is no mind change in the future with data from conjecture e, the new conjecture is equivalent to e; otherwise it is suitably changed, *poisoned*, to make sure that the resulting learner is decisive. This technique was also used in [CK10] to show strongly non-U-shaped learnability.

Finally, for showing classes of languages to be not (strongly) decisively learnable, we adapt a technique known in computability theory as a "priority argument" (note, though, that we do not deal with oracle computations). We use this technique to reprove that decisiveness is a restriction to **TxtGEx**-learning (as shown in [BCM+08]), and then use a variation of the proof to show that strongly decisive learning is a restriction to decisive learning.

Due to space constraints, we cannot give all proofs in this version of the paper. The full version of the paper can be found at `http://arxiv.org/abs/1404.7527`.

2 Mathematical Preliminaries

Unintroduced notation follows [Rog67], a textbook on computability theory.

$\mathbb{N}$ denotes the set of natural numbers, $\{0, 1, 2, \ldots\}$. The symbols $\subseteq$, $\subset$, $\supseteq$, $\supset$ respectively denote the subset, proper subset, superset and proper superset relation between sets; $\setminus$ denotes set difference. $\emptyset$ and λ denote the empty set and the empty sequence, respectively. The quantifier $\forall^\infty x$ means "for all but finitely many x". With dom and range we denote, respectively, domain and range of a given function.

Whenever we consider tuples of natural numbers as input to a function, it is understood that the general coding function $\langle \cdot, \cdot \rangle$ is used to code the tuples into a single natural number. We similarly fix a coding for finite sets and sequences, so that we can use those as input as well. For finite sequences, we suppose that for any $\sigma \subseteq \tau$ we have that the code number of σ is at most the code number of τ. We let $\mathbb{Seq}$ denote the set of all (finite) sequences, and $\mathbb{Seq}_{\leq t}$ the (finite) set of all sequences of length at most t using only elements $\leq t$.

If a function f is not defined for some argument x, then we denote this fact by $f(x)\uparrow$, and we say that f on x *diverges*; the opposite is denoted by $f(x)\downarrow$, and we say that f on x *converges*. If f on x converges to p, then we denote this fact by $f(x)\downarrow = p$. We let $\mathfrak{P}$ denote the set of all partial functions $\mathbb{N} \to \mathbb{N}$ and $\mathfrak{R}$ the set of all total such functions.

$\mathcal{P}$ and $\mathcal{R}$ denote, respectively, the set of all partial computable and the set of all total computable functions (mapping $\mathbb{N} \to \mathbb{N}$).

We let φ be any fixed acceptable programming system for $\mathcal{P}$ (an acceptable programming system could, for example, be based on a natural programming language such as C or Java, or on Turing machines). Further, we let φ_p denote the partial computable function computed by the φ-program with code number p. A set $L \subseteq \mathbb{N}$ is *computably enumerable (ce)* iff it is the domain of a computable function. Let $\mathcal{E}$ denote the set of all **ce** sets. We let W be the mapping such that $\forall e : W(e) = \mathrm{dom}(\varphi_e)$. For each e, we write W_e instead of $W(e)$. W is, then, a mapping from $\mathbb{N}$ onto $\mathcal{E}$. We say that e is an index, or program, (in W) for W_e.

We let Φ be a Blum complexity measure associated with φ (for example, for each e and x, $\Phi_e(x)$ could denote the number of steps that program e takes on input x before terminating). For all e and t we let $W_e^t = \{x \leq t \mid \Phi_e(x) \leq t\}$ (note that a complete description for the finite set W_e^t is computable from e and t). The symbol # is pronounced *pause* and is used to symbolize "no new input data" in a text. For each (possibly infinite) sequence q with its range contained in $\mathbb{N} \cup \{\#\}$, let $\mathrm{content}(q) = (\mathrm{range}(q) \setminus \{\#\})$. By using an appropriate coding, we assume that ? and # can be handled by computable functions. For any function T and all i, we use $T[i]$ to denote the sequence $T(0), \ldots, T(i-1)$ (the empty sequence if $i = 0$ and undefined, if any of these values is undefined).

2.1   Learning Criteria

In this section we formally introduce our setting of learning in the limit and associated learning criteria. We follow [Köt09] in its "building-blocks" approach for defining learning criteria.

A *learner* is a partial computable function $h \in \mathcal{P}$. A *language* is a **ce** set $L \subseteq \mathbb{N}$. Any total function $T : \mathbb{N} \to \mathbb{N} \cup \{\#\}$ is called a *text*. For any given language L, a *text for L* is a text T such that $\mathrm{content}(T) = L$. Initial parts of this kind of text is what learners usually get as information.

An *interaction operator* is an operator β taking as arguments a function h (the learner) and a text T, and that outputs a function p. We call p the *learning sequence* (or *sequence of hypotheses*) of h given T. Intuitively, β defines how a learner can interact with a given text to produce a sequence of conjectures.

We define the interaction operators **G**, **Psd** (partially set-driven learning, [SR84]) and **Sd** (set-driven learning, [WC80]) as follows. For all learners h, texts T and all i,

$$\mathbf{G}(h, T)(i) = h(T[i]);$$
$$\mathbf{Psd}(h, T)(i) = h(\text{content}(T[i]), i);$$
$$\mathbf{Sd}(h, T)(i) = h(\text{content}(T[i])).$$

Thus, in set-driven learning, the learner has access to the set of all previous data, but not to the sequence as in **G**-learning. In partially set-driven learning, the learner has the set of data and the current iteration number.

Successful learning requires the learner to observe certain restrictions, for example convergence to a correct index. These restrictions are formalized in our next definition.

A *learning restriction* is a predicate δ on a learning sequence and a text. We give the important example of explanatory learning (**Ex**, [Gol67]) defined such that, for all sequences of hypotheses p and all texts T,

$$\mathbf{Ex}(p, T) \Leftrightarrow p \text{ total } \wedge [\exists n_0 \forall n \geq n_0 : p(n) = p(n_0) \wedge W_{p(n_0)} = \text{content}(T)].$$

Furthermore, we formally define the restrictions discussed in Section 1 in Figure 4 (where we implicitly require the learning sequence p to be total, as in **Ex**-learning; note that this is a technicality without major importance).

$$\mathbf{Conv}(p, T) \Leftrightarrow [\forall i : \text{content}(T[i+1]) \subseteq W_{p(i)} \Rightarrow p(i) = p(i+1)];$$
$$\mathbf{Caut}(p, T) \Leftrightarrow [\forall i, j : W_{p(i)} \subset W_{p(j)} \Rightarrow i < j];$$
$$\mathbf{NU}(p, T) \Leftrightarrow [\forall i, j, k : i \leq j \leq k \ \wedge \ W_{p(i)} = W_{p(k)} = \text{content}(T) \Rightarrow W_{p(j)} = W_{p(i)}];$$
$$\mathbf{Dec}(p, T) \Leftrightarrow [\forall i, j, k : i \leq j \leq k \ \wedge \ W_{p(i)} = W_{p(k)} \Rightarrow W_{p(j)} = W_{p(i)}];$$
$$\mathbf{SNU}(p, T) \Leftrightarrow [\forall i, j, k : i \leq j \leq k \ \wedge \ W_{p(i)} = W_{p(k)} = \text{content}(T) \Rightarrow p(j) = p(i)];$$
$$\mathbf{SDec}(p, T) \Leftrightarrow [\forall i, j, k : i \leq j \leq k \ \wedge \ W_{p(i)} = W_{p(k)} \Rightarrow p(j) = p(i)];$$
$$\mathbf{SMon}(p, T) \Leftrightarrow [\forall i, j : i < j \Rightarrow W_{p(i)} \subseteq W_{p(j)}];$$
$$\mathbf{Mon}(p, T) \Leftrightarrow [\forall i, j : i < j \Rightarrow W_{p(i)} \cap \text{content}(T) \subseteq W_{p(j)} \cap \text{content}(T)];$$
$$\mathbf{WMon}(p, T) \Leftrightarrow [\forall i, j : i < j \wedge \text{content}(T[j]) \subseteq W_{p(i)} \Rightarrow W_{p(i)} \subseteq W_{p(j)}].$$

Fig. 4. Definitions of learning restrictions

A variant on decisiveness is *syntactic decisiveness*, **SynDec**, a technically useful property defined as follows.

$$\mathbf{SynDec}(p, T) \Leftrightarrow [\forall i, j, k : i \leq j \leq k \ \wedge \ p(i) = p(k) \Rightarrow p(j) = p(i)].$$

We combine any two sequence acceptance criteria δ and δ' by intersecting them; we denote this by juxtaposition (for example, all the restrictions given in Figure 4 are meant to be always used together with **Ex**). With **T** we denote the always true sequence acceptance criterion (no restriction on learning).

A *learning criterion* is a tuple $(\mathcal{C}, \beta, \delta)$, where $\mathcal{C}$ is a set of learners (the admissible learners), β is an interaction operator and δ is a learning restriction;

we usually write $\mathcal{C}\mathbf{Txt}\beta\delta$ to denote the learning criterion, omitting $\mathcal{C}$ in case of $\mathcal{C} = \mathcal{P}$. We say that a learner $h \in \mathcal{C}$ $\mathcal{C}\mathbf{Txt}\beta\delta$-*learns* a language L iff, for all texts T for L, $\delta(\beta(h,T),T)$. The set of languages $\mathcal{C}\mathbf{Txt}\beta\delta$-learned by $h \in \mathcal{C}$ is denoted by $\mathcal{C}\mathbf{Txt}\beta\delta(h)$. We write $[\mathcal{C}\mathbf{Txt}\beta\delta]$ to denote the set of all $\mathcal{C}\mathbf{Txt}\beta\delta$-learnable classes (learnable by some learner in $\mathcal{C}$).

3   Delayable Learning Restrictions

In this section we present technically useful results which show that learners can always be assumed to be in some normal form. We will later always assume our learners to be in the normal form established by Corollary 7, the main result of this section. We start with the definition of *delayable*. Intuitively, a learning criterion δ is delayable iff the output of a hypothesis can be arbitrarily (but not indefinitely) delayed.

Definition 1. Let $\vec{R}$ be the set of all non-decreasing $r : \mathbb{N} \to \mathbb{N}$ with infinite limit inferior, i.e. for all m we have $\forall^\infty n : r(n) \geq m$.

A learning restriction δ is *delayable* iff, for all texts T and T' with content$(T) =$ content(T'), all p and all $r \in \vec{R}$, if $(p,T) \in \delta$ and $\forall n :$ content$(T[r(n)]) \subseteq$ content$(T'[n])$, then $(p \circ r, T') \in \delta$. Intuitively, as long as the learner has at least as much data as was used for a given conjecture, then the conjecture is permissible. Note that this condition holds for $T = T'$ if $\forall n : r(n) \leq n$.

Note that the intersection of two delayable learning criteria is again delayable and that *all* learning restrictions considered in this paper are delayable.

As the name suggests, we can apply *delaying tricks* (tricks which delay updates of the conjecture) in order to achieve fast computation times in each iteration (but of course in the limit we still spend an infinite amount of time). This gives us equally powerful but total learners, as shown in the next theorem. While it is well-known that, for many learning criteria, the learner can be assumed total, this theorem explicitly formalizes conditions under which totality can be assumed (note that there are also natural learning criteria where totality cannot be assumed, such as consistent learning [JORS99]).

Theorem 2. For any delayable learning restriction δ, we have $[\mathbf{Txt}\mathbf{G}\delta] = [\mathcal{R}\mathbf{Txt}\mathbf{G}\delta]$.

Next we define another useful property, which can always be assumed for delayable learning restrictions.

Definition 3. A *locking sequence for a learner h on a language L* is any finite sequence σ of elements from L such that $h(\sigma)$ is a correct hypothesis for L and, for sequences τ with elements from L, $h(\sigma \diamond \tau) = h(\sigma)$[BB75]. It is well known that every learner h learning a language L has a locking sequence on L. We say that a learning criterion I *allows for strongly locking learning* iff, for each I-learnable class of languages $\mathcal{L}$ there is a learner h such that h I-learns $\mathcal{L}$ and, for each $L \in \mathcal{L}$ and any text T for L, there is an n such that $T[n]$ is a locking sequence of h on L (we call such a learner h *strongly locking*).

With this definition we can give the following theorem.

Theorem 4. Let δ be a delayable learning criterion. Then $\mathcal{R}\mathbf{TxtG}\delta\mathbf{Ex}$ allows for strongly locking learning.

Next we define semantic and pseudo-semantic restrictions introduced in [Köt14]. Intuitively, semantic restrictions allow for replacing hypotheses by equivalent ones; pseudo-sematic restrictions allow the same, as long as no new mind changes are introduced.

Definition 5. For all total functions $p \in \mathfrak{P}$, we let

$$\mathrm{Sem}(p) = \{p' \in \mathfrak{P} \mid \forall i : W_{p(i)} = W_{p'(i)}\};$$
$$\mathrm{Mc}(p) = \{p' \in \mathfrak{P} \mid \forall i : p'(i) \neq p'(i+1) \Rightarrow p(i) \neq p(i+1)\}.$$

A sequence acceptance criterion δ is said to be a *semantic restriction* iff, for all $(p,q) \in \delta$ and $p' \in \mathrm{Sem}(p)$, $(p',q) \in \delta$.

A sequence acceptance criterion δ is said to be a *pseudo-semantic restriction* iff, for all $(p,q) \in \delta$ and $p' \in \mathrm{Sem}(p) \cap \mathrm{Mc}(p)$, $(p',q) \in \delta$.

We note that the intersection of two (pseudo-) semantic learning restrictions is again (pseudo-) semantic. All learning restrictions considered in this paper are pseudo-semantic, and all except **Conv**, **SNU**, **SDec** and **Ex** are semantic.

The next lemma shows that, for every pseudo-semantic learning restriction, learning can be done syntactically decisively.

Lemma 6. Let δ be a pseudo-semantic learning criterion. Then we have

$$[\mathcal{R}\mathbf{TxtG}\delta] = [\mathcal{R}\mathbf{TxtGSynDec}\delta].$$

As **SynDec** is a delayable learning criterion, we get the following corollary by taking Theorems 2 and 4 and Lemma 6 together. We will always assume our learners to be in this normal form in this paper.

Corollary 7. Let δ be pseudo-semantic and delayable. Then $\mathbf{TxtG}\delta\mathbf{Ex}$ allows for strongly locking learning by a syntactically decisive total learner.

Fulk showed that any **TxtGEx**-learner can be (effectively) turned into an equivalent learner with many useful properties, including strongly locking learning [Ful90]. One of the properties is called *order-independence*, meaning that on any two texts for a target language the learner converges to the same hypothesis. Another property is called *rearrangement-independence*, where a learner h is rearrangement-independent if there is a function f such that, for all sequences σ, $h(\sigma) = f(\mathrm{content}(\sigma), |\sigma|)$ (intuitively, rearrangement independence is equivalent to the existence of a partially set-driven learner for the same language). We define the collection of all the properties which Fulk showed a learner can have to be the *Fulk normal form* as follows.

Definition 8. We say a **TxtGEx**-learner h is in *Fulk normal form* if $(1) - (5)$ hold.

(1) h is order-independent.
(2) h is rearrangement-independent.
(3) If h **TxtGEx**-learns a language L from some text, then h **TxtGEx**-learns L.
(4) If there is a locking sequence of h for some L, then h **TxtGEx**-learns L.
(5) For all $\mathcal{L} \in$ **TxtGEx**(h), h is strongly locking on $\mathcal{L}$.

The following theorem is somewhat weaker than what Fulk states himself.

Theorem 9 ([Ful90, Theorem 13]). Every **TxtGEx**-learnable set of languages has a **TxtGEx**-learner in Fulk normal form.

4   Full-Information Learning

In this section we consider various versions of cautious learning and show that all of our variants are either no restriction to learning, or equivalent to conservative learning as is shown in Figure 5.

Additionally, we will show that every cautious **TxtGEx**-learnable language is conservative **TxtGEx**-learnable which implies that [**TxtGConvEx**], [**TxtGWMonEx**] and [**TxtGCautEx**] are equivalent. Last, we will separate these three learning criteria from strongly decisive **TxtGEx**-learning and show that [**TxtGSDecEx**] is a proper superset.

Theorem 10. We have that any conservative learner can be assumed cautious and strongly decisive, i.e.

$$[\mathbf{TxtGConvEx}] = [\mathbf{TxtGConvSDecCautEx}].$$

Proof. Let $h \in \mathcal{R}$ and $\mathcal{L}$ be such that h **TxtGConvEx**-learns $\mathcal{L}$. We define, for all σ, a set $M(\sigma)$ as follows

$$M(\sigma) = \{\tau \mid \tau \subseteq \sigma \ \wedge \ \forall x \in \text{content}(\tau) : \Phi_{h(\tau)}(x) \leq |\sigma|\}.$$

We let

$$\forall \sigma : h'(\sigma) = h(\max(M(\sigma))).$$

Let T be a text for a language $L \in \mathcal{L}$. We first show that h' **TxtGEx**-learns L from the text T. As h **TxtGConvEx**-learns L, there are n and e such that $\forall n' \geq n : h(T[n]) = h(T[n']) = e$ and $W_e = L$. Thus, there is $m \geq n$ such that $\forall x \in \text{content}(T[n]) : \Phi_{h(T[n])}(x) \leq m$ and therefore $\forall m' \geq m : h'(T[m]) = h'(T[m']) = e$.

Next we show that h' is strongly decisive and conservative; for that we show that, with every mind change, there is a new element of the target included in the conjecture which is currently not included but is included in all future conjectures; it is easy to see that this property implies both caution and strong decisiveness. Let i and i' be such that $\max(M(T[i'])) = T[i]$. This implies that

$$\text{content}(T[i]) \subseteq W_{h'(T[i'])}.$$

Let $j' > i'$ such that $h'(T[i']) \neq h'(T[j'])$. Then there is $j > i$ such that $\max(M(T[j'])) = T[j]$ and therefore

$$\text{content}(T[j]) \subseteq W_{h'(T[j'])}.$$

Note that in the following diagram j could also be between i and i'.

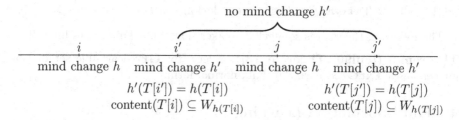

As h is conservative and $\text{content}(T[i]) \subseteq W_{h(T[i])}$, there exists ℓ such that $i < \ell < j$ and $T(\ell) \notin W_{h(T[i])}$. Then we have $\forall n \geq j' : T(\ell) \in W_{h'(T[n])}$ as $T(\ell) \in W_{h'(T[j'])}$.

Obviously h' is conservative as it only outputs (delayed) hypotheses of h (and maybe skip some) and h is conservative. ∎

In the following we consider three new learning restrictions. The learning restriction **Caut$_{\text{Fin}}$** means that the learner never returns a hypothesis for a finite set that is a proper subset of a previous hypothesis. **Caut$_\infty$** is the same restriction for infinite hypotheses. With **Caut$_{\text{Tar}}$** the learner is not allowed to ever output a hypothesis that is a proper superset of the target language that is learned.

Definition 11.

$$\text{Caut}_{\text{Fin}}(p, T) \Leftrightarrow [\forall i < j : W_{p(j)} \subset W_{p(i)} \Rightarrow W_{p(j)} \text{ is infinite}]$$
$$\text{Caut}_\infty(p, T) \Leftrightarrow [\forall i < j : W_{p(j)} \subset W_{p(i)} \Rightarrow W_{p(j)} \text{ is finite}]$$
$$\text{Caut}_{\text{Tar}}(p, T) \Leftrightarrow [\forall i : \neg(\text{content}(T) \subset W_{p(i)})]$$

The proof of the following theorem is essentially the same as given in [OSW86] to show that cautious learning is a proper restriction of **TxtGEx**-learning, we now extend it to strongly decisive learning. Note that a different extension was given in [BCM⁺08] (with an elegant proof exploiting the undecidability of the halting problem), pertaining to *behaviorally correct* learning. The proof in [BCM⁺08] as well as our proof would also carry over to the combination of these two extensions.

Theorem 12. *There is a class of languages that is* **TxtGSDecMonEx**-*learnable, but not* **TxtGCautEx**-*learnable.*

The following theorem contradicts a theorem given as an exercise in [OSW86] (Exercise 4.5.4B).

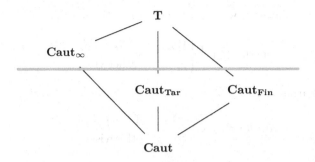

Fig. 5. Relation of different variants of cautious learning. A black line indicates inclusion (bottom to top); all and only the black lines meeting the gray line are proper inclusions.

Theorem 13. For $\delta \in \{\textbf{Caut}, \textbf{Caut}_{\textbf{Tar}}, \textbf{Caut}_{\textbf{Fin}}\}$ we have

$$[\textbf{Txt\,G}\delta\textbf{Ex}] = [\textbf{Txt\,GConvEx}].$$

From the definitions of the learning criteria we have $[\textbf{Txt\,GConvEx}] \subseteq [\textbf{Txt\,GWMonEx}]$. Using Theorem 13 and the equivalence of weakly monotone and conservative learning (using $\textbf{G}$) [KS95, JS98], we get the following.

Corollary 14. We have

$$[\textbf{Txt\,GConvEx}] = [\textbf{Txt\,GWMonEx}] = [\textbf{Txt\,GCautEx}].$$

Using Corollary 14 and Theorem 10 we get that weakly monotone **Txt\,GEx**-learning is included in strongly decisive **Txt\,GEx**-learning. Theorem 12 shows that this inclusion is proper.

Corollary 15. We have

$$[\textbf{Txt\,GWMonEx}] \subset [\textbf{Txt\,GSDecEx}].$$

The next theorem is the last theorem of this section and shows that forbidding to go down to strict *infinite* subsets of previously conjectures sets is no restriction.

Theorem 16. We have

$$[\textbf{Txt\,GCaut}_{\infty}\textbf{Ex}] = [\textbf{Txt\,GEx}].$$

Proof. Obviously we have $[\textbf{Txt\,GCaut}_{\infty}\textbf{Ex}] \subseteq [\textbf{Txt\,GEx}]$. Thus, we have to show that $[\textbf{Txt\,GEx}] \subseteq [\textbf{Txt\,GCaut}_{\infty}\textbf{Ex}]$. Let $\mathcal{L}$ be a set of languages and h be a learner such that h **Txt\,GEx**-learns $\mathcal{L}$ and h is strongly locking on $\mathcal{L}$ (see Corollary 7). We define, for all σ and t, the set M_σ^t such that

$$M_\sigma^t = \{\tau \mid \tau \in \text{Seq}(W_{h(\sigma)}^t \cup \text{content}(\sigma)) \;\wedge\; |\tau \diamond \sigma| \leq t\}.$$

Using the S-m-n Theorem we get a function $p \in \mathcal{R}$ such that

$$\forall \sigma : W_{p(\sigma)} = \text{content}(\sigma) \bigcup_{t \in \mathbb{N}} \begin{cases} W_{h(\sigma)}^t, & \text{if } \forall \rho \in M_\sigma^t : h(\sigma \diamond \rho) = h(\sigma); \\ \emptyset, & \text{otherwise.} \end{cases}$$

We define a learner h' as

$$\forall \sigma : h'(\sigma) = \begin{cases} p(\sigma), & \text{if } h(\sigma) \neq h(\sigma^-); \\ h'(\sigma^-), & \text{otherwise.} \end{cases}$$

We will show now that the learner h' **TxtGCaut$_\infty$Ex**-learns $\mathcal{L}$. Let an $L \in \mathcal{L}$ and a text T for L be given. As h is strongly locking there is n_0 such that for all $\tau \in \text{Seq}(L)$, $h(T[n_0] \diamond \tau) = h(T[n_0])$ and $W_{h(T[n_0])} = L$. Thus we have, for all $n \geq n_0$, $h'(T[n]) = h'(T[n_0])$ and $W_{h'(T[n_0])} = W_{p(T[n_0])} = W_{h(T[n_0])} = L$. To show that the learning restriction **Caut$_\infty$** holds, we assume that there are $i < j$ such that $W_{h'(T[j])} \subset W_{h'(T[i])}$ and $W_{h'(T[j])}$ is infinite. W.l.o.g. j is the first time that h' returns the hypothesis $W_{h'(T[j])}$. Let τ be such that $T[i] \diamond \tau = T[j]$. From the definition of the function p we get that content$(T[j]) \subseteq W_{h'(T[j])} \subseteq W_{h'(T[i])}$. Thus, content$(\tau) \subseteq W_{h'(T[i])} = W_{p(T[i])}$ and therefore $W_{p(T[i])}$ is finite, a contradiction to the assumption that $W_{h'(T[j])}$ is infinite. $\square$

The following theorem can be shown with a priority argument. The detailed proof is about eight pages long, following some ideas given in [BCM+08] for proving the second inequality, but adapted to the priority argument.

Theorem 17. We have

$$[\text{TxtGSDecEx}] \subset [\text{TxtGDecEx}] \subset [\text{TxtGEx}].$$

5   Set-Driven Learning

In this section we give theorems regarding set-driven learning. For this we build on the result that set-driven learning can always be done conservatively [KS95].

First we show that any conservative set-driven learner can be assumed to be cautious and syntactically decisive, an important technical lemma.

Lemma 18. We have

$$[\text{TxtSdEx}] = [\text{TxtSdConvSynDecEx}].$$

In other words, every set-driven learner can be assumed syntactically decisive.

The following Theorem is the main result of this section, showing that set-driven learning can be done not just conservatively, but also strongly decisively and cautiously *at the same time*.

Theorem 19. We have

$$[\text{TxtSdEx}] = [\text{TxtSdConvSDecCautEx}].$$

6   Monotone Learning

In this section we show the hierarchies regarding monotone and strongly monotone learning, simultaneously for the settings of **G** and **Sd** in Theorems 20 and 21. With Theorems 22 and 23 we establish that monotone learnabilty implies strongly decisive learnability.

Theorem 20. There is a language $\mathcal{L}$ that is **TxtSdMonWMonEx**-learnable but not **TxtGSMonEx**-learnable, i.e.

$$[\textbf{TxtSdMonWMonEx}]\backslash[\textbf{TxtGSMonEx}] \neq \emptyset.$$

Theorem 21. There is $\mathcal{L}$ such that $\mathcal{L}$ is **TxtSdWMonEx**-learnable but not **TxtGMonEx**-learnable.

The following theorem is an extension of a theorem from [BCM+08], where the theorem has been shown for decisive learning instead of strongly decisive learning.

Theorem 22. Let $\mathbb{N} \in \mathcal{L}$ and $\mathcal{L}$ be **TxtGEx**-learnable. Then, we have $\mathcal{L}$ is **TxtGSDecEx**-learnable.

Theorem 23. We have that any monotone **TxtGEx**-learnable class of languages is strongly decisive learnable, while the converse does not hold, i.e.

$$[\textbf{TxtGMonEx}] \subset [\textbf{TxtGSDecEx}].$$

Proof. Let $h \in \mathcal{R}$ be a learner and $\mathcal{L} = \textbf{TxtGMonEx}(h)$. We distinguish the following two cases. We call $\mathcal{L}$ *dense* iff it contains a superset of every finite set.

Case 1: $\mathcal{L}$ is dense. We will show now that h **TxtGSMonEx**-learns the class $\mathcal{L}$. Let $L \in \mathcal{L}$ and T be a text for L. Suppose there are i and j with $i < j$ such that $W_{h(T[i])} \nsubseteq W_{h(T[j])}$. Thus, we have $W_{h(T[i])}\backslash W_{h(T[j])} \neq \emptyset$. Let $x \in W_{h(T[i])}\backslash W_{h(T[j])}$. As $\mathcal{L}$ is dense there is a language $L' \in \mathcal{L}$ such that content$(T[j]) \cup \{x\} \in L'$. Let T' be a text for L' and T'' be such that $T'' = T[j] \diamond T'$. Obviously, T'' is a text for L'. We have that $x \in W_{h(T''[i])}$ but $x \notin W_{h(T''[j])}$ which is a contradiction as h is monotone. Thus, h **TxtGSMonEx**-learns $\mathcal{L}$, which implies that h **TxtGWMonEx**-learns $\mathcal{L}$. Using Corollary 15 we get that $\mathcal{L}$ is **TxtGSDecEx**-learnable.

Case 2: $\mathcal{L}$ is not dense. Thus, $\mathcal{L}' = \mathcal{L} \cup \mathbb{N}$ is **TxtGEx**-learnable. Using Theorem 22 $\mathcal{L}'$ is **TxtGSDecEx**-learnable and therefore so is $\mathcal{L}$.

Note that $[\textbf{TxtGSDecEx}] \subseteq [\textbf{TxtGMonEx}]$ does not hold as in *Case 1* with Corollary 15 a proper subset relation is used. $\square$

References

[Ang80]   Angluin, D.: Inductive inference of formal languages from positive data. Information and Control 45, 117–135 (1980)

[BB75]   Blum, L., Blum, M.: Toward a mathematical theory of inductive inference. Information and Control 28, 125–155 (1975)

[BCM+08]   Baliga, G., Case, J., Merkle, W., Stephan, F., Wiehagen, W.: When unlearning helps. Information and Computation 206, 694–709 (2008)

[CK10]   Case, J., Kötzing, T.: Strongly non-U-shaped learning results by general techniques. In: Proc. of COLT 2010, pp. 181–193 (2010)

[CM11]   Case, J., Moelius, S.: Optimal language learning from positive data. Information and Computation 209, 1293–1311 (2011)

[Ful90]   Fulk, M.: Prudence and other conditions on formal language learning. Information and Computation 85, 1–11 (1990)

[Gol67]   Gold, E.: Language identification in the limit. Information and Control 10, 447–474 (1967)

[Jan91]   Jantke, K.: Monotonic and non-monotonic inductive inference of functions and patterns. In: Dix, J., Jantke, K.P., Schmitt, P.H. (eds.) NIL 1990. LNCS, vol. 543, pp. 161–177. Springer, Heidelberg (1991)

[JORS99]   Jain, S., Osherson, D., Royer, J., Sharma, A.: Systems that Learn: An Introduction to Learning Theory, 2nd edn. MIT Press, Cambridge (1999)

[JS98]   Jain, S., Sharma, A.: Generalization and specialization strategies for learning r.e. languages. Annals of Mathematics and Artificial Intelligence 23, 1–26 (1998)

[Köt09]   Kötzing, T.: Abstraction and Complexity in Computational Learning in the Limit. PhD thesis, University of Delaware (2009), http://pqdtopen.proquest.com/#viewpdf?dispub=3373055

[Köt14]   Kötzing, T.: A solution to Wiehagen's thesis. In: Proc. of STACS (Symposium on Theoretical Aspects of Computer Science), pp. 494–505 (2014)

[KS95]   Kinber, E., Stephan, F.: Language learning from texts: Mind changes, limited memory and monotonicity. Information and Computation 123, 224–241 (1995)

[LZ93]   Lange, S., Zeugmann, T.: Monotonic versus non-monotonic language learning. In: Brewka, G., Jantke, K.P., Schmitt, P.H. (eds.) NIL 1991. LNCS (LNAI), vol. 659, pp. 254–269. Springer, Heidelberg (1993)

[OSW82]   Osherson, D., Stob, M., Weinstein, S.: Learning strategies. Information and Control 53, 32–51 (1982)

[OSW86]   Osherson, D., Stob, M., Weinstein, S.: Systems that Learn: An Introduction to Learning Theory for Cognitive and Computer Scientists. MIT Press, Cambridge (1986)

[Rog67]   Rogers, H.: Theory of Recursive Functions and Effective Computability. McGraw Hill, New York (1967); Reprinted by MIT Press, Cambridge (1987)

[SR84]   Schäfer-Richter, G.: Über Eingabeabhängigkeit und Komplexität von Inferenzstrategien. PhD thesis, RWTH Aachen (1984)

[WC80]   Wexler, K., Culicover, P.: Formal Principles of Language Acquisition. MIT Press, Cambridge (1980)

[Wie91]   Wiehagen, R.: A thesis in inductive inference. In: Dix, J., Schmitt, P.H., Jantke, K.P. (eds.) NIL 1990. LNCS, vol. 543, pp. 184–207. Springer, Heidelberg (1991)

On the Role of Update Constraints and Text-Types in Iterative Learning

Sanjay Jain[1,*], Timo Kötzing[2], Junqi Ma[1], and Frank Stephan[1,3,**]

[1] Department of Computer Science, National University of Singapore,
Singapore 117417, Republic of Singapore
`sanjay@comp.nus.edu.sg`, `ma.junqi@nus.edu.sg`
[2] Friedrich-Schiller University, Jena, Germany
`timo.koetzing@uni-jena.de`
[3] Department of Mathematics, National University of Singapore,
Singapore 119076, Republic of Singapore
`fstephan@comp.nus.edu.sg`

Abstract. The present work investigates the relationship of iterative learning with other learning criteria such as decisiveness, caution, reliability, non-U-shapedness, monotonicity, strong monotonicity and conservativeness. Building on the result of Case and Moelius that iterative learners can be made non-U-shaped, we show that they also can be made cautious and decisive. Furthermore, we obtain various special results with respect to one-one texts, fat texts and one-one hypothesis spaces.

1 Introduction

Iterative learning is the most common variant of learning in the limit which addresses memory constraints: the memory of the learner on past data is just its current hypothesis. Due to the padding lemma, this memory is still not void, but finitely many data can be memorised in the hypothesis. However, one subfield of the study of iterative learning considers therefore the usage of class-preserving one-one hypothesis spaces which limit this type of coding during the learning process. Other ways to limit it is to control the amount and types of updates; such constraints also aim for other natural properties of the conjectures: For example, updates have to be motivated by inconsistent data observed (syntactic conservativeness), semantic updates have to be motivated by inconsistent data observed (semantic conservativeness), updates cannot repeat semantically abandoned conjectures (decisiveness), updates cannot go from correct to incorrect hypotheses (non-U-shapedness), conjectures cannot be proper supersets of the language to be learnt (cautiousness) or conjectures have to contain all the data observed so far (consistency). There is already a quite comprehensive body of work on how iterativeness relates with various combinations of these constraints [CK10, GL04, JMZ13, JORS99, Köt09, LG02, LG03, LZ96, LZZ08], however various important questions remained unsolved. A few years ago, Case and Moelius

---

* Supported by NUS grants C252-000-087-001 and R146-000-181-112.
** Supported in part by NUS grant R146-000-181-112.

P. Auer et al. (Eds.): ALT 2014, LNAI 8776, pp. 55–69, 2014.

[CM08b] obtained a breakthrough result by showing that iterative learners can be made non-U-shaped. The present work improves this result by showing that they can also be made decisive — this stands in contrast to the case of the usual non-iterative framework where decisiveness is a real restriction in learning [BCMSW08]. Further results complete the picture and also include the role of hypothesis spaces and text-types in iterative learning.

We completely characterise the relationship of the iterative learning criteria with the different restrictions as given in the diagramme in Figure 1. A line indicates a previously known inclusion. A gray box around criteria indicates equality of these criteria, as found in this work.

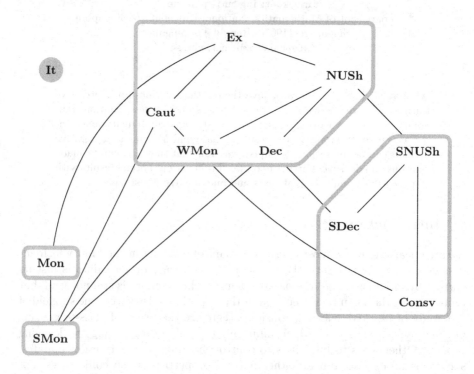

Fig. 1. Relation of criteria combined with iterative learning

The learning criteria investigated in the present work are quite natural. Conservativeness, consistency, cautiousness and decisiveness are natural constraints studied for a long time [Ang80, OSW86]; these criteria require that conjectures contain the data observed (consistency) or that mind changes are based on evidence that the prior hypothesis is incorrect (conservativeness); a lot of work has been undertaken using the assumption that learners are both, consistent and conservative. Monotonicity constraints play an important role in various fields like monotonic versus non-monotonic logic and this is reflected in inductive inference by considering the additional requirement that new hypotheses should be at least as general as the previous ones [Jan91, LZ93]. The fundamental notion of iterative learning is one of the first memory-constraints to be investigated in inductive inference and has been widely studied [LG02, LG03, LZ96, OSW86]; the beauty of this criterion is that the memory limitation comes rather indirectly, as for finitely many steps the memory can be enhanced by padding; after that, however, the learner has to converge and to ignore new data unless it gives enough evidence to undertake a mind change. Osherson, Stob and Weinstein [OSW82] formalised decisiveness as a notion where a learner never semantically returns to an abandoned hypothesis; they left it as an open problem whether the notion of decisiveness is restrictive; it took about two decades until the problem was solved [BCMSW08]. The search for this solution and also the parallels to developmental psychology motivated to study the related notion of non-U-shapedness where a non-U-shaped learner never abandons a correct hypothesis for an incorrect one and later (in a U-shaped way) returns to a correct hypothesis. The study of this field turned out to be quite fruitful and productive and we also consider decisive and non-U-shaped learning and its variants in this paper.

Taking this into account, we believe that the criteria investigated are natural and deserve to be studied; the restrictions on texts which we investigated are motivated from the fact that in the case of memory limitations (like enforced by iterativeness), the learners cannot keep track of which information has been presented before and therefore certain properties of the text (like every datum appearing exactly once or every datum appearing infinitely often) can be exploited by the learner during the learning process. In some cases these exploitations only matter when the restrictions on the hypothesis space make the iterativeness-constraint stricter, as they might rule out padding. Such a restriction is quite natural, as padding is a way to permit finite calculations to go into the update process and thereby bypass the basic idea behind the notion of iterativeness; this is reflected in the finding that the relations between the learning criteria differ for iterative learning in general and iterative learning using a class-preserving one-one hypothesis space.

Due to space restrictions some proofs are omitted. The full paper is available as Technical Report TRA7/14, School of Computing, National University of Singapore.

2   Mathematical Preliminaries

Unintroduced notation follows the textbook of Rogers [Rog67] on recursion theory. The set of natural numbers is denoted by $\mathbb{N} = \{0, 1, 2, \ldots\}$. The symbols $\subseteq, \subset, \supseteq, \supset$ respectively denote the subset, proper subset, superset and proper superset relation between sets. The symbol $\emptyset$ denotes both the empty set and the empty sequence.

With dom and range we denote, respectively, domain and range of a given function. We sometimes denote a partial function f of $n > 0$ arguments $x_1, \ldots, x_n$ in lambda notation (as in Lisp) as $\lambda x_1, \ldots, x_n . f(x_1, \ldots, x_n)$. For example, with $c \in \mathbb{N}$, $\lambda x . c$ is the constantly c function of one argument.

We let $\langle x, y \rangle = \frac{(x+y)(x+y+1)}{2} + x$ be Cantor's Pairing function which is an invertible, order-preserving function from $\mathbb{N} \times \mathbb{N} \to \mathbb{N}$. Whenever we consider tuples of natural numbers as input to a function, it is understood that the general coding function $\langle \cdot, \cdot \rangle$ is used to code the tuples into a single natural number. We similarly fix a coding for finite sets and sequences, so that we can use those as input as well.

If a function f is not defined for some argument x, then we denote this fact by $f(x)\uparrow$ and we say that f on x *diverges*; the opposite is denoted by $f(x)\downarrow$ and we say that f on x *converges*. If f on x converges to p, then we denote this fact by $f(x)\downarrow = p$.

$\mathcal{P}$ and $\mathcal{R}$ denote, respectively, the set of all partial recursive and the set of all recursive functions (mapping $\mathbb{N} \to \mathbb{N}$). We let φ be any fixed acceptable numbering for $\mathcal{P}$ (an acceptable numbering could, for example, be based on a natural programming language such as C or Java). Further, we let φ_p denote the partial-recursive function computed by the φ-program with code number p. A set $L \subseteq \mathbb{N}$ is *recursively enumerable (r.e.)* iff it is the domain of a partial recursive function. We let $\mathcal{E}$ denote the set of all r.e. sets. We let W be the mapping such that $\forall e : W_e = \mathrm{dom}(\varphi_e)$. W is, then, a mapping from $\mathbb{N}$ *onto* $\mathcal{E}$. We say that e is an index, or program, (in W) for W_e. Let $W_{e,s}$ denote W_e enumerated in s steps in some uniform way to enumerate all the W_e's. We let pad be a 1–1 padding function such that for all e and finite sets D, $W_{\mathrm{pad}(e,D)} = W_e$.

The special symbol ? is used as a possible hypothesis (meaning "no change of hypothesis"). The symbol # stands for a pause, that is, for "no new input data in the text". For each (possibly infinite) sequence q with its range contained in $\mathbb{N} \cup \{\#\}$, let $\mathrm{content}(q) = (\mathrm{range}(q) \setminus \{\#\})$. By using an appropriate coding, we assume that ? and # can be handled by recursive functions.

For any function f and all i, we use $f[i]$ to denote the sequence $f(0), \ldots, f(i-1)$ (the empty sequence if $i = 0$ and undefined, if one of these values is undefined).

3   Learning Criteria

In this section we formally introduce our setting of learning in the limit and associated learning criteria. We follow [Köt09] in its "building-blocks" approach for defining learning criteria.

A *learner* is a partial function from $\mathbb{N}$ to $\mathbb{N} \cup \{?\}$. A *language* is a r.e. set $L \subseteq \mathbb{N}$. Any total function $T : \mathbb{N} \to \mathbb{N} \cup \{\#\}$ is called a *text*. For any given language L, a *text for L* is a text T such that $\mathrm{content}(T) = L$. Initial parts of this kind of text is what learners usually get as information. We let σ and τ range over initial segments of texts. Concatenation of two initial segments σ and τ is denoted by $\sigma \diamond \tau$. For a given set of texts F, we let $\mathbf{Txt}^F(L)$ denote the set of all texts in F for L.

An *interaction operator* is an operator β taking as arguments a function M (the learner) and a text T, and that outputs a function p. We call p the *learning sequence* (or *sequence of hypotheses*) of M given T. Intuitively, β defines how a learner can interact with a given text to produce a sequence of conjectures.

We define the sequence generating operators $\mathbf{G}$ and $\mathbf{It}$ (corresponding to the learning criteria discussed in the introduction) as follows. For all learners M, texts T and all i,

$$\mathbf{G}(M, T)(i) = M(T[i]);$$

$$\mathbf{It}(M, T)(i) = \begin{cases} M(\emptyset), & \text{if } i = 0; \\ M(\mathbf{It}(M, T)(i - 1), T(i - 1)), & \text{otherwise;} \end{cases}$$

where $M(\emptyset)$ denotes the *initial conjecture* made by M. Thus, in iterative learning, the learner has access to the previous conjecture, but not to all previous data as in $\mathbf{G}$-learning. With any iterative learner M we associate a learner M^* such that

$$M^*(\emptyset) = M(\emptyset) \text{ and}$$
$$\forall \sigma, x : M^*(\sigma \diamond x) = M(M^*(\sigma), x).$$

Intuitively, M^* on a sequence σ returns the hypothesis which M makes after being fed the sequence σ in order. Note that, for all texts T, $\mathbf{G}(M^*, T) = \mathbf{It}(M, T)$. We let $M(T)$ (respectively $M^*(T)$) denote $\lim_{n \to \infty} M(T[n])$ (respectively, $\lim_{n \to \infty} M^*(T[n])$) if it exists.

Successful learning requires the learner to observe certain restrictions, for example convergence to a correct index. These restrictions are formalised in our next definition.

A *learning restriction* is a predicate δ on a learning sequence and a text. We give the important example of explanatory learning ($\mathbf{Ex}$, [Gol67]) and that of vacillatory learning ($\mathbf{Fex}$, [CL82, OW82, Cas99]) defined such that, for all sequences of hypotheses p and all texts T,

$$\mathbf{Ex}(p, T) \Leftrightarrow [\exists n_0 \forall n \geq n_0 : p(n) = p(n_0) \wedge W_{p(n_0)} = \text{content}(T)];$$
$$\mathbf{Fex}(p, T) \Leftrightarrow [\exists n_0 \exists \text{ finite } D \subset \mathbb{N}$$
$$\forall n \geq n_0 : p(n) \in D \wedge \forall e \in D : W_e = \text{content}(T)].$$

Furthemore, we formally define the restrictions discussed in Section 1 in Figure 2. We combine any two sequence acceptance criteria δ and δ' by intersecting them; we denote this by juxtaposition (for example, all the restrictions given in Figure 2 are meant to be always used together with $\mathbf{Ex}$).

For any set of texts F, interaction operator β and any (combination of) learning restrictions δ, $\mathbf{Txt}^F \beta \delta$ is a *learning criterion*. A learner M $\mathbf{Txt}^F \beta \delta$-*learns* all languages in the class

$$\mathbf{Txt}^F \beta \delta(M) = \{L \in \mathcal{E} \mid \forall T \in \mathbf{Txt}(L) \cap F : \delta(\beta(M, T), T)\}$$

and we use $\mathbf{Txt} \beta \delta$ to denote the set of all $\mathbf{Txt} \beta \delta$-learnable classes (learnable by some learner). Note that we omit the superscript F whenever F is the set of all texts.

In some cases, we consider learning using an explicitly given particular hypothesis space $(H_e)_{e \in \mathbb{N}}$ instead of the usual acceptable numbering $(W_e)_{e \in \mathbb{N}}$. For this, one replaces W_e by H_e in the respective definitions of learning as above.

$\mathbf{Consv}(p, T) \Leftrightarrow [\forall i : \text{content}(T[i+1]) \subseteq W_{p(i)} \Rightarrow p(i) = p(i+1)];$

$\mathbf{Caut}(p, T) \Leftrightarrow [\forall i, j : W_{p(i)} \subset W_{p(j)} \Rightarrow i < j];$

$\mathbf{NUSh}(p, T) \Leftrightarrow [\forall i, j, k : i \leq j \leq k \ \wedge \ W_{p(i)} = W_{p(k)} = \text{content}(T) \Rightarrow W_{p(j)} = W_{p(i)}];$

$\mathbf{Dec}(p, T) \Leftrightarrow [\forall i, j, k : i \leq j \leq k \ \wedge \ W_{p(i)} = W_{p(k)} \Rightarrow W_{p(j)} = W_{p(i)}];$

$\mathbf{SNUSh}(p, T) \Leftrightarrow [\forall i, j, k : i \leq j \leq k \ \wedge \ W_{p(i)} = W_{p(k)} = \text{content}(T) \Rightarrow p(j) = p(i)];$

$\mathbf{SDec}(p, T) \Leftrightarrow [\forall i, j, k : i \leq j \leq k \ \wedge \ W_{p(i)} = W_{p(k)} \Rightarrow p(j) = p(i)];$

$\mathbf{SMon}(p, T) \Leftrightarrow [\forall i, j : i < j \Rightarrow W_{p(i)} \subseteq W_{p(j)}];$

$\mathbf{Mon}(p, T) \Leftrightarrow [\forall i, j : i < j \Rightarrow W_{p(i)} \cap \text{content}(T) \subseteq W_{p(j)} \cap \text{content}(T)];$

$\mathbf{WMon}(p, T) \Leftrightarrow [\forall i, j : i < j \wedge \text{content}(T[j]) \subseteq W_{p(i)} \Rightarrow W_{p(i)} \subseteq W_{p(j)}].$

Fig. 2. Definitions of learning restrictions

4   Plain-Text Learning

In this section we first show that, for iterative learning, the convergence restrictions **Ex** and **Fex** allow for learning the same sets of languages. After that we give the necessary theorems establishing the diagramme given in Figure 1.

Theorem 1. TxtItFex = TxtItEx.

Next we give separating theorems for monotone learning and first show that there is a class which can be learnt iteratively by a learner which is strongly decisive, conservative, monotone and cautious while on the other hand, there is no learner which, even non-iteratively, learns the same class strongly monotonically.

Theorem 2. TxtItSDecConsvMonCautEx ⊄ TxtGSMonEx.

Proof. Let $L_0 = \{0, 2, 4, \ldots\}$ and for all n, $L_{n+1} = \{2m \mid m \leq n\} \cup \{2n+1\}$. Let $\mathcal{L} = \{L_n : n \in \mathbb{N}\}$. Let e be a recursive function computing an r.e. index for L_n: $W_{e(n)} = L_n$. Let $M \in \mathcal{P}$ be the iterative learner which memorises a single state in its conjecture (using padding) and has the following state transition diagramme (an edge labeled $\frac{x}{e}$ means that the edge indicates a state transition on input x with conjecture output e).

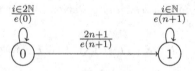

Clearly, M is a **TxtItSDecConsvMonCautEx**-learner for $\mathcal{L}$. It is known that $\mathcal{L}$ is not strongly monotonically learnable.   □

Note that one can modify this protocol such that M only memorises the state; however, M then abstains from repeating correct conjectures and one has to modify the learnability criterion such that outputting a special symbol for repeating the last (correct) conjecture is allowed. The next result shows that there is a class of languages which can be learnt by an iterative learner which is strongly decisive, conservative and cautious; on the other hand, there is no learner, even non-iterative one, that learns the class monotonically.

Theorem 3. TxtItSDecConsvCautEx $\not\subseteq$ TxtGMonEx.

Proof. We consider $L_0 = \{0, 2, 4, \ldots\}$ and, for all n, $L_{2n+1} = \{2m \mid m \leq n\} \cup \{4n+1\}$ and $L_{2n+2} = \{2m \mid m \leq n+1\} \cup \{4n+1, 4n+3\}$. We let $\mathcal{L} = \{L_n \mid n \in \mathbb{N}\}$.

Let e be a recursive function such that, for all n, $W_{e(n)} = L_n$. Let $M \in \mathcal{P}$ be the iterative learner using state transitions as given by the following diagramme.

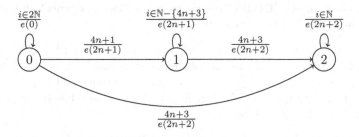

Clearly, M fulfills all the desired requirements for **TxtItSDecConsvCautEx**-learning $\mathcal{L}$. One can show that every learner of $\mathcal{L}$ outputs on some text for some L_{2n+2} hypotheses for L_0, L_{2n+1} and L_{2n+2} (in that order, with possibly other hypotheses in between) and is therefore not learning monotonically. $\square$

The next result shows that there is a class of languages which is simultaneously iteratively, monotonically, decisively, weakly monotonically and cautiously learnable, but not iteratively strongly non-U-shapedly learnable.

Theorem 4. TxtItMonDecWMonCautEx $\not\subseteq$ TxtItSNUShEx.

The next result shows that there is an iteratively and strongly monotonically learnable class which does not have any iterative learner which is strongly non-U-shaped, that is, which never revises a correct hypothesis. The proof uses the notion of a join which is defined as $A \oplus B = \{2x : x \in A\} \cup \{2x+1 : x \in B\}$.

Theorem 5. TxtItSMonEx $\not\subseteq$ TxtItSNUShEx.

Proof. Let $M_0, M_1, \ldots$ denote a recursive listing of all partial recursive iterative learning machines. Consider a class $\mathcal{L}$ consisting of the following sets for each $e \in \mathbb{N}$ (where $F(\cdot)$, $G(\cdot)$ are recursively enumerable sets in the parameters described later):

- $\{2e\} \oplus F(e)$
- $\{2e, 2d+1\} \oplus G(e, d)$

– $\{2e, 2d+1\} \oplus \mathbb{N}$

where,

(a) If there exists an s such that $M_e^*(4e \diamond 1 \diamond \# \diamond 3 \diamond \# \diamond 5 \diamond \# \ldots \diamond 2s+1) = M_e^*(4e \diamond 1 \diamond \# \diamond 3 \diamond \# \diamond 5 \diamond \# \ldots \diamond 2s+1 \diamond \# \diamond 2s'+1)$, for all $s' > s$, then $F(e) = \{0, 1, 2, \ldots, s\}$, else $F(e) = \mathbb{N}$.

(b) If $F(e) = \mathbb{N}$ or $\max(F(e)) > d$, then $G(e, d) = \mathbb{N}$. Otherwise, if there exists a $k > d$ such that $M_e^*(4e \diamond 1 \diamond \# \diamond 3 \diamond \# \diamond 5 \diamond \# \ldots \diamond 2 \max(F(e)) + 1 \diamond \# \diamond 4d + 2 \diamond \#^r) = M_e^*(4e \diamond 1 \diamond \# \diamond 3 \diamond \# \diamond 5 \diamond \# \diamond \ldots \diamond 2max(F(e)) + 1 \diamond \# \diamond 4d + 2 \diamond \#^r \diamond \#) \neq M_e^*(4e \diamond 1 \diamond \# \diamond 3 \diamond \# \diamond 5 \diamond \# \ldots \diamond 2max(F(e)) + 1 \diamond \# \diamond 4d + 2 \diamond \#^r \diamond \# \diamond 2k + 1)$ then $G(e, d) = F(e) \cup \{k\}$ for first such k found in some algorithmic search, else $G(e, d) = F(e)$.

Now, the above class is **TxtItSMonEx** learnable, as the learner can remember seeing $4e, 4d+2$ in the input text, if any:

- Having seen only $4e$, the learner outputs a grammar for $\{2e\} \oplus F(e)$;
- Having seen $4e, 4d+2$, the learner outputs a grammar for $\{2e, 2d+1\} \oplus G(e, d)$ until it sees, (after having seen $4e, 4d+2$), two more odd elements bigger than $2d$ in the input, at which point the learner switches to outputting a grammar for $\{2e, 2d+1\} \oplus \mathbb{N}$.

It is easy to verify that the above learner will **TxtItSMon** learn $\mathcal{L}$.

Now we show that $\mathcal{L}$ is not **TxtItSNUShEx**-learnable. Suppose by way of contradiction that M_e **TxtItSNUShEx**-learns $\mathcal{L}$. Then the following statements hold:

– There exists an s as described in the definition of $F(e)$ above and thus $F(e)$ is finite, as otherwise M_e does not learn $2e \oplus F(e) = 2e \oplus \mathbb{N}$;

– For $d > \max(F(e))$, there exists a $k > d$ as described in the definition of $G(e, d)$, as otherwise M_e does not learn at least one of $\{2e, 2d+1\} \oplus G(e, d)$ and $\{2e, 2d+1\} \oplus \mathbb{N}$;

– Now the learner M_e has two different hypotheses on the segments $(4e \diamond 1 \diamond \# \diamond 3 \diamond \# \diamond \ldots \diamond 2F(e) + 1 \diamond \# \diamond 2k + 1 \diamond \# \diamond 4d + 2 \diamond \#^r)$ and $(4e \diamond 1 \diamond \# \diamond 3 \diamond \# \diamond \ldots \diamond 2F(e) + 1 \diamond \# \diamond 2k + 1 \diamond \# \diamond 4d + 2 \diamond \#^r \diamond 2k + 1)$ and first of them must be correct hypothesis for $\{2e, 2d+1\} \oplus G(e, d)$, as otherwise the learner M_e does not learn it from the text — $4e \diamond 1 \diamond \# \diamond 3 \diamond \# \diamond \ldots \diamond 2F(e) + 1 \diamond \# \diamond 2k + 1 \diamond \# \diamond 4d + 2 \diamond \#^r \diamond \#^\infty$ — see part (b) in the definition of $G(e, d)$, whereas second is a mind change, after the correct hypothesis by M_e on $\{2e, 2d+1\} \oplus G(e, d)$.

Thus, M_e does not **TxtItSNUShEx**-learn $\mathcal{L}$. □

For our following proofs we will require the notion of a *canny* learner [CM08b].

Definition 6 (Case and Moelius [CM08b]). For all iterative learners M, we say that M is *canny* iff

1. M never outputs ?,

2. for all e, $M(e, \#) = e$ and
3. for all x, τ and σ, if $M^*(\sigma \diamond x) \neq M^*(\sigma)$ then $M^*(\sigma \diamond x \diamond \tau \diamond x) = M^*(\sigma \diamond x \diamond \tau)$.

Case and Moelius [CM08b] showed that, for **TxtItEx**-learning, learners can be assumed to be canny.

Lemma 7 (Case and Moelius [CM08b]). *For all $\mathcal{L} \in$ **TxtItEx** there exists canny iterative learner M such that $\mathcal{L} \subseteq$ **TxtItEx**(M).*

The term "sink-locking" means that on any text for a language to be learnt the learner converges to a *sink*, a correct hypothesis which is not abandoned on any continuation of the text. The following result does not only hold for the case where all texts are allowed but also for the case where only fat texts are allowed (see Section 5).

Theorem 8. *Let $\mathcal{L}$ be sink-lockingly **TxtItEx**-learnable. Then $\mathcal{L}$ is cautiously, conservatively, strongly decisively and weakly monotonically **TxtItEx**-learnable.*

The previous theorem gives us the following immediate corollary which states that a class is iteratively strongly decisive learnable from text iff it is iteratively conservatively learnable from text iff it is iteratively strongly non-U-shaped learnable from text.

Corollary 9. *We have that*

$$\textbf{TxtItSDecEx} \ = \ \textbf{TxtItConsvEx} \ = \ \textbf{TxtItSNUShEx}.$$

Proof. We have that strongly decisive or conservative (iterative) learnability trivially implies strongly non-U-shaped learnability. Using Theorem 8 it remains to show that strongly non-U-shaped learnability implies sink-locking learnability. But this is trivial, as the learner can never converge to a correct conjecture that might possibly be abandoned on the given language, as this would contradict strong non-U-shapedness. □

Case and Moelius [CM08b] showed that **TxtItNUShEx** = **TxtItEx**; we finally show that this proof can be extended to also cover decisiveness, weak monotonicity and caution.

Theorem 10. *We have that*

$$\textbf{TxtItEx} \ = \ \textbf{TxtItDecEx} \ = \ \textbf{TxtItWMonEx} \ = \textbf{TxtItCautEx}.$$

Proof. Suppose M is a canny iterative learner which learns a class $\mathcal{L}$. Below we will construct an iterative learner N which is weakly monotonic and learns $\mathcal{L}$. Let

$$C_M(\sigma) = \{x \in \mathbb{N} \cup \{\#\} : M^*(\sigma \diamond x){\downarrow} = M^*(\sigma){\downarrow}\};$$
$$B_M(\sigma) = \{x \in \mathbb{N} \cup \{\#\} : M^*(\sigma \diamond x){\downarrow} \neq M^*(\sigma){\downarrow}\};$$
$$B_M^\cap(\sigma) = \bigcap_{0 \leq i \leq |\sigma|} B_M(\sigma[i]);$$
$$CB_M(\sigma) = \bigcup_{0 \leq i < |\sigma|} C_M(\sigma[i]) \cap B_M(\sigma).$$

Let P be such that for all σ and m and $x \in \mathbb{N} \cup \{\#\}$, $P(\sigma, m, x)$ iff (i) $x \neq \#$ and (ii) $(\exists w)[M^*(\sigma \diamond w)$ converges in x steps, $W_{M^*(\sigma)}$ enumerates w in x steps, $w \in CB_M(\sigma)$ and $m < w \leq x]$.

Let N be such that $N(\emptyset) = f(\emptyset, 0, \emptyset)$, and for all inputs x, and previous conjecture $f(\sigma, m, \alpha)$, N outputs as follows:

$$
\begin{cases}
\uparrow, & \text{(i) if } M^*(\tau)\uparrow \text{ for some } \tau \in \{\sigma, \sigma \diamond \alpha, \sigma \diamond x, \sigma \diamond \alpha \diamond x\}; \\
f(\sigma \diamond \alpha \diamond x, 0, \emptyset), & \text{(ii) if } \neg \text{ (i) and } (x \in B_M^\cap(\sigma) \text{ or } (x \in CB_M(\sigma) \text{ and } x > m)); \\
f(\sigma, m, \alpha \diamond x), & \text{(iii) if } \neg ((i) \text{ or } (ii)) \text{ and} \\
& \quad x \in CB_M(\sigma \diamond \alpha) \\
f(\sigma, x, \emptyset), & \text{(iv) if } \neg ((i) \text{ or } (ii)) \text{ and} \\
& \quad x \in C_M(\sigma \diamond \alpha) \text{ and } P(\sigma, m, x) \text{ and } \alpha = \emptyset; \\
f(\sigma \diamond \alpha \diamond x, 0, \emptyset), & \text{(v) if } \neg ((i) \text{ or } (ii)) \text{ and} \\
& \quad x \in C_M(\sigma \diamond \alpha) \text{ and } P(\sigma, m, x) \text{ and } \alpha \neq \emptyset; \\
f(\sigma, m, \alpha), & \text{(vi) if } \neg ((i) \text{ or } (ii)) \text{ and} \\
& \quad x \in C_M(\sigma \diamond \alpha) \text{ and } \neg P(\sigma, m, x).
\end{cases}
$$

Here $W_{f(\sigma, m, \alpha)}$ is defined as follows.

1. Enumerate content(σ)
   In the following, if the needed $M^*(\cdot)$ (to compute various parameters), is not defined, then do not enumerate any more.
2. Go to stage 0.
   Stage s:
       Let $A_s = $ content(σ) $\cup W_{M^*(\sigma), s}$
   (a) If there exists an $x \in A_s$ such that $x \in B_M^\cap(\sigma)$, then no more elements are enumerated.
   (b) If there exists an $x \in A_s$ such that $x > m$, and $[x \in CB_M(\sigma)$ or $P(\sigma, m, x)]$, then:
           If for all τ with content(τ) $\subseteq A_s$ and $|\tau| \leq |A_s| + 1$, τ not containing $\#$ and τ starting with a y in $CB_M(\sigma)$: $A_s \subseteq W_{f(\sigma \diamond \tau, 0, \emptyset)}$,
           then enumerate A_s and go to stage $s + 1$;
           otherwise, no more elements are enumerated.
           (basically, this is testing if x satisfies clauses ii, iv or v in the defn of M)
   (c) If both (a) and (b) fail, then enumerate A_s, and go to stage $s + 1$.
   End stage s

It can be easily shown by induction on the length of ρ, that for all input ρ, if $N^*(\rho) = f(\sigma, m, \alpha)$, then $M^*(\rho) = M^*(\sigma \diamond \alpha)$.

Now, for finite languages L iteratively learnt by M, if content(σ) $\subseteq L$ and $L \cap B_M^\cap(\sigma) = \emptyset$, then $W_{M^*(\sigma)} = L$. To see this note that if we construct a sequence τ from σ, by inserting elements of $L - $ content(σ) after the initial segment σ' of σ such that $x \in C_M(\sigma')$, then $M^*(\sigma) = M^*(\tau)$, and content(τ) $= L$; thus, $M^*(\sigma) = M^*(\sigma\#^\infty) = M^*(\tau\#^\infty)$, which must be a grammar for L. Thus for such σ, for content$(\alpha) \subseteq L$, using the fact that M is canny and using reverse induction on the number of mind changes made by M on σ (which is bounded by card(L) due to M being canny), it is easy to verify that $W_{f(\sigma,m,\alpha)}$ would be L.

Given an infinite languages $L \in \mathcal{L}$ and a text T for L, consider the output $f(\sigma_n, m_n, \alpha_n)$ of $N^*(T[n])$. As $M^*(T)$ converges, it holds that $\sigma = \lim_{n\to\infty} \sigma_n$ and $\lim_{n\to\infty} \alpha_n$ would converge. For this paragraph fix this σ and α. If $\alpha \neq \emptyset$, then clearly $m = \lim_{n\to\infty} m_n$ also converges, and as $B_M^{\cap}(\sigma) \cap L = \emptyset$, we also have $W_{M^*(\sigma)} = L$. If $\alpha = \emptyset$, then as $M^*(T) = M^*(\sigma)$, we have that $W_{M^*(\sigma)} = L$ and all but finitely many of the elements of L do not belong to $B_M(\sigma)$. Thus, in this case also $m = \lim_{n\to\infty} m_n$ converges. In both cases, m bounds all the elements of L which are in $B_M(\sigma)$. Thus, $f(\sigma, m, \alpha)$ would be a grammar for L.

We show the weak monotonicity of N. Note that, for all σ, α, m, $W_{f(\sigma,m,\alpha)} \subseteq$ content$(\sigma) \cup W_{M^*(\sigma)}$.

Also, note that $W_{f(\sigma,m,\alpha)} \subseteq W_{f(\sigma,m+1,\alpha')}$ for all $m, \alpha, \sigma, \alpha'$ — (P1).

Now suppose N on input $\rho \diamond x$ and previous conjecture (on input ρ) being $f(\sigma, m, \alpha)$ outputs $f(\sigma \diamond \alpha \diamond x, 0, \emptyset)$. This implies that, $x \in B_M^{\cap}(\sigma)$ or $x > m$ and $(CB_M(\sigma)$ or $P(\sigma, m, x))$ hold.

Case 1: content$(\alpha \diamond x)$ is not contained in $W_{f(\sigma,m,\alpha)}$.

In this case clearly content$(\rho \diamond x) \supseteq$ content$(\sigma \diamond \alpha \diamond x)$ and thus, content$(\rho \diamond x)$ is not contained in $W_{f(\sigma,m,\alpha)}$, so mind change is safe.

Case 2: content$(\alpha \diamond x)$ is contained in $W_{f(\sigma,m,\alpha)}$ and thus in content$(\sigma) \cup W_{M^*(\sigma)}$.

Let s be least such that content$(\alpha \diamond x)$ is contained in A_s as in stage s. Then, the definition of $W_{f(\sigma,m,\alpha)}$ ensures that $W_{f(\sigma,m,\alpha)}$ enumerates $A_t, t \geq s$, only if A_t is contained in $W_{f(\sigma \diamond \alpha \diamond x, 0, \emptyset)}$ (note that the case of $A_t = $ content(σ), already satisfies $A_t \subseteq W_{f(\sigma \diamond \alpha \diamond x, 0, \emptyset)}$).

It follows from the above analysis that either the new input is not contained in the previous conjecture of N, or the previous conjecture is contained in the new conjecture. Thus, N is weakly monotonic.

It follows from the above construction that N is also decisive and cautious. To see this, note that whenever mind change of N falls in Case 1 above, all future conjectures of N (beyond input $\rho \diamond x$) contain content$(\alpha \diamond x)$; thus, N never returns to the conjecture $W_{f(\sigma,m,\alpha)}$, which does not contain content$(\alpha \diamond x)$. On the other hand, the mind changes due to Case 2 or mind changes due to N outputting $f(\sigma, m', \alpha')$ after outputting $f(\sigma, m, \alpha)$, are strongly monotonic (see the discussion in Case 2, as well as property (P1) mentioned above). The theorem follows.   □

5   Learning from Fat-Texts and other Texts

In this section we deal with special kinds of texts. A text is called *fat* iff every datum appears infinitely often in that text. A text T is called *one-one* iff for all $x \in$ content(T), there exists a unique n such that $T(n) = x$. We let fat denote the set of all fat texts and one — one the set of all one-one texts. Standard techniques can be used to show the following result.

Theorem 11. TxtItEx $\subset$ Txt^{fat}ItEx $\subset$ TxtGEx.

The above result shows that iterative learners have not only information-theoretic limitations in that they forget past data and cannot recover them (on normal text), but also computational limitations which cannot be compensated by having fat text. Next we show that fat text always allows for learning conservatively (as well as cautiously and strongly decisively).

Theorem 12. Txt^{fat}ItEx $=$ Txt^{fat}ItConsvEx $=$ Txt^{fat}ItSDecEx.

Proposition 13. *(a) There exists a class of languages which is* **TxtItMonEx,** **TxtItSDecEx, TxtItConsvEx**-*learnable but not* **Txt^{fat}SMonEx**-*learnable.*

(b) There is a class which is **TxtItSDecEx**-*learnable (and therefore also* **TxtItConsvEx**-*learnable) but not* **Txt^{fat}ItMonEx** *or* **Txt$^{\text{one−one}}$ItMonEx**-*learnable.*

Theorem 14. TxtItSMonEx $\not\subseteq$ Txt^{fat}ItSNUShEx.

We next show that learning from one-one texts is equivalent to learning from arbitrary text.

Theorem 15. Txt$^{\text{one−one}}$ItEx $=$ TxtItEx.

Theorem 16. *There exists a class $\mathcal{L}$ which is* **Txt$^{\text{one−one}}$ItFex**-*learnable but not* **Txt$^{\text{one−one}}$Ex**-*learnable. Therefore $\mathcal{L}$ is not* **TxtItEx**-*learnable (and hence not* **TxtItFex**-*learnable).*

Proof. Let $\mathcal{L}$ consist of the languages $L_{e,z}$, $z \leq e$, $e, z \in \mathbb{N}$, where $L_{e,z} = \{(e, x, y) : x = z \text{ or } x + y < |W_e|\}$.

The learner on seeing any input element (e, x, y), outputs a grammar (obtained effectively from (e, x)) for $L_{e,\min(\{e,x\})}$.

If W_e is infinite, then $L_{e,e} = L_{e,z}$ for all $z \leq e$, and thus all the (finitely many) grammars output by the learner are for $L_{e,e}$.

If W_e is finite, then $L_{e,z}$ contains only finitely many elements which are not of the form $(e, z, \cdot)$, and thus on any one-one text for $L_{e,z}$, the learner converges to a grammar for $L_{e,z}$.

We now show that $\mathcal{L}$ is not **TxtEx**-learnable. Suppose otherwise that some learner **TxtEx**-learns $\mathcal{L}$. Then, for $e \geq 2$, W_e is infinite iff the learner has a stabilising sequence [BB75, Ful90] τ on the set $\{(e, x, y) : x, y \in \mathbb{N}\}$ and the largest sum $x + y$ for some (e, x, y) occurring in τ is below $|W_e|$. Thus it would be a Σ_2 condition to check whether W_e is infinite in contradiction to the fact that checking whether W_e is infinite is Π_2 complete. Thus such a learner does not exist. $\square$

Theorem 17. *There exists a class of languages which is iteratively learnable using texts where every element which is maximal so far is marked, but is not* **TxtItEx**-*learnable.*

6   Class Preserving Hypotheses Spaces

A one-one hypothesis space might be considered in order to prevent that an iterative learner cheats by storing information in the hypothesis. A hypothesis space $(H_e)_{e \in \mathbb{N}}$ is called class preserving (for learning $\mathcal{L}$) iff $\{H_e : e \in \mathbb{N}\} = \mathcal{L}$. A learner is class preserving, if the hypothesis space used by it is class preserving. The first result shows that the usage of one-one texts increases the learning power of those iterative learners which are forced to use one-one hypothesis spaces, that is, which cannot store information in the hypothesis during the learning process.

Theorem 18. *There exists a class $\mathcal{L}$ having a one-one class preserving hypothesis space such that the following conditions hold:*

(a) $\mathcal{L}$ can be $\mathbf{Txt}^{\mathrm{one-one}}\mathbf{ItEx}$-learnt using any fixed one-one class preserving hypothesis space for $\mathcal{L}$;

(b) $\mathcal{L}$ cannot be $\mathbf{TxtItEx}$-learnt using any fixed one-one class preserving hypothesis space for $\mathcal{L}$.

In general, the hierarchy $\mathbf{SMonEx} \subseteq \mathbf{MonEx} \subseteq \mathbf{WMonEx}$ holds. The following result shows that this hierarchy is proper and that one can get the separations even in the case that the more general criterion is made stricter by enforcing the use of a one-one hypothesis space.

Theorem 19. *(a) $\mathbf{TxtItWMonEx} \not\subseteq \mathbf{TxtItMonEx}$;*
(b) $\mathbf{TxtItMonEx} \not\subseteq \mathbf{TxtItSMonEx}$.
Here the positive sides can be shown using a one-one class preserving hypothesis space.

Theorem 2 and Theorem 3 show the above result and also provide conservatively learnable families for these separations. We now consider learning by *reliable* learners. A learner is *reliable* if it is total and for any text T, if the learner converges on T to a hypothesis e, then e is a correct grammar for content(T). We denote the reliability constraint on the learner by using $\mathbf{Rel}$ in the criterion name. For the following result, we assume (by definition) that if a learner converges to ? on a text, then it is not reliable. The next result shows that there is exactly one class which has a reliable iterative learner using a one-one class preserving hypothesis space and this is the class FIN $= \{L : L$ is finite$\}$.

Theorem 20. *If $\mathcal{L}$ is $\mathbf{TxtItRelEx}$-learnable using a one-one class preserving hypothesis space then $\mathcal{L}$ must be FIN.*

Theorem 21. *There exists a subclass of FIN which is not $\mathbf{TxtItEx}$-learnable using a one-one class preserving hypothesis space.*

Note that in learning theory without loss of generality one assumes that classes are not empty. The next theorem characterises when a class can be iteratively and reliably learnt using a class preserving hypothesis space: it is the case if and only if the set of canonical indices of the languages in the class is recursively enumerable. Note that the hypothesis space considered here is not one-one and that padding is a natural ingredient of the (omitted) learning algorithm.

Theorem 22. *A class $\mathcal{L}$ has a class-preserving iterative and reliable learner iff it does not contain infinite languages and the set $\{e : D_e \in \mathcal{L}\}$ of its canonical indices is recursively enumerable.*

7   Syntactic versus Semantic Conservativeness

A learner is called *semantically conservative* iff whenever it outputs two indices i, j such that $W_i \neq W_j$ and i is output before j then the hypothesis j is based on some observed data not contained in W_i. This notion coincides with syntactic conservative learning in the case of standard explanatory learning; however, in the special case of iterative learning, it is more powerful than the usual notion of conservative learning.

Theorem 23. *There is a class $\mathcal{L}$ which can be learnt iteratively and strongly monotonically and semantically conservatively but which does not have an iterative and syntactically conservative learner.*

References

[Ang80]   Angluin, D.: Inductive inference of formal languages from positive data. Information and Control 45, 117–135 (1980)

[BB75]   Blum, L., Blum, M.: Toward a mathematical theory of inductive inference. Information and Control 28, 125–155 (1975)

[BCMSW08]   Baliga, G., Case, J., Merkle, W., Stephan, F., Wiehagen, R.: When unlearning helps. Information and Computation 206, 694–709 (2008)

[Cas74]   Case, J.: Periodicity in generations of automata. Mathematical Systems Theory 8, 15–32 (1974)

[Cas94]   Case, J.: Infinitary self-reference in learning theory. Journal of Experimental and Theoretical Artificial Intelligence 6, 3–16 (1994)

[Cas99]   Case, J.: The power of vacillation in language learning. SIAM Journal on Computing 28, 1941–1969 (1999)

[CK10]   Case, J., Kötzing, T.: Strongly non-U-shaped learning results by general techniques. In: Proceedings of COLT (Conference on Computational Learning Theory), pp. 181–193 (2010)

[CL82]   Case, J., Lynes, C.: Machine inductive inference and language identification. In: Proceedings of ICALP (International Colloquium on Automata, Languages and Programming), pp. 107–115 (1982)

[CM08a]   Case, J., Moelius III, S.E.: Optimal language learning. In: Freund, Y., Györfi, L., Turán, G., Zeugmann, T. (eds.) ALT 2008. LNCS (LNAI), vol. 5254, pp. 419–433. Springer, Heidelberg (2008)

[CM08b]   Case, J., Moelius, S.E.: U-shaped, iterative, and iterative-with-counter learning. Machine Learning 72, 63–88 (2008)

[Ful90]   Fulk, M.: Prudence and other conditions on formal language learning. Information and Computation 85, 1–11 (1990)

[Gol67]   Mark Gold, E.: Language identification in the limit. Information and Control 10, 447–474 (1967)

[GL04]   Grieser, G., Lange, S.: Incremental learning of approximations from positive data. Information Processing Letters 89, 37–42 (2004)

[Jan91]   Jantke, K.-P.: Monotonic and non-monotonic inductive inference of
             functions and patterns. In: Dix, J., Schmitt, P.H., Jantke, K.P. (eds.)
             NIL 1990. LNCS, vol. 543, pp. 161–177. Springer, Heidelberg (1991)
[JMZ13]   Jain, S., Moelius, S.E., Zilles, S.: Learning without coding. Theoretical
             Computer Science 473, 124–148 (2013)
[JORS99]   Jain, S., Osherson, D., Royer, J., Sharma, A.: Systems that Learn: An
             Introduction to Learning Theory, 2nd edn. MIT Press, Cambridge (1999)
[Köt09]   Kötzing, T.: Abstraction and Complexity in Computational Learning in
             the Limit. PhD thesis, University of Delaware (2009),
             http://pqdtopen.proquest.com/#viewpdf?dispub=3373055
[Köt14]   Kötzing, T.: A Solution to Wiehagen's Thesis. In: Symposium on Theo-
             retical Aspects of Computer Science (STACS 2014), pp. 494–505 (2014)
[LG02]   Lange, S., Grieser, G.: On the power of incremental learning. Theoretical
             Computer Science 288, 277–307 (2002)
[LG03]   Lange, S., Grieser, G.: Variants of iterative learning. Theoretical Com-
             puter Science 292, 359–376 (2003)
[LZ93]   Lange, S., Zeugmann, T.: Monotonic versus non-monotonic language
             learning. In: Brewka, G., Jantke, K.P., Schmitt, P.H. (eds.) NIL 1991.
             LNCS, vol. 659, pp. 254–269. Springer, Heidelberg (1993)
[LZ96]   Lange, S., Zeugmann, T.: Incremental learning from positive data. Jour-
             nal of Computer and System Sciences 53, 88–103 (1996)
[LZZ08]   Lange, S., Zeugmann, T., Zilles, S.: Learning indexed families of recur-
             sive languages from positive data: a survey. Theoretical Computer Sci-
             ence 397, 194–232 (2008)
[OSW82]   Osherson, D., Stob, M., Weinstein, S.: Learning strategies. Information
             and Control 53, 32–51 (1982)
[OSW86]   Osherson, D., Stob, M., Weinstein, S.: Systems that Learn: An Intro-
             duction to Learning Theory for Cognitive and Computer Scientists. MIT
             Press, Cambridge (1986)
[OW82]   Osherson, D., Weinstein, S.: Criteria of language learning. Information
             and Control 52, 123–138 (1982)
[RC94]   Royer, J., Case, J.: Subrecursive Programming Systems: Complexity and
             Succinctness. Research monograph in Progress in Theoretical Computer
             Science. Birkhäuser, Basel (1994)
[Rog67]   Rogers, H.: Theory of Recursive Functions and Effective Computability.
             McGraw Hill, New York (1967); Reprinted by MIT Press, Cambridge
             (1987)
[Wie91]   Wiehagen, R.: A thesis in inductive inference. *Nonmonotonic and In-
             ductive Logic*. In: Dix, J., Schmitt, P.H., Jantke, K.P. (eds.) NIL 1990.
             LNCS, vol. 543, pp. 184–207. Springer, Heidelberg (1991)

Parallel Learning of Automatic Classes of Languages

Sanjay Jain[1],* and Efim Kinber[2]

[1] School of Computing, National University of Singapore, Singapore 117417
sanjay@comp.nus.edu.sg
[2] Department of Computer Science, Sacred Heart University, Fairfield,
CT 06432-1000, U.S.A.
kinber@sacredheart.edu

Abstract. We introduce and explore a model for parallel learning of families of languages computable by finite automata. In this model, an algorithmic or automatic learner takes on n different input languages and identifies at least m of them correctly. For finite parallel learning, for large enough families, we establish a full characterization of learnability in terms of characteristic samples of languages. Based on this characterization, we show that it is the difference $n - m$, the number of languages which are potentially not identified, which is crucial. Similar results are obtained also for parallel learning in the limit. We consider also parallel finite learnability by finite automata and obtain some partial results. A number of problems for automatic variant of parallel learning remain open.

1 Introduction

In this paper, we define and explore a model for learning *automatic* families of languages *in parallel*. A family of languages is called *automatic* if it is an indexed family, and there is a finite automaton that, given an index v of a language and a string u can solve the membership problem for u in the language indexed by v (study of learnability of automatic classes was initiated in [JLS12]). Our aim is to establish if, under what circumstances, and on what expense, learning several languages from an automatic family in parallel can be more powerful than learning one language at a time. In the past, few approaches to learning in parallel have been suggested. One of them, known as *team inference* involves a finite team of learning machines working in parallel on the same input function or language (see for example [Smi82]). Our approach follows the one suggested for parallel learning recursive functions in [KSVW95]: one learning machine is learning a finite collection of (pairwise distinct) languages (in some sense, this model is a generalization of the model introduced in [AGS89]). A similar approach has recently been utilized in a study of prediction of recursive

* Supported in part by NUS grant numbers C-252-000-087-001, R-146-000-181-112 and R-252-000-534-112.

P. Auer et al. (Eds.): ALT 2014, LNAI 8776, pp. 70–84, 2014.

function values in [BKF11]: one algorithm predicts next values of several different input functions.

We consider learning languages in two different, albeight related settings:

a) Finite learning [Gol67]: a learning machine, after seeing a finite amount of input data, terminates and outputs conjectures for grammars of languages being learnt; for this type of learning, we also consider the case when the learner itself is a finite automaton (see [JLS12]).

b) Learning in the limit [Gol67]: a learning machine outputs a potentially infinite sequence of conjectures, stabilizing on a correct grammar for the target language.

The learners in our model use input *texts* — potentially infinite sequences that contain full positive data in a target language, intermittent with periods of "no data". Both settings, under the name of inductive inference, have a long history, see, for example, [JORS99].

A simple example of the family of three languages, $\{0\}, \{1\}, \{0, 1\}$ (which can be trivially made automatic), shows that finite learning of three languages in parallel might be possible, whereas no learner can learn languages in the family one at a time: the desired parallel learner will just wait when texts for the three input languages will be pairwise distinct (when both 0 and 1 will appear in one of the input texts) and output three correct conjectures; on the other hand, if an individual learner gets on the input a text containing all 0-s and settles on the conjecture $\{0\}$, it will be too late if 1 appears in the input.

However, interestingly, when families of languages are large, finite parallel learning of all input languages has no advantage over finite learning of individual languages: as it follows from one of our results (Theorem 8), if the number of languages in an automatic family is at least 4, and the family is learnable in parallel by a finite learner taking three different input texts, then the family is finitely learnable, one language at a time. Therefore, we consider a more general model of parallel learning, where the potential advantage of parallelism may compensate for lack of precision — so-called (m, n) or *frequency* learning: a learner gets input texts for n different languages and learns at least m of them correctly. This model of learning was first suggested and explored for algorithmic learning of recursive functions in [KSVW95]. The idea of frequency learning stems from a more general idea of (m, n)-computation, which, in the recursion-theoretic setting, means the following: to compute a function, an algorithm takes on n different inputs at a time and outputs correct values on at least m inputs. This idea can be traced to the works by G. Rose [Ros60] and B.A. Trakhtenbrot [Tra64] who suggested frequency computation as a deterministic alternative to traditional probabilistic algorithms using randomization. Since then, this idea has been applied to various settings, from computation by finite automata ([Kin76, ADHP05]) to computation with a small number of bounded queries ([BGK96]).

We explore and, whenever it has been possible, determine what makes automatic classes of languages (m, n)-learnable for various numbers n and $m \leq n$. Whereas, in our general model, it is not possible to identify which m conjectures among n are correct, we also consider the special case of finite learning automatic classes when the learner can identify m correct conjectures.

In the theory of language learning, a prominent role belongs to the so-called *characteristic samples* ([Muk92], see also [Ang80] for the related concept of tell-tale sets): a finite subset D of a language L is called a characteristic sample of L (with respect to the family of languages under consideration) if, for every language L' in the family, $D \subseteq L'$ implies $L' = L$. A family of languages satisfies characteristic sample condition if every language in it has a characteristic sample. Several of our characterizations of (m, n)-learnability are based on suitable variants of the characteristic sample condition. Since in all our settings (m, n)-learning (for $m < n$) turns out to be more powerful than learning individual languages, we study and discover interesting relationships between classes of languages (m, n)-learnable with different parameters m and/or n. In particular, we are concerned with the following questions:

a) does (m, n)-learnability imply $(m + 1, n + 1)$-learnability of a class? (thus, increasing frequency of correct conjectures, while keeping the number of possibly erroneous conjectures the same);

b) does $(m + 1, n + 1)$-learnability imply (m, n)-learnability? (thus, loosing in terms of frequency of correct conjectures, but allowing a smaller number of languages to be learnt in parallel, with the same number of possibly erroneous conjectures);

c) does $(m, n + 1)$-learnability imply (m, n)-learnability? (thus, reducing the number of possibly erroneous conjectures and increasing frequency of correct conjectures at the same time).

For each of our variants of learnability, we obtain either full or partial answers to all the above questions, for large enough families.

The structure of our study of (m, n)-learning is as follows. In the next section, we introduce necessary mathematical preliminaries and notation. In Section 3 we formally define our learning models. In Section 4, we take on the case of finite (m, n)-learning when a learner can identify at least m languages learnt correctly — following [KSVW95], we call (m, n)-learning of this kind *superlearning*. In Theorems 7 and 8, for the classes containing at least $2n + 1 - m$ languages, we give a full characterization for (m, n)-superlearnability in terms of characteristic samples. For large classes of languages, this characterization provides us full positive answers to the above questions a), b), and the negative answer to c). We also address the case when the number of languages in a class to be learnt is smaller than $2n + 1 - m$.

In Section 5 we consider finite (m, n)-learning when a learner cannot tell which m conjectures are correct. For large classes of languages, we again obtain a full characterization of (m, n)-learnability in terms of characteristic samples — albeight somewhat different from the case of superlearnability. This characterization, as in case of superlearnability, provides us answers to the questions a), b), and c). The proofs in this section are quite involved — to obtain necessary results, we developed a technique based on bipartite graphs.

In Section 6, we address finite (m, n)-learning by finite automata — automatic learning. We have not been able to come up with a characterization of this type of learnability, however, we answer positively to the question b) and negatively to the question c). The question a) remains open.

In Section 7 we obtain full positive answers for the questions a) and b) and the negative answer to the question c) for (m, n)-learnability of automatic classes in the limit.

2 Preliminaries

The set of natural numbers, $\{0, 1, 2, \ldots\}$, is denoted by N. We let Σ denote a finite alphabet. The set of all strings over the alphabet Σ is denoted by Σ^*. A language is a subset of Σ^*. The length of a string x is denoted by $|x|$. We let ϵ denote the empty string.

A string $x = x(0)x(1)\ldots x(n-1)$ is identified with the corresponding function from $\{0, 1, \ldots, n-1\}$ to Σ. We assume some canonical ordering of members of Σ. Lexicographic order is then the dictionary order over strings. A string w is length-lexicographically before (or smaller than) string w' (written $w <_{ll} w'$) iff $|w| < |w'|$ or $|w| = |w'|$ and w is lexicographically before w'. Furthermore, $w \leq_{ll} w'$ denotes that either $w = w'$ or $w <_{ll} w'$. For any set of strings S, let $\mathrm{succ}_S(w)$ denote the length-lexicographically least w' such that $w' \in S$ and $w <_{ll} w'$ — if there is no such string, then $\mathrm{succ}_S(w)$ is undefined.

We let $\emptyset, \subseteq$ and $\subset$ respectively denote empty set, subset and proper subset. The cardinality of a set S is denoted by $\mathrm{card}(S)$.

We now define the convolution of two strings $x = x(0)x(1)\ldots x(n-1)$ and $y = y(0)y(1)\ldots y(m-1)$, denoted $conv(x, y)$. Let x', y' be strings of length $\max(\{m, n\})$ such that $x'(i) = x(i)$ for $i < n$, $x'(i) = \#$ for $n \leq i < \max(\{m, n\})$, $y'(i) = y(i)$ for $i < m$, and $y'(i) = \#$ for $m \leq i < \max(\{m, n\})$, where $\# \notin \Sigma$ is a special padding symbol. Thus, x', y' are obtained from x, y by padding the smaller string with $\#$'s. Then, $conv(x, y) = z$, where $|z| = \max(\{m, n\})$ and $z(i) = (x'(i), y'(i))$, for $i < \max(\{m, n\})$. Here, note that z is a string over the alphabet $(\Sigma \cup \{\#\}) \times (\Sigma \cup \{\#\})$. Intuitively, giving a convolution of two strings as input to a machine means giving the two strings in parallel, with the shorter string being padded with $\#$s. The definition of convolution of two strings can be easily generalized to convolution of more than two strings. An n-ary relation R is *automatic*, if $\{conv(x_1, x_2, \ldots, x_n) : (x_1, x_2, \ldots, x_n) \in R\}$ is regular. Similarly, an n-ary function f is automatic if $\{conv(x_1, x_2, \ldots, x_n, y) : f(x_1, x_2, \ldots, x_n) = y\}$ is regular.

A family of languages, $(L_\alpha)_{\alpha \in I}$, over some finite alphabet Σ, is called an *automatic family* if (a) the index set I is regular and (b) the set $\{conv(\alpha, x) : \alpha \in I, x \in L_\alpha\}$ is regular. We often identify an automatic family $(L_\alpha)_{\alpha \in I}$ with the class $\mathcal{L} = \{L_\alpha : \alpha \in I\}$, where the indexing is implicit. An automatic family $(L_\alpha)_{\alpha \in I}$ is 1-1 (or the indexing is 1-1), if for all $\alpha, \beta \in I$, $L_\alpha = L_\beta$ implies $\alpha = \beta$.

It can be shown that any family, relation or function that is first-order definable using other automatic relations or functions is itself automatic.

Lemma 1. *[BG00, KN95] Any relation that is first-order definable from existing automatic relations is automatic.*

We use the above lemma implicitly in our proofs, without explicitly stating so. The example below gives some well-known automatic families.

Example 2. (a) *For any fixed k, the class of all subsets of Σ^* having at most k elements is an automatic family.*
(b) *The class of all finite and cofinite subsets of $\{0\}^*$ is an automatic family.*
(c) *The class of closed intervals, consisting of languages $L_{Conv(\alpha,\beta)} = \{x \in \Sigma^* : \alpha \leq_{lex} x \leq_{lex} \beta\}$ where $\alpha, \beta \in \Sigma^*$, over the alphabet Σ is an automatic family.*

3   Learning Automatic Families

A *text T* is a mapping from N to $\Sigma^* \cup \{\#\}$. The content of a text T, denoted content(T), is $\{T(i) : i \in N\} - \{\#\}$. A text T is for a language L iff content$(T) = L$. Intuitively, $\#$'s denote pauses in the presentation of data. Furthermore, $\#^{\infty}$ is the only text for $\emptyset$.

Let $T[n]$ denote $T(0)T(1)\ldots T(n-1)$, the initial sequence of T of length n. We let σ and τ range over finite initial sequences of texts. The length of σ is denoted by $|\sigma|$. For $n \leq |\sigma|$, $\sigma[n]$ denotes $\sigma(0)\sigma(1)\ldots\sigma(n-1)$. The empty sequence is denoted by Λ. Let content$(\sigma) = \{\sigma(i) : i < |\sigma|\}$.

We now consider learning machines. Since we are considering parallel learning, we directly define learners which take as input n texts. Furthermore, to make it easier to define automatic learners, we define the learners as mapping from the current memory and the new datum, to the new memory and conjecture (see [JLS12]). When one does not have any memory constraints (as imposed, for example, by automatic learning requirement), these learners are equivalent to those defined by Gold [Gol67]. The learner uses some hypothesis space $\{H_\alpha : \alpha \in J\}$ to interpret its hypothesis. We always require (without explicitly stating so) that $\{H_\alpha : \alpha \in J\}$ is a uniformly r.e. class (that is, $\{(x,\alpha) : x \in H_\alpha\}$ is r.e.). Often the hypothesis space is even required to be an automatic family, with the index set J being regular.

Definition 3. (Based on [Gol67, JLS12]) Suppose Σ and Δ are finite alphabets used for languages and memory of learners respectively, where $\# \notin \Sigma$. Suppose J is the index set (over some finite alphabet) for the hypothesis space used by the learner. Let ? be a special symbol not in J. Suppose $0 < n$.

(a) A *learner* (from n-texts) is a mapping from $\Delta^* \times (\Sigma^* \cup \{\#\})^n$ to $\Delta^* \times (J \cup \{?\})^n$.
A learner has an initial memory $mem_0 \in \Delta^*$, and an initial hypotheses $(hyp_1^0, hyp_2^0, \ldots, hyp_n^0) \in (J \cup \{?\})^n$.
(b) Suppose a learner $\mathbf{M}$ with the initial memory mem_0 and the initial hypotheses $hyp_1^0, hyp_2^0, \ldots, hyp_n^0$ is given. Suppose $T_1, T_2, \ldots, T_n$ are n texts. Then the definition of $\mathbf{M}$ is extended to sequences as follows.

$\mathbf{M}(\Lambda, \Lambda, \ldots, \Lambda) = (mem_0, hyp_1^0, hyp_2^0, \ldots, hyp_n^0);$
$\mathbf{M}(T_1[s+1], T_2[s+1], \ldots, T_n[s+1]) = \mathbf{M}(mem, T_1(s), T_2(s), \ldots, T_n(s)),$
where $\mathbf{M}(T_1[s], T_2[s], \ldots, T_n[s]) = (mem, hyp_1, hyp_2, \ldots, hyp_n)$, for some
$(hyp_1, hyp_2, \ldots, hyp_n) \in (J \cup \{?\})^n$ and $mem \in \Delta^*$.

(c) We say that $\mathbf{M}$ converges on $T_1, T_2, \ldots, T_n$ to hypotheses $(\beta_1, \beta_2, \ldots, \beta_n) \in (J \cup \{?\})^n$ (written: $\mathbf{M}(T_1, T_2, \ldots, T_n) \downarrow_{hyp} = (\beta_1, \beta_2, \ldots, \beta_n))$ iff there exists a t such that, for all $t' \geq t$,
$\mathbf{M}(T_1[t'], T_2[t'], \ldots, T_n[t']) \in \Delta^* \times \{(\beta_1, \beta_2, \ldots, \beta_n)\}.$

Intuitively, $\mathbf{M}(T_1[s], T_2[s], \ldots, T_2[s]) = (mem, hyp_1, hyp_2, \ldots, hyp_n)$ means that the memory and the hypotheses of the learner $\mathbf{M}$ after having seen the initial parts $T_1[s], T_2[s], \ldots, T_n[s]$ of the n texts are mem and $hyp_1, hyp_2, \ldots, hyp_n$, respectively.

We call the learner automatic, if the corresponding graph of the learner is automatic. That is, $\{conv(mem, x_1, x_2, \ldots, x_n, newmem, \beta_1, \beta_2, \ldots, \beta_n) : \mathbf{M}(mem, x_1, x_2, \ldots, x_n) = (newmem, \beta_1, \beta_2, \ldots, \beta_n)\}$ is regular.

We can think of a learner as receiving the texts $T_1, T_2, \ldots, T_n$ one element at a time from each of the texts. At each input, the learner updates its previous memory, and outputs a new conjecture (hypothesis) for each of the texts. If the sequence of hypotheses converges to a grammar for content(T), then we say that the learner $\mathbf{TxtEx}$-learns the corresponding text ([Gol67]). Here $\mathbf{Ex}$ denotes "explains", and $\mathbf{Txt}$ denotes learning from text. For parallel (m, n)-learnability, we require that the learner converges to a correct grammar for at least m out of the n input texts. Now we define learnability formally.

Definition 4. (Based on [Gol67, KSVW95])
Suppose $\mathcal{L} = \{L_\alpha : \alpha \in I\}$ is a target class, and $\mathcal{H} = \{H_\beta : \beta \in J\}$ is a hypothesis space. Suppose $0 < m \leq n$.

(a) We say that $\mathbf{M}$ (m, n)-$\mathbf{TxtEx}$-learns the class $\mathcal{L}$ (using $\mathcal{H}$ as the hypothesis space) iff for all n-texts $T_1, T_2, \ldots, T_n$ for distinct languages in $\mathcal{L}$, $\mathbf{M}(T_1, T_2, \ldots, T_n) \downarrow_{hyp} = (\beta_1, \beta_2, \ldots, \beta_n)$ such that for at least m different $i \in \{1, 2, \ldots, n\}$, $\beta_i \in J$ and $H_{\beta_i} = $ content(T_i).

(b) (m, n)-$\mathbf{TxtEx} = \{\mathcal{L} : (\exists \mathbf{M})[\mathbf{M} \ (m, n)$-$\mathbf{TxtEx}$-learns $\mathcal{L}$ using some $\mathcal{H}$ as the hypothesis space]$\}$.

(c) We say that $\mathbf{M}$ (m, n)-$\mathbf{TxtFin}$-learns the class $\mathcal{L}$ (using $\mathcal{H}$ as the hypothesis space) iff for all n-texts $T_1, T_2, \ldots, T_n$ for distinct languages in $\mathcal{L}$, there exists an s such that, for all $s' < s$ and $s'' \geq s$:
(i) $\mathbf{M}(T_1[s'], T_2[s'], \ldots, T_n[s']) \in \Delta^* \times (?, ?, \ldots, ?)$ (where there are n ? in the above).
(ii) $\mathbf{M}(T_1[s''], T_2[s''], \ldots, T_n[s'']) \in \Delta^* \times (\beta_1, \beta_2, \ldots, \beta_n)$, where for at least m distinct $i \in \{1, 2, \ldots, n\}$, $\beta_i \in J$ and $H_{\beta_i} = $ content(T_i).

(d) (m, n)-$\mathbf{TxtFin} = \{\mathcal{L} : (\exists \mathbf{M})[\mathbf{M} \ (m, n)$-$\mathbf{TxtFin}$-learns $\mathcal{L}$ using some $\mathcal{H}$ as the hypothesis space]$\}$.

We drop the reference to "using the hypothesis space $\mathcal{H}$", when the hypothesis space is clear from the context. A hypothesis space $\mathcal{H}$ is said to be *class preserving*

[LZ93] for learning a class $\mathcal{L}$ if $\mathcal{L} = \mathcal{H}$. A hypothesis space $\mathcal{H}$ is said to be *class comprising* [LZ93] for learning a class $\mathcal{L}$ if $\mathcal{L} \subseteq \mathcal{H}$.

For (m, n)-**superTxtEx** or (m, n)-**superTxtFin**-learnability, we require the learner to also identify the (at least m) texts which it has learnt. This is done via allowing additional output (along with the hypotheses) to the learner. Then, for superlearnability, we require that the learner specifies the m-texts which it is able to learn, outputting 1 (along with the corresponding hypothesis) to denote that it has learnt the text and 0 to denote that it is not making any guarantees about its hypothesis.

When we are considering automatic learners (that is, learners, whose graphs are regular [JLS12]), we prefix the learning criterion **TxtEx** or **TxtFin** by **Auto**. For this we also require the hypothesis space used to be an automatic family.

For ease of notation, rather than giving the learner as a mapping from memory and n-input elements to new memory and conjecture, we often just give an informal description of the learner, where for finite learning the learner will output only one conjecture different from $(?, ?, \ldots, ?)$. It will be clear from the context how the formal learners can be obtained from the description.

Definition 5. [Muk92] We say that S is a *characteristic sample* for L with respect to $\mathcal{L}$ iff (a) S is a finite subset of L and (b) for all $L' \in \mathcal{L}$, $S \subseteq L'$ implies $L = L'$.

Using Lemma 1, it is easy to see that testing whether or not a finite set S is a characteristic sample of L with respect to automatic family $\mathcal{L}$ is decidable effectively in S and index for L (since we can express such a property using first-order formula).

The following lemma is used implicitly in several proofs, without explicitly referring to it.

Lemma 6. *Suppose $\mathcal{L}$ and a language $L \in \mathcal{L}$ are given such that L does not have a characteristic sample with respect to $\mathcal{L}$. Then, either (a) there exists $L' \in \mathcal{L}$ such that $L \subset L'$ or (b) for all n, there are $X_n \in \mathcal{L}$, such that X_n are pairwise distinct and $L \cap \{x : x \leq n\} \subseteq X_n \cap \{x : x \leq n\}$.*

4 (m, n)-superTxtFin-Learnability

The next two theorems give a full characterization of (m, n)-**superTxtFin**-learnability for large automatic classes.

Theorem 7. *Suppose $0 < m \leq n$. Suppose $\mathcal{L}$ is automatic, and for all except at most $n - m$ $L \in \mathcal{L}$, there exists a characteristic sample for L with respect to $\mathcal{L}$. Then $\mathcal{L}$ is (m, n)-superTxtFin-learnable.*

Proof. The desired learner $\mathbf{M}$, on any input texts $T_1, \ldots, T_n$, searches for an r such that at least m of content($T_1[r]$), content($T_2[r]$), $\ldots$, content($T_n[r]$) are characteristic samples for some languages in $\mathcal{L}$ (before finding such an r, $\mathbf{M}$ conjectures ? for all the texts). When it finds such an r, $\mathbf{M}$ outputs corresponding grammars for the languages on the corresponding texts and lists them as having been learnt (the conjectures on remaining texts are irrelevant).   ∎

The following result shows that the above result is optimal for large enough classes of languages. For small finite classes, we need some special considerations, as illustrated by Remark 13 below.

Theorem 8. *Suppose $0 < m \leq n$. If $\mathcal{L}$ has at least $2n + 1 - m$ languages, then (m, n)-superTxtFin-learnability of $\mathcal{L}$ implies there are at most $n - m$ languages in $\mathcal{L}$ which do not have characteristic sample with respect to $\mathcal{L}$.*

Proof. (sketch) Suppose by way of contradiction otherwise. Pick at least $n-m+1$ languages in the class which do not have a characteristic sample with respect to $\mathcal{L}$. Let these languages be $A_0, A_1, \ldots, A_{n-m}$.

For each $r \leq n - m$, let $B_r \in \mathcal{L} - \{A_0, A_1, \ldots, A_{n-m}\}$ be a language in $\mathcal{L}$ which is a superset of A_r. If there is no such language, then B_r is taken to be an arbitrary member of $\mathcal{L} - \{A_0, A_1, \ldots, A_{n-m}\}$. The B_i's may not be different from each other.

Note that if B_r is not a superset of A_r, then there exist infinitely many pairwise distinct languages $S_r^w \in \mathcal{L}$, $w \in \Sigma^*$, such that each S_r^w contains $A_r \cap \{x : x \leq_{ll} w\}$ — this follows from the fact that there is no characteristic sample for any A_i in $\mathcal{L}, i \leq n - m$.

Now consider the behaviour of the superlearner on the texts $T_0, T_1, \ldots, T_{n-1}$ for languages $A_0, A_1, \ldots, A_{n-m}, C_{n-m+1}, \ldots, C_{n-1}$, where $C_{m-n+1}, \ldots, C_{n-1}$ are members of $\mathcal{L}$ which are different from $A_r, B_r, r \leq n - m$. Suppose the superlearner outputs its conjecture (different from $(?, ?, \ldots, ?)$) after seeing input $T_0[s], T_1[s], \ldots, T_{n-1}[s]$. As the superlearner identifies at least m languages, it has to identify at least one $A_0, A_1, \ldots, A_{n-m}$, say A_r. Suppose content$(T_r[s]) \subseteq \{x : x \leq_{ll} w\}$ Then one can replace A_r by B_r or by an appropriate one of $S_r^{w'}$, $w' \geq_{ll} w$, which is not among $A_0, A_1, \ldots, A_{n-m}, C_{n-m+1}, \ldots, C_{n-1}$, thus making the superlearner fail.   ∎

The following corollaries easily follow from the above two theorems.

Corollary 9. *Suppose $0 < n$. If an automatic class $\mathcal{L}$ contains at least $n + 1$ languages and is (n, n)-superTxtFin-learnable, then every language in $\mathcal{L}$ has a characteristic sample and, thus, the class is TxtFin-learnable.*

Corollary 10. *Suppose $0 < m \leq n$. If a large enough automatic class $\mathcal{L}$ is (m, n)-superTxtFin-learnable, then it is $(m-1, n-1)$-superTxtFin-learnable and $(m + 1, n + 1)$-superTxtFin-learnable.*

Corollary 11. *Suppose $0 < m < n$. There exists an automatic class $\mathcal{L}$ that is (m, n)-superTxtFin-learnable, but not $(m, n - 1)$-superTxtFin-learnable.*

For $m = 1$, Theorem 8 can be strengthened to

Theorem 12. *Suppose $0 < n$. Suppose $\mathcal{L}$ contains at least n languages L which do not have a characteristic sample with respect to $\mathcal{L}$. Then, $\mathcal{L}$ is not $(1, n)$-superTxtFin-learnable.*

Remark 13. Let $2 \leq m \leq n$.

Let $L_{2r} = \{2r\}$, $L_{2r+1} = \{2r, 2r+1\}$, for $r \leq n - m$.

Let $L_i = \{i\}$, for $2n - 2m + 2 \leq i < 2n - m$.

Let $\mathcal{L} = \{L_i : i < 2n - m\}$.

Now, $\mathcal{L}$ contains $n - m + 1$ languages (L_{2r}, for $r \leq n - m$) which do not have a characteristic sample with respect to $\mathcal{L}$. However, $\mathcal{L}$ is (m,n)-**superTxtFin**-learnable. To see (m,n)-**superTxtFin**-learnability, note that in any collection of n languages from $\mathcal{L}$, there can be at most $n - m$ different $s \leq n - m$ such that the collection contains L_{2s} but not L_{2s+1}. Note that if the collection contains both L_{2s} and L_{2s+1}, then we can identify both of them, from texts, as the languages given as input to (m,n)-**superTxtFin**-learner are supposed to be different. Thus, one can easily (m,n)-**superTxtFin**-learn the class $\mathcal{L}$.

5 (m,n)-TxtFin-Learnability

Our first goal is to find a necessary condition for finite (m,n)-learnability of large automatic classes in terms of characteristic samples. For this, we introduce the concept of a *cut* of a bipartite graph and an important technical lemma.

Definition 14. Suppose $G = (V, E)$ is a bipartite graph, where V_1, V_2 are the two partitions of the vertices. Then,

(a) (V_1', V_2') is called a *cut* of G if G does not contain any edges between $V_1 - V_1'$ and $V_2 - V_2'$. (V_1', V_2') is called a *minimum cut*, if it is a cut which minimizes card$(V_1' \cup V_2')$.

(b) $E' \subseteq E$ is called a *matching* if for all distinct $(v, w), (v', w') \in E'$, $v \neq v'$ and $w \neq w'$. E' is called a *maximum matching* if E' is a matching with maximum cardinality.

Note that cuts are usually defined using edges rather than vertices, however for our purposes it is convenient to define cut sets using vertices. We often write a bipartite graph (V, E) as (V_1, V_2, E), where V_1, V_2 are the two partitions. For example, consider the graph $V = \{a, b, c_1, c_2, \ldots, c_r, d_1, d_2, \ldots, d_k\}$ with $E = \{(a, d_1), (a, d_2), \ldots, (a, d_k), (c_1, b), (c_2, b), \ldots, (c_r, b)\}$. The minimum cut in the graph would be $(\{a\}, \{b\})$. Note also that $\{(a, d_1), (c_1, b)\}$ forms a maximum matching in the graph. Both minimum cut and maximum matching have same cardinality. This is not an accident, and the following lemma can be proven using the Max-Flow-Min-Cut Theorem (by adding a source node, with edge to each vertex in V_1, and a sink node, with edge from each vertex in V_2). For Max-Flow-Min-Cut Theorem and related concepts see, for example, [PS98].

Lemma 15. *For any bipartite graph, the size of the minimum cut is the same as the size of the maximum matching.*

Now we can show that the existence of characteristic samples for all the languages in the class, except at most $n - 1$ ones, (where characteristic samples are relative to the class excluding the $n - 1$ latter languages) is a necessary condition for $(1, n)$-**TxtFin**-learnability. Note that this characteristic sample condition is similar, but different from the one for $(1, n)$-**superTxtFin**-learning.

Theorem 16. *Suppose $\mathcal{L}$ is $(1,n)$-**TxtFin**-learnable.*

Then there exists a subset $\mathcal{S}$ of $\mathcal{L}$ of size at most $n-1$ such that every language in $\mathcal{L} - \mathcal{S}$ has a characteristic sample with respect to $\mathcal{L} - \mathcal{S}$.

Proof. Suppose $\mathbf{M}$ $(1,n)$-**TxtFin**-learns $\mathcal{L}$.

Let $\mathcal{L}' = \{L \in \mathcal{L} : (\exists \text{ distinct } S_w^L \in \mathcal{L} \text{ for each } w \in \Sigma^*)[L \cap \{x : x \leq_{ll} w\} \subseteq S_w^L]\}$.

For each L in $\mathcal{L}'$, $w \in \Sigma^*$, fix S_w^L as in the definition of $\mathcal{L}'$.

Let $\mathcal{L}'' = \{L \in \mathcal{L} - \mathcal{L}' : L \text{ does not have a characteristic sample with respect to } \mathcal{L} - \mathcal{L}'\}$.

Let $\mathcal{L}''' = \{A \in \mathcal{L} - \mathcal{L}' : (\exists L \in \mathcal{L} - \mathcal{L}')[L \subset A]\}$.

Claim. $\text{card}(\mathcal{L}') < n$.

To see the claim, suppose $\mathcal{L}'$ has $\geq n$ languages. Then as input to $\mathbf{M}$, we can give texts $T_1, T_2, \ldots, T_n$ for $C_1, C_2, \ldots, C_n \in \mathcal{L}'$. Suppose $\mathbf{M}$ converges, say, after seeing $T_1[m], T_2[m], \ldots, T_n[m]$, to conjecture $(p_1, p_2, \ldots, p_n)$ (different from $(?, ?, \ldots, ?)$). Then consider texts T_i' extending $T_i[m]$, where T_i' is a text for $E_i = S_{j_i}^{C_i}$, for some j_i such that $S_{j_i}^{C_i} \supseteq \text{content}(T_i[m])$ and (a) p_i is not a grammar for $S_{j_i}^{C_i}$ and (b) $S_{j_i}^{C_i}$ are pairwise distinct for different i. Note that this can be easily ensured. Then, $\mathbf{M}$ fails on input $T_1', T_2', \ldots, T_n'$. This completes the proof of the claim.

Suppose $\text{card}(\mathcal{L}') = n - r$.

Note that every language in $\mathcal{L}''$ has a proper superset in $\mathcal{L}'''$ and every language in $\mathcal{L}'''$ has a proper subset in $\mathcal{L}''$. Consider the bipartite graph G formed by having the vertex set $V_1 = \mathcal{L}''$ and $V_2 = \mathcal{L}'''$, and edge between (L'', L''') iff $L'' \subset L'''$. (If $L \in \mathcal{L}'' \cap \mathcal{L}'''$, then for the purposes of the bipartite graph, we consider corresponding vertex in V_1 and V_2 representing L as different).

Claim. There exists a cut of G of size at most $r - 1$.

Assume by way of contradiction otherwise. Then, by Lemma 15, there exists a matching of size at least r. Let this matching be $(A_1, B_1), \ldots, (A_r, B_r)$. Here, each $A_i \in \mathcal{L}''$ and each $B_i \in \mathcal{L}'''$. A_i's are pairwise distinct, B_i's are pairwise distinct, but A_i's and B_j's might coincide with each other. Assume without loss of generality that if $i < j$, then $A_j \not\subseteq A_i$. Now consider giving the learner input texts $T_1, T_2, \ldots, T_n$ for $A_1, A_2, \ldots, A_r, C_{r+1}, C_{r+2}, \ldots, C_n$, where $C_{r+1}, C_{r+2}, \ldots, C_n$ are distinct members of $\mathcal{L}'$. Suppose M outputs conjecture $(p_1, p_2, \ldots, p_n)$ (different from $(?, ?, \ldots, ?)$) after seeing input $T_1[m], T_2[m], \ldots, T_n[m]$.

Now we define texts $T_1', \ldots, T_n'$ extending $T_1[m], \ldots, T_n[m]$ respectively for languages $E_1, \ldots, E_n$, on which $\mathbf{M}$ fails.

Below we will define $E_1, \ldots, E_r$. Definition of $E_{r+1}, \ldots, E_n$ can be done appropriately as done in the case above when $\mathcal{L}'$ was at least n. Now we define E_j by induction from $j = r$ to 1.

Suppose we have already defined $E_r, \ldots, E_{j+1}$, and are now defining E_j. We will also maintain languages $B_1', B_2', \ldots$, which change over the construction. Initially, let $B_i' = B_i$ for all i. We will have at any stage (by induction) the invariants that (a) $B'_1, \ldots, B'_j, E_{j+1}, \ldots, E_r$ are pairwise distinct, (b) $A_1, A_2, \ldots, A_j, E_{j+1}, \ldots, E_r$ are pairwise distinct, (c) $A_i \subset B'_i$ for $1 \leq i \leq j$.

It is easy to verify that induction hypothesis holds when $j = r$. Note that this implies that $B'_j \neq A_i$, for $1 \leq i \leq j$ (as $A_i \not\supseteq A_j$, for all $1 \leq i \leq j$).

Definition of E_j:

If p_j is a grammar for A_j, then let $E_j = B'_j$ (and other values do not change).

If p_j is not a grammar for A_j, then let $E_j = A_j$. If one of $B'_i = A_j$, for $i < j$, then replace B'_i by B'_j. Other variables do not change value.

It is now easy to verify that the construction maintains the invariants, and thus **M** fails to $(1, n)$-**TxtFin**-learn $\mathcal{L}$. The claim follows.

Now, it is easy to verify that taking $\mathcal{S}$ as $\mathcal{L}'$ unioned with the cut of G as in the claim, satisfies the requirements of the Theorem. ∎

Note that by appropriate modification of the proof, one can also show:

Theorem 17. *Suppose $\mathcal{L}$ is (m, n)-**TxtFin**-learnable and $\mathcal{L}$ contains at least $2n - m + 1$ languages.*

Then, there exists a subset $\mathcal{S}$ of $\mathcal{L}$ of size at most $n - m$ such that every language in $\mathcal{L} - \mathcal{S}$ has a characteristic sample with respect to $\mathcal{L} - \mathcal{S}$.

Now we show that the necessary condition of the previous Theorem is sufficient for (m, n)-**TxtFin**-learning.

Theorem 18. *Suppose $\mathcal{L}$ is automatic. Suppose there exists a subset $\mathcal{S}$ of $\mathcal{L}$ of size at most $n - m$ such that every language in $\mathcal{L} - \mathcal{S}$ has a characteristic sample with respect to $\mathcal{L} - \mathcal{S}$. Then, $\mathcal{L}$ is (m, n)-**TxtFin**-learnable.*

Proof. Let $\mathcal{L}' = \{L \in \mathcal{L} : (\exists \text{ infinitely many distinct } S^L_w \in \mathcal{L}, w \in \Sigma^*)[L \cap \{x : x \leq_{ll} w\} \subseteq S^L_w]\}$.

Note that $\mathcal{L}' \subseteq \mathcal{S}$. Thus, $\mathrm{card}(\mathcal{L}') = n - r \leq n - m$, for some r.

Furthermore, for all $L \in \mathcal{L} - \mathcal{L}'$, there exists a finite subset X of L such that there exist at most finitely many $L' \in \mathcal{L}$ satisfying $X \subseteq L'$. Furthermore, none of the members of $\mathcal{L} - \mathcal{L}'$ are contained in any member of $\mathcal{L}'$.

So the learner **M** behaves as follows on input texts $T_1, T_2, \ldots, T_n$. It first searches for an s such that, for at least r members j of $\{1, 2, \ldots n\}$,

(a) $\mathrm{content}(T_j[s])$ is not contained in any L in $\mathcal{L}'$, and
(b) $\mathrm{content}(T_j[s])$ is contained in at most finitely many of L in $\mathcal{L}$.

Note that there exists such an s, and it can be effectively found, given an automatic numbering of $\mathcal{L}$. Without loss of generality, for ease of notation, from now on we assume that these r members are $\{1, 2, \ldots, r\}$.

This gives us that the corresponding r texts $T_1, T_2, \ldots, T_r$ can only be for languages from $\mathcal{L} - \mathcal{L}'$. Up to card$(\mathcal{S}) - (n - r)$ of these may be from $\mathcal{S} - \mathcal{L}'$, and thus at least $n - $ card$(\mathcal{S})$ are from $\mathcal{L} - \mathcal{S}$.

Let $\mathcal{H} = \{L \in \mathcal{L} : (\exists i : 1 \leq i \leq r)[\text{content}(T_i[s]) \subseteq L]\}$. By (b) above, $\mathcal{H}$ is finite. Arrange elements of $\mathcal{H}$ in a directed graph G, where there is an edge from L to L' iff $L \subset L'$ and no other $L'' \in \mathcal{H}$ satisfies $L \subset L'' \subset L'$. Note that the graph is acyclic. Also, note that there is no path from any $L \in \mathcal{L} - \mathcal{S}$ to another $L' \in \mathcal{L} - \mathcal{S}$ (as this would imply that $L \subset L'$, and thus no characteristic sample for L with respect to $\mathcal{L} - \mathcal{S}$ would exist).

Now let $s' > s$ be such that

(c) for each i, $1 \leq i \leq r$, there exists a (necessarily unique) $L \in \mathcal{H}$ such that content$(T_i[s']) \subseteq L$ but content$(T_i[s']) \not\subseteq L'$ for any other $L' \in \mathcal{H}$ which satisfies $L' \not\supseteq L$ — we assign T_i to the node L in the graph G in this case, and

(d) for each node L in G, at most one T_i is assigned to L.

Note that such s' will eventually be found as the texts $T_1, T_2, \ldots, T_r$ are for different languages from $\mathcal{H}$. Once such s' is found, the learner outputs grammar for L on T_i iff T_i is assigned to the node L. We now claim that the above learner (m, n)-**TxtFin**-learns $\mathcal{L}$. For this, it suffices to show that the learner is correct on at least m of the texts $T_1, T_2, \ldots, T_r$.

(**) We will only count the correctness of the learner for languages in $\mathcal{L} - \mathcal{S}$.

Let G' be just as graph G, except that the texts assigned to nodes may change — T_i is assigned to a node L iff T_i is actually a text for L. Note that each node in G and G' is assigned at most one text.

Note that if T_i is assigned to a node L in G, but to a node L' in G', then $L \subseteq L'$. Now consider the texts T_i on which the learner is wrong. These texts can be divided up into maximal chains of the form $T_{i_1}, T_{i_2}, T_{i_3}, \ldots, T_{i_{j-1}}$, where (i) T_{i_s} is assigned to A_{i_s} in G and $A_{i_{s+1}}$ (which represented content(T_{i_s})) in G', (ii) $A_{i_s} \subset A_{i_{s+1}}$, for $1 \leq s < j$, (iii) no text is assigned to A_{i_1} in G' and no text is assigned to A_{i_j} in G, and (iv) the different maximal chains as above do not have any texts/nodes in the graph in common. Thus, we can consider such maximal chains independently for error computation: each of these chains has at most one member from $\mathcal{L} - \mathcal{S}$ (since members of $\mathcal{L} - \mathcal{S}$ are pairwise not included in each other), and thus contain at least $j - 1$ members from $\mathcal{S} - \mathcal{L}'$. Thus, the learner fails on at most $card(\mathcal{S} - \mathcal{L}')$ texts among $T_1, T_2, \ldots, T_r$. It follows that the learner is correct on at least $r - card(\mathcal{S} - \mathcal{L}')$ many texts from $T_1, T_2, \ldots, T_r$, and thus correct on at least m input texts. ∎

The following corollaries easily follow from the above two theorems.

Corollary 19. *Suppose $0 < m \leq n$. If a (large enough) automatic class $\mathcal{L}$ is (m, n)-**TxtFin**-learnable, then it is $(m - 1, n - 1)$-**TxtFin**-learnable and $(m + 1, n + 1)$-**TxtFin**-learnable.*

Corollary 20. *Suppose $0 < m < n$. There exists an automatic class $\mathcal{L}$ that is (m, n)-**TxtFin**-learnable, but not $(m, n - 1)$-**TxtFin**-learnable.*

6 Automatic (m, n)-Finite Learning

In this section, we consider finite (m, n)-learning by finite automata. Proofs are omitted due to space constraints.

Note that, for automatic learnability, results of the previous section do not hold. This is illustrated by the following result, based on techniques from [JLS12].

Theorem 21. *[JLS12] Let* $\mathcal{L} = \{L : (\exists n)(\exists x \in \Sigma^n)[L = \Sigma^n - \{x\}]\}$. *Then,* $\mathcal{L}$ *is not* **AutoTxtEx***-learnable.*

The proof for the above theorem can be generalized to show that $\mathcal{L}$ is not $(1, k)$-**AutoTxtEx**-learnable. Note that every language in the class has a characteristic sample (the language itself).

In the sequel, without loss of generality assume that all languages have at most one grammar in the hypothesis space (which is automatic). So below equality of languages is equivalent to grammars being the same.

First we show that, for finite automatic classes of languages, (m, n)-**TxtFin**-learnability implies (m, n)-**AutoTxtFin**-learnability.

Theorem 22. *Suppose* $\mathcal{L}$ *is an automatic finite class. Then,* $\mathcal{L}$ *is* (m, n)-**TxtFin***-learnable implies* $\mathcal{L}$ *is* (m, n)-**AutoTxtFin***-learnable.*

Our next goal is to show that, for large enough automatic classes, automatic finite $(m + 1, n + 1)$-learnability implies automatic finite (m, n)-learnability.

Theorem 23. *Let* $0 < m \leq n$. *Suppose* $\mathcal{L}$ *is an infinite automatic class which is* $(m+1, n+1)$-**AutoTxtFin***-learnable. Then* $\mathcal{L}$ *is* (m, n)-**AutoTxtFin***-learnable.*

Remark 24. Suppose $0 < m \leq n$, $\mathcal{L}$ is finite and contains at least $2n + 2 - m$ languages, and $\mathcal{L}$ is $(m + 1, n + 1)$-**AutoTxtFin**-learnable. Then, $\mathcal{L}$ is (m, n)-**AutoTxtFin**-learnable. This holds as by Theorem 17 and Theorem 18, $\mathcal{L}$ is (m, n)-**TxtFin**-learnable and thus by Theorem 22, $\mathcal{L}$ is (m, n)-**AutoTxtFin**-learnable.

We have not been able to prove that automatic finite (m, n)-learnability implies automatic finite $(m+1, n+1)$-learnability (even for large automatic classes). Yet, we can show that automatic finite (m, n)-learnability does not imply even finite $(m, n - 1)$-learnability.

Proposition 25 *Suppose* $r \in N$, *and* $r \geq 1$. *Let* $L_{a^i} = \{a^i\}$, *Let* $L_{b^i} = \{b^i\}$, *and* $L_{c^i} = \{b^i, c^i\}$.
   Let $\mathcal{L} = \{L_{a^i} : i \geq 1\} \cup \{L_{b^i} : 1 \leq i \leq r\} \cup \{L_{c^i} : 1 \leq i \leq r\}$.
   Then, for $m \geq 1$, $\mathcal{L}$ *is* $(m, m + r)$-**AutosuperTxtFin***-learnable, but not* $(m, m + r - 1)$-**TxtFin***-learnable.*

Corollary 26. *For all* m, n *such that* $0 < m \leq n - 1$, *there exists an automatic family which can be* (m, n)-**AutoTxtFin***-learnt but not* $(m, n - 1)$-**AutoTxtFin***-learnt.*

7 (m, n)-TxtEx-Learning

In this section we consider (m, n)-learning of automatic classes in the limit. We show that the number of languages learnable in parallel can be increased or decreased when $n - m$, the number of languages which may be erroneously identified, remains the same. For the rest of this section, we assume that the hypothesis spaces are always automatic. Proofs are omitted due to space constraints.

Theorem 27. *Suppose* $0 < m \leq n$. *Suppose* $\mathcal{L}$ *is an automatic class. If* $\mathcal{L}$ *can be* (m, n)-**TxtEx**-*learnt then* $\mathcal{L}$ *can be* $(m + 1, n + 1)$-**TxtEx**-*learnt.*

Theorem 28. *Suppose* $0 < m \leq n$. *Suppose* $\mathcal{L}$ *is automatic. Suppose* $\mathcal{L}$ *is* $(m + 1, n + 1)$-**TxtEx**-*learnable. Then,* $\mathcal{L}$ *is* (m, n)-**TxtEx**-*learnable.*

Theorem 29. *Suppose* $0 < n$. *There exists an automatic* $\mathcal{L}$ *which can be* $(1, n + 1)$-**TxtEx**-*learnt but not* $(1, n)$-**TxtEx**-*learnt.*

Corollary 30. *For* $0 < m \leq n$, *there exists an automatic* $\mathcal{L}$ *such that* $\mathcal{L}$ *can be* $(m, n + 1)$-**TxtEx**-*learnt but not* (m, n)-**TxtEx**-*learnt.*

8   Conclusion

We defined and explored a model of parallel learning n languages at a time when at least m languages are required to be learnt correctly. Similarly to (m, n)-computation being a deterministic alternative to probabilistic computation based on randomization, our model suggests a deterministic alternative to traditional probablistic learnability of languages (explored, for example, in [Pit89] and [WFK84]; as L. Pitt showed in [Pit89], learning using traditional probability is strongly related to another type of parallel deterministic learning — learning a language by a team). It turns out that, for the finite (m, n)-learnability, the maximum number $n - m$ of languages in the automatic family that do not have characteristic samples is the crucial factor defining learnability (and not the frequency m out of n of correct conjectures — as it follows from our results, increasing frequency not necessarily diminishes learnability of families of languages). Since a family of languages with a larger number of languages without characteristic samples is more topologically complex, the number $n - m$ can be interpreted as a measure of this complexity, and we have shown that there are learnability hierarchies based on this complexity measure.

Several interesting problems remain open. The major problem is finding characterizations — if any — for (m, n)-**AutoTxtFin**-learnability and for (m, n)-**TxtEx**-learnability. It is open whether (m, n)-**AutoTxtFin**-learnability implies $(m + 1, n + 1)$-**AutoTxtFin**-learnability. Another potentially interesting area of research would be finding if and how frequency learnability can help in terms of efficiency of learning.

Acknowledgements. We thank the referees of ALT 2014 for several helpful comments which improved the presentation of the paper.

References

[ADHP05] Austinat, H., Diekert, V., Hertrampf, U., Petersen, H.: Regular frequency computations. Theoretical Computer Science 330, 15–20 (2005)

[AGS89] Angluin, D., Gasarch, W., Smith, C.: Training sequences. Theoretical Computer Science 66, 255–272 (1989)

[Ang80] Angluin, D.: Inductive inference of formal languages from positive data. Information and Control 45, 117–135 (1980)

[BG00] Blumensath, A., Grädel, E.: Automatic structures. In: 15th Annual IEEE Symposium on Logic in Computer Science (LICS), pp. 51–62. IEEE Computer Society (2000)

[BGK96] Biegel, R., Gasarch, W., Kinber, E.: Frequency computation and bounded queries. Theoretical Computer Science 163, 177–192 (1996)

[BKF11] Balodis, K., Kucevalovs, I., Freivalds, R.: Frequency prediction of functions. In: Kotásek, Z., Bouda, J., Černá, I., Sekanina, L., Vojnar, T., Antoš, D. (eds.) MEMICS 2011. LNCS, vol. 7119, pp. 76–83. Springer, Heidelberg (2012)

[Gol67] Gold, E.M.: Language identification in the limit. Information and Control 10(5), 447–474 (1967)

[JLS12] Jain, S., Luo, Q., Stephan, F.: Learnability of automatic classes. Journal of Computer and System Sciences 78(6), 1910–1927 (2012)

[JORS99] Jain, S., Osherson, D., Royer, J., Sharma, A.: Systems that Learn: An Introduction to Learning Theory, 2nd edn. MIT Press, Cambridge (1999)

[Kin76] Kinber, E.: Frequency computations in finite automata. Kibernetika 2, 7–15 (1976) (in Russian); English translation in Cybernetics 12, 179–187

[KN95] Khoussainov, B., Nerode, A.: Automatic presentations of structures. In: Leivant, D. (ed.) LCC 1994. LNCS, vol. 960, pp. 367–392. Springer, Heidelberg (1995)

[KSVW95] Kinber, E., Smith, C., Velauthapillai, M., Wiehagen, R.: On learning multiple concepts in parallel. Journal of Computer and System Sciences 50, 41–52 (1995)

[LZ93] Lange, S., Zeugmann, T.: Language learning in dependence on the space of hypotheses. In: Proceedings of the Sixth Annual Conference on Computational Learning Theory, pp. 127–136. ACM Press (1993)

[Muk92] Mukouchi, Y.: Characterization of finite identification. In: Jantke, K.P. (ed.) AII 1992. LNCS, vol. 642, pp. 260–267. Springer, Heidelberg (1992)

[PS98] Papadimitriou, C.H.S., Steiglitz, K.: Combinatorial Optimization: Algorithms and Complexity. Dover (1998)

[Pit89] Pitt, L.: Probabilistic inductive inference. Journal of the ACM 36, 383–433 (1989)

[Ros60] Rose, G.: An extended notion of computability. Abstracts of International Congress for Logic, Methodology and Philosophy of Science, p. 14 (1960)

[Smi82] Smith, C.: The power of pluralism for automatic program synthesis. Journal of the ACM 29, 1144–1165 (1982)

[Tra64] Trakhtenbrot, B.: On the frequency of computation of functions. Algebra i Logika 2, 25–32 (1964)

[WFK84] Wiehagen, R., Freivalds, R., Kinber, E.: On the power of probabilistic strategies in inductive inference. Theoretical Computer Science 28, 111–133 (1984)

Algorithmic Identification of Probabilities Is Hard

Laurent Bienvenu[1], Benoît Monin[2], and Alexander Shen[3]

[1] Laboratoire Poncelet
laurent.bienvenu@computability.fr
[2] LIAFA
benoit.monin@liafa.univ-paris-diderot.fr
[3] LIRMM
alexander.shen@lirmm.fr; on leave from IITP RAS

Abstract. Suppose that we are given an infinite binary sequence which is random for a Bernoulli measure of parameter p. By the law of large numbers, the frequency of zeros in the sequence tends to p, and thus we can get better and better approximations of p as we read the sequence. We study in this paper a similar question, but from the viewpoint of inductive inference. We suppose now that p is a computable real, and one asks for more: as we are reading more and more bits of our random sequence, we have to eventually guess the exact parameter p (in the form of its Turing code). Can one do such a thing uniformly for all sequences that are random for computable Bernoulli measures, or even for a 'large enough' fraction of them? In this paper, we give a negative answer to this question. In fact, we prove a very general negative result which extends far beyond the class of Bernoulli measures.

1 Introduction

1.1 Learnability of Sequences

The study of learnability of computable sequences is concerned with the following problem. Suppose we have a black box that generates some infinite computable sequence of bits $X = X(0)X(1)X(2), \ldots$ We do not know the program running in the box, and want to guess it looking at finite prefixes

$$X \restriction n = X(0) \ldots X(n-1)$$

for increasing n. There could be different programs that produce the same sequence, and it is enough to guess one of them (since there is no way to distinguish between them looking at the output bits). The more bits we see, the more information we have about the sequence. For example, it is hard to say something about a sequence seeing only its first bit 1, but looking at the prefix

$$110010010000111111011010101000$$

P. Auer et al. (Eds.): ALT 2014, LNAI 8776, pp. 85–95, 2014.

one may observe that this is a prefix of the binary expansion of π, and guess that the machine inside the box does exactly that (though the machine may as well produce the binary expansion of, say, $47627751/15160384$).

The hope is that, as we gain access to more and more bits, we will *eventually* figure out how the sequence X is generated. More precisely, we hope to have a computable function $\mathfrak{A}$ such that for every computable X, the sequence

$$\mathfrak{A}(X \restriction 1), \ \mathfrak{A}(X \restriction 2), \ \mathfrak{A}(X \restriction 3), \ldots$$

converges to a program (=Turing machine) that computes X. This is referred to as *identification in the limit*, and can be understood in two ways:

- Strong success: for every computable X, the above sequence converges to a single program that produces X.
- Weak success: for every computable X, all but finitely many terms of the above sequence are programs that produce X (may be, different ones).

The first type of success is often referred to as *exact* (EX) and the second type as *behaviorally correct* (BC). Either way, such an algorithm $\mathfrak{A}$ does not exist in general. The main obstacle: certain machines are not total (produce only finitely many bits), and distinguishing total machines from non-total ones cannot be done computably. (If we restrict ourselves to some decidable class of total machines, e.g., primitive recursive functions, then exact learning is possible: let $\mathfrak{A}(u)$ be the first machine in the class that is compatible with u.) We refer the reader to [ZZ08] for a detailed survey of learnability of computable functions.

1.2   Learnability of Probability Measures

Recently, Vitanyi and Chater [VC13] proposed to study a related problem. Suppose that instead of a total deterministic machine, the black box contains an *almost total probabilistic machine* M. By "almost total" machine we mean a randomized algorithm that produces an infinite sequence with probability 1. The output distribution of such a machine is a computable probability measure μ_M over the space 2^ω of infinite binary sequences. Again, our ultimate goal is to guess what machine is in the box, i.e., to give a reasonable explanation for the observed sequence X. For example, observing the sequence

$$00011111111000011000000000011111111111111$$

one may guess that M is a probabilistic machine that starts with 0 and then chooses each output bit to be equal to the previous one with probability $4/5$ (so the change happens with probability $1/5$), making all the choices independently.

What should count as a good guess for some observed sequence? Again there is no hope to distinguish between some machine M and another machine M' that has the same output distribution $\mu_{M'} = \mu_M$. So our goal should be to reconstruct the output distribution and not the specific machine.

But even this is too much to ask for. Assume that we have agreed that some machine M is a plausible explanation for some sequence X. Consider another machine M' that starts by tossing a coin and then (depending on the outcome) either generates an infinite sequence of zeros or simulates M'. If X is a plausible output of M, then X is also a plausible output for M', because it may happen (with probability $1/2$) that M' simulates M.

A reasonable formalization of 'good guess' is provided by the theory of algorithmic randomness. As Chater and Vitanyi recall, there is a widely accepted formalization of "plausible outputs" for an almost total probabilistic machine with output distribution μ: the notion of Martin-Löf random sequences with respect to μ. These are the sequences which pass all effective statistical tests for the measure μ, also known as μ-Martin-Löf tests. (We assume that the reader is familiar with algorithmic randomness and Kolmogorov complexity. The most useful references for our purposes are [Gác05] and [LV08].) Having this notion in mind, one could look for an algorithm $\mathfrak{A}$ with the following property:

for every almost total probabilistic machine M with output distribution μ_M, for μ_M-almost all X, the sequence $\mathfrak{A}(X \upharpoonright 1), \mathfrak{A}(X \upharpoonright 2), \mathfrak{A}(X \upharpoonright 3), \ldots$ identifies in the limit an almost total probabilistic machine M' such that X is $\mu_{M'}$-Martin-Löf random.

Note that this requirement uses two machines M and M' (more precisely, their output distributions): the first one is used when we speak about "almost all" X, and the second is used in the definition of Martin-Löf randomness. Here M' may differ from M and, moreover, may be different for different X.

Vitanyi and Chater suggest that this can be achieved in the strongest sense (EX): the guesses $\mathfrak{A}(X \upharpoonright n)$ converge to a single code of some machine M'. The main result of this paper says that even a much weaker goal cannot be achieved.

Let us consider a rather weak notion of success: $\mathfrak{A}$ *succeeds* on X if there exists $c > 0$ such that for all sufficiently large n the guess $\mathfrak{A}(X \upharpoonright n)$ is a machine M' such that X is $\mu_{M'}$-Martin-Löf random with randomness deficiency[1] less than c. So the machines $\mathfrak{A}(X \upharpoonright n)$ may be different, we only require that X is Martin-Löf random (with bounded deficiency) for almost all of them. (If almost all machines $\mathfrak{A}(X \upharpoonright n)$ generate the same distribution and X is Martin-Löf random with respect to this distribution, this condition is guaranteed to be true.)

Moreover, we require $\mathfrak{A}$ to be successful only with some positive probability instead of probability 1, and only for machines from some class: for every machine M from this class of machines, $\mathfrak{A}$ is required to succeed with μ_M-probability at least $\delta > 0$, for some δ independent of M.

Of course, this class should not be too narrow: if it contains only one machine M, the algorithm $\mathfrak{A}$ can always produce a code for this machine. The exact conditions on the class will be discussed in the next section.

The proof of this result is quite involved. In the rest of the paper, we specify which classes of machines are considered, present the proof and discuss the consequences of this result.

---

[1] See below about the version of the randomness deficiency function that we use.

2   Identifying Measures

2.1   Background and Notation

Let us start by providing some notation and background.

We denote by 2^ω the set of infinite binary sequences and by $2^{<\omega}$ the set of finite binary sequences (or *strings*). The length of a string σ is denoted by $|\sigma|$. The n-th element of a sequence $X(0), X(1), \ldots$ is $X(n-1)$ (assuming that the length of X is at least n); the string $X \restriction n = X(0)X(1)\ldots X(n-1)$ is *n-bit prefix of X*. We write $\sigma \preceq X$ if σ is a prefix of X (of some finite length).

The space 2^ω is endowed with the distance d defined by

$$d(X, Y) = 2^{-\min\{n : X(n) \neq Y(n)\}}$$

This distance is compatible with the product topology generated by *cylinders*

$$[\sigma] = \{X \in 2^\omega \; : \; \sigma \preceq X\}$$

A cylinder is both open and closed ($= clopen$). Thus, any finite union of cylinders is also clopen. It is easy to see, by compactness, that the converse holds: every clopen subset of 2^ω is a finite union of cylinders. We say that a clopen set C has *granularity at most n* if it can be written as a finite union of cylinders $[\sigma]$ with all σ's of length at most n. We denote by Γ_n the family of clopen sets of granularity at most n.

The space of Borel probability measures over 2^ω is denoted by $\mathcal{M}(2^\omega)$. It is equipped with the weak topology. Several classical distances are compatible with this topology; for our purposes, it will be convenient to use the distance ρ, constructed as follows: For $\mu, \nu \in \mathcal{M}(2^\omega)$, let $\rho_n(\mu, \nu)$ (for an integer n) be the quantity

$$\rho_n(\mu, \nu) = \max_{C \in \Gamma_n} |\mu(C) - \nu(C)|$$

and then set

$$\rho(\mu, \nu) = \sum_n 2^{-n} \rho_n(\mu, \nu)$$

The *open* (resp. *closed*) *ball $\mathcal{B}$ of center μ and radius r* is the set of measures ν such that $\rho(\mu, \nu) < r$ (resp. $\rho(\mu, \nu) \leq r$). Note that for any ν in this open (resp. closed) ball, if C is a clopen set of granularity at most n, then $|\mu(C) - \nu(C)| < 2^n r$ (resp. $\leq 2^n r$). The distance ρ makes $\mathcal{M}(2^\omega)$ a computable compact metric space; its computable points are called *computable probability measures*. A measure is computable if and only if it is the output distribution of some almost total probabilistic Turing machine (see, e.g., [Gác05]). Since $\mathcal{M}(2^\omega)$ is a computable metric space, one can define partial computable functions from some discrete space $\mathcal{X}$ (such as $\mathbb{N}$) to $\mathcal{M}(2^\omega)$ via type-2 computability: a partial function $f :\subseteq \mathcal{X} \to \mathcal{M}(2^\omega)$ is partial computable if there is an algorithm g that for every input $x \in \mathcal{X}$ enumerates a (finite or infinite) list of rational balls[2] $\mathcal{B}_1, \mathcal{B}_2,\ldots$ in $\mathcal{M}(2^\omega)$ such that $\mathcal{B}_{i+1} \subseteq \mathcal{B}_i$, the radius of $\mathcal{B}_i$ is less than 2^{-i}, and for every x in the domain of f, the list of enumerated balls is infinite and their intersection is the singleton $\{f(x)\}$. (We do not require any specific behavior outside the domain of f.)

Let us introduce two non-standard, but important in this paper, pieces of terminology: having fixed the algorithm g associated to f, we write $\mathrm{err}(f(x)) < \varepsilon$ to mean that the list of balls produced by g on input x contains a ball of radius less than ε (the justification for this notation is that when such a ball is enumerated, should $f(x)$ be defined, we know its value with error at most ε for the distance ρ). When the algorithm g on input x enumerates an empty list of balls, we say that g is *null* on input x.

We denote by K the prefix-free Kolmogorov complexity function. Given a computable measure μ, we call *randomness deficiency of X with respect to μ* the quantity

$$\mathbf{d}(X|\mu) = \sup_n \left[\log \frac{1}{\mu([X \restriction n])} - \mathrm{K}(X \restriction n) \right]$$

It is known that $X \in 2^\omega$ is μ-Martin-Löf random (or μ-random for short) if $\mathbf{d}(X|\mu) < \infty$. This definition is slightly non-standard; to get a more standard one, one has to add μ as the condition (with some precautions). However, the above is enough for our purposes.

We say that two measures μ and ν are *orthogonal* if there is a set having μ-measure 1 and ν-measure 0.

If $\mathcal{B}$ is a ball (open or closed) in $\mathcal{M}(2^\omega)$, with center μ and radius r, we define the *estimated deficiency* of X relative to $\mathcal{B}$ by

$$\mathbf{ed}(X|\mathcal{B}) = \sup_n \left[\log \frac{1}{\mu([X \restriction n]) + 2^n \, r} - \mathrm{K}(X \restriction n) \right]$$

Note that $\mathbf{ed}(X|\mathcal{B})$ is a lower bound for $\mathbf{d}(X|\nu)$ for every $\nu \in \mathcal{B}$: we know that the value of $\nu([X \restriction n])$ does not exceed $\mu([X \restriction n]) + 2^n \, r$ for every ν in the ball $\mathcal{B}$. For a fixed pair (X, μ) we have $\lim_{\mathcal{B} \to \mu} \mathbf{ed}(X|\mathcal{B}) = \mathbf{d}(X|\mu)$: if $\mathbf{d}(X|\mu)$ is large, one of the terms (for some n) is large, and the corresponding term in $\mathbf{ed}(X|\mathcal{B})$ is close to it if $\mathcal{B}$ has small radius and contains μ.

Sometimes in the paper we will use the notation $\mathbf{ed}(X|\mathfrak{A}(\sigma))$. By this we mean the supremum of $\mathbf{ed}(X|\mathcal{B})$ over all balls $\mathcal{B}$ output by $\mathfrak{A}$ on input σ.

The next lemma will be useful in the sequel.

Lemma 1 (Randomness deficiency lemma). *Let $\mathcal{B} \subseteq \mathcal{M}(2^\omega)$ be a ball of center μ (rational measure) and rational radius not exceeding r, and let C be a clopen set of granularity at most n. Then for all $X \in C$:*

$$\mathbf{ed}(X|\mathcal{B}) \geq \log \frac{\mu(X \restriction n)}{\mu(X \restriction n) + 2^n r} - \log \mu(C) - \mathrm{K}(C, \mu, r, n) - O(1)$$

---

[2] We fix some natural dense set of finitely representable measures. Rational balls are balls of rational radius with centers in this set. Such balls can also be finitely represented.

Proof. Knowing C, μ, r, n, one can build a prefix-free machine which associates to every string σ of length n such that $[\sigma] \subseteq C$ a a description of size $-\log \mu(X \restriction n) - \log \mu(C)$, so that indeed

$$\sum_{\sigma} 2^{-\log \mu(X \restriction n) - \log \mu(C)} = \frac{1}{\mu(C)} \sum_{\sigma} \mu(\sigma) = 1$$

where the sums are taken over those σ such that $[\sigma] \subseteq C$. This shows that for every such σ of length n, $K(\sigma) \leq -\log \mu(X \restriction n) - \log \mu(C) + K(C, \mu, r, n) - O(1)$. Applying the definition of **ed**, we get, for all $X \in C$

$$\mathbf{ed}(X | \mathcal{B}) \geq \log \frac{1}{\mu([X \restriction n]) + 2^n r} - K(X \restriction n)$$

$$\geq \log \frac{1}{\mu(X \restriction n) + 2^n r} + \log \mu(X \restriction n) - \log \mu(C) - K(C, \mu, r, n) - O(1)$$

$$\geq \log \frac{\mu(X \restriction n)}{\mu(X \restriction n) + 2^n r} - \log \mu(C) - K(C, \mu, r, n) - O(1)$$

$\square$

2.2   The Main Theorem

Now we return to the formulation of our main result. The *learning algorithm* is a partial computable function $\mathfrak{A} :\subseteq 2^{<\omega} \to \mathcal{M}(2^\omega)$; it gets the prefix $X \restriction n$ of a sequence X and computes (in type-2 sense) some measure $\mathfrak{A}(X \restriction n)$. (Such a computable function can be converted into an algorithm that, given an input string, produces a program that computes the output measure, and vice versa.) We say that $\mathfrak{A}$ *BC-succeeds* on a sequence $X \in 2^\omega$ if $\mathfrak{A}(X \restriction n)$ outputs the same computable measure μ for all sufficiently large n, and X is Martin-Löf random with respect to μ. This is a weaker requirement that exact (EX) success mentioned above: the algorithm is obliged to produce the same measure (for almost all n), but is not obliged to produce the same machine. Our main result, in its weak form, says that this goal cannot be achieved for all sequences that are random with respect to some computable measure:

Theorem 2. *There is no algorithm $\mathfrak{A}$ that BC-succeeds on every sequence X which is random with respect to some computable measure.*

As we have discussed, we prove a stronger version of this result—stronger in three directions.

First, we require the learning algorithm to succeed only on sequences that are random with respect to measures in some restricted class, for example, the class of Bernoulli measures (the main particular case considered by Chater and Vitanyi).

Second, for each measure μ in this class we do not require the algorithm to succeed on all sequences X that are μ-Martin-Löf random: it is enough that it succeeds with some fixed positive μ-probability (a weaker condition).

Finally, the notion of success on a sequence X is now weaker: we do not require that the algorithm produces (for all sufficiently long inputs) some specific measure, asking only that it gives 'good explanations' for the observed sequence from some point on. More specifically, we say that an algorithm $\mathfrak{A}$ *BD-succeeds* (BD stands for 'bounded deficiency') on some X, if for some c and for all sufficiently large n the measure $\mathfrak{A}(X \restriction n)$ is defined and X is random with deficiency at most c with respect to this measure. Clearly BC-success implies BD-success. (Note that in our definition the randomness deficiency depends only on the measure but not on the algorithm that computes it.)

We now are ready to state our main result in its strong form.

Theorem 3. *Let $\mathcal{M}_0$ be a subspace of $\mathcal{M}(2^\omega)$ with the following properties*:

- $\mathcal{M}_0$ *is effectively closed, i.e., one can enumerate a sequence of open balls in $\mathcal{M}(2^\omega)$ whose union is the complement of $\mathcal{M}_0$.*
- $\mathcal{M}_0$ *is recursively enumerable, i.e., one can enumerate the open balls in $\mathcal{M}(2^\omega)$ which intersect $\mathcal{M}_0$.*
- *every non-empty open subset of $\mathcal{M}_0$ (i.e., a non-empty intersection of an open set in $\mathcal{M}(2^\omega)$ with $\mathcal{M}_0$) contains infinitely many pairwise orthogonal computable measures.*

and let $\delta > 0$. Then there is no algorithm $\mathfrak{A}$ such that for every computable $\mu \in \mathcal{M}_0$, the μ-measure of sequences X on which $\mathfrak{A}$ BD-succeeds is at least δ.

The notion of an recursively enumerable closed set is standard in computable analysis, see [Wei00, Definition 5.1.1].

Note that the hypotheses on the class $\mathcal{M}_0$ are not very restrictive: many standard classes of probability measures have these properties. Bernoulli measures B_p (independent trials with success probability p, where p is a parameter in $[0, 1]$) are an obvious example; so there is no algorithm that can learn all Bernoulli measures (not to speak about all Markov chains). Let us give another interesting example: for every parameter $p \in [0, 1]$, consider measure μ_p associated to the stochastic process which generates a binary sequence bit-by-bit as follows: the first bit is 1, and the conditional probability of 1 after $\sigma 10^k$ is $p/(k+1)$. The class $\mathcal{M}(2^\omega) = \{\mu_p : p \in [0, 1]\}$ satisfies the hypotheses of the theorem.

Note also that these hypotheses are not added for convenience: although they might not be optimal, they cannot be outright removed. If we do not require compactness, then the class of Bernoulli measures B_p with *rational* parameter p would qualify, but it is easy to see that this class admits an algorithm which correctly identifies each of the measures in the class with probability 1. The third condition is important, too. Consider the measures B_0 and B_1 concentrated on the sequences $0000\ldots$ and $1111\ldots$ respectively. Then the class $\mathcal{M}_0 = \{pB_0 + (1-p)B_1 \mid p \in [0,1]\}$ is indeed effectively compact, but it is obvious that there is an algorithm that succeeds with probability 1 for all measures of that class (in the most strong sense: the first bit determines the entire sequence). For the second condition we do not have a counterexample showing that it is really needed, but it is true for all the natural classes (it is guaranteed to be true if $\mathcal{M}_0$ has a computable dense sequence).

3   The Proof of the Main Theorem

The rest of the paper is devoted to proving Theorem 3. Fix a subset $\mathcal{M}_0$ of $\mathcal{M}(2^\omega)$ satisfying the hypotheses of the theorem, and some $\delta > 0$. In the sequel, by "success" we always mean BD-success.

For every algorithm $\mathfrak{A}$ we consider the set of sequences on which it succeeds. We say that $\mathfrak{A}$ is δ-*good* if this success set has μ-probability at least δ for every $\mu \in \mathcal{M}_0$. We need to show that δ-good algorithms do not exist.

Let us introduce some useful notation. First, let

$$\textsc{Succ}(\mathfrak{A}, c, n) = \big\{ X \in 2^\omega \ : \ (X \restriction n) \in \mathrm{dom}(\mathfrak{A}) \wedge \mathbf{d}(X | \mathfrak{A}(X \restriction n)) \le c \big\}$$

be the set of X on which $\mathfrak{A}$ achieves "local success" on the prefix of length n for randomness deficiency c. The success set is then $\bigcup_c \bigcup_N \bigcap_{n \ge N} \textsc{Succ}(\mathfrak{A}, c, n)$.

According to our type-2 definition, the algorithm computing $\mathfrak{A}$ produces (for each input string) a finite or infinite sequence of balls (we assume that i-th ball has radius at most 2^{-i}). We will write '$\mathcal{B} \in \mathfrak{A}(\sigma)$' to signify that on input σ this algorithm enumerates the ball $\mathcal{B}$ at some point. For any function $f : 2^{<\omega} \to [0,1]$ converging to 0, we define the set $\textsc{Prec}(\mathfrak{A}, f, n)$ of points X which are 'precise enough' in the sense that $\mathfrak{A}(X \restriction n)$ almost outputs a measure:

$$\textsc{Prec}(\mathfrak{A}, f, n) = \{ X \in 2^\omega \ : \mathrm{err}(\mathfrak{A}(X \restriction n)) < f(X \restriction n) \}$$

(notice that $\textsc{Prec}(\mathfrak{A}, f, n)$ is a clopen set because the membership of X in $\textsc{Prec}(\mathfrak{A}, f, n)$ is determined fully by the first n bits of X). The specific choice of f (how 'precise' should be the output measure) is discussed later.

In contrast to $\textsc{Prec}$, we define the following "nullity" sets:

$$\textsc{Null}(\mathfrak{A}, N) = \big\{ X \in 2^\omega \ : \ \mathfrak{A}(X \restriction n) \text{ is null for every } n \ge N \big\}.$$

Proposition 4 (Nullity amplification). *Assume that $\mathfrak{A}$ is a δ-good algorithm, N is an integer, $\eta \ge 0$ is a real number and $\mathcal{B}$ is an open ball intersecting $\mathcal{M}_0$ such that $\mu(\textsc{Null}(\mathfrak{A}, N)) \ge \eta$ for all $\mu \in \mathcal{B} \cap \mathcal{M}_0$. Then there is a non-empty ball $\mathcal{B}' \subseteq \mathcal{B}$ intersecting $\mathcal{M}_0$, an integer $N' \ge N$ and a δ-good algorithm $\mathfrak{A}'$ such that $\mu(\textsc{Null}(\mathfrak{A}', N')) \ge \eta + \delta/2$ for all $\mu \in \mathcal{B}' \cap \mathcal{M}_0$.*

This proposition clearly shows that there can be no δ-good algorithm: if there was one, one could construct by induction (taking for the base case $\eta = 0$, $N = 0$, and $\mathcal{B} = $ any ball intersecting $\mathcal{M}_0$) a sequence of δ-good algorithms $\mathfrak{A}_i$, a non-increasing sequence of balls $\mathcal{B}_i$ intersecting $\mathcal{M}_0$, and a non-decreasing sequence of integers N_i such that $\mu(\textsc{Null}(\mathfrak{A}_i, N_i)) \ge \delta + i \cdot (\delta/2)$ for every $\mu \in \mathcal{B}_i \cap \mathcal{M}_0$, which gives a contradiction for large i. Thus, all we need to do is to prove this proposition.

Proof. Fix $\mathfrak{A}$, N, η and $\mathcal{B}$ as in the hypotheses of the proposition. For $m \ge N$, define a decreasing sequence of effectively open sets $\mathcal{U}_m$ by

$$\mathcal{U}_m = \{ \mu \mid (\exists n > m)\, (\mu(\textsc{Prec}(\mathfrak{A}, f, n)) > 1 - \eta - \delta/2) \}.$$

The first step of this proof consists in showing that if f is carefully chosen to tend to 0 fast enough, then only finitely many of the $\mathcal{U}_m$ can be dense in $\mathcal{B} \cap \mathcal{M}_0$.

The way we do this is by proving the following fact: if $\mathcal{U}_m$ is dense in $\mathcal{B} \cap \mathcal{M}_0$ for some m, then for every $\mathcal{B}' \subseteq \mathcal{B}$ intersecting $\mathcal{M}_0$, one can effectively find $\mathcal{B}'' \subseteq \mathcal{B}'$ intersecting $\mathcal{M}_0$ such that for all $\mu \in \mathcal{B}''$, $\mu(\mathrm{SUCC}(\mathfrak{A}, n, n)) < 7\delta/8$ for some $n \geq m \geq N$.

This would yield a contradiction since this would allow us to construct a computable sequence of decreasing balls $\mathcal{B}_m$, all intersecting $\mathcal{M}_0$, where all $\mu \in \mathcal{B}_m$ would be such that $\mu(\mathrm{SUCC}(\mathfrak{A}, n, n)) < 7\delta/8$ for some $n \geq m$, and thus the intersection of the $\mathcal{B}_m$ would be a computable measure μ^* – belonging to $\mathcal{M}_0$ by closedness of $\mathcal{M}_0$ – for which the success set of $\mathfrak{A}$ has μ^*-measure at most $7\delta/8$, a contradiction.

The definition of f on strings of length n will depend on a "large enough" parameter $s = s(n)$ which we will define later as a computable function of n. Suppose s has already been chosen. We shall first define in terms of s an important auxiliary computable function L. It is computed as follows. For a given n, let $\varepsilon = \min(2^{-n} \cdot \delta/4, r)$ where r is the radius of $\mathcal{B}$.

First, we effectively find $k(\varepsilon)$ rational balls $\mathcal{D}_1, \mathcal{D}_2, \cdots \mathcal{D}_{k(\varepsilon)}$, all intersecting $\mathcal{M}_0$, whose union covers $\mathcal{M}_0$ and for any ball of radius at least ε, one of the $\mathcal{D}_i$ is contained in this ball. (To do this, enumerate all balls with rational center and radius smaller than $\varepsilon/3$. By effective compactness of the space of measures $\mathcal{M}(2^\omega)$ and since $\mathcal{M}_0$ is effectively closed, one can find a finite number of them, call them $\mathcal{D}_1, \mathcal{D}_2, \ldots, \mathcal{D}_{k(\varepsilon)}$, which cover $\mathcal{M}_0$ entirely. Now, let $\mathcal{A}$ be a ball of radius at least ε intersecting $\mathcal{M}_0$ and μ its center. Since μ is at distance $\varepsilon/3$ of some measure $\nu \in \mathcal{M}_0$. But the $\mathcal{D}_i$'s cover $\mathcal{M}_0$, so ν belongs to some ball $\mathcal{D}_i$, and by the triangular inequality, every member of $\mathcal{D}_i$ is at distance at most $2\varepsilon/3$ of μ, hence $\mathcal{D}_i$ is contained in $\mathfrak{A}$).

Then, inside each ball $\mathcal{D}_i$, we effectively find 2^s rational measures $\xi_1^{(i)}, \ldots, \xi_{2^s}^{(i)}$ and pairwise disjoint clopen sets $V_1^{(i)}, \ldots, V_{2^s}^{(i)}$ such that $\xi_j^{(i)}(V_j^{(i)}) > 1 - \delta/8$.

To see that this can be done, observe that the conditions '$\xi_1, \ldots, \xi_s \in \mathcal{B}$', 'the V_i are disjoint', and '$\xi_i(V_i) > 1 - \varepsilon$ for all i' are all Σ_1^0-conditions. Therefore, all we need to argue is that such measures and clopen sets exist. By our assumption on $\mathcal{M}_0$, let $\xi_1, \ldots, \xi_s$ be pairwise orthogonal measures inside $\mathcal{B}$. By definition, this means that for every pair (i, j) with $i \neq j$, there exists a set $S_{i,j} \subseteq 2^\omega$ such that $\xi_i(S_{i,j}) = 1$ and $\xi_j(S_{i,j}) = 0$. For each i, let $S_i = \bigcap_{j \neq i} S_{i,j}$. One can easily check that $\xi_i(S_i) = 1$ for all i and $\xi_i(S_j) = 0$ when $i \neq j$. The measure of a set is the infimum of the measures of open sets covering it. Therefore, for each i there is an open set U_i covering S_i such that $\xi_j(U_i) \leq 2^{-s-1}\varepsilon$ for $i \neq j$ (and of course, $\xi_i(U_i) = 1$ for all i). Now we use the fact that the measure of an open set is the supremum of the measures of the clopen sets it contains. Therefore, for each i there exists a clopen set $U_i' \subseteq U_i$ such that $\xi_i(U_i') \geq 1 - \varepsilon/2$ (and of course $\xi_i(U_j') \leq 2^{-s-1}\varepsilon$ for $i \neq j$). Now $V_i = U_i' \setminus \bigcup_{j \neq i} U_j'$ for each i is a clopen set of ξ_i-measure at least $1 - \varepsilon/2 - 2^s \cdot 2^{-s-1}\varepsilon = 1 - \varepsilon$. The pairwise disjointness of the V_i is clear from their definition.

Compute the maximum of the granularities of all the clopen sets $V_j^{(i)}$ for $i \leq k(\varepsilon)$ and $j \leq 2^s$ and denote this maximum by $L(n)$.

Suppose now that for every non-empty $\mathcal{B}' \subseteq \mathcal{B}$ intersecting $\mathcal{M}_0$, there exists some $\mu \in \mathcal{B}'$ and some n,

$$\mu(\text{PREC}(\mathfrak{A}, f, n)) > 1 - \eta - \delta/2$$

for some measure $\mu \in \mathcal{B}$ and some $n \geq N$. Set again $\varepsilon = \min(2^{-n} \cdot \delta/4, r)$ and compute a family $\mathcal{D}_1, \mathcal{D}_2, \cdots \mathcal{D}_{k(\varepsilon)}$ intersecting $\mathcal{M}_0$ and whose union covers $\mathcal{M}_0$ so that for any ball $\mathcal{B}$ of radius at least ε, there is some $\mathcal{D}_i \subseteq \mathcal{B}$.

Recall that $\text{PREC}(\mathfrak{A}, f, n)$ is a clopen set of granularity n. Thus, if $\rho(\nu, \mu) < 2^{-n} \cdot \delta/4$, then $\nu(\text{PREC}(\mathfrak{A}, f, n)) > 1 - \eta - \delta/2 - \delta/4 = 1 - \eta - 3\delta/4$. And thus, by definition of the $\mathcal{D}_i$, there exists i such that for all $\nu \in \mathcal{D}_i$, $\nu(\text{PREC}(\mathfrak{A}, f, n)) > 1 - \eta - 3\delta/4$. Moreover, such an i can be found effectively knowing $\text{PREC}(\mathfrak{A}, f, n)$ and δ. Fix such an i and set $\mathcal{D} = \mathcal{D}_i$.

Now consider the behaviour of the algorithm $\mathfrak{A}$ on all possible strings σ of length n. On some of these strings, the algorithm does not achieve precision $f(\sigma)$; we ignore such strings. On some others, $\mathfrak{A}(\sigma)$ achieves precision $f(\sigma)$ and thus returns a sequence containing some ball $\mathcal{A}$ of radius less than $f(\sigma)$. Call $\mathcal{A}_1, ..., \mathcal{A}_t$ all such balls (obtained by some $\mathfrak{A}(\sigma)$ with σ of length n). Note that $t \leq 2^n$. Let $\alpha_1, ..., \alpha_t$ be the centers of these balls, and consider their average $\beta = (1/t) \sum_{i \leq t} \alpha_i$. Since the V_i are disjoint and there are 2^s-many of them, by the pigeonhole principle, there exists some j such that $\beta(V_j) \leq 2^{-s}$, and thus $\alpha_i(V_j) \leq t \cdot 2^{-s} \leq 2^{n-s}$ for all i. Fix such a j and set $V = V_j$, and $\xi = \xi_j$.

Recalling that the granularity of V is at most $L(n)$, we can apply the randomness deficiency lemma, we have for all $X \in V$:

$$\mathbf{ed}(X | \mathfrak{A}(X \restriction n)) \geq \log \frac{\alpha_i(X \restriction L(n))}{\alpha_i(X \restriction L(n)) + 2^{L(n)} f(X \restriction n)} - \log \alpha_i(V)$$
$$- K(V, n, s(n)) - O(1)$$

where α_i is the center of the ball of radius $f(X \restriction n)$ enumerated by $\mathfrak{A}(X \restriction n)$. And this finally tells us how the function f should be defined: we require that $2^{L(n)} f(X \restriction n)$ is smaller than $\alpha_i(X \restriction L(n))$, so as to make constant the first term of the right-hand-side. It seems to be a circular definition, but it is not the case: we can *define* $f(\sigma)$ to be the first rational q we find such that $\mathfrak{A}(\sigma)$ enumerates a ball of radius at most q and such that the center α of this ball is such that $\alpha(\sigma) > 2^{L(|\sigma|)} q$. This makes f a partial computable function, which is fine for our construction. Note also that $f(\sigma)$ can be undefined if $\mathfrak{A}(\sigma)$ is a measure γ such that $\gamma(\sigma) = 0$, but we need not worry about this case because it automatically makes the algorithm fail on σ (because the γ-deficiency of any extension of σ is infinite).

It remains to evaluate the Kolmogorov complexity of V. What we need to observe that $K(V)$ can be computed from $\text{PREC}(\mathfrak{A}, f, n)$, which, being a clopen set of granularity at most n, has complexity at most $2^{n+O(1)}$. Indeed, knowing this set, one can compute the open set of measures ν such that $\nu(\text{PREC}(\mathfrak{A}, f, n)) > 1 - \eta - 3\delta/4$ and effectively find a ball $\mathcal{D}$ as above. Then, from $\mathcal{D}$, the sequence of clopen sets $V_1, \ldots, V_{2^s}$ can be effectively computed. Moreover, to choose the V

as above, we need to know β, hence the sequence of measures $\alpha_1, \ldots \alpha_t$. But these can also be found knowing $\mathrm{PREC}(\mathfrak{A}, f, n)$, by definition of the latter. Thus we have established that $K(V) \leq 2^{n+O(1)}$.

Plugging all these complexity estimates in the above expression, we get

$$\mathbf{ed}(X|\mathfrak{A}(X \restriction n)) \geq s(n) - n - K(s(n)) - O(1) \tag{1}$$
$$\geq s(n) - 2\log(s(n)) - n - O(1) \tag{2}$$

Thus, by taking $s(n) = 2n + d$ for some large enough constant d, we get that

$$\mathbf{ed}(X|\mathfrak{A}(X \restriction n)) > n$$

for all $X \in V$. But the clopen set V has ξ-measure at least $1 - \delta/8$, so by definition of the $\mathcal{A}_i$, $\mathfrak{A}$ returns a ξ-inconsistent answer for deficiency level n on a set of ξ-measure at least $1 - \eta - 3\delta/4 - \delta/8$ of strings of length n. Note that this is a Σ_1^0-property of ξ, so we can in fact effectively find a ball $\mathcal{B}''$ intersecting $\mathcal{M}_0$ on which this happens. For every $\nu \in \mathcal{B}''$, $\mathfrak{A}(\sigma)$ is null on a set of strings of ν-measure at least η (by assumption) and is inconsistent on a set of measure at least $1 - \eta - 7\delta/8$, so $\mathrm{SUCC}(\mathfrak{A}, n, n)$ has a ν-measure of at most $7\delta/8$, which is the contradiction we wanted.

Now, we have reached our first goal which was to show that some $\mathcal{U}_{N'}$ is not dense in $\mathcal{B} \cap \mathcal{M}_0$ for some N'. Note that the $\mathcal{U}_m$ are non-increasing so this further means that there is a ball $\mathcal{B}' \subseteq \mathcal{B}$ such that $\mathcal{B} \cap \mathcal{M}_0$ does not intersect any of the $\mathcal{U}_m$ for $m \geq N'$. By definition, this means that on any measure ν of that ball $\mathcal{B}'$, the algorithm does not reach precision $f(\sigma)$ on a set of strings σ of ν-measure at least $\eta + \delta/2$. Thus, it suffices to consider the algorithm $\mathfrak{A}'$ which on any input σ does the following: it runs $\mathfrak{A}(\sigma)$ until $\mathfrak{A}(\sigma)$ reaches precision $f(\sigma)$. If this never happens, $\mathfrak{A}'(\sigma)$ remains null. If it does, then $\mathfrak{A}'(\sigma)$ returns the same list of balls as $\mathfrak{A}(\sigma)$. Clearly the algorithm $\mathfrak{A}'$ is δ-good since for every σ in the domain of $\mathfrak{A}$, $\mathfrak{A}'(\sigma) = \mathfrak{A}(\sigma)$. But by construction our new algorithm $\mathfrak{A}'$ is such that $\nu(\mathrm{NULL}(\mathfrak{A}', N')) \geq \eta + \delta/2$ for all $\nu \in \mathcal{B}'$. This finishes the proof. □

Acknowledgements. This publication was made possible through the support of a grant from the John Templeton Foundation. The opinions expressed in this publication are those of the authors and do not necessarily reflect the views of the John Templeton Foundation.

References

[Gác05] Gács, P.: Uniform test of algorithmic randomness over a general space. Theoretical Computer Science 341(1-3), 91–137 (2005)

[LV08] Li, M., Vitányi, P.: An introduction to Kolmogorov complexity and its applications, 3rd edn. Texts in Computer Science. Springer, New York (2008)

[VC13] Vitanyi, P., Chater, N.: Algorithmic identification of probabilities (2013), http://arxiv.org/abs/1311.7385

[Wei00] Weihrauch, K.: Computable analysis. Springer, Berlin (2000)

[ZZ08] Zeugmann, T., Zilles, S.: Learning recursive functions: a survey. Theoretical Computer Science 397, 4–56 (2008)

Learning Boolean Halfspaces with Small Weights from Membership Queries

Hasan Abasi[1], Ali Z. Abdi[2], and Nader H. Bshouty[1]

[1] Department of Computer Science
Technion, Haifa, Israel
[2] Convent of Nazareth School
Grade 11, Abas 7, Haifa, Israel

Abstract. We consider the problem of proper learning a Boolean Half-space with integer weights $\{0, 1, \ldots, t\}$ from membership queries only. The best known algorithm for this problem is an adaptive algorithm that asks $n^{O(t^5)}$ membership queries where the best lower bound for the number of membership queries is n^t [4].

In this paper we close this gap and give an adaptive proper learning algorithm with two rounds that asks $n^{O(t)}$ membership queries. We also give a non-adaptive proper learning algorithm that asks $n^{O(t^3)}$ membership queries.

1 Introduction

We study the problem of the learnability of boolean halfspace functions from membership queries [2, 1]. Boolean halfspace is a function $f = [w_1 x_1 + \cdots + w_n x_n \geq u]$ from $\{0, 1\}^n$ to $\{0, 1\}$ where the *weights* $w_1, \ldots, w_n$ and the *threshold* u are integers. The function is 1 if the arithmetic sum $w_1 x_1 + \cdots + w_n x_n$ is greater or equal to u and zero otherwise. In the *membership query* model [2, 1] the learning algorithm has access to a *membership oracle* $\mathcal{O}_f$, for some *target function* f. The oracle can receive an assignment $a \in \{0, 1\}^n$ from the algorithm and returns $f(a)$. A *proper learning algorithm* for a class of functions C is an algorithm that has access to $\mathcal{O}_f$ where $f \in C$ asks membership queries and returns a function g *in* C that is equivalent to f.

The problem of learning classes from membership queries only were motivated from many problems in different areas such as computational biology that arises in whole-genome (DNA) shotgun sequencing [8, 5, 11], DNA library screening [15], multiplex PCR method of genome physical mapping [13], linkage discovery problems of artificial intelligence [11], chemical reaction problem [3, 6, 7] and signature coding problem for the multiple access adder channels [9].

Another scenario that motivate the problem of learning Halfspaces is the following. Given a set of n similar looking objects of unknown weights (or any other measure), but from some class of weights W. Suppose we have a scale (or a measure instrument) that can only indicate whether the weight of any set of objects exceeds some unknown fixed threshold (or capacity). How many weighing do one needs in order to find the weights (or all possible weights) of the objects.

P. Auer et al. (Eds.): ALT 2014, LNAI 8776, pp. 96–110, 2014.

In this paper we study the problem of proper learnability of boolean halfspace functions with $t + 1$ different non-negative weights $W = \{0, 1, \ldots, t\}$ from membership queries. The best known algorithm for this problem is an adaptive algorithm that asks $n^{O(t^5)}$ membership queries where the best lower bound for the number of membership queries is n^t [4].

In this paper we close the above gap and give an adaptive proper learning algorithm with two rounds that asks $n^{O(t)}$ membership queries. We also give a non-adaptive proper learning algorithm that asks $n^{O(t^3)}$ membership queries. All the algorithms in this paper runs in time that is linear in the membership query complexity.

Extending such result to non-positive weights is impossible. In [4] Abboud et. al. showed that in order to learn boolean Halfspace functions with weights $W = \{-1, 0, 1\}$, we need at least $O(2^{n-o(n)})$ membership queries. Therefore the algorithm that asks all the 2^n queries in $\{0, 1\}^n$ is optimal for this case. Shevchenko and Zolotykh [16] studied halfspace function over the domain $\{0, 1, \ldots, k - 1\}^n$ when n is fixed and no constraints on the coefficients. They gave the lower bound $\Omega(\log^{n-2} k)$ for learning this class from membership queries. Hegedüs [14] prove the upper bound $O(\log^n k / \log \log k)$. For fixed n Shevchenko and Zolotykh [17] gave a polynomial time algorithm (in $\log k$) for this class. One of the reviewers noted that applying Theorem 3 in [14], the upper bound $O(\log^{n-2} k)$ for the teaching dimension of a halfspace, [12], gives the upper bound $O(\log^{n-1} k / \log \log k)$.

This paper is organized as follows. In Section 2 we give some definitions and preliminary results. In Section 3 we show that any boolean halfspace with polynomially bounded coefficients can be expressed as an Automaton of polynomial size. A result that will be used in Section 4. In Section 4 we give the two round learning algorithm and the non-adaptive algorithm.

2   Definitions and Preliminary Results

In this section we give some definitions and preliminary results that will be used throughout the paper

2.1   Main Lemma

In this subsection we prove two main results that will be frequently used in this paper

For integers $t < r$ we denote $[t] := \{1, 2, \ldots, t\}$, $[t]_0 = \{0, 1, \ldots, t\}$ and $[t, r] = \{t, t + 1, \ldots, r\}$.

We first prove the following

Lemma 1. *Let* $w_1, \ldots, w_m \in [-t, t]$ *where at least one* $w_j \notin \{-t, 0, t\}$ *and*

$$\sum_{i=1}^{m} w_i = r \in [-t + 1, t - 1].$$

There is a permutation $\phi : [m] \to [m]$ *such that for every* $j \in [m]$, $W_j :=$ $\sum_{i=1}^{j} w_{\phi(i)} \in [-t+1, t-1]$.

Proof. Since there is j such that $w_j \in [-t+1, t-1] \backslash \{0\}$ we can take $\phi(1) = j$. Then $W_1 = w_j \in [-t+1, t-1]$. If there is j_1, j_2 such that $w_{j_1} = t$ and $w_{j_2} = -t$ we set $\phi(2) = j_1$, $\phi(3) = j_2$ if $W_1 < 0$ and $\phi(2) = j_2$, $\phi(3) = j_1$ if $W_1 > 0$. We repeat the latter until there are either no more t or no more $-t$ in the rest of the elements.

Assume that we have chosen $\phi(1), \ldots, \phi(k-1)$ such that $W_j \in [-t+1, t-1]$ for $j \in [k-1]$. We now show how to determine $\phi(k)$ so that $W_k \in [-t+1, t-1]$. If $W_{k-1} = \sum_{i=1}^{k-1} w_{\phi(i)} > 0$ and there is $q \notin \{\phi(1), \ldots, \phi(k-1)\}$ such that $w_q < 0$ then we take $\phi(k) := q$. Then $W_k = W_{k-1} + w_q \in [-t+1, t-1]$. If $W_{k-1} < 0$ and there is $q \notin \{\phi(1), \ldots, \phi(k-1)\}$ such that $w_q > 0$ then we take $\phi(k) := q$. Then $W_k = W_{k-1} + w_q \in [-t+1, t-1]$. If for every $q \notin \{\phi(1), \ldots, \phi(k-1)\}$, $w_q > 0$ (resp. $w_q < 0$) then we can take an arbitrary order of the other elements and we get $W_{k-1} < W_k < W_{k+1} < \cdots < W_m = r$ (resp. $W_{k-1} > W_k > W_{k+1} > \cdots > W_m = r$). If $W_{k-1} = 0$ then there must be $q \notin \{\phi(1), \ldots, \phi(k-1)\}$ such that $w_q \in [-t+1, t-1]$. This is because not both t and $-t$ exist in the elements that are not assigned yet. We then take $\phi(k) := q$.

This completes the proof. $\qquad\qquad\qquad\qquad\qquad\qquad\qquad\qquad\qquad\qquad$ $\square$

We now prove the first main lemma

Lemma 2. *Let* $w_1, \ldots, w_m \in [-t, t]$ *and*

$$\sum_{i=1}^{m} w_i = r \in [-t+1, t-1].$$

There is a partition $S_1, S_2, \ldots, S_q$ *of* $[m]$ *such that*

1. *For every* $j \in [q-1]$, $\sum_{i \in S_j} w_i = 0$.
2. $\sum_{i \in S_q} w_i = r$.
3. *For every* $j \in [q]$, $|S_j| \leq 2t - 1$.
4. *If* $r \neq 0$ *then* $|S_q| \leq 2t - 2$.

Proof. If $w_1, \ldots, w_m \in \{-t, 0, t\}$ then r must be zero, and the number of non-zero elements is even and half of them are equal to t and the other half are equal to $-t$. Then we can take $S_i = \{-t, t\}$ or $S_i = \{0\}$ for all i. Therefore we may assume that at least one $w_j \notin \{-t, 0, t\}$.

By Lemma 1 we may assume w.l.o.g (by reordering the elements) that such that $W_j := \sum_{i=1}^{j} w_i \in [-t+1, t-1]$ for all $j \in [m]$. Let $W_0 = 0$. Consider $W_0, W_1, W_2, \ldots, W_{2t-1}$. By the pigeonhole principle there is $0 \leq j_1 < j_2 \leq 2t-1$ such that $W_{j_2} = W_{j_1}$ and then $W_{j_2} - W_{j_1} = \sum_{i=j_1+1}^{j_2} w_i = 0$. We then take $S_1 = \{j_1 + 1, \ldots, j_2\}$. Notice that $|S_1| = j_2 - j_1 \leq 2t - 1$.

Since $\sum_{i \notin S_1} w_i = r$ we can repeat the above to find $S_2, S_3, \cdots$. This can be repeated as long as $|[m] \backslash (S_1 \cup S_2 \cup \cdots \cup S_h)| \geq 2t - 1$. This proves $1 - 3$.

We now prove 4. If $g := |[m]\backslash(S_1 \cup S_2 \cup \cdots \cup S_h)| < 2t - 1$ then define $S_{h+1} = [m]\backslash(S_1 \cup S_2 \cup \cdots \cup S_h)$ and we get 4 for $q = h+1$. If $g = 2t-1$ then $W_0 = 0, W_1, W_2, \ldots, W_{2t-1} = r$ and since $r \neq 0$ we must have $0 \leq j_1 < j_2 \leq 2t - 1$ and $j_2 - j_1 < 2t - 1$ such that $W_{j_2} = W_{j_1}$. Then define $S_{h+1} = \{j_1 + 1, \ldots, j_2\}$, $S_{h+2} = [m]\backslash(S_1 \cup S_2 \cup \cdots \cup S_{h+1})$ and $q = h + 2$. Then $|S_{h+2}| \leq 2t - 2$, $\sum_{i \in S_{h+1}} w_i = W_{j_2} - W_{j_1} = 0$ and $\sum_{i \in S_{h+2}} w_i = r$. □

The following example shows that the bound $2t - 2$ for the size of set in Lemma 2 is tight. Consider the $2t - 2$ elements $w_1 = w_2 = \cdots = w_{t-1} = t$ and $w_t = w_{t+1} = \cdots = w_{2t-2} = -(t - 1)$. The sum of any subset of elements is distinct. By adding the element $w_{2t-1} = -(t - 1)$ it is easy to show that the bound $2t - 1$ in the lemma is also tight.

We now prove the second main lemma

Lemma 3. *Let* $(w_1, v_1), \ldots, (w_m, v_m) \in [-t, t]^2$ *and*

$$\sum_{i=1}^{m}(w_i, v_i) = (r, s) \in [-t+1, t-1]^2.$$

There is $M \subseteq [m]$ *such that*

1. $\sum_{i \in M}(w_i, v_i) = (r, s).$
2. $|M| \leq 8t^3 - 4t^2 - 2t + 1.$

Proof. Since $w_1, \ldots, w_m \in [-t, t]$ and $\sum_{i=1}^{m} w_i = r \in [-t+1, t-1]$, by Lemma 2, there is a partition $S_1, \ldots, S_q$ of $[m]$ that satisfies the conditions $1 - 4$ given in the lemma. Let $V_j = \sum_{i \in S_j} v_i$ for $j = 1, \ldots, q$. We have

$$V_j \in [-t|S_j|, t|S_j|] \subseteq [-t(2t - 1), t(2t - 1)] \subset [-2t^2, 2t^2]$$

for $j = 1, \ldots, q$ and

$$\sum_{i=1}^{q-1} V_i = s - V_q \in [-2t^2 + 1, 2t^2 - 1].$$

If $s - V_q = 0$ then for $M = S_q$ we have $|M| = |S_q| \leq 2t - 1 \leq 8t^3 - 4t^2 - 2t + 1$ and

$$\sum_{i \in M}(w_i, v_i) = \sum_{i \in S_q}(w_i, v_i) = (r, V_q) = (r, s).$$

Therefore we may assume that $s - V_q \neq 0$.

Consider $V_1, V_2, \ldots, V_{q-1}$. By 4 in Lemma 2 there is a set $Q \subseteq [q - 1]$ of size at most $2(2t^2) - 2 = 4t^2 - 2$ such that $\sum_{i \in Q} V_i = s - V_q$. Then for

$$M = S_q \cup \bigcup_{i \in Q} S_i$$

we have
$$|M| \leq (2t - 1) + (4t^2 - 2)(2t - 1) = 8t^3 - 4t^2 - 2t + 1$$
and
$$\sum_{i \in M} (w_i, v_i) = \sum_{i \in S_q} (w_i, v_i) + \sum_{j \in Q} \sum_{i \in S_j} (w_i, v_i)$$
$$= (r, V_q) + \sum_{j \in Q} (0, V_j)$$
$$= (r, V_q) + (0, s - V_q) = (r, s).$$

$\square$

2.2   Boolean Functions

For a boolean function $f(x_1, \ldots, x_n) : \{0, 1\}^n \to \{0, 1\}$, $1 \leq i_1 < i_2 < \cdots < i_k \leq n$ and $\sigma_1, \ldots, \sigma_k \in \{0, 1\}$ we denote by

$$f|_{x_{i_1} = \sigma_1, x_{i_2} = \sigma_2, \cdots, x_{i_k} = \sigma_k}$$

the function f when fixing the variables x_{i_j} to σ_j for all $j \in [k]$. For $a \in \{0, 1\}^n$ we denote by $a|_{x_{i_1} = \sigma_1, x_{i_2} = \sigma_2, \cdots, x_{i_k} = \sigma_k}$ the assignment a where each a_{i_j} is replaced by σ_j for all $j \in [k]$. We note here (and throughout the paper) that $f|_{x_{i_1} = \sigma_1, x_{i_2} = \sigma_2, \cdots, x_{i_k} = \sigma_k}$ is a function from $\{0, 1\}^n \to \{0, 1\}$ with the same variables $x_1, \ldots, x_n$ as of f. Obviously

$$f|_{x_{i_1} = \sigma_1, x_{i_2} = \sigma_2, \cdots, x_{i_k} = \sigma_k}(a) = f(a|_{x_{i_1} = \sigma_1, x_{i_2} = \sigma_2, \cdots, x_{i_k} = \sigma_k}).$$

When $\sigma_1 = \cdots = \sigma_k = \xi$ and $S = \{x_{i_1}, \ldots, x_{i_k}\}$ we denote

$$f|_{S \leftarrow \xi} = f|_{x_{i_1} = \xi, x_{i_2} = \xi, \cdots, x_{i_k} = \xi}.$$

In the same way we define $a|_{S \leftarrow \xi}$. We denote by $0^n = (0, 0, \ldots, 0) \in \{0, 1\}^n$ and $1^n = (1, 1, \ldots, 1) \in \{0, 1\}^n$. For two assignments $a \in \{0, 1\}^k$ and $b \in \{0, 1\}^j$ we denote by $ab \in \{0, 1\}^{k+j}$ the concatenation of the two assignments.

For two assignments $a, b \in \{0, 1\}^n$ we write $a \leq b$ if for every i, $a_i \leq b_i$. A boolean function $f : \{0, 1\}^n \to \{0, 1\}$ is *monotone* if for every two assignments $a, b \in \{0, 1\}^n$, if $a \leq b$ then $f(a) \leq f(b)$. Recall that every monotone boolean function f has a unique representation as a reduced monotone DNF [1]. That is, $f = M_1 \vee M_2 \vee \cdots \vee M_s$ where each *monomial* M_i is an ANDs of input variables and for every monomial M_i there is a unique assignment $a^{(i)} \in \{0, 1\}^n$ such that $f(a^{(i)}) = 1$ and for every $j \in [n]$ where $a_j^{(i)} = 1$ we have $f(a^{(i)}|_{x_j=0}) = 0$. We call such assignment a *minterm* of the function f. Notice that every monotone DNF can be uniquely determined by its minterms.

We say that x_i is *relevant* in f if $f|_{x_i=0} \neq f|_{x_i=1}$. Obviously, if f is monotone then x_i is relevant in f if there is an assignment a such that $f(a|_{x_i=0}) = 0$ and $f(a|_{x_i=1}) = 1$. We say that a is a *semiminterm* of f if for every $a_i = 1$ either $f(a|_{x_i=0}) = 0$ or x_i is not relevant in f.

For two assignments $a, b \in \{0,1\}^n$ we define the *distance* between a and b as $wt(a+b)$ where wt is the Hamming weight and $+$ is the bitwise exclusive or of assignments. The set $B(a, d)$ is the set of all assignments that are of distance at most d from $a \in \{0,1\}^n$.

2.3 Symmetric and Nonsymmetric

We say that a boolean function f is *symmetric* in x_i and x_j if for any $\xi_1, \xi_2 \in \{0,1\}$ we have $f|_{x_i=\xi_1,x_j=\xi_2} \equiv f|_{x_i=\xi_2,x_j=\xi_1}$. Obviously, this is equivalent to $f|_{x_i=0,x_j=1} \equiv f|_{x_i=1,x_j=0}$. We say that f is *nonsymmetric* in x_i and x_j if it is not symmetric in x_i and x_j. This is equivalent to $f|_{x_i=0,x_j=1} \not\equiv f|_{x_i=1,x_j=0}$. We now prove

Lemma 4. *Let f be a monotone function and $1 \le i < j \le n$. Then f is nonsymmetric in x_i and x_j if and only if there is a minterm a of f such that $a_i + a_j = 1$ (one is 0 and the other is 1) where $f(a|_{x_i=0,x_j=1}) \ne f(a|_{x_i=1,x_j=0})$.*

Proof. Since f is nonsymmetric in x_i and x_j we have $f|_{x_i=0,x_j=1} \not\equiv f|_{x_i=1,x_j=0}$ and therefore there is an assignment a' such that

$$f|_{x_i=0,x_j=1}(a') \ne f|_{x_i=1,x_j=0}(a').$$

Suppose w.l.o.g. $f|_{x_i=0,x_j=1}(a') = 0$ and $f|_{x_i=1,x_j=0}(a') = 1$. Take a minterm $a \le a'$ of $f|_{x_i=1,x_j=0}$. Notice that $a_i = a_j = 0$. Otherwise we can flip them to 0 without changing the value of the function $f|_{x_i=1,x_j=0}$ and then a is not a minterm. Then $f|_{x_i=1,x_j=0}(a) = 1$ and since $a \le a'$, $f|_{x_i=0,x_j=1}(a) = 0$.

We now prove that $b = a|_{x_i=1,x_j=0}$ is a minterm of f. Since $b|_{x_i=0} = a|_{x_i=0,x_j=0} < a|_{x_i=0,x_j=1}$ we have $f(b|_{x_i=0}) < f(a|_{x_i=0,x_j=1}) = f|_{x_i=0,x_j=1}(a) = 0$ and therefore $f(b|_{x_i=0}) = 0$. For any $b_k = 1$ where $k \ne i$, since a is a minterm for $f|_{x_i=1,x_j=0}$, we have $f(b|_{x_k=0}) = f|_{x_i=1,x_j=0}(a|_{x_k=0}) = 0$. Therefore b is a minterm of f.   □

We write $x_i \sim_f x_j$ when f is symmetric in x_i and x_j and call $\sim_f$ the symmetric relation of f. The following folklore result is proved for completeness

Lemma 5. *The relation $\sim_f$ is an equivalence relation.*

Proof. Obviously, $x_i \sim_f x_i$ and if $x_i \sim_f x_j$ then $x_j \sim_f x_i$. Now if $x_i \sim_f x_j$ and $x_j \sim_f x_k$ then $f|_{x_i=\xi_1,x_j=\xi_2,x_k=\xi_3} \equiv f|_{x_i=\xi_2,x_j=\xi_1,x_k=\xi_3} \equiv f|_{x_i=\xi_2,x_j=\xi_3,x_k=\xi_1} \equiv f|_{x_i=\xi_3,x_j=\xi_2,x_k=\xi_1}$ and therefore $x_i \sim_f x_k$.   □

2.4 Properties of Boolean Halfspaces

A *Boolean Halfspace* function is a boolean function $f : \{0,1\}^n \to \{0,1\}$, $f = [w_1 x_1 + w_2 x_2 + \cdots + w_n x_n \ge u]$ where $w_1, \ldots, w_n, u$ are integers, defined as $f(x_1, \ldots, x_n) = 1$ if $w_1 x_1 + w_2 x_2 + \cdots + w_n x_n \ge u$ and 0 otherwise. The numbers w_i, $i \in [n]$ are called the *weights* and u is called the *threshold*. The class HS is the class of all Boolean Halfspace functions. The class HS_t is the class of all Boolean

Halfspace functions with weights $w_i \in [t]_0$ and the class $\mathrm{HS}_{[-t,t]}$ is the class of all Boolean Halfspace functions with weights $w_i \in [-t, t]$. Obviously, the functions $f \in \mathrm{HS}_t$ are monotone. The representation of the above Boolean Halfspaces are not unique. For example, $[3x_1 + 2x_2 \geq 2]$ is equivalent to $[x_1 + x_2 \geq 1]$. We will assume that

$$\text{There is an assignment } a \in \{0,1\}^n \text{ such that } w_1 a_1 + \cdots + w_n a_n = b \quad (1)$$

Otherwise we can replace b by the minimum integer $w_1 a_1 + \cdots + w_n a_n$ where $f(a) = 1$ and get an equivalent function. Such a is called a *strong assignment* of f. If f is monotone, a is strong assignment and minterm of f then a is called a *strong minterm*.

The following lemma follows from the above definitions

Lemma 6. *Let* $f \in \mathrm{HS}_t$. *We have*

1. *If* a *is strong assignment of* f *then* a *is semiminterm of* f.
2. *If all the variables in* f *are relevant then any semiminterm of* f *is a minterm of* f.

We now prove

Lemma 7. *Let* $f = [w_1 x_1 + w_2 x_2 + \cdots + w_n x_n \geq u] \in \mathrm{HS}_t$. *Then*

1. *If* $w_1 = w_2$ *then* f *is symmetric in* x_1 *and* x_2.
2. *If* f *is symmetric in* x_1 *and* x_2 *then there are* w_1' *and* w_2' *such that* $|w_1' - w_2'| \leq 1$ *and* $f \equiv [w_1' x_1 + w_2' x_2 + w_3 x_3 \cdots + w_n x_n \geq u] \in \mathrm{HS}_t$.

Proof. If $w_1 = w_2$ then for any assignment $z = (z_1, z_2, \ldots, z_n)$ we have $w_1 z_1 + w_2 z_2 + \cdots + w_n z_n = w_1 z_2 + w_2 z_1 + \cdots + w_n z_n$. Therefore, $f(0, 1, x_3, \ldots, x_n) \equiv f(1, 0, x_3, \ldots, x_n)$.

Suppose $w_1 > w_2$. It is enough to show that $f \equiv g := [(w_1 - 1)x_1 + (w_2 + 1)x_2 + w_3 x_3 \cdots + w_n x_n \geq u]$. Obviously, $f(x) = g(x)$ when $x_1 = x_2 = 1$ or $x_1 = x_2 = 0$. If $f(0, 1, x_3, \ldots, x_n) \equiv f(1, 0, x_3, \ldots, x_n)$ then $w_1 + w_3 x_3 + w_4 x_4 + \cdots + w_n x_n \geq u$ if and only if $w_2 + w_3 x_3 + w_4 x_4 + \cdots + w_n x_n \geq u$ and therefore $w_1 + w_3 x_3 + w_4 x_4 + \cdots + w_n x_n \geq u$ if and only if $(w_1 - 1) + w_3 x_3 + w_4 x_4 + \cdots + w_n x_n \geq u$ if and only if $(w_2 + 1) + w_3 x_3 + w_4 x_4 + \cdots + w_n x_n \geq u$. $\square$

We now prove

Lemma 8. *Let* $f \in \mathrm{HS}_t$. *Let* a *be any assignment such that* $f(a) = 1$ *and* $f(a|_{x_i=0}) = 0$ *for some* $i \in [n]$. *There is a strong assignment of* f *in* $B(a, 2t - 2)$.

Proof. Let $f = [w_1 x_1 + \cdots + w_n x_n \geq u]$. Since $f(a) = 1$ and $f|_{x_i=0}(a) = 0$, $a_i = 1$ and we have $w_1 a_1 + w_2 a_2 + \cdots + w_n a_n = u + u'$ where $t - 1 \geq u' \geq 0$. If $u' = 0$ then $a \in B(a, 2t - 2)$ is a strong assignment. So we may assume that $u' \neq 0$.

By (1) there is an assignment b where $w_1 b_1 + w_2 b_2 + \cdots + w_n b_n = u$. Therefore $w_1(b_1 - a_1) + w_2(b_2 - a_2) + \cdots + w_n(b_n - a_n) = -u'$. Since $w_i(b_i - a_i) \in [-t, t]$, by

Lemma 2 there is $S \subseteq [n]$ of size at most $2t - 2$ such that $\sum_{i \in S} w_i(b_i - a_i) = -u'$. Therefore

$$u = -u' + (u + u') = \sum_{i \in S} w_i(b_i - a_i) + \sum_{i=1}^{n} w_i a_i = \sum_{i \in S} w_i b_i + \sum_{i \notin S} w_i a_i.$$

Thus the assignment c where $c_i = b_i$ for $i \in S$ and $c_i = a_i$ for $i \notin S$ is a strong assignment of f and $c \in B(a, 2t - 2)$. □

The following will be used to find the relevant variables

Lemma 9. *Let* $f \in HS_t$. *Suppose* x_k *is relevant in* f. *Let* a *be any assignment such that* $a_k = 1$, $f(a) = 1$ *and* $f(a|_{x_j=0}) = 0$ *for some* $j \in [n]$. *There is* $c \in B(a, 2t - 2)$ *such that* $c_k = 1$, $f(c) = 1$ *and* $f(c|_{x_k=0}) = 0$.

Proof. Let $f = [w_1 x_1 + \cdots + w_n x_n \geq u]$. Since $f(a) = 1$ and $f(a|_{x_j=0}) = 0$ we have $a_j = 1$ and $w_1 a_1 + w_2 a_2 + \cdots + w_n a_n = u + u'$ where $t - 1 \geq u' \geq 0$. Let b a minterm of f such that $b_k = 1$. Since b is a minterm we have $w_1 b_1 + w_2 b_2 + \cdots + w_n b_n = u + u''$ where $t - 1 \geq u'' \geq 0$ and since $f(b|_{x_k=0}) = 0$ we also have $u'' - w_k < 0$. If $u'' = u'$ then we may take $c = a$. Therefore we may assume that $u'' \neq u'$.

Hence $\sum_{i=1, i \neq k}^{n} w_i(b_i - a_i) = u'' - u' \in [-t+1, t-1] \setminus \{0\}$. By Lemma 2 there is $S \subseteq [n] \setminus \{k\}$ of size at most $2t - 2$ such that $\sum_{i \in S} w_i(b_i - a_i) = u'' - u'$. Therefore

$$u + u'' = \sum_{i \in S} w_i(b_i - a_i) + \sum_{i=1}^{n} w_i a_i = \sum_{i \in S} w_i b_i + \sum_{i \notin S} w_i a_i.$$

Thus the assignment c where $c_i = b_i$ for $i \in S$ and $c_i = a_i$ for $i \notin S$ satisfies $c_k = a_k = 1$ and $c \in B(a, 2t - 2)$. Since $\sum_{i=1, i \neq k}^{n} w_i c_i = u + u'' - b_k < u$ we have $f(c|_{x_k=0}) = 0$. □

The following will be used to find the order of the weights

Lemma 10. *Let* $f \in HS_t$ *be nonsymmetric in* x_1 *and* x_2. *For any minterm* a *of* f *of weight at least 2 there is* $b \in B(a, 2t + 1)$ *such that* $b_1 + b_2 = 1$ *and* $f|_{x_1=0, x_2=1}(b) \neq f|_{x_1=1, x_2=0}(b)$.

Proof. Let $f = [w_1 x_1 + \cdots + w_n x_n \geq u]$. Assume w.l.o.g $w_1 > w_2$. By Lemma 4 there is a minterm $c = (1, 0, c_3, \ldots, c_n)$ such that $f(c) = 1$ and $f(0, 1, c_3, \ldots, c_n) = 0$. Then $W_1 := w_1 + w_3 c_3 + \cdots + w_n c_n = u + v$ where $0 \leq v \leq t - 1$ and $W_2 := w_2 + w_3 c_3 + \cdots + w_n c_n = u - z$ where $1 \leq z \leq t-1$. In fact $-z = v - w_1 + w_2$. Since a is a minterm we have $W_3 := w_1 a_1 + \cdots + w_n a_n = u + h$ where $0 \leq h \leq t-1$. It is now enough to find $b \in B(a, 2t - 2)$ such that either

1. $b_1 = 1$, $b_2 = 0$ and $w_1 b_1 + \cdots + w_n b_n = u + v$, or
2. $b_1 = 0$, $b_2 = 1$ and $w_1 b_1 + \cdots + w_n b_n = u - z$.

This is because if $b_1 = 1$, $b_2 = 0$ and $w_1b_1 + \cdots + w_nb_n = u + v$ (the other case is similar) then $f(1, 0, b_2, \ldots, b_n) = 1$ and since $w_1 \cdot 0 + w_2 \cdot 1 + w_3 \cdot a_3 \cdots + w_na_n = u + v - w_1 + w_2 = u - z$ we have $f(0, 1, b_2, \ldots, b_n) = 0$.

We now have four cases

Case I. $a_1 = 1$ and $a_2 = 0$: Then $W_1 - W_3 = w_3(c_3 - a_3) + \cdots + w_n(c_n - a_n) = v - h \in [-t+1, t-1]$. By Lemma 2 there is $S \subseteq [3, n]$ of size at most $2t - 1$ such that $\sum_{i \in S} w_i(c_i - a_i) = v - h$. Therefore

$$u + v = v - h + W_3 = \sum_{i \in S} w_i(c_i - a_i) + \sum_{i=1}^{n} w_ia_i = \sum_{i \in S} w_ic_i + \sum_{i \notin S} w_ia_i.$$

Now define b to be $b_i = c_i$ for $i \in S$ and $b_i = a_i$ for $i \notin S$. Since $1, 2 \notin S$ $b_1 = a_1 = 1$ and $b_2 = a_2 = 0$. Since $b \in B(a, 2t - 1) \subset B(a, 2t + 1)$ and b satisfies 1. the result follows for this case.

Case II. $a_1 = 0$ and $a_2 = 1$: Since a is of weight at least 2, we may assume w.l.o.g that $a_3 = 1$. Since a is a minterm $f(a) = 1$ and $f(a|_{x_3=0}) = 0$ and therefore for $a' = a|_{x_3=0}$ we have $W_4 := w_1a'_1 + w_2a'_2 + \cdots + w_na'_n = u - h'$ where $1 \le h' \le t-1$. Then $W_2 - W_4 = \sum_{i=3}^{n} w_i(c_i - a'_i) = h' - z \in [-t+1, t-1]$. By Lemma 2 there is $S \subseteq [3, n]$ of size at most $2t - 1$ such that $\sum_{i \in S} w_i(c_i - a'_i) = h' - z$. Therefore

$$u - z = h' - z + W_4 = \sum_{i \in S} w_i(c_i - a'_i) + \sum_{i=1}^{n} w_ia'_i = \sum_{i \in S} w_ic_i + \sum_{i \notin S} w_ia'_i.$$

Now define b to be $b_i = c_i$ for $i \in S$ and $b_i = a'_i$ for $i \notin S$. Since $1, 2 \notin S$ $b_1 = a'_1 = 0$ and $b_2 = a'_2 = 1$. Since $b \in B(a', 2t - 1) \subset B(a, 2t + 1)$ and b satisfies 2. the result follows for this case.

Case III. $a_1 = 1$ and $a_2 = 1$: Since a is a minterm $f(a) = 1$ and $f(a|_{x_1=0}) = 0$ and therefore for $a' = a|_{x_1=0}$ we have $W_4 := w_1a'_1 + w_2a'_2 + \cdots + w_na'_n = u - h'$ where $1 \le h' \le t - 1$. We now proceed exactly as in Case II.

Case IV. $a_1 = 0$ and $a_2 = 0$: Since a is of weight at least 2 we may assume w.l.o.g that $a_3 = 1$. Since a is a minterm $f(a) = 1$ and $f(a|_{x_3=0}) = 0$ and therefore for $a' = a|_{x_3=0}$ we have $W_4 := a'_1w_1 + a'_2w_2 + \cdots + a'_nw_n = u - h'$ where $1 \le h' \le t - 1$. If $f(a'|_{x_2=1}) = 0$ then proceed as in Case II to get $b \in B(a, 2t + 1)$ that satisfies 2. If $f(a'|_{x_1=1}) = 1$ then proceed as in Case I. Now the case where $f(a'|_{x_2=1}) = 1$ and $f(a'|_{x_1=1}) = 0$ cannot happen since $w_1 > w_2$. $\square$

The following will be used for the non-adaptive algorithm

Lemma 11. *Let $f, g \in HS_t$ be such that $f \not\approx g$. For any minterm b of f there is $c \in B(b, 8t^3 + O(t^2))$ such that $f(c) + g(c) = 1$.*

Proof. Let $f = [w_1x_1 + \cdots + w_nx_n \ge u]$ and $g = [w'_1x_1 + \cdots + w'_nx_n \ge u']$. Since $f \not\approx g$, there is $a' \in \{0, 1\}^n$ such that $f(a') = 1$ and $g(a') = 0$. Let $a \le a'$ be a minterm of f. Then $f(a) = 1$ and since $a \le a'$ we also have $g(a) = 0$. Therefore $w_1a_1 + \cdots + w_na_n = u + r$ where $0 \le r \le t-1$ and $w'_1a_1 + \cdots + w'_na_n = u' - s$ for some integer $s \geq 1$. Since b is a minterm of f we have $w_1 b_1 + \cdots + w_n b_n = u + r'$ where $0 \leq r' \leq t - 1$. If $g(b) = 0$ then take $c = b$. Otherwise, if for some $b_i = 1$, $g(b|_{x_i=0}) = 1$ then take $c = b|_{x_i=0}$. Therefore we may assume that b is also a minterm of g. Thus $w_1' b_1 + \cdots + w_n' b_n = u + s'$ where $0 \leq s' \leq t - 1$.

Consider the sequence Z_i, $i = 1, \ldots, n+s-1$ where $Z_i = (w_i(a_i - b_i), w_i'(a_i - b_i))$ for $i = 1, \ldots, n$ and $Z_i = (0, 1)$ for $i = n+1, \ldots, n+s-1$. Then

$$\sum_{i=1}^{n+s-1} Z_i = (r - r', -1 - s') \in [-t, t]^2.$$

By Lemma 3 there is a set $S \subseteq [n+s-1]$ of size $8t^3 + O(t^2)$ such that $\sum_{i \in S} Z_i = (r - r', -1 - s')$. Therefore, there is a set $T \subseteq [n]$ of size at most $8t^3 + O(t^2)$ such that $\sum_{i \in T} Z_i = (r - r', -\ell - 1 - s')$ for some $\ell > 0$. Therefore

$$\sum_{i \in T} w_i(a_i - b_i) = r - r' \quad \text{and} \quad \sum_{i \in T} w_i'(a_i - b_i) = -\ell - 1 - s'.$$

Define c such that $c_i = a_i$ for $i \in T$ and $c_i = b_i$ for $i \notin T$. Then

$$\sum_{i=1}^{n} w_i c_i = u + r \geq u \quad \text{and} \quad \sum_{i=1}^{n} w_i' c_i = u' - \ell - 1 < u'.$$

Therefore $f(c) = 1$ and $g(c) = 0$. This gives the result.   □

3   Boolean Halfspace and Automata

In this section we show that functions in $\mathrm{HS}_{[-t,t]}$ has an automaton representation of $poly(n, t)$ size.

Lemma 12. *Let $f_1, f_2, \ldots, f_k \in \mathrm{HS}_{[-t,t]}$ and $g : \{0,1\}^k \to \{0,1\}$. Then $g(f_1, \ldots, f_k)$ can be represented with an Automaton of size $(2t)^k n^{k+1}$.*

Proof. Let $f_i = [w_{i,1} x_1 + \cdots + w_{i,n} x_n \geq u_i]$, $i = 1, \ldots, k$. Define the following automaton: The alphabet of the automaton is $\{0,1\}$. The states are $S \subseteq [n]_0 \times [-tn, tn]^k$. The automaton has $n+1$ levels. States in level i are connected only to states in level $i+1$ for all $i \in [n]_0$. We denote by S_i the states in level i. We also have $S_i \subseteq \{i\} \times [-tn, tn]^k$ so the first entry of the state indicates the level that the state belongs to. The state $(0, (0, 0, \ldots, 0))$ is the initial state and is the only state in level 0. That is $S_0 = \{(0, (0, 0, \ldots, 0))\}$. We now show how to connect states in level i to states in level $i+1$. Given a state $s = (i, (W_1, W_2, \ldots, W_k))$ in S_i. Then the transition function for this state is

$$\delta((i, (W_1, W_2, \ldots, W_k)), 0) = (i + 1, (W_1, W_2, \ldots, W_k))$$

and

$$\delta((i, (W_1, W_2, \ldots, W_k)), 1) = (i+1, (W_1 + w_{1,i+1}, W_2 + w_{2,i+1}, \ldots, W_k + w_{k,i+1})).$$

The accept states (where the output of the automaton is 1) are all the states $(n, (W_1, \ldots, W_k))$ where $g([W_1 \geq u_1], [W_2 \geq u_2], \ldots, [W_n \geq u_n]) = 1$. Here $[W_i \geq u_i] = 1$ if $W_i \geq u_i$ and zero otherwise. All other states are nonaccept states (output 0).

We now claim that the above automaton is equivalent to $g(f_1, \ldots, f_k)$. The proof is by induction on n. The claim we want to prove is that the sub-automaton that starts from state $s = (i, (W_1, W_2, \ldots, W_k))$ computes a function g_s that is equivalent to the function $g(f_1^i, \ldots, f_k^i)$ where $f_j^i = [w_{j,i+1}x_{i+1} + \cdots + w_{j,n}x_n \geq u_j - W_j]$. This immediately follows from the fact that

$$g_s|_{x_{i+1}=0} \equiv g_{\delta(s,0)}, \quad \text{and} \quad g_s|_{x_{i+1}=1} \equiv g_{\delta(s,1)}.$$

It remains to prove the result for level n. The claim is true for the states at level n because

$$g(f_1^n, \ldots, f_k^n) = g([0 \geq u_1 - W_1], \ldots, [0 \geq u_n - W_n])$$
$$= g([W_1 \geq u_1], [W_2 \geq u_2], \ldots, [W_n \geq u_n]).$$

This completes the proof. $\square$

Now the following will be used in the sequel

Lemma 13. *Let* $f_1, f_2 \in \mathrm{HS}_{[-t,t]}$. *There is an algorithm that runs in time* $O(t^2 n^3)$ *and decides whether* $f_1 \equiv f_2$. *If* $f_1 \not\equiv f_2$ *then the algorithm finds an assignment* a *such that* $f_1(a) \neq f_2(a)$.

Proof. We build an automaton for $f_1 + f_2$. If there is no accept state then $f_1 \equiv f_2$. If there is, then any path from the start state to an accept state defines an assignment a such that $f_1(a) \neq f_2(a)$.

The time complexity follows from Lemma 12. $\square$

4   Two Rounds and Non-adaptive Algorithm

In this section we give a two rounds algorithm for learning HS_t that uses $n^{O(t)}$ membership queries.

The algorithm is in Figure 1. Let $f = [w_1 x_1 + w_2 x_2 + \cdots + w_n x_n \geq u]$ be the target function. Let

$$A_m = \bigcup_{i,j=0}^{n} B(0^i 1^{n-i-j} 0^j, m).$$

In Lemma 14 below we show that by querying all the assignments in A_{2t-1} the algorithm can find all the relevant variables of the target function f. In the first step in round 1 the algorithm finds the relevant variables X of f. In Lemma 16 below we show that by querying all the assignments in A_{4t-1} the algorithm can find all the nonsymmetric pairs of variables and therefore the order of the corresponding weights. This is the second step in round 1. The algorithm define a set Y of all the pairs (x_j, x_k) where $w_j < w_k$ and then order all the weights.

By the end of the first round, the algorithm knows the relevant variables and the order of the weights of the target function.

In the second round, the algorithm defines the set $\mathcal{F}$ of all possible functions in HS_t that has weights with the order that was found in round 1. As we will see below $|\mathcal{F}| = n^{O(t)}$. For every two non-equivalent $f_1, f_2 \in \mathcal{F}$ the algorithm finds an assignment a such that $f_1(a) \neq f_2(a)$. Then either $f(a) \neq f_1(a)$ in which case f_1 is removed from $\mathcal{F}$, or $f(a) \neq f_2(a)$ in which case f_2 is removed from $\mathcal{F}$. All the functions remain in $\mathcal{F}$ are equivalent to the target function f.

Learning HalfSpace $f = [w_1 x_1 + w_2 x_2 + \cdots + w_n x_n \geq u]$

Round 1.
$X \leftarrow \emptyset, Y \leftarrow \emptyset$
Ask Membership Queries for all the assignments in
$$A_{4t-1} := \bigcup_{i,j=0}^{n} B(0^i 1^{n-i-j} 0^j, 4t-1)$$
For $k = 1, \ldots, n$
   If there is $a \in A_{2t-2}$ such that $a_k = 1$, $a|_{x_k=0} \in A_{2t-1}$ and $f(a) \neq f(a|_{x_k=0})$
   then $X \leftarrow X \cup \{x_k\}$.
   /* X contains all the relevant variables. If $x_i \notin X$ then $w_i = 0$ */
For each $x_j, x_k \in X$ do
   If there is $b \in A_{4t-1}$ such that $b_j + b_k = 1$
       and $f|_{x_j=0, x_k=1}(b) = 1$, $f|_{x_j=1, x_k=0}(b) = 0$ then $Y \leftarrow Y \cup \{(x_j, x_k)\}$
   /* $(x_j, x_k) \in Y$ iff x_j, x_k are nonsymmetric and $w_j < w_k$ */
Find an order $x_{j_1}, \ldots, x_{j_r}$ of all the elements of X such that
   $(x_{j_{i+1}}, x_{j_i}) \notin Y$ for all $i = 1, \ldots, r-1$
   /* Here we have $w_{j_1} \leq w_{j_2} \leq \cdots \leq w_{j_r}$ */

Round 2.
$D \leftarrow \emptyset$
$\mathcal{F} = $ All possible functions $g := [w_1' x_1 + w_2' x_2 + \cdots + w_n' x_n \geq u']$ where
   $1 \leq w_{j_1}' \leq w_{j_2}' \leq \cdots \leq w_{j_r}' \leq t$, all other $w_j' = 0$ and $u' \in [nt]$
For every two non-equivalent functions $f_1, f_2 \in \mathcal{F}$
   Find a such that $f_1(a) \neq f_2(a)$
   $D \leftarrow D \cup \{a\}$.
Ask Membership Queries for all the assignments in D.
For every $g \in \mathcal{F}$ and $a \in D$ if $g(a) \neq f(a)$ then $\mathcal{F} \leftarrow \mathcal{F} \backslash \{g\}$
Output($\mathcal{F}$)

Fig. 1. Two Rounds Algorithm

We now present a complete analysis. Let $f = [w_1 x_1 + \ldots + w_n x_n \geq u]$. If $f(0^n) = 1$ then $f \equiv 1$. If there is a minterm of weight one, i.e., $f(a) = 1$ for some $a \in B(0^n, 1)$, then $0 \leq u \leq t$ and then all the minterms of f are of weight at most t. In this case we can find all the minterms in one round by asking membership queries for all the assignments in $B(0, t)$ (all other assignments gives 0), finding all the relevant variables and the nonsymmetric variables and move to the second round. This case is handled separately to avoid unnecessary complications in the following analysis. Therefore we may assume that all the minterms of f are of weight at least two.

Consider the set

$$A_m = \bigcup_{i,j=0}^{n} B(0^i 1^{n-i-j} 0^j, m).$$

we now prove

Lemma 14. *Let $f \in HS_t$. The variable x_k is relevant in f if and only if there is $a \in A_{2t-2}$ such that $a_k = 1$, $a|_{x_k=0} \in A_{2t-1}$ and $f(a) \neq f(a|_{x_k=0})$.*

Proof. If x_k is relevant in f then $f \not\equiv 0, 1$ and therefore $f(0^n) = 0$ and $f(1^n) = 1$. Therefore there is an element a in the following sequence

$$0^n, 0^{k-1} 1 0^{n-k}, 0^{k-1} 1^2 0^{n-k-1}, \ldots, 0^{k-1} 1^{n-k+1}, 0^{k-2} 1^{n-k+2}, \ldots, 01^{n-1}, 1^n$$

and $j \in [n]$ such that $f(a) = 1$ and $f(a|_{x_j=0}) = 0$. Notice that $a_k = 1$ and therefore by Lemma 9 there is $c \in B(a, 2t - 2)$ such that $c_k = 1$, $f(c) = 1$ and $f(c|_{x_k=0}) = 0$. Since $c|_{x_k=0} \in B(a, 2t - 1)$, the result follows.   □

Therefore from the assignments in A_{2t-1} one can determine the relevant variables in f. This implies that we may assume w.l.o.g that all the variables are relevant.

We now show

Lemma 15. *If all the variables in $f \in HS_t$ are relevant then there is a strong minterm $a \in A_{2t-2}$ of f.*

Proof. Follows from Lemma 8 and Lemma 6.   □

Lemma 16. *Let $f \in HS_t$ and suppose all the variables in f are relevant. Suppose f is nonsymmetric in x_j and x_k. There is $b \in A_{4t-1}$ such that $b_1 + b_2 = 1$ and $f|_{x_j=0,x_k=1}(b) \neq f|_{x_j=1,x_k=0}(b)$.*

Proof. By Lemma 15 there is a minterm $a \in A_{2t-2}$ of f. Since $wt(a) > 1$, by Lemma 10 there is $b \in B(a, 2t - 1)$ such that $b_j + b_k = 1$ and $f|_{x_j=0,x_k=1}(b) \neq f|_{x_j=1,x_k=0}(b)$. Since $b \in B(a, 2t + 1) \subseteq A_{4t-1}$ the result follows.   □

Therefore from the assignments in A_{4t-1} one can find a permutation ϕ of the variables in f such that $f\phi = [w_1' x_1 + w_2' x_2 + \cdots + w_n' x_n \geq u]$ and $w_1' \leq w_2' \leq \cdots \leq w_n'$.

This completes the first round. We now may assume w.l.o.g that $f = [w_1 x_1 + \cdots + w_n x_n \geq u]$ and $1 \leq w_1 \leq w_2 \leq \cdots \leq w_n \leq t$ and all the variables are relevant. The goal of the second round is to find $w_i \in [1, t]$ and $u \in [0, nt]$. Since we know that $1 \leq w_1 \leq w_2 \leq \cdots \leq w_n \leq t$ we have

$$\binom{n+t-1}{t-1} nt \leq n^{t+1}$$

choices of $w_1, \ldots, w_n$. That is, at most n^{t+1} possible functions in HS_t. For every two such functions f_1, f_2 we use Lemma 13 to find out if $f_1 \equiv f_2$ and if not to find an assignment a such that $f_1(a) \neq f_2(a)$. This takes time

$$\binom{n^{t+1}}{2} t^2 n^3 \leq n^{2t+7}.$$

Let D the set of all such assignments. Then $|D| \leq n^{2t+2}$. In the second round we ask membership queries with all the assignments in D.

Now notice that if $f_1(a) \neq f_2(a)$ then either $f(a) \neq f_1(a)$ or $f(a) \neq f_2(a)$. This shows that the assignments in B eliminates all the functions that are not equivalent to the target and all the remaining functions are equivalent to the target.

Now using Lemma 11 one can replace the set D by $B(b, 8t^3 + O(t^2))$ for any minterm b of f. Lemma 11 shows that for any $f_1 \not\equiv f_2$ there is $a \in B(b, 8t^3 + O(t^2))$ such that $f_1(a) \neq f_2(a)$. This shows that the two rounds can be made into one round and therefore changes the algorithm to a non-adaptive algorithm.

References

[1] Aigner, M.: Combinatorial Search. Wiley Teubner Series on Applicable Theory in Computer Science. Teubner, Stuttgart (1988)

[2] Angluin, D.: Queries and Concept Learning. Machine Learning 2(4), 319–342 (1987)

[3] Alon, A., Asodi, V.: Learning a Hidden Subgraph. SIAM J. Discrete Math. 18(4), 697–712 (2005)

[4] Abboud, E., Agha, N., Bshouty, N.H., Radwan, N., Saleh, F.: Learning Threshold Functions with Small Weights Using Membership Queries. In: COLT 1999. pp. 318–322 (1999)

[5] Alon, N., Beigel, R., Kasif, S., Rudich, S., Sudakov, B.: Learning a Hidden Matching. SIAM J. Comput. 33(2), 487–501 (2004)

[6] Angluin, D., Chen, J.: Learning a Hidden Hypergraph. Journal of Machine Learning Research 7, 2215–2236 (2006)

[7] Angluin, D., Chen, J.: Learning a hidden graph using $O(\log n)$ queries per edge. J. Comput. Syst. Sci. 74(4), 546–556 (2008)

[8] Beigel, R., Alon, N., Kasif, S., Serkan Apaydin, M., Fortnow, L.: An optimal procedure for gap closing in whole genome shotgun sequencing. In: RECOMB 2001, pp. 22–30 (2001)

[9] Biglieri, E., Gyorfi, L.: Multiple Access Channels: Theory and Practice. IOS Press (2007)

[10] Bshouty, N.H.: Exact Learning from Membership Queries: Some Techniques, Results and New Directions. In: Jain, S., Munos, R., Stephan, F., Zeugmann, T. (eds.) ALT 2013. LNCS, vol. 8139, pp. 33–52. Springer, Heidelberg (2013)

[11] Choi, S.-S., Kim, J.H.: Optimal query complexity bounds for finding graphs. Artif. Intell. 174(9-10), 551–569 (2010)

[12] Chirkov, A.Y., Zolotykh, N.Y.: On the number of irreducible points in polyhedra arXiv:1306.4289

[13] Grebinski, V., Kucherov, G.: Reconstructing a Hamiltonian Cycle by Querying the Graph: Application to DNA Physical Mapping. Discrete Applied Mathematics 88(1-3), 147–165 (1998)
[14] Hegedüs, T.: Generalized teaching dimensions and the query complexity of learning. In: Proceedings of the 8th Annual ACM Conference on Computational Learning Theory (COLT 1995), pp. 108–117. ACM Press, New York (1995)
[15] Ngo, H.Q., Du., D.-Z.: A Survey on Combinatorial Group Testing Algorithms with Applications to DNA Library Screening. DIMACS Series in Discrete Mathematics and Theoretical Computer Science
[16] Shevchenko, V.N., Zolotykh, N.Y.: Lower Bounds for the Complexity of Learning Half-Spaces with Membership Queries. In: Richter, M.M., Smith, C.H., Wiehagen, R., Zeugmann, T. (eds.) ALT 1998. LNCS (LNAI), vol. 1501, pp. 61–71. Springer, Heidelberg (1998)
[17] Zolotykh, N.Y., Shevchenko, V.N.: Deciphering threshold functions of k-valued logic. Discrete Analysis and Operations Research. Novosibirsk 2(3), 18–23 (1995); English transl.: Korshunov, A.D. (ed.): Operations Research and Discrete Analysis, pp. 321–326. Kluwer Ac. Publ., Netherlands (1997)

On Exact Learning Monotone DNF
from Membership Queries

Hasan Abasi[1], Nader H. Bshouty[1], and Hanna Mazzawi[2]

[1] Department of Computer Science
Technion, Haifa, Israel
[2] IBM Research - Haifa, Israel

Abstract. In this paper, we study the problem of learning a monotone DNF with at most s terms of size (number of variables in each term) at most r (s term r-MDNF) from membership queries. This problem is equivalent to the problem of learning a general hypergraph using hyperedge-detecting queries, a problem motivated by applications arising in chemical reactions and genome sequencing.

We first present new lower bounds for this problem and then present deterministic and randomized adaptive algorithms with query complexities that are almost optimal. All the algorithms we present in this paper run in time linear in the query complexity and the number of variables n. In addition, all of the algorithms we present in this paper are asymptotically tight for fixed r and/or s.

1 Introduction

We consider the problem of learning a monotone DNF with at most s monotone terms, where each monotone term contains at most r variables (s term r-MDNF) from membership queries [1]. This is equivalent to the problem of learning a general hypergraph using hyperedge-detecting queries, a problem that is motivated by applications arising in chemical reaction and genome sequencing.

1.1 Learning Hypergraph

A hypergraph is $H = (V, E)$ where V is the set of vertices and $E \subseteq 2^V$ is the set of edges. The dimension of the hypergraph H is the cardinality of the largest set in E. For a set $S \subseteq V$, the *edge-detecting queries* $Q_H(S)$ is answered "Yes" or "No", indicating whether S contains all the vertices of at least one edge of H. Our learning problem is equivalent to learning a hidden hypergraph of dimension r with s edges using edge-detecting queries [4].

This problem has many applications in chemical reactions and genome sequencing. In chemical reactions, we are given a set of chemicals, some of which react and some which do not. When multiple chemicals are combined in one test tube, a reaction is detectable if and only if at least one set of the chemicals in the tube reacts. The goal is to identify which sets react using as few experiments as possible. The time needed to compute which experiments to do is a

P. Auer et al. (Eds.): ALT 2014, LNAI 8776, pp. 111–124, 2014.
© Springer International Publishing Switzerland 2014

secondary consideration, though it is polynomial for the algorithms we present [5]. See [18, 11, 3, 2, 13, 4, 21, 5, 15] for more details and other applications.

1.2   Previous Results

In [5], Angluin and Chen presented a deterministic optimal adaptive learning algorithm for learning s-term 2-MDNF with n variables. They also gave a lower bound of $\Omega((2s/r)^{r/2} + rs \log n)$ for learning the class of s-term r-MDNF when $r < s$. In [4], Angluin and Chen gave a randomized algorithm for s-term r-uniform MDNF (the size of each term is *exactly* r) that asks $O(2^{4r} s \cdot poly(r, \log n))$ membership queries. For s-term r-MDNF where $r \le s$, they gave a randomized learning algorithm that asks $O(2^{r+r^2/2} s^{1+r/2} \cdot poly(\log n))$ membership queries.

Literature has also addressed learning some subclasses of s-term 2-MDNF. Those classes have specific applications to genome sequencing. See [18, 11, 3, 2, 13, 4, 21, 5, 15]. In this paper we are interested in learning the class of all s-term r-MDNF formulas for any r and s.

1.3   Our Results

In this paper, we distinguish between two cases: $s \ge r$ and $s < r$.

For $s < r$, we first prove the lower bound $\Omega((r/s)^{s-1} + rs \log n)$. We then give three algorithms. Algorithm I is a deterministic algorithm that asks $O(r^{s-1} + rs \log n)$ membership queries. Algorithm II is a deterministic algorithm that asks $O(s \cdot N((s-1; r); sr) + rs \log n)$ membership queries where $N((s-1; r); sr)$ is the size of $(sr, (s-1, r))$-cover free family (see Subsection 2.2 for the definition of cover free) that can be constructed in time linear in its size. Bshouty and Gabizon showed in [10] that a $(sr, (s-1, r))$-cover free family of size $(r/s)^{s-1+o(1)}$ can be constructed in linear time and therefore algorithm II is almost optimal. Algorithm III is a randomized algorithm that asks

$$O\left(\binom{s+r}{s}\sqrt{sr}\log(sr) + rs \log n\right) = O\left(\left(\frac{r}{s}\right)^{s-1+o(1)} + rs \log n\right)$$

membership queries. This algorithm is almost optimal.

For the case $s \ge r$, Angluin and Chen, [5], gave the lower bound $\Omega((2s/r)^{r/2} + rs \log n)$. We give two algorithms that are almost tight. The first algorithm, Algorithm IV, is a deterministic algorithm that asks $(crs)^{r/2+1.5} + rs \log n$ membership queries for some constant c. The second algorithm, Algorithm V, is a randomized algorithm that asks $(c's)^{r/2+0.75} + rs \log n$ membership queries for some constant c'.

All the algorithms we present in this paper run in time linear in the query complexity and n. Additionally, all the algorithms we describe in this paper are asymptotically tight for fixed r and s.

The following table summarizes our results. We have removed the term $rs \log n$ from all the bounds to be able to fit this table in this page. Det. and Rand. stands for deterministic algorithm and randomized algorithm, respectively.

r,s	Lower Bound	Algorithm	Rand./Det.	Upper Bound
$r > s$, $rs \log n + \left(\frac{r}{s}\right)^{s-1}$		Alg. I	Det.	$rs \log n + r^{s-1}$
		Alg. II	Det.	$rs \log n + \left(\frac{r}{s}\right)^{s+o(s)}$
		Alg. III	Rand.	$rs \log n + (\log r)\sqrt{s}e^s \left(\frac{r}{s}+1\right)^s$
$r \leq s$, $rs \log n + \left(\frac{2s}{r}\right)^{r/2}$		Alg. IV	Det.	$rs \log n + s^{r/2+o(r)}$
		Alg. IV	Rand.	$rs \log n + \sqrt{r}(3e)^r (\log s)s^{r/2+1}$

2   Definitions and Notations

For a vector w, we denote by w_i the ith entry of w. For a positive integer j, we denote by $[j]$ the set $\{1, 2, \ldots, j\}$.

Let $f(x_1, x_2, \ldots, x_n)$ be a Boolean function from $\{0,1\}^n$ to $\{0,1\}$. For an assignment $a \in \{0,1\}^n$ we say that f is ξ in a (or a is ξ in f) if $f(a) = \xi$. We say that a is zero in x_i if $a_i = 0$. For a set of variables S, we say that a is zero in S if for every $x_i \in S$, a is zero in x_i. Denote $X_n = \{x_1, \ldots, x_n\}$.

For a Boolean function $f(x_1, \ldots, x_n)$, $1 \leq i_1 < i_2 < \cdots < i_k \leq n$ and $\sigma_1, \ldots, \sigma_k \in \{0,1\}$ we denote by

$$f|_{x_{i_1}=\sigma_1, x_{i_2}=\sigma_2, \cdots, x_{i_k}=\sigma_k}$$

the function f when fixing the variables x_{i_j} to σ_j for all $j \in [k]$. We denote by $a|_{x_{i_1}=\sigma_1, x_{i_2}=\sigma_2, \cdots, x_{i_k}=\sigma_k}$ the assignment a where each a_{i_j} is replaced by σ_j for all $j \in [k]$. Note that

$$f|_{x_{i_1}=\sigma_1, x_{i_2}=\sigma_2, \cdots, x_{i_k}=\sigma_k}(a) = f(a|_{x_{i_1}=\sigma_1, x_{i_2}=\sigma_2, \cdots, x_{i_k}=\sigma_k}).$$

When $\sigma_1 = \cdots = \sigma_k = \xi$ and $S = \{x_{i_1}, \ldots, x_{i_k}\}$, we denote

$$f|_{x_{i_1}=\sigma_1, x_{i_2}=\sigma_2, \cdots, x_{i_k}=\sigma_k}$$

by $f|_{S \leftarrow \xi}$. In the same way, we define $a|_{S \leftarrow \xi}$. We denote by $1^n = (1, 1, \ldots, 1) \in \{0,1\}^n$.

For two assignments $a, b \in \{0,1\}^n$, we write $a \leq b$ if for every i, $a_i \leq b_i$. A Boolean function $f : \{0,1\}^n \to \{0,1\}$ is *monotone* if for every two assignments $a, b \in \{0,1\}^n$, if $a \leq b$ then $f(a) \leq f(b)$. Recall that every monotone Boolean function f has a unique representation as a reduced monotone DNF [1]. That is, $f = M_1 \vee M_2 \vee \cdots \vee M_s$ where each *monomial* M_i is an ANDs of input variables, and for every monomial M_i there is a unique assignment $a^{(i)} \in \{0,1\}^n$ such that $f(a^{(i)}) = 1$ and for every $j \in [n]$ where $a_j^{(i)} = 1$ we have $f(a^{(i)}|_{x_j=0}) = 0$. We call such assignment a *minterm* of the function f. Notice that every monotone DNF can be uniquely determined by its minterms [1]. That is, $a \in \{0,1\}^n$ is a minterm of f iff $M := \wedge_{a_i=1}x_i$ is a term in f.

For a monotone DNF, $f(x_1, x_2, \ldots, x_n) = M_1 \vee M_2 \vee \cdots \vee M_s$, and a variable x_i, we say that x_i is *t-frequent* if it appears in more than or equal to t terms. A monotone DNF f is called *read k monotone* DNF, if none of its variables is $k+1$-frequent. An s term r-MDNF is a monotone DNF with at most s monotone terms, where each monotone term contains at most r variables.

2.1   Learning Model

Consider a *teacher* (or a black box) that has a *target function* $f : \{0,1\}^n \to \{0,1\}$ that is s-term r-MDNF. The teacher can answer *membership queries*. That is, when receiving $a \in \{0,1\}^n$ it returns $f(a)$. A *learning algorithm* is an algorithm that can ask the teacher membership queries. The goal of the learning algorithm is to *exactly learn* (exactly find) f with minimum number of membership queries and optimal time complexity.

In our algorithms, for a function f we will denote by MQ_f the oracle that answers the membership queries. That is, for $a \in \{0,1\}^n$, $MQ_f(a) = f(a)$.

2.2   Cover Free Families

The problem $(n, (s,r))$-*cover free family*, $((n, (s,r))$-CFF$)$, [17], is equivalent to the following problem: A $(n, (s,r))$-cover free family is a set $A \subseteq \{0,1\}^n$ such that for every $1 \leq i_1 < i_2 < \cdots < i_d \leq n$ where $d = s + r$ and every $J \subseteq [d]$ of size $|J| = s$ there is $a \in A$ such that $a_{i_k} = 0$ for all $k \in J$ and $a_{i_j} = 1$ for all $j \notin J$. Denote by $N((s;r);n)$ the minimum size of such set. The lower bounds in [22] are

$$N((s;r);n) \geq \Omega \left(\frac{(s+r)}{\log \binom{s+r}{s}} \binom{s+r}{s} \log n \right).$$

It is known that a set of random

$$m = O \left(\sqrt{\min(r,s)} \binom{s+r}{s} \left((s+r)\log n + \log \frac{1}{\delta} \right) \right) \tag{1}$$

vectors $a^{(i)} \in \{0,1\}^n$, where each $a_j^{(i)}$ is 1 with probability $r/(s+r)$, is a $(n, (s,r))$-cover free family with probability at least $1 - \delta$.

In [9, 8], Bshouty gave a deterministic construction of $(n, (s,r))$-CFF of size

$$C := \min((2e)^s r^{s+1}, (2e)^r s^{r+1}) \log n$$
$$= \binom{s+r}{r} 2^{\min(s \log s, r \log r)(1+o(1))} \log n \tag{2}$$

that can be constructed in time $C \cdot n$. Fomin et. al. in [16] gave a construction of size

$$D := \binom{s+r}{r} 2^{O\left(\frac{r+s}{\log \log (r+s)} \right)} \log n \tag{3}$$

that can be constructed in time $D \cdot n$. The former bound, (2), is better than the latter when $s \geq r \log r \log \log r$ or $r \geq s \log s \log \log s$. We also note that the former bound, (2), is almost optimal, i.e.,

$$\binom{s+r}{r}^{1+o(1)} \log n,$$

when $r = s^{\omega(1)}$ or $r = s^{o(1)}$ and the latter bound, (3), is almost optimal when

$$o(s \log \log s \log \log \log s) = r = \omega \left(\frac{s}{\log \log s \log \log \log s} \right).$$

Recently, Bshouty and Gabizon, [10], gave a linear time almost optimal construction of $(n, (s, r))$-CFF for any r and s. The size of the $(n, (r, s))$-CFF in [10] is

$$\binom{s+r}{r}^{1+o(1)} \log n \tag{4}$$

where the $o(1)$ is with respect to $\min(r, s)$.

3   Lower Bounds

In this section, we prove some lower bounds.

3.1   General Lower Bound

In this section, we prove that the information theoretic lower bound for learning a class C from membership queries is also a lower bound for any randomized learning algorithm. We believe it is a folklore result, but we could not find the proof in the literature. We first state the following information-theoretic lower bound for deterministic learning algorithm,

Lemma 1. *Let C be any class of Boolean function. Then any deterministic learning algorithm for C must ask at least $\log|C|$ membership queries.*

We now prove,

Lemma 2. *Let C be any class of boolean function. Then any Monte Carlo (and therefore, Las Vegas) randomized learning algorithm that learns C with probability at least $3/4$ must ask at least $\log|C| - 1$ membership queries.*

Proof. Let $\mathcal{A}$ be a randomized algorithm that for every $f \in C$ and an oracle MQ_f that answers membership queries for f, asks m membership queries and satisfies

$$\mathbf{Pr}_s\left[\mathcal{A}(MQ_f, s) = f\right] \geq \frac{3}{4}$$

where $s \in \{0,1\}^N$ is chosen randomly uniformly for some large N. Here $\mathcal{A}(MQ_f, s)$ is the output of the algorithm $\mathcal{A}$ when running with the membership oracle MQ_f for f and random seed s. Consider the random variable $X_f(s)$ that is 1 if $\mathcal{A}(MQ_f, s) = f$ and 0, otherwise. Then for every f, $\mathbf{E}_s[X_f] \geq 3/4$. Therefore, for random uniform $f \in C$

$$3/4 \leq \mathbf{E}_f[\mathbf{E}_s[X_f]] = \mathbf{E}_s[\mathbf{E}_f[X_f(s)]].$$

and by Markov Bound for at least $1/2$ of the elements $s \in \{0,1\}^N$ we have $\mathbf{E}_f[X_f(s)] \geq 1/2$. Let $S \subseteq \{0,1\}^N$ be the set of such elements. Then $|S| \geq 2^N/2$.

Let $s_0 \in S$ and $C_{s_0} \subseteq C$ the class of functions f where $X_f(s_0) = 1$. Then $|C_{s_0}| \geq |C|/2$ and $\mathcal{A}(MQ_f, s_0)$ is a deterministic algorithm that learns the class C_{s_0}. Using the information theoretic lower bound for deterministic algorithm, we conclude that $\mathcal{A}(MQ_f, s_0)$ must ask at least

$$m \geq \log |C_{s_0}| = \log(1/2) + \log |C|$$

membership queries.   □

Specifically, since the number of s-term r-MDNF is

$$\binom{\binom{n}{r}}{s}$$

we have,

Corollary 1. *Any Monte Carlo (and therefore Las Vegas) randomized learning algorithm for the class of s-term r-MDNF must ask on average at least $rs \log n$ membership queries.*

3.2   Two Lower Bounds

In this section, we give two lower bounds. The first is from [4] and the second follows using the same techniques used in [12].

In [4], Angluin and Chen proved,

Theorem 1. *Let r and s be integers. Let k and ℓ be two integers such that*

$$\ell \leq r, \quad s \geq \binom{k}{2}\ell + 1.$$

Any (Monte Carlo) randomized learning algorithm for the class of s-term r-MDNF must ask at least

$$k^\ell - 1$$

membership queries.

Specifically, when $s \gg r$ we have the lower bound

$$\Omega\left(\left(\frac{2s}{r}\right)^{r/2}\right)$$

membership queries. Also, for any integer λ where

$$\binom{\lambda}{2}r + 1 \leq s < \binom{\lambda+1}{2}r$$

we have the lower bound $\lambda^r - 1$.

We now prove the following lower bound.

Theorem 2. *Let r and s be integers and ℓ and t be two integers such that*

$$\ell - \left\lfloor \frac{\ell}{t} \right\rfloor \leq r, \qquad \left\lfloor \frac{\ell}{t} \right\rfloor \leq s - 1.$$

Any (Monte Carlo) randomized learning algorithm for the class of s-term r-MDNF must ask at least $t^{\lfloor \ell/t \rfloor}$ membership queries.

Specifically, for $r \gg s$ we have the lower bound

$$\left(\frac{r}{s} \right)^{s-1}.$$

and for any constant integer λ and $\lambda s \leq r < (\lambda + 1)s$ we have the lower bound

$$(\lambda + 1)^{s-1}.$$

Proof. Let $m = \lfloor \ell/t \rfloor$. Consider the monotone terms $M_j = x_{(j-1)t+1} \cdots x_{jt}$ for $j = 1, 2, \ldots, m$. Define $M_{i,k}$ where $i = 1, \ldots, m$ and $k = 1, \ldots, t$ the monotone term M_i without the variable $x_{(i-1)t+k}$. Let $M_{k_1,k_2,\ldots,k_m} = M_{1,k_1} M_{2,k_2} \cdots M_{m,k_m}$. Let $f = M_1 \vee M_2 \vee \cdots \vee M_m$ and $g = M_1 \vee M_2 \vee \cdots \vee M_m \vee M_{k_1,k_2,\ldots,k_m}$. It is easy to see that f and g are s-term r-MDNF. The only way we can distinguish between the two hypothesis f and g is by guessing an assignment that is 1 in all its first mt entries except for the entire $k_1, t + k_2, 2t + k_3, \ldots, (m-1)t + k_m$. That is, by guessing $k_1, k_2, \ldots, k_m$. This takes an average of t^m guesses. Since both f and g are s-term r-MDNF, the result follows. $\square$

For $r \gg s$, we choose $\ell = r$ and t such that $\lfloor \ell/t \rfloor = s - 1$. Since $s - 1 = \lfloor \ell/t \rfloor \geq \ell/t - 1$, we have $t \geq r/s$ and the result follows.

For $\lambda s \leq r < (\lambda + 1)s$, proving the lower bound for $r = \lambda s$ is sufficient. Take $t = \lambda + 1$ and $\ell = (\lambda + 1)s - 1$.

4   Optimal Algorithms for Monotone DNF

In this section, we present the algorithms (Algorithm I-V) that learn the class of s-term r-MDNF. We first give a simple algorithm that learns one term. We then give three algorithms (Algorithm I-III) for the case $r > s$ and two algorithms (Algorithm IV-V) for the case $s \geq r$.

4.1   Learning One Monotone Term

In this section, we prove the following result.

Lemma 3. *Let $f(x) = M_1 \vee M_2 \vee \cdots \vee M_s$ be the target function where each M_i is a monotone term of size at most r. Suppose $g(x) = M_1 \vee M_2 \vee \cdots \vee M_{s'}$ and $h(x) = M_{s'+1} \vee M_{s'+2} \vee \cdots \vee M_s$. If a is an assignment such that $g(a) = 0$ and $h(a) = 1$, then a monotone term in $h(x)$ can be found with*

$$O\left(r \log \frac{n}{r} \right)$$

membership queries.

Proof. First notice that since g is monotone, for any $b \leq a$ we have $g(b) = 0$. Our algorithm finds a minterm $b \leq a$ of f and therefore b is a minterm of h.

First, if the number of ones in a is $2r$, then we can find a minterm by flipping each bit in a that does change the value of f and get a minterm. This takes at most $2r$ membership queries.

If the number of ones in a is $w > 2r$, then we divide the entries of a that are equal to 1 into $2r$ disjoint sets $S_1, S_2, \ldots, S_{2r}$ where for every i, the size of S_i is either $\lfloor w/(2r) \rfloor$ or $\lceil w/(2r) \rceil$. Now for $i = 1, 2, \ldots, 2r$, we flip all the entries of S_i in a to zero and ask a membership query. If the function is one, we keep those entries 0. Otherwise we set them back to 1 and proceed to $i + 1$. At the end of this procedure, at most r sets are not flipped. Therefore, at least half of the bits in a are flipped to zero using $2r$ membership queries. Therefore, the number of membership queries we need to get a minterm is $2r \log(n/2r) + 2r$.   □

We will call the above procedure **Find-Term**. See Figure 1.

Find-Term(f, g, h, a)
$g(a) = 0$ and $h(a) = 1$.
1) If the number of ones in a, $wt(a)$, is less than or equal $2r$ then
    For $i = 1$ to n
        If $a_i = 1$ and $f(a|_{x_i=0}) = 1$ then $a \leftarrow a|_{x_i=0}$.
    Output(a).
2) Let $S = \{x_i \mid a_i = 1\}$
3) Partition $S = S_1 \cup S_2 \cup \cdots \cup S_{2r}$ such that $|S_i| \in \{\lfloor w/(2r) \rfloor$ or $\lceil w/(2r) \rceil\}$.
4) For $i = 1$ to $2r$
        If $f(a|_{S_i \leftarrow 0}) = 1$ then $a \leftarrow a|_{S_i \leftarrow 0}$.
5) Goto 1.

Fig. 1. Algorithm Find-Term for finding a new term in f

4.2 The case $r > s$

In this section, we present three algorithms, two deterministic and one randomized. We start with the deterministic algorithm.

Deterministic Algorithm: Consider the class s-term r-MDNF. Let f be the target function. Given $s - \ell$ monotone terms $h := M_1 \vee M_2 \vee \cdots \vee M_{s-\ell}$ that are known to the learning algorithm to be in f. The learning algorithm goal is to find a new monotone term. In order to find a new term we need to find an assignment a that is zero in $M_1 \vee M_2 \vee \cdots \vee M_{s-\ell}$ and 1 in the function f. Then by the procedure **Find-Term** in Subsection 4.1, we get a new term in $O(r \log n)$ additional membership queries.

To find such an assignment, we present three algorithms:

Algorithm I: (Exhaustive Search) choose a variable from each M_i and set it to zero and set all the other variables to 1. The set of all such assignments is denoted by A. If f is 1 in some $a \in A$, then find a new term using **Find-Term**. We now show.

Lemma 4. *If $f \not\equiv h$, then* **Algorithm I** *finds a new term in $r^{s-\ell} + O(r \log n)$ membership queries.*

Proof. Since the number of variables in each term in $h := M_1 \vee M_2 \vee \cdots \vee M_{s-\ell}$ is at most r the number of assignments in A is at most $r^{s-\ell}$. Since we choose one variable from each term in h and set it to zero, all the assignments in A are zero in h. We now show that one of the assignments in A must be 1 in f, and therefore a new term can be found.

Let b be an assignment that is 1 in f and zero in h. Such assignment exists because otherwise $f \Rightarrow h$ and since $h \Rightarrow f$ we get $f \equiv h$. Since $h(b) = 0$ there is at least one variable x_{j_i} in each M_i that is zero in b. Then the assignment $a := 1^n|_{x_{j_1}=0,\ldots,x_{j_{s-\ell}}=0}$ is in A and $h(a) = 0$. Since $a \geq b$ we also have $f(a) = 1$.

The number of queries in this algorithm is

$$\sum_{\ell=1}^{s} O\left(r^{s-\ell} + r \log n\right) = O(r^{s-1} + rs \log n).$$

$\square$

We now present the second algorithm. Recall that $X_n = \{x_1, \ldots, x_n\}$.

Algorithm II

1) Let V be the set of variables that appear in $M_1 \vee M_2 \vee \cdots \vee M_{s-\ell}$.
2) Take a $(|V|, (s-\ell, r))$-CFF A over the variables V.
3) For each $a \in A$
    3.1) Define an assignment a' that is a_i in x_i for every $x_i \in V$
        and 1 in x_i for every $x_i \in X_n \backslash V$.
    3.2) If $M_1 \vee M_2 \vee \cdots \vee M_{s-\ell}$ is 0 in a' and f is one in a'
        then find a new term using **Find-Term**

Fig. 2. Algorithm II for the case $r > s$

We now show,

Lemma 5. *If $f \not\equiv h$, then* **Algorithm II** *finds a new term in $N((s-\ell; r); (s-\ell)r) + O(r \log n)$ membership queries.*

Proof. Let $h := M_1 \vee M_2 \vee \cdots \vee M_{s-\ell}$. Let b be an assignment that is 1 in f and zero in h. Since $h(b) = 0$, there is at least one variable x_{j_i} in each M_i that is zero in b. Consider the set $U = \{x_{j_i} | i = 1, \ldots, s-\ell\}$. Since $f(b) = 1$ there is a new term M in f that is one in b. That is, all of its variables are one in b.

Let W be the set of all variables in M. Since A is $(|V|, (s-\ell, r))$-CFF and since $|U \cup (W \cap V)| \le s - \ell + r$ there is an assignment $a \in A$ that is 0 in each variable in U and is one in each variable in $W \cap V$. Since a' is also 0, in each variable in U we have $h(a') = 0$. Since a' is one in each variable in $W \cap V$ and one in each variable $W \backslash V$, we have $M(a') = 1$ and therefore $f(a') = 1$.

The number of queries in Algorithm II is

$$\sum_{\ell=1}^{s-1} N((s - \ell; r); (s - \ell)r) + r \log n = O(sN((s-1; r); sr) + rs \log n).$$

This completes the proof. $\square$

Randomized Algorithm: Our third algorithm, Algorithm III, is a randomized algorithm. It is basically Algorithm II where an $(rs, (s-1, r))$-CFF A is randomly constructed, as in (1). Notice that an $(rs, (s-1, r))$-CFF is also an $(|V|, (s-\ell, r))$-CFF, so it can be used in every round of the algorithm. The algorithm fails if there is a new term that has not been found and this happens if and only if A is not $(rs, (s-1, r))$-CFF. So the failure probability is δ. By (1), this gives a Monte Carlo randomized algorithm with query complexity

$$O\left(\sqrt{s} \binom{s+r}{s} \left(r \log r + \log \frac{1}{\delta}\right) + rs \log n\right).$$

4.3   The case $r < s$

In this section, we present two algorithms. Algorithm IV is deterministic and Algorithm V is randomized. We start with the deterministic algorithm.

Deterministic Algorithm: In this section, we present Algorithm IV, used when $r < s$. For this case, we prove the following.

Theorem 3. *There is a deterministic learning algorithm for the class of s-term r-MDNF that asks*

$$O\left((3e)^r (rs)^{r/2+1.5} + rs \log n\right),$$

membership queries.

Before proving this theorem, we first prove learnability in simpler settings. We prove the following.

Lemma 6. *Let $f(x_1, x_2, \ldots, x_n) = M_1 \vee \cdots \vee M_s$ be the target s-term r-MDNF. Suppose the learning algorithm knows some of the terms, $h = M_1 \vee M_2 \vee \cdots \vee M_{s-\ell}$ and knows that $M_{s-\ell+1}$ is of size r'. Suppose that h is a read k monotone DNF. Then, there exists an algorithm that finds a new term (not necessarily $M_{s-\ell+1}$) using*

$$O\left(N((r'k; r'); sr)\right) + r \log n,$$

membership queries.

LearnRead(MQ_f, s, ℓ, r')
1) Let V be the set of variables that appear in h.
2) Let A be a $(|V|, (r'k, r'))$-CFF over the variables V.
3) For each $a \in A$
      3.1) Let $a' \in \{0,1\}^n$ where a' is a_i in each $x_i \in V$,
          and one in each $x_i \in X_n \backslash V$.
      3.2) $X \leftarrow \emptyset$.
      3.3) For each M_i, $i = 1, \ldots, s - \ell$ such that $M_i(a') = 1$ do
          Take any variable x_j in M_i and set $X \leftarrow X \cup \{x_j\}$
      3.4) Set $a'' \leftarrow a'|_{X \leftarrow 0}$.
      3.5) If $f(a'') = 1$ and $h(a'') = 0$
          then find a new term using **Find-Term**.

Fig. 3. Finding a new term in read k

Proof. Consider the algorithm in Figure 3.

Let V be the set of variables that appear in h. Let $M := M_{s-\ell+1}$. Let U be the set of variables in M and $W = U \cap V$. Each variable in W can appear in at most k terms in h. Let w.l.o.g $h' := M_1 \vee \cdots \vee M_t$ be those terms. Notice that $t \leq |W|k \leq r'k$. In each term M_i, $i \leq t$ one can choose a variable x_{j_i} that is not in W. This is because, if all the variable in M_i are in W, then $M \Rightarrow M_i$ and then f is not reduced MDNF.

Let $Z = \{x_{j_i} | i = 1, \ldots, t\}$. Since $|Z| \leq t \leq r'k$ and $|U| \leq r'$ there is $a \in A$ that is 0 in every variable in Z and is 1 in every variable in U. Now notice that a' in step 3.1 in the algorithm is the same as a over the variables in Z and therefore $h'(a') = 0$. Also a' is the same as a over the variables in U and therefore $M(a') = 1$. Now notice that since $M_i(a') = 0$ for $i \leq t$, in step 3.4 in the algorithm we only flip a'_i that correspond to variables in the terms M_i, $i > t$. The set of variables in each other term M_i, $i > t$ is disjoint with U. Therefore if for some $i > t$, $M_i(a') = 1$ then setting any variable x_j in M_i that is one in a' to zero will not change the values $M(a') = 1$ and (from monotonicity) $h'(a') = 0$. Eventually, we will have an assignment a'' that satisfies $h(a'') = 0$ and $M(a'') = 1$ which implies $f(a'') = 1$. □

In the following lemma, we remove the restriction on h.

Lemma 7. *Let* $f(x_1, x_2, \ldots, x_n) = M_1 \vee \cdots \vee M_s$ *be the target s-term r-MDNF. Suppose some of the terms, $h = M_1 \vee M_2 \vee \ldots \vee M_{s-\ell}$, are already known to the learning algorithm. Then, for any integer d, there exists an algorithm that finds a new term using*

$$O\left(\sum_{i=1}^{r} \binom{r\sqrt{ds}}{i} N(((r-i)\sqrt{s/d}; (r-i)); rs) + r \log n\right),$$

membership queries.

Proof. Consider the algorithm in Figure 4.

Learn(s, ℓ)
1) Let S be the set of $\sqrt{s/d}$-frequent variables in h.
2) For every $R \subseteq S$ of size $|R| \leq r$ do
    2.1) Define $A \in (\{0,1\} \cup X_n)^n$ that is 1 in R and 0 in $S \backslash R$
       and $A_i = x_i$ for every $x_i \notin S$.
    2.2) Run **LearnRead**$(MQ_{f(A)}, s, \ell, r - |R|)$ to find a''.
3) Use $a''|_{R \leftarrow 1, S \backslash R \leftarrow 0}$ to find a new term using **Find-Term**.

Fig. 4. Finding a new term

First note that in step 2.2, $f(A)$ is considered in **LearnRead** as a function in all the variables X_n. Note also that the oracle $MQ_{f(A)}$ can be simulated by MQ_f, since $f(A)(a) = f(a|_{R \leftarrow 1, S \backslash R \leftarrow 0})$.

Let W be the set of variables that appear in $M := M_{s-\ell+1}$ and $R = S \cap W$. Note that A is zero in all $S \backslash R$ and 1 in R and therefore $f(A)$ is now a read $\sqrt{s/d}$ and $M(A)$ contains at most $|W \backslash R| \leq r - |R|$ variables. Therefore, when we run **LearnRead**$(MQ_{f(A)}, s, \ell, r - |R|)$ we find an assignment a'' that is 1 in $M(A)$ and zero in $f(A)$ and then $a''|_{R \leftarrow 1, S \backslash R \leftarrow 0}$ is one in f and zero in h.

We now find the number of queries. By the Pigeon hole principle, there are at most $|S| \leq r\sqrt{ds}$ that are $\sqrt{s/d}$-frequent. The number of sets $R \subseteq S$ of size i is $\binom{r\sqrt{ds}}{i}$. For each set, we run **LearnRead**$(MQ_{f(A)}, s, \ell, r - |R|)$ that by Lemma 6 asks $N(((r - i)\sqrt{s/d}; (r - i)); rs)$ queries. This implies the result. $\qquad\square$

We now prove Theorem 3.

Proof. We choose $d = r$. Then by the construction (2), we have

$$\binom{r\sqrt{ds}}{i} N(((r - i)\sqrt{s/d}; (r - i)); rs) \leq \left(\frac{er\sqrt{rs}}{i}\right)^i (2e)^{r-i} \left(\frac{(r - i)\sqrt{s}}{\sqrt{r}}\right)^{r-i+3}$$

$$\leq e^r 2^{r-i} (\sqrt{rs})^{r+3} \left(\frac{r}{i}\right)^i \left(\frac{r - i}{r}\right)^{r-i+3}$$

$$\leq e^r 2^{r-i} \binom{r}{i} (\sqrt{rs})^{r+3}.$$

and therefore

$$\sum_{i=1}^{r} \binom{r\sqrt{ds}}{i} N(((r - i)\sqrt{s/d}; (r - i)); rs) \leq (3e)^r (rs)^{r/2+1.5}.$$

Using the result in (4) with $d = 1$ we get the bound

$$s^{r/2+o(r)}$$

$\qquad\square$

Randomized Algorithm: In this section, we give a randomized algorithm for the case $s > r$.

The randomized algorithm is the same as the deterministic one, except that each CFF is constructed randomly, as in (1) with probability of success $1 - \delta/s$. Since the algorithm in Lemma 7 is running s times, the probability of success of the algorithm is at least $1 - \delta$. We choose $d = 1$ in Lemma 7 and get

$$\binom{r\sqrt{ds}}{i} N(((r-i)\sqrt{s/d}; (r-i)); rs)$$

$$\leq \left(\frac{er\sqrt{s}}{i}\right)^i \sqrt{r}(e(\sqrt{s}+1))^{r-i} \left(2s\log rs + \log\frac{s}{\delta}\right).$$

$$\leq e^r 2^{r-i} \left(\frac{r}{i}\right)^i \sqrt{r} s^{r/2}(s\log s + \log(1/\delta))$$

and therefore number of queries used in the algorithm is

$$\sum_{i=1}^{r} \binom{r\sqrt{ds}}{i} N(((r-i)\sqrt{s/d}; (r-i)); rs) \leq \sqrt{r}(3e)^r s^{r/2}(s\log s + \log(1/\delta)).$$

5   Conclusion and Open Problems

In this paper, we gave an almost optimal adaptive exact learning algorithms for the class of s-term r-MDNF. When r and s are fixed, the bounds are asymptotically tight. Some gaps occur between the lower bounds and upper bounds. For $r \geq s$, the gap is c^s for some constant c and for $r \leq s$ the gap is $r^{r/2}$. It is interesting to close these gaps. Finding a better deterministic construction of CFF will give better deterministic algorithms.

Another challenging problem is finding tight bounds for non-adaptive learning of this class.

References

[1] Angluin, D.: Queries and Concept Learning. Machine Learning 2(4), 319–342 (1987)
[2] Alon, N., Asodi, V.: Learning a Hidden Subgraph. SIAM J. Discrete Math. 18(4), 697–712 (2005)
[3] Alon, N., Beigel, R., Kasif, S., Rudich, S., Sudakov, B.: Learning a Hidden Matching. SIAM J. Comput. 33(2), 487–501 (2004)
[4] Angluin, D., Chen, J.: Learning a Hidden Hypergraph. Journal of Machine Learning Research 7, 2215–2236 (2006)
[5] Angluin, D., Chen, J.: Learning a Hidden Graph using $O(\log n)$ Queries per Edge. J. Comput. Syst. Sci. 74(4), 546–556 (2008)
[6] Alon, N., Moshkovitz, D., Safra, S.: Algorithmic Construction of Sets for k-Restrictions. ACM Transactions on Algorithms 2(2), 153–177 (2006)

[7] Bshouty, N.H.: Exact Learning from Membership Queries: Some Techniques, Results and New Directions. In: Jain, S., Munos, R., Stephan, F., Zeugmann, T. (eds.) ALT 2013. LNCS, vol. 8139, pp. 33–52. Springer, Heidelberg (2013)

[8] Bshouty, N.H.: Linear time Constructions of some d-Restriction Problems. CoRR abs/1406.2108 (2014)

[9] Bshouty, N.H.: Testers and their Applications. Electronic Collouium on Computational Complexity (ECCC) 19, 11 (2012); ITCS 2014, pp. 327–352

[10] Bshouty, N.H., Gabizon, A.: Almost Optimal Cover-Free Family (in preperation)

[11] Beigel, R., Alon, N., Kasif, S., Serkan Apaydin, M., Fortnow, L.: An Optimal procedure for gap Closing in whole Genome Shotgun Sequencing. In: RECOMB 2001, pp. 22–30 (2001)

[12] Bshouty, N.H., Goldman, S.A., Hancock, T.R., Matar, S.: Asking Questions to Minimize Errors. J. Comput. Syst. Sci. 52(2), 268–286 (1996)

[13] Bouvel, M., Grebinski, V., Kucherov, G.: Combinatorial Search on Graphs Motivated by Bioinformatics Applications: A Brief Survey. In: Kratsch, D. (ed.) WG 2005. LNCS, vol. 3787, pp. 16–27. Springer, Heidelberg (2005)

[14] Bshouty, N.H., Hellerstein, L.: Attribute-Efficient Learning in Query and Mistake-bound Models. In: COLT, pp. 235–243 (1996)

[15] Chang, H., Chen, H.-B., Fu, H.-L., Shi, C.-H.: Reconstruction of hidden graphs and threshold group testing. J. Comb. Optim. 22(2), 270–281 (2011)

[16] Fomin, F.V., Lokshtanov, D., Saurabh, S.: Efficient Computation of Representative Sets with Applications in Parameterized and Exact Algorithms. In: SODA 2014, pp. 142–151 (2014)

[17] Kautz, W.H., Singleton, R.C.: Nonrandom binary superimposed codes. IEEE Trans. Inform. Theory 10(4), 363–377 (1964)

[18] Grebinski, V., Kucherov, G.: Reconstructing a Hamiltonian Cycle by Querying the Graph: Application to DNA Physical Mapping. Discrete Applied Mathematics 88(1-3), 147–165 (1998)

[19] Kleitman, D.J., Spencer, J.: Families of k-independent sets. Discrete Mathematics 6(3), 255–262 (1972)

[20] Naor, M., Schulman, L.J., Srinivasan, A.: Splitters and Near-optimal Derandomization. In: FOCS 1995, pp. 182–191 (1995)

[21] Reyzin, L., Srivastava, N.: Learning and Verifying Graphs Using Queries with a Focus on Edge Counting. In: Hutter, M., Servedio, R.A., Takimoto, E. (eds.) ALT 2007. LNCS (LNAI), vol. 4754, pp. 285–297. Springer, Heidelberg (2007)

[22] Stinson, D.R., Wei, R., Zhu, L.: Some New Bounds for Cover free Families. Journal of Combinatorial Theory, Series A 90(1), 224–234 (2000)

Learning Regular Omega Languages

Dana Angluin[1,*] and Dana Fisman[2,**]

[1] Yale University, New Haven, Connecticut, USA
[2] University of Pennsylvania, Philadelphia, Pennsylvania, USA

Abstract. We provide an algorithm for learning an unknown regular set of infinite words, using membership and equivalence queries. Three variations of the algorithm learn three different canonical representations of omega regular languages, using the notion of families of DFAs. One is of size similar to $L_\$$, a DFA representation recently learned using L^* [7]. The second is based on the syntactic FORC, introduced in [14]. The third is introduced herein. We show that the second can be exponentially smaller than the first, and the third is at most as large as the first two, with up to a quadratic saving with respect to the second.

1 Introduction

The L^* algorithm learns an unknown regular language in polynomial time using membership and equivalence queries [2]. It has proved useful in many areas including AI, neural networks, geometry, data mining, verification and many more. Some of these areas, in particular verification, call for an extension of the algorithm to regular ω-languages, i.e. regular languages over infinite words.

Regular ω-languages are the main means to model reactive systems and are used extensively in the theory and practice of formal verification and synthesis. The question of learning regular ω-languages has several natural applications in this context. For instance, a major critique of *reactive-system synthesis*, the problem of synthesizing a reactive system from a given temporal logic formula, is that it shifts the problem of implementing a system that adheres to the specification in mind to formulating a temporal logic formula that expresses it. A potential customer of a computerized system may find it hard to specify his requirements by means of a temporal logic formula. Instead, he might find it easier to provide good and bad examples of ongoing behaviors (or computations) of the required system, or classify a given computation as good or bad — a classical scenario for interactive learning of an unknown language using membership and equivalence queries.

Another example, concerns *compositional reasoning*, a technique aimed to improve scalability of verification tools by reducing the original verification task into subproblems. The simplification is typically based on the assume-guarantee reasoning principles and requires identifying adequate environment assumptions

---

* Research of this author was supported by US NSF grant CCF-0916389.
** Research of this author was supported by US NSF grant CCF-1138996.

P. Auer et al. (Eds.): ALT 2014, LNAI 8776, pp. 125–139, 2014.

for components. A recent approach to the automatic derivation of assumptions uses L^* [5, 1, 17] and a model checker for the different component playing the role of the teacher. Using L^* allows learning only *safety* properties (a subset of ω-regular properties that state that something bad hasn't happened and can be expressed by automata on finite words). To learn *liveness* and *fairness* properties, we need to extend L^* to the full class of regular ω-languages — a problem considered open for many years [11].

The first issue confronted when extending to ω-languages is how to cope with infinite words? Some finite representation is needed. There are two main approaches for that: one considers only finite prefixes of infinite computations and the other considers ultimately periodic words, i.e., words of the form uv^ω where v^ω stands for the infinite concatenation of v to itself. It follows from McNaughton's theorem [15] that two ω-regular languages are equivalent if they agree on the set of ultimately periodic words, justifying their use for representing examples.

Work by de la Higuera and Janodet [6] gives positive results for polynomially learning in the limit *safe* regular ω-languages from *prefixes*, and negative results for learning any strictly subsuming class of regular ω-languages from prefixes. A regular ω-language L is *safe* if for all $w \notin L$ there exists a prefix u of w such that any extension of u is not in L. This work is extended in [8] to learning bi-ω languages from subwords.

Saoudi and Yokomori [19] consider ultimately periodic words and provide an algorithm for learning in the limit the class of *local* ω-languages and what they call *recognizable* ω-languages. An ω-language is said to be *local* if there exist $I \subseteq \Sigma$ and $C \subseteq \Sigma^2$ such that $L = I\Sigma^\omega - \Sigma^*C\Sigma^\omega$. An ω-language is referred to as *recognizable* [19] if it is recognizable by a deterministic automaton all of whose states are accepting.

Maler and Pnueli [13] provide an extension of the L^* algorithm, using ultimately periodic words as examples, to the class of regular ω-languages which are recognizable by both deterministic Büchi and deterministic co-Büchi automata. This is the subset for which the straightforward extension of right-congruence to infinite words gives a Myhill-Nerode characterization [20]. Generalizing this to wider classes calls for finding a Myhill-Nerode characterization for larger classes of regular ω-languages. This direction of research was taken in [10, 14] and is one of the starting points of our work.

In fact the full class of regular ω-languages can be learned using the result of Calbrix, Nivat and Podelski [4]. They define for a given ω-language L the set $L_\$ = \{u\$v \mid u \in \Sigma^*, v \in \Sigma^+, uv^\omega \in L\}$ and show that $L_\$$ is regular by constructing an NFA and a DFA accepting it. Since DFAs are canonical for regular languages, it follows that a DFA for $L_\$$ is a canonical representation of L. Such a DFA can be learned by the L^* algorithm provided the teacher's counter examples are ultimately periodic words, given e.g. as a pair (u, v) standing for uv^ω — a quite reasonable assumption that is common to the other works too. This DFA can be converted to a Büchi automaton recognizing it. This approach was studied and implemented by Farzan et al. [7]. For a Büchi automaton with m states, Calbrix et al. provide an upper bound of $2^m + 2^{2m^2+m}$ on the size of a DFA for $L_\$$.

So the full class of regular ω-languages can be learnt using membership and equivalence queries, yet not very efficiently. We thus examine an alternative canonical representation of the full class of regular ω-languages. Maler and Staiger [14] show that regular ω-languages can be represented by a family of right congruences (FORC, for short). With a given ω-language they associate a particular FORC, the *syntactic* FORC, which they show to be the coarsest FORC recognizing the language. We adapt and relax the notion of FORC to families of DFAs (FDFA, for short). We show that the syntactic FORC can be factorially smaller than $L_\$$. That is, there exists a family of languages L_n for which the syntactic FDFA is of size $O(n)$ and the minimal DFA for $L_\$$ is of size $\Omega(n!)$. We then provide a third representation, the *recurrent* FDFA. We show that the recurrent FDFA is at most as large as both the syntactic FDFA and an FDFA corresponding to $L_\$$, with up to a quadratic saving with respect to the syntactic FDFA.

We provide a learning algorithm L^ω that can learn an unknown regular ω-language using membership and equivalence queries. The learned representations use the notion of families of DFAs (FDFAs). Three variations of the algorithm can learn the three canonical representations: the periodic FDFA (the FDFA corresponding to $L_\$$), the syntactic FDFA (the FDFA corresponding to the syntactic FORC) and the recurrent FDFA. The running time of the three learning algorithms is polynomial in the size of the periodic FDFA.

2   Preliminaries

Let Σ be a finite set of symbols. The set of finite words over Σ is denoted Σ^*, and the set of infinite words, termed ω-words, over Σ is denoted Σ^ω. A *language* is a set of finite words, that is, a subset of Σ^*, while an ω-language is a set of ω-words, that is, a subset of Σ^ω. Throughout the paper we use u, v, x, y, z for finite words, w for ω-words, a, b, c for letters of the alphabet Σ, and i, j, k, l, m, n for natural numbers. We use $[i..j]$ for the set $\{i, i+1, \ldots, j\}$. We use $w[i]$ for the i-th letter of w and $w[i..k]$ for the subword of v starting at the i-th letter and ending at the k-th letter, inclusive.

An *automaton* is a tuple $M = \langle \Sigma, Q, q_0, \delta \rangle$ consisting of a finite alphabet Σ of symbols, a finite set Q of states, an initial state q_0 and a transition function $\delta : Q \times \Sigma \to 2^Q$. A run of an automaton on a finite word $v = a_1 a_2 \ldots a_n$ is a sequence of states $q_0, q_1, \ldots, q_n$ such that $q_{i+1} \in \delta(q_i, a_{i+1})$. A run on an infinite word is defined similarly and results in an infinite sequence of states. The transition function can be extended to a function from $Q \times \Sigma^*$ by defining $\delta(q, \lambda) = q$ and $\delta(q, av) = \delta(\delta(q, a), v)$ for $q \in Q$, $a \in \Sigma$ and $v \in \Sigma^*$. We often use $M(v)$ as a shorthand for $\delta(q_0, v)$ and $|M|$ for the number of states in Q. A transition function is *deterministic* if $\delta(q, a)$ is a singleton for every $q \in Q$ and $a \in \Sigma$, in which case we use $\delta(q, a) = q'$ rather than $\delta(q, a) = \{q'\}$.

By augmenting an automaton with an acceptance condition α, obtaining a tuple $\langle \Sigma, Q, q_0, \delta, \alpha \rangle$, we get an *acceptor*, a machine that accepts some words and rejects others. An acceptor accepts a word, if one of the runs on that word is accepting. For finite words the acceptance condition is a set $F \subseteq Q$ and a run on v is accepting if it ends in an accepting state, i.e. if $\delta(q_0, v) \in F$. For infinite words, there are many acceptance conditions in the literature, here we mention three: Büchi, co-Büchi and Muller. Büchi and co-Büchi acceptance conditions are also a set $F \subseteq Q$. A run of a Büchi automaton is accepting if it visits F infinitely often. A run of a co-Büchi is accepting if it visits F only finitely many times. A Muller acceptance condition is a map $\tau : 2^Q \to \{+, -\}$. A run of a Muller automaton is accepting if the set S of states visited infinitely often along the run is such that $\tau(S) = +$. The set of words accepted by an acceptor A is denoted $[\![A]\!]$.

We use three letter acronyms to describe acceptors. The first letter is D or N: D if the transition relation is deterministic and N if it is not. The second letter is one of $\{F,B,C,M\}$: F if this is an acceptor over finite words, B, C, M if it is an acceptor over infinite words with Büchi, co-Büchi or Muller acceptance condition, respectively. The third letter is always A for acceptor. For finite words DFAs and NFAs have the same expressive power. For infinite words the theory is much more involved. For instance, DBAs are weaker than NBAs, DMAs are as expressive as NMAs, and NBAs are as expressive as DMAs. A language is said to be *regular* if it is accepted by a DFA. An ω-language is said to be *regular* if it is accepted by a DMA.

An equivalence relation $\sim$ on Σ^* is a *right-congruence* if $x \sim y$ implies $xv \sim yv$ for every $x, y, v \in \Sigma^*$. The *index* of $\sim$, denoted $|\sim|$ is the number of equivalence classes of $\sim$. Given a language L its *canonical right congruence* $\sim_L$ is defined as follows: $x \sim_L y$ iff $\forall v \in \Sigma^*$ we have $xv \in L \iff yv \in L$. We use $[\sim]$ to denote the equivalence classes of the right-congruence $\sim$ (instead of the more common notation $\Sigma^*/\sim$). For a word $v \in \Sigma^*$ the notation $[v]$ is used for the class of $\sim$ in which v resides.

A right congruence $\sim$ can be naturally associated with an automaton $M_\sim = \langle \Sigma, Q, q_0, \delta \rangle$ as follows: the set of states Q are the equivalence classes of $\sim$. The initial state q_0 is the equivalence class $[\lambda]$. The transition function δ is defined by $\delta([u], \sigma) = [u\sigma]$. Similarly, given an automaton $M = \langle \Sigma, Q, q_0, \delta \rangle$ we can naturally associate with it a right congruence as follows: $x \sim_M y$ iff $\delta(q_0, x) = \delta(q_0, y)$. The Myhill-Nerode Theorem states that a language L is regular iff $\sim_L$ is of finite index. Moreover, if L is accepted by a DFA A then $\sim_A$ refines $\sim_L$. Finally, the index of $\sim_L$ gives the size of the minimal DFA for L.

For ω-languages, the right congruence $\sim_L$ is defined similarly, by quantifying over ω-words. That is, $x \sim_L y$ iff $\forall w \in \Sigma^\omega$ we have $xw \in L \iff yw \in L$. Given a deterministic automaton M we can define $\sim_M$ exactly as for finite words. However, for ω-regular languages, right-congruence alone does not suffice to obtain a "Myhill-Nerode" characterization. As an example consider the language $L_1 = \Sigma^* a^\omega$. We have that $\sim_{L_1}$ consists of just one equivalence class, but obviously an acceptor recognizing L_1 needs more than a single state.

3   Canonical Representations of Regular ω-Languages

As mentioned in the introduction, the language $L_\$ = \{u\$v \mid u \in \Sigma^*, v \in \Sigma^+, uv^\omega \in L\}$ provides a canonical representation for a regular ω-language L. As the upper bound of going from a given Büchi automaton of size m to $L_\$$, is quite large $(2^m + 2^{2m^2+m})$ we investigate other canonical representations.

Second Canonical Representation - Syntactic FORC. Searching for a notion of right congruence adequate for regular ω-languages was the subject of many works (c.f. [21, 12, 9, 3, 14]). In the latest of these [14] Maler and Staiger proposed the notion of a *family of right-congruences* or FORC.

Definition 1 (FORC, Recognition by FORC [14]). *A family of right congruences (in short* FORC*) is a pair* $\mathcal{R} = (\sim, \{\approx^u\}_{u\in[\sim]})$ *such that*

1. $\sim$ *is a right congruence,*
2. $\approx^u$ *is a right congruence for every* $u \in [\sim]$*, and*
3. $x \approx^u y$ *implies* $ux \sim uy$ *for every* $u, x, y \in \Sigma^*$*.*

An ω-language L *is recognized by a* FORC $\mathcal{R} = (\sim, \approx^u)$ *if it can be written as a union of sets of the form* $[u]([v]_u)^\omega$ *such that* $uv \sim_L u$.[1]

Definition 2 (Syntactic FORC [14]). *Let* $x, y, u \in \Sigma^*$*, and* L *be a regular* ω-language. *We use* $x \approx^u_s y$ *iff* $ux \sim_L uy$ *and* $\forall v \in \Sigma^*$ *if* $uxv \sim_L u$ *then* $u(xv)^\omega \in L \iff u(yv)^\omega \in L$*. The syntactic* FORC *of* L *is* $(\sim_L, \{\approx^u_s\}_{u\in[\sim_L]})$*.*

Theorem 1 (Minimality of the Syntactic FORC [14]). *An* ω-language *is regular iff it is recognized by a finite* FORC*. Moreover, for every regular* ω-language, *its syntactic* FORC *is the coarsest* FORC *recognizing it.*

Moving to Families of DFAs. We have seen in the preliminaries how a right congruence defines an automaton, and that the latter can be augmented with an acceptance criterion to get an acceptor for regular languages. In a similar way, we would like to define a family of automata, and augment it with an acceptance criterion to get an acceptor for regular ω-languages.

Definition 3 (Family of DFAs (FDFA)). *A family of* DFAs $\mathcal{F} = (M, \{A_q\})$ *over an alphabet* Σ *consists of a leading automaton* $M = (\Sigma, Q, q^0, \delta)$ *and progress* DFAs $A_q = (\Sigma, S_q, s^0_q, \delta_q, F_q)$ *for each* $q \in Q$*.*

Note that the definition of FDFA, does not impose the third requirement in the definition of FORC. If needed this condition can be imposed by the progress DFAs themselves.[2]

---

[1] The original definition of recognition by a FORC requires also that the language L be saturated by $\mathcal{R}$. An ω-language L is saturated by $\mathcal{R}$ if for every u, v s.t. $uv \sim u$ it holds that $[u]([v]_u)^\omega \subseteq L$. It is shown in [14] that for finite FORCs, covering and saturation coincide. Thus, the definition here only requires that L is covered by $\mathcal{R}$.

[2] In [10] Klarlund also suggested the notion of family of DFAs. However, that work did require the third condition in the definition of FORC to hold.

Definition 4 (Syntactic FDFA). *Let L be a regular ω-language, and let M be the automaton corresponding to $\sim_L$. For every equivalence class $[u]$ of $\sim_L$ let A_S^u be the DFAs corresponding to $\approx_S^u$, where the accepting states are the equivalence classes $[v]$ of $\approx_S^u$ for which $uv \sim_L u$ and $uv^\omega \in L$. We use $\mathcal{F}_S$ to denote the FDFA $(M, \{A_S^u\})$, and refer to it as the* syntactic *FDFA.*[3]

The following is a direct consequence of Theorem 1 and Definitions 1 and 4.

Proposition 1. *Let L be an ω-language and $\mathcal{F}_S = (M, \{A_S^u\})$ the syntactic FDFA. Let $w \in \Sigma^\omega$ be an ultimately periodic word. Then $w \in L$ iff there exists u and v such that $w = uv^\omega$, $uv \sim_L u$ and $v \in [\![A_S^{\tilde{u}}]\!]$ where $\tilde{u} = M(u)$.*

To get an understanding of the subtleties in the definition of $\approx_S^u$ we consider the following simpler definition of a right congruence for the progress automata, and the corresponding FDFA. It is basically the FDFA version of $L_\$$.

Definition 5 (Periodic FDFA). *Let $x, y, u \in \Sigma^*$ and L be an ω-language. We use $x \approx_P^u y$ iff $\forall v \in \Sigma^*$ we have $u(xv)^\omega \in L \iff u(yv)^\omega \in L$. Let M be the automaton corresponding to $\sim_L$. For every equivalence class $[u]$ of $\sim_L$ let A_P^u be the DFA corresponding to $\approx_P^u$ where the accepting states are the equivalence classes $[v]$ of $\approx_S^u$ for which $uv^\omega \in L$. We use $\mathcal{F}_P$ to denote the FDFA $(M, \{A_P^u\})$, and refer to it as the* periodic *FDFA.*[4]

It is not hard to see that the following proposition holds.

Proposition 2. *Let L be a regular ω-language and $\mathcal{F}_P = (M, \{A_S^u\})$ the periodic FDFA. Let $w \in \Sigma^\omega$ be an ultimately periodic word. Then $w \in L$ iff there exists u and v such that $w = uv^\omega$, $uv \sim_L u$ and $v \in [\![A_P^{\tilde{u}}]\!]$ where $\tilde{u} = M(u)$.*

Maler and Staiger show that the syntactic FORC is the coarsest FORC. They do not compare its size with that of other representations. Below we show that it can be factorially more succinct than the periodic FORC, and the same arguments can be used to show that the syntactic FORC is factorially more succinct than the DFA for $L_\$$. Intuitively, the reason is that $\mathcal{F}_P$ pertinaciously insists on finding every period of u, while $\mathcal{F}_S$ may not accept a valid period, if it accepts some repetition of it. For instance, take $L_2 = (aba + bab)^\omega$. Then A_P^λ accepts ab as this is a valid period, yet A_S^λ rejects it, since $\lambda \not\sim_L ab$ but it does accept its repetition $ababab$. This flexibility is common to all acceptance conditions used in the theory of ω-automata (Büchi, Muller, etc.) but is missing from $L_\$$ and $\mathcal{F}_P$. And as the following example shows, it can make a very big difference.

Theorem 2. *There exists a family of languages L_n whose syntactic FDFA has $O(n)$ states but the periodic FDFA has at least $n!$ states.*

Proof. Consider the languages L_n over the alphabet $\Sigma_n = [0..n]$ described by the DBA $\mathcal{B}$ in Fig. 1 on the left.[5] The leading automaton $\mathcal{L}$ looks like $\mathcal{B}$ but has

---

[3] The syntactic FDFA is well defined since, as shown in [14], $uv^\omega \in L$ implies $[u][v]^\omega \subseteq L$.

[4] It is easy to see that $uv^\omega \in L$ implies $[u][v]^\omega \subseteq L$. Thus the periodic FDFA is well defined.

[5] The automata for the languages L_n are the deterministic version of the automata for a family M_n introduced by Michel [16] to prove there exists an NBA with $O(n)$ states whose complement cannot be recognized by an NBA with fewer than $n!$ states.

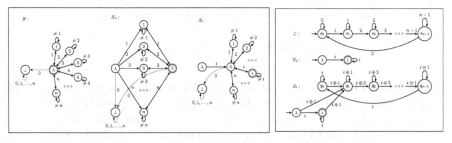

Fig. 1. On the left, Büchi automaton $\mathcal{B}$ for L_n, and the syntactic FDFA for L_n. On the right, leading automaton $\mathcal{L}$ for T_n, the syntactic and recurrent progress automaton for i, $\mathcal{S}_i$ and $\mathcal{R}_i$.

no accepting states. The syntactic progress DFA for the empty word is described by $\mathcal{S}_\lambda$ (in the middle), the syntactic progress DFA for any $i \in [1..n]$ is given by $\mathcal{S}_i$ (on the right), and the syntactic progress DFA for $\bot$ is the trivial DFA accepting the empty language.

We now show that the progress automaton for λ in the periodic FDFA requires at least $(n + 1)!$ states. The idea of the proof is as follows. Given a word v we use f_v to denote the function from $[0..n]$ to $[0..n]$ that satisfies $f(i) = \delta(i, v)$. We show that with each permutation $\pi = (i_0 \, i_1 \ldots \, i_n)$ of $[0..n]$ we can associate a word v_π (of length at most $3n$) such that $f_{v_\pi} = \pi$ (i.e. $f_{v_\pi}(k) = i_k$ for every $k \in [0..n]$). Let V be the set of all such words. We then show that for each $v_1, v_2 \in V$ we can find a word y such that $v_1 y^\omega \in L_n$ iff $v_2 y^\omega \notin L$. Since V is of size $(n + 1)!$ any DFA with fewer than $(n + 1)!$ states is bound to make a mistake.

We now show that we can associate with each π the promised word v_π. With $\pi_0 = (0 \; 1 \; \ldots n)$ we associate the word λ. It is known that one can get from any permutation π to any other permutation π' by a sequence of at most n transpositions (transformations switching exactly two elements). It thus suffices to provide for any permutations $\pi = (i_0 \, i_1 \ldots \, i_n)$ and $\pi' = (i'_0 \, i'_1 \ldots \, i'_n)$ differing in only two elements, a word u such that $f_{v_\pi u} = \pi'$. Suppose π and π' differ in indices j and k. If both $i_j \neq 0$ and $i_k \neq 0$ then the word

Table 1. Distinguishing word y

Case	Condition	y
1	$i_0 = 0, j_0 \neq 0$	$j_0 0 j_0$
2	$i_0 \neq 0, j_0 = 0$	$i_0 0 i_0$
3	$i_0 \neq 0, j_0 \neq 0, i_0 \neq j_0$	$i_0 0 i_0 j_0 j_0$
4	$i_0 = j_0 = 0, i_k \neq j_k, i_k = k$	$k j_k 0 j_k$
5	$i_0 = j_0 = 0, i_k \neq j_k, j_k = k$	$k i_k 0 i_k$
6	$i_0 = j_0 = 0, i_k \neq j_k, i_k \neq k, j_k \neq k$	$k j_k 0 j_k i_k$
7	$i_0 = j_0 \neq 0, i_k = k$	$i_0 k j_k 0 j_k$
8	$i_0 = j_0 \neq 0, j_k = k$	$i_0 k i_k 0 i_k$
9	$i_0 = j_0 \neq 0, i_k \neq k, j_k \neq k$	$i_0 k j_k 0 j_k i_j$

$i_j i_k i_j$ will take state i_j to i_k and state i_k to i_j and leave all other states unchanged. If $i_j = 0$ the word i_k does the job, and symmetrically if $i_k = 0$ we choose i_j. We have thus shown that with each permutation π we can associate a word v_π such that $f_{v_\pi} = \pi$.

We now show that for each two such words v_1, v_2 we can find a differentiating word y. Let $f_{v_1} = (i_0 \, i_1 \ldots i_n)$ and $f_{v_2} = (j_0 \, j_1 \, \ldots \, j_n)$. Table 1 explains how we choose y. In the first three cases we get $f_{v_1 y}(0) = 0$ and $f_{v_2 y}(0) = \bot$ or vice versa.

In the rest of the cases we get $f_{v_1 y}(0) = k$, $f_{v_1 y}(k) = 0$ and $f_{v_2 y}(0) = k$, $f_{v_2 y}(k) = \bot$ or vice versa. Thus $f_{(v_1 y)^2} = f_{v_1 y}^2(0) = 0$ and $f_{(v_2 y)^n} = f_{v_2 y}^n(0) = \bot$ for any $n \geq 2$, or vice versa. Thus $(v_1 y)^\omega \in L$ iff $(v_2 y)^\omega \notin L$. $\square$

Families of FDFAs as Acceptors. Families of automata are not an operational acceptor. The answer to whether a given ultimately periodic word $w \in \Sigma^\omega$ is accepted by the FDFA relies on the existence of a decomposition of w into uv^ω, but it is not clear how to find such a decomposition. We would like to use families of automata as acceptors for pairs of words, such that (u, v) being accepted implies uv^ω is. We can try defining acceptance as follows.

Definition 6 (FDFA Exact Acceptance). *Let $\mathcal{F} = (M, \{A_u\})$ be a FDFA and u, v finite words. We say that $(u, v) \in [\![\mathcal{F}]\!]_{\mathrm{E}}$ if $v \in [\![A_{\tilde{u}}]\!]$ where $\tilde{u} = M(u)$.*

Since our goal is to use families of automata as acceptors for regular ω-languages, and an ultimately periodic ω-word w may be represented by different pairs (u, v) and (x, y) such that $w = uv^\omega = xy^\omega$ (where $u \neq x$ and/or $v \neq y$) it makes sense to require the representation to be *saturated*, in the following sense.

Definition 7 (Saturation). *A language L of pairs of finite words is said to be saturated if for every u, v, x, y such that $uv^\omega = xy^\omega$ we have $(u, v) \in L \iff (x, y) \in L$.*

Calbrix et al. [4] have showed that (1) $L_\$$ is saturated, and (2) a regular language of pairs K is saturated iff it is $L_\$$ for some regular ω-language L. It is thus not surprising that the periodic family is saturated as well.

Proposition 3. *Let L be an ω-language and $\mathcal{F}_\mathrm{P}$ and $\mathcal{F}_\mathrm{S}$ the corresponding periodic and syntactic FDFAs. Then $[\![\mathcal{F}_\mathrm{P}]\!]_{\mathrm{E}}$ is saturated.*

The language $[\![\mathcal{F}_\mathrm{S}]\!]_{\mathrm{E}}$ on the other hand, is not necessarily saturated. Consider $L_3 = a^\omega + ab^\omega$. Let $x = aa$, $y = a$, $u = a$, $v = a$. It can be seen that although $xy^\omega = uv^\omega$ we have $(aa, a) \in [\![\mathcal{F}_\mathrm{S}]\!]_{\mathrm{E}}$ yet $(a, a) \notin [\![\mathcal{F}_\mathrm{S}]\!]_{\mathrm{E}}$. The reason is that, in order to be smaller, the syntactic family does not insist on finding every possible legitimate period v of u (e.g. period a of a in this example). Instead, it suffices in finding a repetition of it v^k, starting from some u so that reading uv on the leading automaton takes us back to the state we got to after reading u.

Given a right congruence $\sim$ of finite index and a periodic word w we say that (x, y) is a *factorization* of w with respect to $\sim$ if $w = xy^\omega$ and $xy \sim x$. If w is given by a pair (u, v) so that $w = uv^\omega$ we can define the *normalized factorization* of (u, v) as the pair (x, y) such that (x, y) is a factorization of uv^ω, $x = uv^i$, $y = v^j$ and $0 \leq i < j$ are the smallest for which $uv^i \sim_L uv^{i+j}$. Since $\sim$ is of finite index, there must exist such i and j such that $i + j < |\!\sim\!| + 1$. If we base our acceptance criteria on the normalized factorization, we achieve that $[\![\mathcal{F}_\mathrm{S}]\!]_{\mathrm{N}}$ is saturated as well.

Definition 8 (FDFA Normalized Acceptance). *Let $\mathcal{F} = (M, \{A_u\})$ be an* FDFA, *and u, v finite words. We say that $(u, v) \in [\![\mathcal{F}]\!]_{\mathbb{N}}$ if $y \in A_{M(x)}$ where (x, y) is the normalized factorization of (x, y) with respect to $\sim_M$.*

Proposition 4. *Let L be an ω-language and $\mathcal{F}_P$ and $\mathcal{F}_S$ the corresponding periodic and syntactic families. Then $[\![\mathcal{F}_P]\!]_{\mathbb{N}}$ and $[\![\mathcal{F}_S]\!]_{\mathbb{N}}$ are saturated.*

Proof. We show that given an ultimately periodic word w, for any x, y such that $w = xy^\omega$, $(x, y) \in [\![\mathcal{F}_S]\!]_{\mathbb{N}}$ iff $w \in L$. This shows that $[\![\mathcal{F}_S]\!]_{\mathbb{N}}$ is a saturated acceptor of L. Assume towards contradiction that $w = uv^\omega$, $w \in L$ yet $(u, v) \notin [\![\mathcal{F}_S]\!]_{\mathbb{N}}$. Let (x, y) be the normalized factorization of (u, v) with respect to $\sim_L$. We have $xy \sim_L x$, $x = uv^i$ and $y = v^j$ for some i, j. Let $\tilde{x}$ be $M(xy^j)$. Let $\tilde{y} = A_{\tilde{x}}(y)$. Thus $\tilde{y}$ is not an accepting state. Meaning $\tilde{x}\tilde{y}^\omega \notin L$.

On the other hand we have that $uv^\omega \in L$ and $uv^\omega = uv^i v^\omega = xv^\omega$. Since $\tilde{x} \sim_L x$ we get that $\tilde{x}y^\omega \in L$ and since $y^\omega = (y^j)^\omega$ that $\tilde{x}(y^j)^\omega \in L$. Since $y \approx_S^{\tilde{x}} \tilde{y}$ and $\tilde{x}y \sim_L \tilde{x}$ it follows that $\tilde{x}\tilde{y}^\omega \notin L$. Contradiction.

Similar arguments show that $[\![\mathcal{F}_P]\!]_{\mathbb{N}}$ as well is saturated. $\square$

New Canonical Representation - The Recurrent FDFA. We note that there is some redundancy in the definition of the syntactic FDFA: the condition that $ux \sim_L uy$ can be checked on the leading automaton rather than refine the definitions of the $\approx_S^u$'s. We thus propose the following definition of right congruence, and corresponding FDFA.

Definition 9 (Recurrent FDFA). *Let $x, y, u \in \Sigma^*$ and L be an ω-language. We use $x \approx_R^u y$ iff $\forall v \in \Sigma^*$ if $uxv \sim_L u$ and $u(xv)^\omega \in L$ then $uyv \sim_L u$ and $u(yv)^\omega \in L$. We use $\mathcal{F}_R$ to denote the FDFA $(M, \{A_R^u\})$ where the accepting states of A_R^u are those v for which $uv \sim_L u$ and $uv^\omega \in L$. We refer to $\mathcal{F}_R$ as the recurrent FDFA.*

Note that the proof of Proposition 4 did not make use of the additional requirement $ux \sim_L uy$ of $\approx_S^u$. The same arguments thus show that the recurrent FDFA is a saturated acceptor of L.

Proposition 5. *Let L be a regular ω-language and $\mathcal{F}_R = (M, \{A_R^u\})$ be its recurrent FDFA. Then $[\![\mathcal{F}_R]\!]_{\mathbb{N}}$ is saturated and is an acceptor of L.*

It follows from the definitions of $\approx_S^u$ and $\approx_R^u$ and $\approx_P^u$ that (a) $\approx_P^u$ refines $\approx_R^u$ (b) $\approx_S^u$ refines $\approx_R^u$ and (c) if $|\sim_L| = n$ and $|\approx_R^u| = m$ then $\approx_S^u$ is of size at most nm. Thus there is at most a quadratic size reduction in the recurrent FDFA, with respect to the syntactic FDFA. We show a matching lower bound.

Proposition 6. *There exists a family of languages T_n such that the size of the syntactic FDFA for T_n is $\Theta(n^2)$ and the size of the recurrent FDFA is $\Theta(n)$.*

Proof. Consider the alphabet $\Sigma_n = \{a_0, a_1, \ldots, a_{n-1}\}$. Let L_i abbreviate a_i^+. Let U_i be the set of ultimately periodic words $(L_0 L_1 \ldots L_{n-1})^* (L_0 L_1 \ldots L_i) a_i^\omega$. Finally let T_n be the union $U_0 \cup U_1 \cup \ldots U_{n-1}$. In Figure 1 on the right we show its

---

Algorithm 1. The Learner L^ω

---

1 Initialize the leading table $\mathcal{T} = (S, \tilde{S}, E, T)$ with $S = \tilde{S} = \{\lambda\}$ and $E = \{(\lambda, \lambda)\}$.
2 $CloseTable(\mathcal{T}, \text{ENT}_1, \text{DFR}_1)$ and let $M = Aut_1(\mathcal{T})$.
3 **forall** $u \in \tilde{S}$ **do**
4 $\quad$ Initialize the table for u, $\mathcal{T}_u = (S_u, \tilde{S}_u, E_u, T_u)$, with $S_u = \tilde{S}_u = E_u = \{\lambda\}$.
5 $\quad$ $CloseTable(\mathcal{T}_u, \text{ENT}_2^u, \text{DFR}_2^u)$ and let $A_u = Aut_2(\mathcal{T}_u)$.
6 Let (a, u, v) be the teacher's response on the equivalence query $\mathcal{H} = (M, \{A_u\})$.
7 **while** $a = $ "no" **do**
8 $\quad$ Let (x, y) be the normalized factorization of (u, v) with respect to M.
9 $\quad$ Let $\tilde{x}$ be $M(x)$.
10 $\quad$ **if** $\text{MQ}(x, y) \neq \text{MQ}(\tilde{x}, y)$ **then**
11 $\quad\quad$ $E = E \cup FindDistinguishingExperiment(x, y)$.
12 $\quad\quad$ $CloseTable(\mathcal{T}, \text{ENT}_1, \text{DFR}_1)$ and let $M = Aut_1(\mathcal{T})$.
13 $\quad\quad$ **forall** $u \in \tilde{S}$ **do**
14 $\quad\quad\quad$ $CloseTable(\mathcal{T}_u, \text{ENT}_2^u, \text{DFR}_2^u)$ and let $A_u = Aut_2(\mathcal{T}_u)$.
15 $\quad$ **else**
16 $\quad\quad$ $E_{\tilde{x}} = E_{\tilde{x}} \cup FindDistinguishingExperiment(\tilde{x}, y)$.
17 $\quad\quad$ $CloseTable(\mathcal{T}_{\tilde{x}}, \text{ENT}_2^{\tilde{x}}, \text{DFR}_2^{\tilde{x}})$ and let $A_x = Aut_2(\mathcal{T}_{\tilde{x}})$.
18 $\quad$ Let (a, u, v) be the teacher's response on equivalence query $\mathcal{H} = (M, \{A_u\})$.
19 **return** $\mathcal{H}$

---

leading DFA and the syntactic and recurrent progress DFAs for state i. (The sink state is not shown, and $\oplus, \ominus$ are plus and minus modulo n.) The total number of states for the recurrent FDFA is $(n + 1) + 3n + 1$ and for the syntactic FDFA it is $(n + 1) + n(n + 3) + (n + 1)$. $\square$

We observe that the recurrent family may not produce a minimal result. Working with the normalized acceptance criterion, we have that a progress DFA P_u for leading state u should satisfy $[u]([\![P_u]\!] \cap C_u) = L \cap [u]C_u$ where $C_u = \{v \mid uv \sim_L u\}$. Thus, in learning P_u we have *don't cares* for all the words that are not in C_u. Minimizing a DFA with don't cares is an NP-hard problem [18]. The recurrent FDFA chooses to treat all don't cares as rejecting.

4 Learning ω-regular Languages via Families of DFAs

In the previous section we have provided three canonical representations of regular ω-languages as families of DFAs. The L^* algorithm provides us an efficient way to learn a DFA for an unknown regular language. Have we reduced the problem to using L^* for the different DFAs of the family? Not quite. This would be true if we had oracles answering membership and equivalence for the languages of the leading and progress DFAs. But the question we consider assumes we have oracles for answering membership and equivalence queries for the unknown regular ω-language. Specifically, the membership oracle, given a pair (u, v) answers whether $uv^\omega \in L$, and the equivalence oracle answers whether a given FDFA $\mathcal{F}$, satisfies $[\![\mathcal{F}]\!]_\mathbb{N} = L$ and returns a counterexample if not. The counterexample is in the format (a, u, v) where a is one of the strings "yes" or "no", and if it is "no" then uv^ω is in $(L \setminus [\![\mathcal{F}]\!]_\mathbb{N}) \cup ([\![\mathcal{F}]\!]_\mathbb{N} \setminus L)$.

We use a common scheme for learning the three families ($\mathcal{F}_\mathrm{P}$, $\mathcal{F}_\mathrm{S}$ and $\mathcal{F}_\mathrm{R}$) under the normalized acceptance criteria, see Alg. 1. This is a simple modification of the L^* algorithm to learn an unknown DFA using membership and equivalence queries [2]. We first explain the general scheme. Then we provide the necessary details for obtaining the learning algorithm for each of the families, and prove correctness.

Auxiliary Procedures. The algorithm makes use of the notion of an *observation table*. An observation table is a tuple $\mathcal{T} = (S, \tilde{S}, E, T)$ where S is a prefix-closed set of strings, E is a set of experiments trying to differentiate the S strings, and $T : S \times E \to D$ stores in entry $T(s, e)$ an element in some domain D. Some criterion should be given to determine when two strings $s_1, s_2 \in S$ should be considered distinct (presumably by considering the contents of the respective rows of the table). The component $\tilde{S}$ is the subset of strings in S considered distinct. A table is *closed* if S is prefix closed and for every $s \in \tilde{S}$ and $a \in \Sigma$ we have $sa \in S$.

The procedure *CloseTable* thus uses two sub-procedures ENT and DFR to fulfill its task. Procedure ENT is used to fill in the entries of the table. This procedure invokes a call to the membership oracle. The procedure DFR is used to determine which rows of the table should be differentiated. Closing the leading table is done using ENT$_1$ and DFR$_1$. Closing the progress table for u is done using ENT$_2$ and DFR$_2$. (This is where the algorithms for the different families differ.)

A closed table can be transformed into an automaton by identifying the automaton states with $\tilde{S}$, the initial state with the empty string, and for every letter $a \in \Sigma$ defining the transition $\delta(s_1, a) = s_2$ iff $s_2 \in \tilde{S}$ is the representative of $s_1 a$. By designating certain states as accepting, e.g. those for which $T(s, \lambda) = d_*$ for some designated $d_* \in D$, we get a DFA. Procedures $Aut_1(\mathcal{T})$ and $Aut_2(\mathcal{T})$ are used for performing this transformation, for the leading and progress tables respectively.

The Main Scheme. The algorithm starts by initializing and closing the leading table (lines 1-2), and the respective progress tables (lines 3-5) and asking an equivalence query about the resulting hypothesis. The algorithm then repeats the following loop (lines 7-18) until the equivalence query returns "yes".

If the equivalence query returns a counter example (u, v) the learner first obtains the normalized factorization (x, y) of (u, v) with respect to its current leading automaton (line 8). It then checks whether membership queries for (x, y) and $(\tilde{x}, y)$, where $\tilde{x}$ is the state M arrives at after reading x, return different results. If so, it calls the procedure *FindDistinguishingExperiment* to find a distinguishing experiment to add to the leading table (line 11). It then closes the leading table and all the progress tables (lines 12-14) and obtains a new hypothesis $\mathcal{H}$ (line 18).

If membership queries for (x, y) and $(\tilde{x}, y)$ return the same results, it calls the procedure *FindDistinguishingExperiment* to find a distinguishing experiment in the progress automaton for $\tilde{x}$ (line 16). It then closes this table (line 17) and obtains a new hypothesis (line 18).

It is clear that if the learning algorithm halts, its output is correct. We discuss the time complexity at the end of the section.

Specializing for the Periodic, Syntactic and Recurrent Families. We now turn to provide the details for specializing L^ω to learn the different families $\mathcal{F}_P$, $\mathcal{F}_S$ and $\mathcal{F}_R$.

The different learning algorithms differ in the content they put in the progress tables (i.e. procedure ENT$_2$), in the criterion for differentiating rows in a progress table (i.e. procedure DFR$_2$), the states they choose to be accepting (i.e. procedure Aut_2) and the way they find a distinguishing experiment (i.e. procedure *FindDistinguishingExperiment*). The details of the latter are given within the respective proofs of correctness.

For learning the leading automaton, which is same in all 3 families, the following procedures: ENT$_1$, DFR$_1$ and Aut_1 are used. For $u \in \Sigma^*$ and $xy^\omega \in \Sigma^\omega$ the procedure ENT$_1(u, xy^\omega)$ returns whether uxy^ω is in the unknown language L. Given two row strings $u_1, u_2 \in S$ the procedure DFR$_1(u_1, u_2)$ returns true, if there exists $w \in E$ s.t. $T(u_1, w) \neq T(u_2, w)$. We use Aut_1 for the procedure transforming the leading table into a DFA with no accepting states.

For the periodic FDFA, given $u, x, v \in \Sigma^*$, we have ENT$_P^u(x, v) = \mathsf{T}$ iff $u(xv)^\omega \in L$, and DFR$_P(x_1, x_2)$ is simply $\exists v \in E$ s.t. $T(x_1, v) \neq T(x_2, v)$. The procedure Aut_P declares a state x as accepting if $T(x, \lambda) = \mathsf{T}$.

Theorem 3. *Calling the learner L^ω with* ENT$_1$,DFR$_1$,Aut_1 *and* ENT$_P$,DFR$_P$,Aut_P *halts and returns the periodic* FDFA.

Proof. We need to show that in each iteration of the while loop at least one state is added to one of the tables. Suppose the returned counter example is (u, v), and its normalized factorization with respect to the current leading automaton M is (x, y). The learner then checks whether membership queries for (x, y) and $(\tilde{x}, y)$ return different results where $\tilde{x} = M(x)$. Let $|x| = n$ and for $i \in [1..n]$ let $s_i = M(x[1..i])$ be the state of the leading automaton reached after reading the first i symbols of x. Then $\tilde{x} = s_n$, and we know that a sequence of membership queries with (x, y), $(s_1x[2..n], y)$, $(s_2x[3..n], y)$, and so on, up to $(s_n, y) = (\tilde{x}, y)$ has different answers for the first and last queries. Thus, a sequential search of this sequence suffices to find a consecutive pair, say $(s_{i-1}x[i..n], y)$ and $(s_ix[i+1..n], y)$, with different answers to membership queries. This shows that the experiment $(x[i+1..n], y)$ distinguishes $s_{i-1}x[i]$ from s_i in the leading table, though $\delta(s_{i-1}, x[i]) = s_i$, so that adding it, there will be at least one more state in the leading automaton.

If membership queries for (x, y) and $(\tilde{x}, y)$ return same answers, we look for an experiment that will distinguish a new state in the progress table of $\tilde{x}$. Let $\tilde{y} = M_{\tilde{x}}(y)$. Let $|y| = n$ and for $i \in [1..n]$ let $s_i = A_{\tilde{x}}(y[1..i])$ be the state reached by $A_{\tilde{x}}$ after reading the first i symbols of y. Thus $s_n = \tilde{y}$. Consider the sequence (λ, y), $(s_1, y[2..n])$, $(s_2, y[3..n])$, up to (s_n, λ). Then we know the entry for first and last return different results, though they should not. We can thus find, in an analogous manner to the first case, a suffix y' of y that is a differentiating experiment for the progress table for $\tilde{x}$. □

For the syntactic FDFA, given $u, x, v \in \Sigma^*$, the procedure $\text{ENT}_S^u(x, v)$ returns a pair $(m, c) \in \{\mathsf{T}, \mathsf{F}\} \times \{\mathsf{T}, \mathsf{F}\}$ such that $m = \mathsf{T}$ iff $u(xv)^\omega \in L$ and $c = \mathsf{T}$ iff $uxv \sim_L u$. Given two rows x_1 and x_2 in the progress table corresponding to leading state u, the procedure $\text{DFR}_S^u(x_1, x_2)$ returns true if either $M(x_1) \neq M(x_2)$, or there exists an experiment $v \in E_u$ for which $T(u_1, v) = (m_1, c_1)$, $T(u_2, v) = (m_2, c_2)$ and $(c_1 \vee c_2) \wedge (m_1 \neq m_2)$. The procedure Aut_S declares a state x as accepting if $T(x, \lambda) = (\mathsf{T}, \mathsf{T})$.

Theorem 4. *Calling the learner L^ω with $\text{ENT}_1, \text{DFR}_1, Aut_1$ and $\text{ENT}_S^u, \text{DFR}_S^u, Aut_S$ halts and returns the syntactic FDFA.*

Proof. Again we need to show that each iteration of the while loop creates a state. The proof for the first part is same as in Thm. 3. For the second part, let (x, y) be the normalized factorization of the given counter example w.r.t to current leading automaton. We can consider the sequence of experiments (λ, y), $(s_1, y[2..n])$, $(s_2, y[3..n])$, up to (s_n, λ) as we did in the periodic case. Now, however, the fact that two experiments (x_1, v), (x_2, v) differ in the answer to the membership query does not guarantee they would get distinguished, as this fact might be hidden if for both $M(ux_iv) \neq M(u)$. Let (m_0, c_0), $(m_1, c_1), \ldots, (m_n, c_n)$ be the respective results for the entry query. We know that $m_0 \neq m_n$. Also, we know that $c_0 = \mathsf{T}$ since we chose x and y so that $M(xy) = M(x)$. Let i be the smallest for which $m_{i-1} \neq m_i$. If all the c_j's for $j \leq i$ are T, we can find a distinguishing experiment as in the case of periodic FDFA. Otherwise let k be the smallest for which $c_k = \mathsf{F}$. Then $M(xs_{k-1}y[k..n]) = M(x)$ but $M(xs_ky[k+1..n]) \neq M(x)$. Therefore $y[k+1..n]$ distinguishes $s_{k-1}y[k]$ from s_k and so we add it to the experiments $E_{\tilde{x}}$ of the progress table for $\tilde{x}$. □

For the recurrent FDFA, given $u, x, v \in \Sigma^*$ the query $\text{ENT}_R^u(x, v)$ is same as $\text{ENT}_S^u(x, v)$. The criterion for differentiating rows, is more relaxed though. Given two rows x_1 and x_2 in the progress table corresponding to leading state u, the procedure $\text{DFR}_R^u(x_1, x_2)$ returns true if there exists an experiment v for which $T(x_1, v) = (\mathsf{T}, \mathsf{T})$ and $T(x_2, v) \neq (\mathsf{T}, \mathsf{T})$ or vice versa. The procedure Aut_R also declares a state x as accepting if $T(x, \lambda) = (\mathsf{T}, \mathsf{T})$.

Theorem 5. *Calling the learner L^ω with $\text{ENT}_1, \text{DFR}_1, Aut_1$ and $\text{ENT}_R^u, \text{DFR}_R^u, Aut_R$ halts and returns the recurrent FDFA.*

Proof. The first part is same as in the proof of Theorem 3. For the second part, let (x, y) be the normalized factorization of (u, v) with respect to M. Let $\tilde{x}$ be

$M(x)$ and let $\tilde{y}$ be $A_{\tilde{x}}(\tilde{y})$. As in the proof of Theorem 4, consider the sequence of experiments (λ, y), $(s_1, y[2..n])$, $(s_2, y[3..n])$ up to (s_n, λ) and the respective entries (m_0, c_0), $(m_1, c_1), \ldots, (m_n, c_n)$ in the table $T_{\tilde{x}}$. We know that out of (m_0, c_0) and (m_n, c_n) one is (T, T) and the other one is not. Therefore for some i we should have that (m_i, c_i) is (T, T) and (m_{i-1}, c_{i-1}) is not, or vice versa. Thus, the experiment $y[i + 1..n]$ distinguishes $s_{i-1}y[i]$ from s_i. □

Starting with a Given Leading Automaton. In [10] Klarlund has shown that while the syntactic FORC is the coarsest FORC recognizing a certain language, it is not necessarily the minimal one. That is, taking a finer (bigger) leading congruence may yield smaller progress congruences. In particular, he showed a family of languages K_n where $|\sim_{K_n}| = 1$, and its syntactic progress DFA is of size $O(n^n)$, but it has an FDFA with n states in the leading automaton and n states in each of the progress DFAs — thus the total size is $O(n^2)$. The family K_n over the alphabet $\Sigma_n = \{a_1, a_2, \ldots, a_n\} \cup \{\bar{b} \mid \bar{b} \in \{0, 1\}^n\}$ accepts all words where at some point a_i appears infinitely often, all other a_j's stop appearing, and the number of 1's in the i-th track between two occurrences of a_i is exactly n. It can be seen that the recurrent FDFA will also have $O(n^n)$ states. The family that has a total of $O(n)$ states has a leading automaton K with n states, remembering which letter among the a_i's was the last to occur.

We can change L^ω so that it starts with a given leading automaton, and proceeds exactly as before. The resulting algorithm may end up refining the leading automaton if necessary. If we apply it to learn K_n by giving it K as the leading automaton, the learnt syntactic/recurrent families would have $O(n^2)$ states as well.

Time Complexity. In each iteration of the while loop, i.e. in processing each counter example, at least one new state is added either to the leading automaton or to one of the progress automata. If the leading automaton is learned first, we are guaranteed that we have not distinguished more states than necessary, and so, since each operation of the while loop is polynomial in the size of the learned family, the entire procedure will run in time polynomial in the size of the learned family. However, it can be that we will unnecessarily add states to a progress automaton, since the leading automaton has not been fully learned yet, in which case the progress automaton may try to learn the exact periods as does the periodic family. At a certain point the leading automaton will be exact and the size of that progress automaton will shrink as necessary. But the worse case time complexity for all three families is thus polynomial in the size of the periodic family, rather than the size of the learned family.

Acknowledgment. We would like to thank Oded Maler and Ludwig Staiger for referring us to [4], and Daniel Neider for referring us to [7].

References

[1] Alur, R., Cerný, P., Madhusudan, P., Nam, W.: Synthesis of interface specifications for Java classes. In: POPL, pp. 98–109 (2005)

[2] Angluin, D.: Learning regular sets from queries and counterexamples. Inf. Comput. 75(2), 87–106 (1987)

[3] Arnold, A.: A syntactic congruence for rational omega-languages. Theor. Comput. Sci. 39, 333–335 (1985)

[4] Calbrix, H., Nivat, M., Podelski, A.: Ultimately periodic words of rational w-languages. In: Main, M.G., Melton, A.C., Mislove, M.W., Schmidt, D., Brookes, S.D. (eds.) MFPS 1993. LNCS, vol. 802, pp. 554–566. Springer, Heidelberg (1994)

[5] Cobleigh, J.M., Giannakopoulou, D., Păsăreanu, C.S.: Learning assumptions for compositional verification. In: Garavel, H., Hatcliff, J. (eds.) TACAS 2003. LNCS, vol. 2619, pp. 331–346. Springer, Heidelberg (2003)

[6] de la Higuera, C., Janodet, J.-.C.: Inference of [omega]-languages from prefixes. Theor. Comput. Sci. 313(2), 295–312 (2004)

[7] Farzan, A., Chen, Y.-F., Clarke, E.M., Tsay, Y.-K., Wang, B.-Y.: Extending automated compositional verification to the full class of omega-regular languages. In: Ramakrishnan, C.R., Rehof, J. (eds.) TACAS 2008. LNCS, vol. 4963, pp. 2–17. Springer, Heidelberg (2008)

[8] Jayasrirani, M., Humrosia Begam, M., Thomas, D.G., Emerald, J.D.: Learning of bi-languages from factors. Machine Learning Research Jour. 21, 139–144 (2012)

[9] Jürgensen, H., Thierrin, G.: On-languages whose syntactic monoid is trivial. International Journal of Parallel Programming 12(5), 359–365 (1983)

[10] Klarlund, N.: A homomorphism concept for omega-regularity. In: 8th Inter. Conf. on Computer Science Logic (CSL), pp. 471–485 (1994)

[11] Leucker, M.: Learning meets verification. In: de Boer, F.S., Bonsangue, M.M., Graf, S., de Roever, W.-P. (eds.) FMCO 2006. LNCS, vol. 4709, pp. 127–151. Springer, Heidelberg (2007)

[12] Lindner, R., Staiger, L.: Eine Bemerkung über nichtkonstantenfreie sequentielle operatoren. Elektronische Informationsverarbeitung und Kybernetik 10(4), 195–202 (1974)

[13] Maler, O., Pnueli, A.: On the learnability of infinitary regular sets. Inf. Comput. 118(2), 316–326 (1995)

[14] Maler, O., Staiger, L.: On syntactic congruences for omega-languages. Theor. Comput. Sci. 183(1), 93–112 (1997)

[15] McNaughton, R.: Testing and generating infinite sequences by a finite automaton. Information and Control 9, 521–530 (1966)

[16] Michel, M.: Complementation is much more difficult with automata on infinite words. In: Manuscript, CNET (1988)

[17] Nam, W., Madhusudan, P., Alur, R.: Automatic symbolic compositional verification by learning assumptions. FMSD 32(3), 207–234 (2008)

[18] Pfleeger, C.F.: State reduction in incompletely specified finite-state machines. IEEE Transactions on Computers 22(12), 1099–1102 (1973)

[19] Saoudi, A., Yokomori, T.: Learning local and recognizable omega-languages and monadic logic programs. In: EUROCOLT. LNCS, vol. 1121, pp. 50–59. Springer (1993)

[20] Staiger, L.: Finite-state omega-languages. J. Comput. Syst. Sci. 27(3), 434–448 (1983)

[21] Trakhtenbrot, B.: Finite automata and monadic second order logic. Siberian Math. J., 103–131 (1962)

Selecting Near-Optimal Approximate State Representations in Reinforcement Learning

Ronald Ortner[1], Odalric-Ambrym Maillard[2], and Daniil Ryabko[3]

[1] Montanuniversitaet Leoben, Austria
[2] The Technion, Israel
[3] Inria Lille-Nord Europe, équipe SequeL, France, and Inria Chile
rortner@unileoben.ac.at, odalric-ambrym.maillard@ens-cachan.org,
daniil@ryabko.net

Abstract. We consider a reinforcement learning setting introduced in [5] where the learner does not have explicit access to the states of the underlying Markov decision process (MDP). Instead, she has access to several models that map histories of past interactions to states. Here we improve over known regret bounds in this setting, and more importantly generalize to the case where the models given to the learner do not contain a true model resulting in an MDP representation but only approximations of it. We also give improved error bounds for state aggregation.

1 Introduction

Inspired by [3], in [5] a reinforcement learning setting has been introduced where the learner does not have explicit information about the state space of the underlying Markov decision process (MDP). Instead, the learner has a set of *models* at her disposal that map histories (i.e., observations, chosen actions and collected rewards) to states. However, only some models give a correct MDP representation. The first regret bounds in this setting were derived in [5]. They recently have been improved in [6] and extended to infinite model sets in [7]. Here we extend and improve the results of [6] as follows. First, we do not assume anymore that the model set given to the learner contains a true model resulting in an MDP representation. Instead, models will only approximate an MDP. Second, we improve the bounds of [6] with respect to the dependence on the state space.

For discussion of potential applications and related work on learning state representations in POMDPs (like predictive state representations), we refer to [5–7]. Here we only would like to mention the recent work [2] that considers a similar setting, however is mainly interested in the question whether the true model will be identified in the long run, a question we think is subordinate to that of minimizing the regret, which means fast learning of optimal behavior.

1.1 Setting

For each time step $t = 1, 2, \ldots$, let $\mathcal{H}_t := \mathcal{O} \times (\mathcal{A} \times \mathcal{R} \times \mathcal{O})^{t-1}$ be the set of histories up to time t, where $\mathcal{O}$ is the set of observations, $\mathcal{A}$ a finite set of

P. Auer et al. (Eds.): ALT 2014, LNAI 8776, pp. 140–154, 2014.
© Springer International Publishing Switzerland 2014

actions, and $\mathcal{R} = [0,1]$ the set of possible rewards. We consider the following reinforcement learning problem: The learner receives some initial observation $h_1 = o_1 \in \mathcal{H}_1 = \mathcal{O}$. Then at any time step $t > 0$, the learner chooses an action $a_t \in \mathcal{A}$ based on the current history $h_t \in \mathcal{H}_t$, and receives an immediate reward r_t and the next observation o_{t+1} from the unknown environment. Thus, h_{t+1} is the concatenation of h_t with (a_t, r_t, o_{t+1}).

State Representation Models. A *state-representation model* ϕ is a function from the set of histories $\mathcal{H} = \bigcup_{t \geq 1} \mathcal{H}_t$ to a finite set of states $\mathcal{S}$. A particular role will be played by state-representation models that induce a *Markov decision process (MDP)*. An MDP is defined as a decision process in which at any discrete time t, given action a_t, the probability of immediate reward r_t and next observation o_{t+1}, given the past history h_t, only depends on the current observation o_t i.e., $P(o_{t+1}, r_t | h_t a_t) = P(o_{t+1}, r_t | o_t, a_t)$, and this probability is also independent of t. Observations in this process are called *states* of the environment. We say that a state-representation model ϕ is a *Markov model* of the environment, if the process $(\phi(h_t), a_t, r_t), t \in \mathbb{N}$ is an MDP. Note that such an MDP representation needs not be unique. In particular, we assume that we obtain a Markov model when mapping each possible history to a unique state. Since these states are not visited more than once, this model is not very useful from the practical point of view, however. In general, an MDP is denoted as $M(\phi) = (\mathcal{S}_\phi, \mathcal{A}, r, p)$, where $r(s,a)$ is the mean reward and $p(s'|s,a)$ the probability of a transition to state $s' \in \mathcal{S}_\phi$ when choosing action $a \in \mathcal{A}$ in state $s \in \mathcal{S}_\phi$.

We assume that there is an *underlying* true Markov model ϕ° that gives a finite and *weakly communicating* MDP, that is, for each pair of states $s, s' \in \mathcal{S}^\circ := \mathcal{S}_{\phi^\circ}$ there is a $k \in \mathbb{N}$ and a sequence of actions $a_1, \ldots, a_k \in \mathcal{A}$ such that the probability of reaching state s' when starting in state s and taking actions $a_1, \ldots, a_k$ is positive. In such a weakly communicating MDP we can define the *diameter* $D := D(\phi^\circ) := D(M(\phi^\circ))$ to be the expected minimum time it takes to reach any state starting from any other state in the MDP $M(\phi^\circ)$, cf. [4]. In finite state MDPs, the *Poisson equation* relates the average reward ρ_π of any policy π to the single step mean rewards and the transition probabilities. That is, for each *policy* π that maps states to actions, it holds that

$$\rho_\pi + \lambda_\pi(s) = r(s, \pi(s)) + \sum_{s' \in \mathcal{S}^\circ} p(s'|s, \pi(s)) \cdot \lambda_\pi(s'), \tag{1}$$

where λ_π is the so-called *bias* vector of π, which intuitively quantifies the difference in accumulated rewards when starting in different states. Accordingly, we are sometimes interested in the *span* of the bias vector λ of an optimal policy defined as $\text{span}(\lambda) := \max_{s \in \mathcal{S}^\circ} \lambda(s) - \min_{s' \in \mathcal{S}^\circ} \lambda(s')$. In the following we assume that rewards are bounded in $[0,1]$, which implies that $\text{span}(\lambda)$ is upper bounded by D, cf. [1, 4].

Problem Setting. Given a finite set of models Φ (not necessarily containing a Markov model), we want to construct a strategy that performs as well as the algorithm that knows the underlying true Markov model ϕ°, including its rewards and transition probabilities. For that purpose we define for the Markov model $\phi°$ the *regret* of any strategy at time T, cf. [1, 3, 4], as

$$\Delta(\phi°, T) := T\rho^*(\phi°) - \sum_{t=1}^{T} r_t,$$

where r_t are the rewards received when following the proposed strategy and $\rho^*(\phi°)$ is the average optimal reward in $\phi°$, i.e. $\rho^*(\phi°) := \rho^*(M(\phi°)) := \rho(M(\phi°), \pi_{\phi°}^*) := \lim_{T\to\infty} \frac{1}{T}\mathbb{E}\left[\sum_{t=1}^{T} r_t(\pi_{\phi°}^*)\right]$ where $r_t(\pi_{\phi°}^*)$ are the rewards received when following the optimal policy $\pi_{\phi°}^*$ on $M(\phi°)$. Note that for weakly communicating MDPs the average optimal reward does not depend on the initial state.

We consider the case when Φ is finite and the learner has no knowledge of the correct approximation errors of each model in Φ. Thus, while for each model $\phi \in \Phi$ there is an associated $\epsilon = \epsilon(\phi) \geq 0$ which indicates the aggregation error (cf. Definition 1 below), this ϵ is unknown to the learner for each model.

We remark that we cannot expect to perform as well as the unknown underlying Markov model, if the model set only provides approximations. Thus, if the best approximation has error ϵ we have to be ready to accept respective error of order ϵD per step, cf. the lower bound provided by Theorem 2 below.

Overview. We start with explicating our notion of approximation in Section 2, then introduce our algorithm in Section 3, present our regret bounds in Section 4, and conclude with the proofs in Section 5. Due to space constraints some proofs can only be found in the extended version [9].

2   Preliminaries: MDP Approximations

Approximations. Before we give the precise notion of *approximation* we are going to use, first note that in our setting the transition probabilities $p(h'|h, a)$ for any two histories $h, h' \in \mathcal{H}$ and an action a are well-defined. Then given an arbitrary model ϕ and a state $s' \in \mathcal{S}_\phi$, we can define the aggregated transition probabilities $p^{\mathrm{agg}}(s'|h, a) := \sum_{h':\phi(h')=s'} p(h'|h, a)$. Note that the true transition probabilities under $\phi°$ are then given by $p(s'|s, a) := p^{\mathrm{agg}}(s'|h, a)$ for $s = \phi°(h)$ and $s' \in \mathcal{S}°$.

Definition 1. *A model ϕ is said to be an ϵ-approximation of the true model $\phi°$ if: (i) for all histories h, h' with $\phi(h) = \phi(h')$ and all actions a*

$$\left|r(\phi°(h), a) - r(\phi°(h'), a)\right| < \epsilon, \text{ and } \left\|p(\cdot|\phi°(h), a) - p(\cdot|\phi°(h'), a)\right\|_1 < \tfrac{\epsilon}{2}, \quad (2)$$

and (ii) there is a surjective mapping $\alpha : \mathcal{S}° \to \mathcal{S}_\phi$ such that for all histories h and all actions a it holds that

$$\sum_{\dot{s}'\in\mathcal{S}_\phi} \left|p^{\mathrm{agg}}(\dot{s}'|h, a) - \sum_{s'\in\mathcal{S}°:\alpha(s')=\dot{s}'} p^{\mathrm{agg}}(s'|h, a)\right| < \tfrac{\epsilon}{2}. \quad (3)$$

Intuitively, condition (2) assures that the approximate model aggregates only histories that are mapped to similar states under the true model. Complementary, condition (3) guarantees that the state space under the approximate model resembles the true state space.[1] Note that any model will be an ϵ-approximation of the underlying true model ϕ° for sufficiently large ϵ.

A particular natural case are approximation models ϕ which also satisfy

$$\forall h, h' \in \mathcal{H} : \ \phi^\circ(h) = \phi^\circ(h') \implies \phi(h) = \phi(h').$$

That is, intuitively, states in $\mathcal{S}^\circ$ are aggregated to meta-states in $\mathcal{S}_\phi$, and (3) holds trivially.

We may carry over our definition of ϵ-approximation to MDPs. This will turn out useful, since each approximate model can be interpreted as an MDP approximation, cf. Section 5.1 below.

Definition 2. *An MDP $\bar{M} = (\bar{\mathcal{S}}, \mathcal{A}, \bar{r}, \bar{p})$ is an ϵ-approximation of another MDP $M = (\mathcal{S}, \mathcal{A}, r, p)$ if there is a surjective function $\alpha : \mathcal{S} \to \bar{\mathcal{S}}$ such that for all s in $\mathcal{S}$:*

$$\left|\bar{r}(\alpha(s), a) - r(s, a)\right| < \epsilon, \ \text{and} \ \sum_{\dot{s}' \in \bar{\mathcal{S}}} \left|\bar{p}(\dot{s}'|\alpha(s), a) - \sum_{s':\alpha(s')=\dot{s}'} p(s'|s, a)\right| < \epsilon. \quad (4)$$

Error Bounds for ϵ-Approximations. The following is an error bound on the error made by an ϵ-approximation. It generalizes bounds of [8] from ergodic to communicating MDPs. For a proof see Appendix A of [9].

Theorem 1. *Let M be a communicating MDP and $\bar{M}$ be an ϵ-approximation of M. Then*

$$\left|\rho^*(M) - \rho^*(\bar{M})\right| \leq \epsilon(D(M) + 1).$$

The following is a matching lower bound on the error by aggregation. This is an improvement over the results in [8], which only showed that the error approaches 1 when the diameter goes to infinity.

Theorem 2. *For each $\epsilon > 0$ and each $2 < D < \frac{4}{\epsilon}$ there are MDPs M, $\bar{M}$ such that $\bar{M}$ is an ϵ-approximation of M, M has diameter $D(M) = D$, and*

$$\left|\rho^*(M) - \rho^*(\bar{M})\right| > \tfrac{1}{56}\epsilon D(M).$$

Proof. Consider the MDP M shown in Figure 1 (left), where the (deterministic) reward in states s_0, s_0' is 0 and 1 in state s_1. We assume that $0 < \varepsilon := \frac{\epsilon}{2} < \delta := \frac{2}{D}$. Then the diameter $D(M)$ is the expected transition time from s_0' to s_0 and equal to $\frac{2}{\delta} = D$. Aggregating states s_0, s_0' gives the MDP $\bar{M}$ on the right hand side of Figure 1. Obviously, $\bar{M}$ is an ϵ-approximation of M. It is straightforward to check that the stationary distribution μ (of the only policy) in M is $(\mu(s_0), \mu(s_0'), \mu(s_1)) = \left(\frac{\delta}{3\varepsilon+4\delta}, \frac{\varepsilon+\delta}{3\varepsilon+4\delta}, \frac{2\varepsilon+2\delta}{3\varepsilon+4\delta}\right)$, while the stationary distribution in $\bar{M}$ is $(\frac{1}{2}, \frac{1}{2})$. Thus, the difference in average reward is

$$\left|\rho^*(M) - \rho^*(\bar{M})\right| = \tfrac{2\varepsilon+2\delta}{3\varepsilon+4\delta} - \tfrac{1}{2} \ = \ \tfrac{\varepsilon}{2(3\varepsilon+4\delta)} \ > \ \tfrac{\varepsilon}{14\delta} = \tfrac{1}{56}\epsilon D(M). \qquad \square$$

---

[1] The allowed error in the conditions for the transition probabilities is chosen to be $\frac{\epsilon}{2}$ so that the total error with respect to the transition probabilities is ϵ. This matches the respective condition for MDP approximations in Definition 2, cf. also Section 5.1.

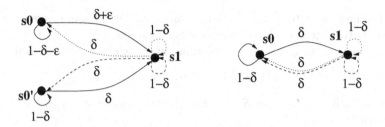

Fig. 1. The MDPs M (left) and $\bar{M}$ (right) in the proof of Theorem 2. Solid, dashed, and dotted arrows indicate different actions.

Theorems 1 and 2 compare the optimal policies of two different MDPs, however it is straightforward to see from the proofs that the same error bounds hold when comparing on some MDP M the optimal average reward $\rho^*(M)$ to the average reward when applying the optimal policy of an ϵ-approximation $\bar{M}$ of M. Thus, when we approximate an MDP M by an ϵ-approximation $\bar{M}$, the respective error of the optimal policy of $\bar{M}$ on M can be of order $\epsilon D(M)$ as well. Hence, we cannot expect to perform below this error if we only have an ϵ-approximation of the true model at our disposal.

3   Algorithm

The OAMS algorithm (shown in detail as Algorithm 1) we propose for the setting introduced in Section 1 is a generalization of the OMS algorithm of [6]. Application of the original OMS algorithm to our setting would not work, since OMS compares the collected rewards of each model to the reward it would receive if the model were Markov. Models not giving sufficiently high reward are identified as non-Markov and rejected. In our case, there may be no Markov model in the set of given models Φ. Thus, the main difference to OMS is that OAMS for each model estimates and takes into account the possible approximation error with respect to a closest Markov model.

OAMS proceeds in episodes $k = 1, 2, \ldots$, each consisting of several runs $j = 1, 2, \ldots$. In each run j of some episode k, starting at time $t = t_{kj}$, OAMS chooses a policy π_{kj} applying the *optimism in face of uncertainty* principle twice.

Plausible Models. First, OAMS considers for each model $\phi \in \Phi$ a set of *plausible* MDPs $\mathcal{M}_{t,\phi}$ defined to contain all MDPs with state space $\mathcal{S}_\phi$ and with rewards r^+ and transition probabilities p^+ satisfying

$$\left| r^+(s,a) - \widehat{r}_t(s,a) \right| \leq \tilde{\epsilon}(\phi) + \sqrt{\frac{\log(48 S_\phi A t^3/\delta)}{2 N_t(s,a)}}, \tag{5}$$

$$\left\| p^+(\cdot|s,a) - \widehat{p}_t(\cdot|s,a) \right\|_1 \leq \tilde{\epsilon}(\phi) + \sqrt{\frac{2 S_\phi \log(48 S_\phi A t^3/\delta)}{N_t(s,a)}}, \tag{6}$$

where $\tilde{\epsilon}(\phi)$ is the estimate for the approximation error of model ϕ (cf. below), $\widehat{p}_t(\cdot|s,a)$ and $\widehat{r}_t(s,a)$ are respectively the empirical state-transition probabilities and the mean reward at time t for taking action a in state $s \in \mathcal{S}_\phi$, $S_\phi := |\mathcal{S}_\phi|$ denotes the number of states under model ϕ, $A := |\mathcal{A}|$ is the number of actions, and $N_t(s, a)$ is the number of times action a has been chosen in state s up to time t. (If a hasn't been chosen in s so far, we set $N_t(s, a)$ to 1.) The inequalities (5) and (6) are obviously inspired by Chernov bounds that would hold with high probability in case the respective model ϕ is Markov, cf. also Lemma 1 below.

Optimistic MDP for Each Model ϕ. In line 4, the algorithm computes for each model ϕ a so-called optimistic MDP $M_t^+(\phi) \in \mathcal{M}_{t,\phi}$ and an associated optimal policy $\pi_{t,\phi}^+$ on $M_t^+(\phi)$ such that the average reward $\rho(M_t^+(\phi), \pi_{t,\phi}^+)$ is maximized. This can be done by extended value iteration (EVI) [4]. Indeed, if $r_t^+(s, a)$ and $p_t^+(s'|s, a)$ denote the optimistic rewards and transition probabilities of $M_t^+(\phi)$, then EVI computes optimistic state values $\mathbf{u}_{t,\phi}^+ = (u_{t,\phi}^+(s))_s \in \mathbb{R}^{S_\phi}$ such that (cf. [4])

$$\widehat{\rho}_t^+(\phi) := \min_{s \in \mathcal{S}_\phi} \left\{ r_t^+(s, \pi_{t,\phi}^+(s)) + \sum_{s'} p_t^+(s'|s, \pi_{t,\phi}^+(s)) \, u_{t,\phi}^+(s') - u_{t,\phi}^+(s) \right\} \quad (7)$$

is an approximation of $\rho^*(M_t^+(\phi))$, that is,

$$\widehat{\rho}_t^+(\phi) \geq \rho^*(M_t^+(\phi)) - 2/\sqrt{t}. \quad (8)$$

Optimistic Model Selection. In line 5, OAMS chooses a model $\phi_{kj} \in \Phi$ with corresponding MDP $M_{kj} = M_t^+(\phi_{kj})$ and policy $\pi_{kj} := \pi_{t,\phi_{kj}}^+$ that maximizes the average reward penalized by the term $\mathbf{pen}(\phi, t)$ defined as

$$\mathbf{pen}(\phi, t) := 2^{-j/2} \left(\left(\lambda(\mathbf{u}_{t,\phi}^+) \sqrt{2S_\phi} + \frac{3}{\sqrt{2}} \right) \sqrt{S_\phi A \log\left(\frac{48 S_\phi A t^3}{\delta} \right)} \right) \quad (9)$$

$$+ \lambda(\mathbf{u}_{t,\phi}^+) \sqrt{2 \log(\frac{24 t^2}{\delta})} \right) + 2^{-j} \lambda(\mathbf{u}_{t,\phi}^+) + \tilde{\epsilon}(\phi) \left(\lambda(\mathbf{u}_{t,\phi}^+) + 3 \right),$$

where we define $\lambda(\mathbf{u}_{t,\phi}^+) := \max_{s \in \mathcal{S}_\phi} u_{t,\phi}^+(s) - \min_{s \in \mathcal{S}_\phi} u_{t,\phi}^+(s)$ to be the empirical value span of the optimistic MDP $M_t^+(\phi)$. Intuitively, the penalization term is an upper bound on the per-step regret of the model ϕ in the run to follow in case ϕ is chosen, cf. eq. (35) in the proof of the main theorem. Similar to the REGAL algorithm of [1] this shall prefer simpler models (i.e., models having smaller state space and smaller value span) to more complex ones.

Termination of Runs and Episodes. The chosen policy π_{kj} is then executed until either (i) run j reaches the maximal length of 2^j steps, (ii) episode k terminates when the number of visits in some state has been doubled (line 12), or (iii) the executed policy π_{kj} does not give sufficiently high rewards (line 9). That is, at any time t in run j of episode k it is checked whether the total reward in the current run is at least $\ell_{kj} \rho_{kj} - \mathbf{lob}_{kj}(t)$, where $\ell_{kj} := t - t_{kj} + 1$ is the (current) length of run j in episode k, and $\mathbf{lob}_{kj}(t)$ is defined as

$$\mathbf{lob}_{kj}(t) := \left(\lambda_{kj}^+ \sqrt{2S_{kj}} + \frac{3}{\sqrt{2}} \right) \sum_{s \in \mathcal{S}_{kj}} \sum_{a \in \mathcal{A}} \sqrt{v_{kj}(s, a) \log\left(\frac{48 S_{kj} A t_{kj}^3}{\delta} \right)}$$

$$+ \lambda_{kj}^+ \sqrt{2 \ell_{kj} \log\left(\frac{24 t_{kj}^2}{\delta} \right)} + \lambda_{kj}^+ + \tilde{\epsilon}(\phi_{kj}) \ell_{kj} (\lambda_{kj}^+ + 3), \quad (10)$$

Algorithm 1. Optimal Approximate Model Selection (OAMS)

---

input set of models Φ, confidence parameter $\delta \in (0,1)$, precision parameter $\epsilon_0 \in (0,1)$
1: Let t be the current time step, and set $\tilde{\epsilon}(\phi) := \epsilon_0$ for all $\phi \in \Phi$.
2: **for** episodes $k = 1, 2, \ldots$ **do**
3:    **for** runs $j = 1, 2, \ldots$ **do**
4:       $\forall \phi \in \Phi$, use EVI to compute an optimistic MDP $M_t^+(\phi)$ in $\mathcal{M}_{t,\phi}$ (the set of *plausible* MDPs defined via the confidence intervals (5) and (6) for the estimates so far), a (near-)optimal policy $\pi_{t,\phi}^+$ on $M_t^+(\phi)$ with approximate average reward $\widehat{\rho}_t^+(\phi)$, and the empirical value span $\lambda(\mathbf{u}_{t,\phi}^+)$.
5:       Choose model $\phi_{kj} \in \Phi$ such that

$$\phi_{kj} = \operatorname*{argmax}_{\phi \in \Phi} \left\{ \widehat{\rho}_t^+(\phi) - \mathbf{pen}(\phi, t) \right\}. \tag{11}$$

6:       Set $t_{kj} := t$, $\rho_{kj} := \widehat{\rho}_t^+(\phi_{kj})$, $\pi_{kj} := \pi_{t,\phi_{kj}}^+$, and $\mathcal{S}_{kj} := \mathcal{S}_{\phi_{kj}}$.
7:       **for** 2^j steps **do**
8:          Choose action $a_t := \pi_{kj}(s_t)$, get reward r_t, observe next state $s_{t+1} \in \mathcal{S}_{kj}$.
9:          **if** the total reward collected so far in the current run is less than

$$(t - t_{kj} + 1)\rho_{kj} - \mathbf{lob}_{kj}(t), \tag{12}$$

          then
10:            $\tilde{\epsilon}(\phi_{kj}) := 2\tilde{\epsilon}(\phi_{kj})$
11:            Terminate current episode.
12:          **else if** $\sum_{j'=1}^{j} v_{kj'}(s_t, a_t) = N_{t_{k1}}(s_t, a_t)$ **then**
13:            Terminate current episode.
14:          **end if**
15:       **end for**
16:    **end for**
17: **end for**

---

where $\lambda_{kj}^+ := \lambda(\mathbf{u}_{t_{kj},\phi_{kj}}^+)$, $S_{kj} := S_{\phi_{kj}}$, and $v_{kj}(s, a)$ are the (current) state-action counts of run j in episode k. That way, OAMS assumes each model to be Markov, as long as it performs well. We will see that $\mathbf{lob}_{kj}(t)$ can be upper bounded by $\ell_{kj}\mathbf{pen}(\phi_{kj}, t_{kj})$, cf. eq. (35) below.

Guessing the Approximation Error. The algorithm tries to guess for each model ϕ the correct approximation error $\epsilon(\phi)$. In the beginning the guessed value $\tilde{\epsilon}(\phi)$ for each model $\phi \in \Phi$ is set to the precision parameter ϵ_0, the best possible precision we aim for. Whenever the reward test fails for a particular model ϕ, it is likely that $\tilde{\epsilon}(\phi)$ is too small and it is therefore doubled (line 10).

4   Regret Bounds

The following upper bound on the regret of OAMS is the main result of this paper.

Theorem 3. *There are $c_1, c_2, c_3 \in \mathbb{R}$ such that in each learning problem where the learner is given access to a set of models Φ not necessarily containing the* *true model ϕ°, the regret $\Delta(\phi^\circ, T)$ of OAMS (with parameters δ, ϵ_0) with respect to the true model ϕ° after any $T \geq SA$ steps is upper bounded by*

$$c_1 \cdot DSA(\log(\tfrac{1}{\epsilon_0})\log T + \log^2 T) + c_2 \cdot DA(\log^{3/2} T)\sqrt{S_\phi S \log(\tfrac{1}{\epsilon_0})T}$$

$$+c_3 \cdot D(\log(\tfrac{1}{\epsilon_0}) + T)\max\{\epsilon_0, \epsilon(\phi)\}$$

with probability at least $1 - \delta$, where $\phi \in \Phi$ is an $\epsilon(\phi)$-approximation of the true underlying Markov model ϕ°, $D := D(\phi^\circ)$, and $S := \sum_{\phi \in \Phi} S_\phi$.

As already mentioned, by Theorem 2 the last term in the regret bound is unavoidable when only considering models in Φ. Note that Theorem 3 holds for *all* models $\phi \in \Phi$. For the best possible bound there is a payoff between the size S_ϕ of the approximate model and its precision $\epsilon(\phi)$.

When the learner knows that Φ contains a Markov model ϕ°, the original OMS algorithm of [6] can be employed. In case when the total number $S = \sum_\phi S_\phi$ of states over all models is large, i.e., $S > D^2|\Phi|S^\circ$, we can improve on the state space dependence of the regret bound given in [6] as follows. The proof (found in Appendix H of [9]) is a simple modification of the analysis in [6] that exploits that by (11) the selected models cannot have arbitrarily large state space.

Theorem 4. *If Φ contains a Markov model ϕ°, with probability at least $1 - \delta$ the regret of OMS is bounded by $\tilde{O}(D^2 S^\circ A\sqrt{|\Phi|T})$.*

Discussion. Unfortunately, while the bound in Theorem 3 is optimal with respect to the dependence on the horizon T, the improvement on the state space dependence that we could achieve in Theorem 4 for OMS is not as straightforward for OAMS and remains an open question just as the optimality of the bound with respect to the other appearing parameters. We note that this is still an open question even for learning in MDPs (without additionally selecting the state representation) as well, cf. [4].

Another direction for future work is the extension to the case when the underlying true MDP has continuous state space. In this setting, the models have the natural interpretation of being discretizations of the original state space. This could also give improvements over current regret bounds for continuous reinforcement learning as given in [10]. Of course, the most challenging goal remains to generate suitable state representation models algorithmically instead of assuming them to be given. However, at the moment it is not even clear how to deal with the case when an infinite set of models is given.

5   Proof of Theorem 3

The proof is divided into three parts and follows the lines of [6], now taking into account the necessary modifications to deal with the approximation error. First, in Section 5.1 we deal with the error of ϵ-approximations. Then in Section 5.2, we show that all state-representation models ϕ which are an $\epsilon(\phi)$-approximation of a Markov model pass the test in (12) on the rewards collected so far with high probability, provided that the estimate $\tilde{\epsilon}(\phi) \geq \epsilon(\phi)$. Finally, in Section 5.3 we use this result to derive the regret bound of Theorem 3.

5.1   Error Bounds for ϵ-Approximate Models

We start with some observations about the empirical rewards and transition probabilities our algorithm calculates and employs for each model ϕ. While the estimated rewards $\widehat{r}$ and transition probabilities $\widehat{p}$ used by the algorithm do in general not correspond to some underlying true values, the expectation values of $\widehat{r}$ and $\widehat{p}$ are still well-defined, given the history $h \in \mathcal{H}$ so far. Indeed, consider some $h \in \mathcal{H}$ with $\phi(h) = \dot{s} \in \mathcal{S}_\phi$, $\phi^\circ(h) = s \in \mathcal{S}^\circ$, and an action a, and assume that the estimates $\widehat{r}(\dot{s}, a)$ and $\widehat{p}(\cdot|\dot{s}, a)$ are calculated from samples when action a was chosen after histories $h_1, h_2, \ldots, h_n \in \mathcal{H}$ that are mapped to the same state $\dot{s}$ by ϕ. (In the following, we will denote the states of an approximation ϕ by variables with dot, such as $\dot{s}$, $\dot{s}'$, etc., and states in the state space $\mathcal{S}^\circ$ of the true Markov model ϕ° without a dot, such as s, s', etc.) Since rewards and transition probabilities are well-defined under ϕ°, we have

$$\mathbb{E}[\widehat{r}(\dot{s}, a)] = \tfrac{1}{n} \sum_{i=1}^{n} r(\phi^\circ(h_i), a), \text{ and } \mathbb{E}[\widehat{p}(\dot{s}'|\dot{s}, a)] = \tfrac{1}{n} \sum_{i=1}^{n} \sum_{h':\phi(h')=\dot{s}'} p(h'|h_i, a). \quad (13)$$

Since ϕ maps the histories $h, h_1, \ldots, h_n$ to the same state $\dot{s} \in \mathcal{S}_\phi$, the rewards and transition probabilities in the states $\phi^\circ(h), \phi^\circ(h_1), \ldots, \phi^\circ(h_n)$ of the true underlying MDP are ϵ-close, cf. (2). It follows that for $s = \phi^\circ(h)$ and $\dot{s} = \phi(h)$

$$\left| \mathbb{E}[\widehat{r}(\dot{s}, a)] - r(s, a) \right| = \left| \tfrac{1}{n} \sum_{i=1}^{n} \left(r(\phi^\circ(h_i), a) - r(\phi^\circ(h), a) \right) \right| < \epsilon(\phi). \quad (14)$$

For the transition probabilities we have by (3) for $i = 1, \ldots, n$

$$\sum_{\dot{s}' \in \mathcal{S}_\phi} \left| p^{\mathrm{agg}}(\dot{s}'|h_i, a) - \sum_{s' \in \mathcal{S}^\circ : \alpha(s')=\dot{s}'} p^{\mathrm{agg}}(s'|h_i, a) \right| < \tfrac{\epsilon(\phi)}{2}. \quad (15)$$

Further, all h_i as well as h are mapped to $\dot{s}$ by ϕ so that according to (2) and recalling that $s = \phi^\circ(h)$ we have for $i = 1, \ldots, n$

$$\sum_{\dot{s}' \in \mathcal{S}_\phi} \left| \sum_{s' \in \mathcal{S}^\circ : \alpha(s')=\dot{s}'} p^{\mathrm{agg}}(s'|h_i, a) - \sum_{s' \in \mathcal{S}^\circ : \alpha(s')=\dot{s}'} p(s'|s, a) \right|$$

$$\leq \sum_{s' \in \mathcal{S}^\circ} \left| p^{\mathrm{agg}}(s'|h_i, a) - p(s'|s, a) \right| < \tfrac{\epsilon(\phi)}{2}. \quad (16)$$

By (15) and (16) for $i = 1, \ldots, n$

$$\sum_{\dot{s}' \in \mathcal{S}_\phi} \left| p^{\mathrm{agg}}(\dot{s}'|h_i, a) - \sum_{s' \in \mathcal{S}^\circ : \alpha(s')=\dot{s}'} p(s'|s, a) \right| < \epsilon(\phi), \quad (17)$$

so that from (13) and (17) we can finally bound

$$\sum_{\dot{s}' \in \mathcal{S}_\phi} \left| \mathbb{E}[\widehat{p}(\dot{s}'|\dot{s}, a)] - \sum_{s' \in \mathcal{S}^\circ : \alpha(s')=\dot{s}'} p(s'|s, a) \right|$$

$$\leq \tfrac{1}{n} \sum_{i=1}^{n} \sum_{\dot{s}' \in \mathcal{S}_\phi} \left| p^{\mathrm{agg}}(\dot{s}'|h_i, a) - \sum_{s' \in \mathcal{S}^\circ : \alpha(s')=\dot{s}'} p(s'|s, a) \right| < \epsilon(\phi). \quad (18)$$

Thus, according to (14) and (18) the ϵ-approximate model ϕ gives rise to an MDP $\bar{M}$ on $\mathcal{S}_\phi$ with rewards $\bar{r}(\dot{s}, a) := \mathbb{E}[\hat{r}(\dot{s}, a)]$ and transition probabilities $\bar{p}(\dot{s}'|\dot{s}, a) := \mathbb{E}[\hat{p}(\dot{s}'|\dot{s}, a)]$ that is an ϵ-approximation of the true MDP $M(\phi^\circ)$. Note that $\bar{M}$ actually depends on the history so far.

The following lemma gives some basic confidence intervals for the estimated rewards and transition probabilities. For a proof sketch see Appendix B of [9].

Lemma 1. *Let t be an arbitrary time step and $\phi \in \Phi$ be the model employed at step t. Then the estimated rewards $\hat{r}$ and transition probabilities $\hat{p}$ satisfy for all $\dot{s}, \dot{s}' \in \mathcal{S}_\phi$ and all $a \in \mathcal{A}$*

$$\hat{r}(\dot{s}, a) - \mathbb{E}[\hat{r}(\dot{s}, a)] \leq \sqrt{\frac{\log(48 S_\phi A t^3/\delta)}{N_t(\dot{s}, a)}},$$

$$\left\| \hat{p}(\cdot|\dot{s}, a) - \mathbb{E}[\hat{p}(\cdot|\dot{s}, a)] \right\|_1 \leq \sqrt{\frac{2 S_\phi \log(48 S_\phi A t^3/\delta)}{N_t(\dot{s}, a)}},$$

each with probability at least $1 - \frac{\delta}{24 t^2}$.

The following is a consequence of Theorem 1, see Appendix C of [9] for a detailed proof.

Lemma 2. *Let ϕ° be the underlying true Markov model leading to MDP $M = (\mathcal{S}^\circ, \mathcal{A}, r, p)$, and ϕ be an ϵ-approximation of ϕ°. Assume that the confidence intervals given in Lemma 1 hold at step t for all states $\dot{s}, \dot{s}' \in \mathcal{S}_\phi$ and all actions a. Then the optimistic average reward $\hat{\rho}_t^+(\phi)$ defined in (7) satisfies*

$$\hat{\rho}_t^+(\phi) \geq \rho^*(M) - \epsilon(D(M) + 1) - \frac{2}{\sqrt{t}}.$$

5.2   Approximate Markov Models Pass the Test in (12)

Assume that the model $\phi_{kj} \in \Phi$ employed in run j of episode k is an $\epsilon_{kj} := \epsilon(\phi_{kj})$-approximation of the true Markov model. We are going to show that ϕ_{kj} will pass the test (12) on the collected rewards with high probability at any step t, provided that $\tilde{\epsilon}_{kj} := \tilde{\epsilon}(\phi_{kj}) \geq \epsilon_{kj}$.

Lemma 3. *For each step t in some run j of some episode k, given that $t_{kj} = t'$ the chosen model ϕ_{kj} passes the test in (12) at step t with probability at least $1 - \frac{\delta}{6 t'^2}$ whenever $\tilde{\epsilon}_{kj}(\phi_{kj}) \geq \epsilon(\phi_{kj})$.*

Proof. In the following, $\dot{s}_\tau := \phi_{kj}(h_\tau)$ and $s_\tau := \phi^\circ(h_\tau)$ are the states at time step τ under model ϕ_{kj} and the true Markov model ϕ°, respectively.

Initial Decomposition. First note that at time t when the test is performed, we have $\sum_{\dot{s} \in \mathcal{S}_{kj}} \sum_{a \in \mathcal{A}} v_{kj}(\dot{s}, a) = \ell_{kj} = t - t' + 1$, so that

$$\ell_{kj} \rho_{kj} - \sum_{\tau=t'}^t r_\tau = \sum_{\dot{s} \in \mathcal{S}_{kj}} \sum_{a \in \mathcal{A}} v_{kj}(\dot{s}, a) \left(\rho_{kj} - \hat{r}_{t':t}(\dot{s}, a) \right),$$

where $\widehat{r}_{t':t}(\dot{s}, a)$ is the empirical average reward collected for choosing a in $\dot{s}$ from time t' to the current time t in run j of episode k.

Let $r_{kj}^+(\dot{s}, a)$ be the rewards and $p_{kj}^+(\cdot | \dot{s}, a)$ the transition probabilities of the optimistic model $M_{t_{kj}}^+(\phi_{kj})$. Noting that $v_{kj}(\dot{s}, a) = 0$ when $a \neq \pi_{kj}(\dot{s})$, we get

$$\ell_{kj}\rho_{kj} - \sum_{\tau=t'}^{t} r_\tau = \sum_{\dot{s},a} v_{kj}(\dot{s}, a)\big(\widehat{\rho}_{kj}^+(\phi_{kj}) - r_{kj}^+(\dot{s}, a)\big) \tag{19}$$

$$+ \sum_{\dot{s},a} v_{kj}(\dot{s}, a)\big(r_{kj}^+(\dot{s}, a) - \widehat{r}_{t':t}(\dot{s}, a)\big). \tag{20}$$

We continue bounding the two terms (19) and (20) separately.

Bounding the Reward Term (20). Recall that $r(s, a)$ is the mean reward for choosing a in s in the true Markov model ϕ°. Then we have at each time step $\tau = t', \ldots, t$ with probability at least $1 - \frac{\delta}{12t'^2}$

$$r_{kj}^+(\dot{s}_\tau, a) - \widehat{r}_{t':t}(\dot{s}_\tau, a) = \big(r_{kj}^+(\dot{s}_\tau, a) - \widehat{r}_{t'}(\dot{s}_\tau, a)\big)$$
$$+ \big(\widehat{r}_{t'}(\dot{s}_\tau, a) - \mathbb{E}[\widehat{r}_{t'}(\dot{s}_\tau, a)]\big) + \big(\mathbb{E}[\widehat{r}_{t'}(\dot{s}_\tau, a)] - r(s_\tau, a)\big)$$
$$+ \big(r(s_\tau, a) - \mathbb{E}[\widehat{r}_{t':t}(\dot{s}_\tau, a)]\big) + \big(\mathbb{E}[\widehat{r}_{t':t}(\dot{s}_\tau, a)] - \widehat{r}_{t':t}(\dot{s}_\tau, a)\big)$$
$$\leq \tilde{\epsilon}_{kj} + 2\sqrt{\frac{\log(48 S_{kj} A t'^3 / \delta)}{2 N_{t'}(\dot{s}, a)}} + 2\epsilon_{kj} + \sqrt{\frac{\log(48 S_{kj} A t'^3 / \delta)}{2 v_{kj}(\dot{s}, a)}}, \tag{21}$$

where we bounded the first term in the decomposition by (5), the second term by Lemma 1, the third and fourth according to (14), and the fifth by an equivalent to Lemma 1 for the rewards collected so far in the current run. In summary, with probability at least $1 - \frac{\delta}{12t'^2}$ we can bound (20) as

$$\sum_{\dot{s},a} v_{kj}(\dot{s}, a)\big(r_{kj}^+(\dot{s}, a) - \widehat{r}_{t':t}(\dot{s}, a)\big) \leq 3\tilde{\epsilon}_{kj}\ell_{kj} + \frac{3}{\sqrt{2}} \sum_{\dot{s},a} \sqrt{v_{kj}(\dot{s}, a) \log\big(\frac{48 S_{kj} A t'^3}{\delta}\big)}, \tag{22}$$

where we used the assumption that $\tilde{\epsilon}_{kj} \geq \epsilon_{kj}$ as well as $v_{kj}(\dot{s}, a) \leq N_{t'}(\dot{s}, a)$.

Bounding the Bias Term (19). First, notice that we can use (7) to bound

$$\sum_{\dot{s},a} v_{kj}(\dot{s}, a)\big(\widehat{\rho}_{kj}^+(\phi_{kj}) - r_{kj}^+(\dot{s}, a)\big) \leq \sum_{\dot{s},a} v_{kj}(\dot{s}, a)\Big(\sum_{\dot{s}'} p_{kj}^+(\dot{s}' | \dot{s}, a)\, u_{kj}^+(\dot{s}') - u_{kj}^+(\dot{s})\Big),$$

where $u_{kj}^+(\dot{s}) := u_{t_{kj}, \phi_{kj}}^+(\dot{s})$ are the state values given by EVI. Further, since the transition probabilities $p_{kj}^+(\cdot | \dot{s}, a)$ sum to 1, this is invariant under a translation of the vector $\mathbf{u}_{kj}^+$. In particular, defining $w_{kj}(\dot{s}) := u_{kj}^+(\dot{s}) - \frac{1}{2}\big(\min_{\dot{s}\in S_{kj}} u_{kj}^+(\dot{s}) + \max_{\dot{s}\in S_{kj}} u_{kj}^+(\dot{s})\big)$, so that $\|\mathbf{w}_{kj}\|_\infty = \lambda_{kj}^+/2$, we can replace $\mathbf{u}_{kj}^+$ with $\mathbf{w}_{kj}$, and (19) can be bounded as

$$\sum_{\dot{s},a} v_{kj}(\dot{s}, a)\big(\widehat{\rho}_{kj}^+(\phi_{kj}) - r_{kj}^+(\dot{s}, a)\big)$$

$$\leq \sum_{\dot{s},a} \sum_{\tau=t'}^{t} \mathbb{1}\big\{(\dot{s}_\tau, a_\tau) = (\dot{s}, a)\big\}\Big(\sum_{\dot{s}'\in S_{kj}} p_{kj}^+(\dot{s}' | \dot{s}_\tau, a)\, w_{kj}(\dot{s}') - w_{kj}(\dot{s}_\tau)\Big). \tag{23}$$

Now we decompose for each time step $\tau = t', \ldots, t$

$$\sum_{\dot{s}' \in \mathcal{S}_{kj}} p_{kj}^+(\dot{s}'|\dot{s}_\tau, a)\, w_{kj}(\dot{s}') - w_{kj}(\dot{s}_\tau) =$$

$$\sum_{\dot{s}' \in \mathcal{S}_{kj}} \left(p_{kj}^+(\dot{s}'|\dot{s}_\tau, a) - \widehat{p}_{t'}(\dot{s}'|\dot{s}_\tau, a) \right) w_{kj}(\dot{s}') \tag{24}$$

$$+ \sum_{\dot{s}' \in \mathcal{S}_{kj}} \left(\widehat{p}_{t'}(\dot{s}'|\dot{s}_\tau, a) - \mathbb{E}[\widehat{p}_{t'}(\dot{s}'|\dot{s}_\tau, a)] \right) w_{kj}(\dot{s}') \tag{25}$$

$$+ \sum_{\dot{s}' \in \mathcal{S}_{kj}} \left(\mathbb{E}[\widehat{p}_{t'}(\dot{s}'|\dot{s}_\tau, a)] - \sum_{s':\alpha(s')=\dot{s}'} p(s'|s_\tau, a) \right) w_{kj}(\dot{s}') \tag{26}$$

$$+ \sum_{\dot{s}' \in \mathcal{S}_{kj}} \sum_{s':\alpha(s')=\dot{s}'} p(s'|s_\tau, a)\, w_{kj}(\dot{s}') - w_{kj}(\dot{s}_\tau) \tag{27}$$

and continue bounding each of these terms individually.

Bounding (24): Using $\|\mathbf{w}_{kj}\|_\infty = \lambda_{kj}^+/2$, (24) is bounded according to (6) as

$$\sum_{\dot{s}' \in \mathcal{S}_{kj}} (p_{kj}^+(\dot{s}'|\dot{s}_\tau, a) - \widehat{p}_{t'}(\dot{s}'|\dot{s}_\tau, a)) w_{kj}(\dot{s}') \leq \left\| p_{kj}^+(\cdot|\dot{s}_\tau, a) - \widehat{p}_{t'}(\cdot|\dot{s}_\tau, a) \right\|_1 \|\mathbf{w}_{kj}\|_\infty$$

$$\leq \frac{\tilde{\epsilon}_{kj}\lambda_{kj}^+}{2} + \frac{\lambda_{kj}^+}{2}\sqrt{\frac{2S_{kj}\log(48S_{kj}At'^3/\delta)}{N_{t'}(s,a)}}. \tag{28}$$

Bounding (25): Similarly, by Lemma 1 with probability at least $1 - \frac{\delta}{24t'^2}$ we can bound (25) at all time steps τ as

$$\sum_{\dot{s}' \in \mathcal{S}_{kj}} \left(\widehat{p}_{t'}(\dot{s}'|\dot{s}_\tau, a) - \mathbb{E}[\widehat{p}_{t'}(\dot{s}'|\dot{s}_\tau, a)] \right) w_{kj}(\dot{s}') \leq \frac{\lambda_{kj}^+}{2}\sqrt{\frac{2S_{kj}\log(48S_{kj}At'^3/\delta)}{N_{t'}(s,a)}}. \tag{29}$$

Bounding (26): By (18) and using that $\|\mathbf{w}_{kj}\|_\infty = \lambda_{kj}^+/2$, we can bound (26) by

$$\sum_{\dot{s}' \in \mathcal{S}_{kj}} \left(\mathbb{E}[\widehat{p}_{t'}(\dot{s}'|\dot{s}_\tau, a)] - \sum_{s':\alpha(s')=\dot{s}'} p(s'|s_\tau, a) \right) w_{kj}(\dot{s}') < \frac{\epsilon_{kj}\lambda_{kj}^+}{2}. \tag{30}$$

Bounding (27): We set $w'(s) := w_{kj}(\alpha(s))$ for $s \in \mathcal{S}^\circ$ and rewrite (27) as

$$\sum_{\dot{s}' \in \mathcal{S}_{kj}} \sum_{s':\alpha(s')=\dot{s}'} p(s'|s_\tau, a)\, w_{kj}(\dot{s}') - w_{kj}(\dot{s}_\tau) = \sum_{s' \in \mathcal{S}^\circ} p(s'|s_\tau, a)\, w'(s') - w'(s_\tau). \tag{31}$$

Summing this term over all steps $\tau = t', \ldots, t$, we can rewrite the sum as a martingale difference sequence, so that Azuma-Hoeffding's inequality (e.g., Lemma 10 of [4]) yields that with probability at least $1 - \frac{\delta}{24t'^3}$

$$\sum_{\tau=t'}^{t} \sum_{s' \in \mathcal{S}^\circ} p(s'|s_\tau, a)\, w'(s') - w'(s_\tau) = \sum_{\tau=t'}^{t} \left(\sum_{s'} p(s'|s_\tau, a)\, w'(s') - w'(s_{\tau+1}) \right)$$

$$+ w'(s_{t+1}) - w'(s_{t'}) \leq \lambda_{kj}^+ \sqrt{2\ell_{kj}\log(\frac{24t'^3}{\delta})} + \lambda_{kj}^+, \tag{32}$$

since the sequence $X_\tau := \sum_{s'} p(s'|s_\tau, a)\, w'(s') - w'(s_{\tau+1})$ is a martingale difference sequence with $|X_t| \le \lambda_{kj}^+$.

Wrap-up. Summing over the steps $\tau = t', \ldots, t$, we get from (23), (27), (28), (29), (30), (31), and (32) that with probability at least $1 - \frac{\delta}{12 t'^2}$

$$\sum_{\dot{s}, a} v_{kj}(\dot{s}, a)\big(\widehat{\rho}_{kj}^+(\phi_{kj}) - r_{kj}^+(\dot{s}, a)\big) \;\le\; \tilde{\epsilon}_{kj} \lambda_{kj}^+ \ell_{kj}$$

$$+\lambda_{kj}^+ \sum_{\dot{s}, a} \sqrt{2 v_{kj}(\dot{s}, a)\, S_{kj} \log\Big(\tfrac{48 S_{kj} At'^3}{\delta}\Big)} + \lambda_{kj}^+ \sqrt{2\ell_{kj} \log\Big(\tfrac{24 t'^2}{\delta}\Big)} + \lambda_{kj}^+, \quad (33)$$

using that $v_{kj}(\dot{s}, a) \le N_{t'}(\dot{s}, a)$ and the assumption that $\epsilon_{kj} \le \tilde{\epsilon}_{kj}$. Combining (20), (22), and (33) gives the claimed lemma.   □

Summing Lemma 3 over all episodes gives the following lemma, for a detailed proof see Appendix D of [9].

Lemma 4. *With probability at least $1 - \delta$, for all runs j of all episodes k the chosen model ϕ_{kj} passes all tests, provided that $\tilde{\epsilon}_{kj}(\phi_{kj}) \ge \epsilon(\phi_{kj})$.*

5.3   Preliminaries for the Proof of Theorem 3

We start with some auxiliary results for the proof of Theorem 3. Lemma 5 bounds the bias span of the optimistic policy, Lemma 6 deals with the estimated precision of ϕ_{kj}, and Lemma 7 provides a bound for the number of episodes. For proofs see Appendix E, F, and G of [9].

Lemma 5. *Assume that the confidence intervals given in Lemma 1 hold at some step t for all states $\dot{s} \in S_\phi$ and all actions a. Then for each ϕ, the set of plausible MDPs $\mathcal{M}_{t,\phi}$ contains an MDP $\widetilde{M}$ with diameter $D(\widetilde{M})$ upper bounded by the true diameter D, provided that $\tilde{\epsilon}(\phi) \ge \epsilon(\phi)$. Consequently, the respective bias span $\lambda(\mathbf{u}_{t,\phi}^+)$ is bounded by D as well.*

Lemma 6. *If all chosen models ϕ_{kj} pass all tests in run j of episode k whenever $\tilde{\epsilon}(\phi_{kj}) \ge \epsilon(\phi_{kj})$, then $\tilde{\epsilon}(\phi) \le \max\{\epsilon_0, 2\epsilon(\phi)\}$ always holds for all models ϕ.*

Lemma 7. *Assume that all chosen models ϕ_{kj} pass all tests in run j of episode k whenever $\tilde{\epsilon}(\phi_{kj}) \ge \epsilon(\phi_{kj})$. Then the number of episodes K_T after any $T \ge SA$ steps is upper bounded as $K_T \le SA \log_2\big(\tfrac{2T}{SA}\big) + \sum_{\phi:\epsilon(\phi)>\epsilon_0} \log_2\big(\tfrac{\epsilon(\phi)}{\epsilon_0}\big)$.*

5.4   Bounding the Regret (Proof of Theorem 3)

Now we can finally turn to showing the regret bound of Theorem 3. We will assume that all chosen models ϕ_{kj} pass all tests in run j of episode k whenever $\tilde{\epsilon}(\phi_{kj}) \ge \epsilon(\phi_{kj})$. According to Lemma 4 this holds with probability at least $1 - \delta$.

Let $\phi_{kj} \in \Phi$ be the model that has been chosen at time t_{kj}, and consider the last but one step t of run j in episode k. The regret Δ_{kj} of run j in episode k with respect to $\rho^* := \rho^*(\phi^\circ)$ is bounded by

$$\Delta_{kj} := (\ell_{kj} + 1)\rho^* - \sum_{\tau=t_{kj}}^{t+1} r_\tau \leq \ell_{kj}(\rho^* - \rho_{kj}) + \rho^* + \ell_{kj}\rho_{kj} - \sum_{\tau=t_{kj}}^{t} r_\tau,$$

where as before $\ell_{kj} := t - t_{kj} + 1$ denotes the length of run j in episode k up to the considered step t. By assumption the test (12) on the collected rewards has been passed at step t, so that

$$\Delta_{kj} \leq \ell_{kj}(\rho^* - \rho_{kj}) + \rho^* + \mathbf{lob}_{kj}(t), \tag{34}$$

and we continue bounding the terms of $\mathbf{lob}_{kj}(t)$.

Bounding the Regret with the Penalization Term. Since we have $v_{kj}(\dot{s}, a) \leq N_{t_{k1}}(\dot{s}, a)$ for all $\dot{s} \in \mathcal{S}_{kj}, a \in \mathcal{A}$ and also $\sum_{\dot{s},a} v_{kj}(\dot{s}, a) = \ell_{kj} \leq 2^j$, by Cauchy-Schwarz inequality $\sum_{\dot{s},a} \sqrt{v_{kj}(\dot{s}, a)} \leq 2^{j/2}\sqrt{S_{kj}A}$. Applying this to the definition (10) of $\mathbf{lob}_{kj}$, we obtain from (34) and by the definition (9) of the penalty term that

$$\Delta_{kj} \leq \ell_{kj}(\rho^* - \rho_{kj}) + \rho^* + 2^{j/2}\left(\lambda_{kj}^+\sqrt{2S_{kj}} + \tfrac{3}{\sqrt{2}}\right)\sqrt{S_{kj}A\log\left(\tfrac{48S_{kj}At_{kj}^3}{\delta}\right)}$$
$$+ \lambda_{kj}^+\sqrt{2\ell_{kj}\log\left(\tfrac{24t_{kj}^2}{\delta}\right)} + \lambda_{kj}^+ + \tilde{\epsilon}(\phi_{kj})\ell_{kj}(\lambda_{kj}^+ + 3)$$
$$\leq \ell_{kj}(\rho^* - \rho_{kj}) + \rho^* + 2^j\mathbf{pen}(\phi_{kj}, t_{kj}). \tag{35}$$

The Key Step. Now, by definition of the algorithm and Lemma 2, for any approximate model ϕ we have

$$\rho_{kj} - \mathbf{pen}(\phi_{kj}, t_{kj}) \geq \widehat{\rho}_{t_{kj}}^+(\phi) - \mathbf{pen}(\phi, t_{kj}) \tag{36}$$
$$\geq \rho^* - (D+1)\epsilon(\phi) - \mathbf{pen}(\phi, t_{kj}) - 2t_{kj}^{-1/2},$$

or equivalently $\rho^* - \rho_{kj} + \mathbf{pen}(\phi_{kj}, t_{kj}) \leq \mathbf{pen}(\phi, t_{kj}) + (D+1)\epsilon(\phi) + 2t_{kj}^{-1/2}$. Multiplying this inequality with 2^j and noting that $\ell_{kj} \leq 2^j$ then gives

$$\ell_{kj}(\rho^* - \rho_{kj}) + 2^j\mathbf{pen}(\phi_{kj}, t_{kj}) \leq 2^j\mathbf{pen}(\phi, t_{kj}) + 2^j(D+1)\epsilon(\phi) + 2^{j+1}t_{kj}^{-1/2}.$$

Combining this with (35), we get by application of Lemma 5, i.e., $\lambda(\mathbf{u}_{t_{kj},\phi}^+) \leq D$, and the definition of the penalty term (9) that

$$\Delta_{kj} \leq \rho^* + 2^{j/2}\left(\left(D\sqrt{2S_\phi} + \tfrac{3}{\sqrt{2}}\right)\sqrt{S_\phi A\log\left(\tfrac{48S_\phi At_{kj}^3}{\delta}\right)} + D\sqrt{2\log(\tfrac{24t_{kj}^2}{\delta})}\right)$$
$$+ D + 2^j\tilde{\epsilon}(\phi)(D+3) + 2^j(D+1)\epsilon(\phi) + 2^{j+1}t_{kj}^{-1/2}.$$

By Lemma 6 and using that $2t_{kj} \geq 2^j$ (so that $2^{j+1}t_{kj}^{-1/2} \leq 2\sqrt{2} \cdot 2^{j/2}$) we get

$$\Delta_{kj} \leq \rho^* + 2^{j/2}\left(\left(D\sqrt{2S_\phi} + \tfrac{3}{\sqrt{2}}\right)\sqrt{S_\phi A\log\left(\tfrac{48S_\phi At_{kj}^3}{\delta}\right)} + D\sqrt{2\log(\tfrac{24t_{kj}^2}{\delta})}\right)$$
$$+ D + \tfrac{3}{2} \cdot 2^j \max\{\epsilon_0, 2\epsilon(\phi)\}(D + \tfrac{7}{3}) + 2\sqrt{2} \cdot 2^{j/2}. \tag{37}$$

Summing over Runs and Episodes. Let J_k be the total number of runs in episode k, and let K_T be the total number of episodes up to time T. Noting that $t_{kj} \leq T$ and summing (37) over all runs and episodes gives

$$\Delta(\phi^\circ, T) = \sum_{k=1}^{K_T} \sum_{j=1}^{J_k} \Delta_{kj} \leq (\rho^* + D) \sum_{k=1}^{K_T} J_k + \tfrac{3}{2} \max\{\epsilon_0, 2\epsilon(\phi)\}(D + \tfrac{7}{3}) \sum_{k=1}^{K_T} \sum_{j=1}^{J_k} 2^j$$

$$+ \left((D\sqrt{2S_\phi} + \tfrac{3}{\sqrt{2}})\sqrt{S_\phi A \log(\tfrac{48 S_\phi A T^3}{\delta})} + D\sqrt{2 \log(\tfrac{24 T^2}{\delta})} + 2\sqrt{2} \right) \sum_{k=1}^{K_T} \sum_{j=1}^{J_k} 2^{j/2}.$$

As shown in Section 5.2 of [6], $\sum_k J_k \leq K_T \log_2(2T/K_T)$, $\sum_k \sum_j 2^j \leq \sqrt{2(T + K_T)}$ and $\sum_k \sum_j 2^{j/2} \leq \sqrt{2K_T \log_2(2T/K_T)(T + K_T)}$, and we may conclude the proof applying Lemma 7 and some minor simplifications. □

Acknowledgments. This research was funded by the Austrian Science Fund (FWF): P 26219-N15, the European Community's FP7 Program under grant agreements n° 270327 (CompLACS) and 306638 (SUPREL), the Technion, the Ministry of Higher Education and Research of France, Nord-Pas-de-Calais Regional Council, and FEDER (Contrat de Projets Etat Region CPER 2007-2013).

References

[1] Bartlett, P.L., Tewari, A.: REGAL: A regularization based algorithm for reinforcement learning in weakly communicating MDPs. In: Proc. 25th Conf. on Uncertainty in Artificial Intelligence, UAI 2009, pp. 25–42. AUAI Press (2009)

[2] Hallak, A., Castro, D.D., Mannor, S.: Model selection in Markovian processes. In: 19th ACM SIGKDD Int'l Conf. on Knowledge Discovery and Data Mining, KDD 2013, pp. 374–382. ACM (2013)

[3] Hutter, M.: Feature Reinforcement Learning: Part I: Unstructured MDPs. J. Artificial General Intelligence 1, 3–24 (2009)

[4] Jaksch, T., Ortner, R., Auer, P.: Near-optimal regret bounds for reinforcement learning. J. Mach. Learn. Res. 11, 1563–1600 (2010)

[5] Maillard, O.A., Munos, R., Ryabko, D.: Selecting the state-representation in reinforcement learning. Adv. Neural Inf. Process. Syst. 24, 2627–2635 (2012)

[6] Maillard, O.A., Nguyen, P., Ortner, R., Ryabko, D.: Optimal regret bounds for selecting the state representation in reinforcement learning. In: Proc. 30th Int'l Conf. on Machine Learning, ICML 2013. JMLR Proc., vol. 28, pp. 543–551 (2013)

[7] Nguyen, P., Maillard, O.A., Ryabko, D., Ortner, R.: Competing with an infinite set of models in reinforcement learning. In: Proc. 16th Int'l Conf. on Artificial Intelligence and Statistics, AISTATS 2013. JMLR Proc., vol. 31, pp. 463–471 (2013)

[8] Ortner, R.: Pseudometrics for state aggregation in average reward markov decision processes. In: Hutter, M., Servedio, R.A., Takimoto, E. (eds.) ALT 2007. LNCS (LNAI), vol. 4754, pp. 373–387. Springer, Heidelberg (2007)

[9] Ortner, R., Maillard, O.A., Ryabko, D.: Selecting Near-Optimal Approximate State Representations in Reinforcement Learning. Extended version, http://arxiv.org/abs/1405.2652

[10] Ortner, R., Ryabko, D.: Online Regret Bounds for Undiscounted Continuous Reinforcement Learning. Adv. Neural Inf. Process. Syst. 25, 1772–1780 (2012)

Policy Gradients for CVaR-Constrained MDPs

L.A. Prashanth

INRIA Lille - Nord Europe, Team SequeL, France
prashanth.la@inria.fr

Abstract. We study a risk-constrained version of the stochastic shortest path (SSP) problem, where the risk measure considered is Conditional Value-at-Risk (CVaR). We propose two algorithms that obtain a locally risk-optimal policy by employing four tools: stochastic approximation, mini batches, policy gradients and importance sampling. Both the algorithms incorporate a CVaR estimation procedure, along the lines of [3], which in turn is based on Rockafellar-Uryasev's representation for CVaR and utilize the likelihood ratio principle for estimating the gradient of the sum of one cost function (objective of the SSP) and the gradient of the CVaR of the sum of another cost function (constraint of the SSP). The algorithms differ in the manner in which they approximate the CVaR estimates/necessary gradients - the first algorithm uses stochastic approximation, while the second employs mini-batches in the spirit of Monte Carlo methods. We establish asymptotic convergence of both the algorithms. Further, since estimating CVaR is related to rare-event simulation, we incorporate an importance sampling based variance reduction scheme into our proposed algorithms.

1 Introduction

Risk-constrained Markov decision processes (MDPs) have attracted a lot of attention recently in the reinforcement learning (RL) community (cf. [8, 18, 14, 19]). However, unlike previous works that focused mostly on variance of the return as a measure of risk, we consider Conditional Value-at-Risk (CVaR) as a risk measure. CVaR has the form of a conditional expectation, where the conditioning is based on a constraint on Value-at-Risk (VaR).

The aim in this paper is to find a *risk-optimal* policy in the context of a stochastic shortest path (SSP) problem. A risk-optimal policy is one that minimizes the sum of one cost function (see $G^\theta(s^0)$ in (1)), while ensuring that the conditional expectation of the sum of another cost function (see $C^\theta(s^0)$ in (1)) given some confidence level, stays bounded, i.e., the solution to the following risk-constrained problem: For a given $\alpha \in (0, 1)$ and $K_\alpha > 0$,

$$\min_{\theta \in \Theta} \mathbb{E}\underbrace{\left[\sum_{m=0}^{\tau-1} g(s_m, a_m) \,\big|\, s_0 = s^0\right]}_{G^\theta(s^0)} \text{ subject to } \mathrm{CVaR}_\alpha \underbrace{\left[\sum_{m=0}^{\tau-1} c(s_m, a_m) \,\big|\, s_0 = s^0\right]}_{C^\theta(s^0)} \le K_\alpha.$$

$$(1)$$

In the above, s^0 is the starting state and the actions $a_0, \ldots, a_{\tau-1}$ are chosen according to a randomized policy π_θ governed by θ. Further, $g(s, a)$ and $c(s, a)$ are cost functions

P. Auer et al. (Eds.): ALT 2014, LNAI 8776, pp. 155–169, 2014.

that take a state s and an action a as inputs and τ is the first passage time to the recurrent state of the underlying SSP (see Section 2 for a detailed formulation). In [8], a similar problem is considered in a finite horizon MDP, though under a strong separability assumption for the cost function $c(s, a)$.

The problem (1) is motivated by applications in finance and energy markets. For example, consider a portfolio reallocation problem where the aim is to find an investment strategy that achieves a targeted asset allocation. The portfolio is composed of assets (e.g. stocks) and the gains obtained by buying or selling assets is stochastic and depends on the market situation. A *risk-averse* investor would prefer a investment strategy that alters the mix of assets in the portfolio that **(i)** quickly achieves the target asset allocation (modeled by the objective in (1)), and **(ii)** minimizes the worst-case losses incurred (modeled by the CVaR constraint in (1)). Another problem of interest, as outlined in [8], is in the re-insurance business. The insurance companies collect premiums for providing coverage, but run the risk of heavy payouts due to catastrophic events and this problem can be effectively cast into the framework of a risk-constrained SSP.

Solving the risk-constrained problem (1) is challenging due to two reasons:
(i) Finding a globally risk-optimal policy is intractable even for a simpler case when the risk is defined as the variance of the return of an MDP (see [12]). The risk-constrained MDP that we consider is more complicated in comparison, since CVaR is a conditional expectation, with the conditioning governed by an event that bounds a probability.
(ii) For the sake of optimization of the CVaR-constrained MDP that we consider in this paper, it is required to estimate both VaR/CVaR of the total cost ($C^\theta(s^0)$ in (1)) as well as its gradient.
(iii) Since VaR/CVaR concerns the tail of the distribution of the total cost, a variance reduction technique is required to speed up the estimation procedure.
We avoid the first problem by proposing a policy gradient scheme that is proven to converge to a locally risk-optimal policy. The second problem is alleviated using two principled approaches: stochastic approximation [15, 3] and mini-batch [2] procedures for estimating VaR/CVaR and policy gradients using likelihood ratios [4]. The final problem is solved by incorporating an importance sampling scheme.

The contributions of this paper are summarized as follows:
(I) First, using the representation of CVaR (and also VaR) as the solution of a certain convex optimization problem by Rockafellar and Uryasev [16], we develop a stochastic approximation procedure, along the lines of [3], for estimating the CVaR of a policy for an SSP. In addition, we also propose a mini-batch variant to estimate CVaR. Mini-batches are in the spirit of Monte Carlo methods and have been proposed in [2] under a different optimization context for stochastic proximal gradient algorithms.
(II) Second, we develop two novel policy gradient algorithms for finding a (locally) risk-optimal policy of the CVaR-constrained SSP. The first algorithm is a four timescale stochastic approximation scheme that **(a)** on the fastest two timescales, estimates VaR/CVaR and uses the policy-gradient principle with likelihood ratios to estimate the gradient of the total cost $G^\theta(s^0)$ as well as CVaR of another cost sum $C^\theta(s^0)$; **(b)** updates the policy parameter in the negative descent direction on the intermediate timescale and performs dual ascent for the Lagrange multiplier on the slowest timescale.

On the other hand, the second algorithm operates on two timescales as it employs mini-batches to estimate the CVaR as well as the necessary gradients.

(III) Third, we adapt our proposed algorithms to incorporate importance sampling (IS). This is motivated by the fact that when the confidence level α is close to 1, estimating VaR as well as CVaR takes longer as the interesting samples used to estimate CVaR come from the tail of the distribution of the total cost $C^\theta(s^0)$ random variable. We provide a non-trivial adaptation of the IS scheme proposed in [11] to our setting. Unlike [11] which requires the knowledge of transition dynamics, we use the randomized policies to derive sampling ratios for the IS procedure.

The rest of the paper is organized as follows: In Section 2 we formalize the CVaR-constrained SSP and in Section 3 describe the structure of our proposed algorithms. In Section 4 we present the first algorithm based on stochastic approximation and in Section 5 we present the mini-batch variant. In Section 6, we sketch the convergence of our algorithms and later in Section 7 describe the importance sampling variants of our algorithms. In Section 8, we review relevant previous works. Finally, in Section 9 we provide the concluding remarks.

2   Problem Formulation

In this section, we first introduce VaR/CVaR risk measures, then formalize the stochastic shortest path problem and subsequently define the CVaR-constrained SSP.

2.1   Background on VaR and CVaR

For any random variable X, we define the VaR at level $\alpha \in (0, 1)$ as

$$\text{VaR}_\alpha(X) := \inf \{\xi \mid \mathbb{P}(X \leq \xi) \geq \alpha\}.$$

If the distribution of X is continuous, then VaR is the lowest solution to $\mathbb{P}(X \leq \xi) = \alpha$. VaR as a risk measure has several drawbacks, which precludes using standard stochastic optimization methods. This motivated the definition of coherent risk measures in [1]. A risk measure is coherent if it is convex, monotone, positive homogeneous and translation equi-variant. CVaR is one popular risk measure defined by

$$\text{CVaR}_\alpha(X) := \mathbb{E}[X \mid X \geq \text{VaR}_\alpha(X)].$$

Unlike VaR, the above is a coherent risk measure.

2.2   Stochastic Shortest Path (SSP)

We consider a SSP with a finite state space $\mathcal{S} = \{0, 1, \ldots, r\}$, where 0 is a special cost-free and absorbing terminal state. The set of feasible actions in state $s \in \mathcal{S}$ is denoted by $\mathcal{A}(s)$. A transition from state s to s' under action $a \in \mathcal{A}(s)$ occurs with probability $p_{ss'}(a)$ and incurs the following costs: $g(s, a)$ and $c(s, a)$, respectively.

A policy specifies how actions are chosen in each state. A *stationary* randomized policy $\pi(\cdot|s)$ maps any state s to a probability vector on $\mathcal{A}(s)$. As is standard in policy gradient algorithms, we parameterize the policy and assume that the policy is continuously differentiable in the parameter $\theta \in \mathbb{R}^d$. Since a policy π is identifiable by its parameter θ, we use them interchangeably in this paper.

As defined in [5], a proper policy is one which ensures that there is a positive probability that the terminal state 0 will be reached, starting from any initial state, after utmost r transitions. This in turn implies the states $1, \ldots, r$ are transient. We assume that class of parameterized policies considered, i.e., $\{\pi_\theta \mid \theta \in \Theta\}$, is proper. We assume that Θ is a compact and convex subset of $\mathbb{R}^d$.

2.3   CVaR-Constrained SSP

As outlined earlier, the risk-constrained objective is:

$$
\min_{\theta \in \Theta} \mathbb{E}\underbrace{\left[\sum_{m=0}^{\tau-1} g(s_m, a_m) \,\middle|\, s_0 = s^0\right]}_{G^\theta(s^0)} \text{ subject to } \mathrm{CVaR}_\alpha \underbrace{\left[\sum_{m=0}^{\tau-1} c(s_m, a_m) \,\middle|\, s_0 = s^0\right]}_{C^\theta(s^0)} \leq K_\alpha.
$$

where τ denotes the first visiting time to terminal state 0, i.e., $\tau = \min\{m \mid s_m = 0\}$. The actions $a_0, \ldots, a_{\tau-1}$ are chosen according to the randomized policy π_θ. Further, α and K_α are constants that specify the confidence level and constraint bound for CVaR, respectively.

Using the standard trick of Lagrangian relaxation for constrained optimization problems, we convert (1) to the following unconstrained problem:

$$
\max_\lambda \min_\theta \left[\mathcal{L}^{\theta,\lambda}(s^0) := G^\theta(s^0) + \lambda\big(\mathrm{CVaR}_\alpha(C^\theta(s^0)) - K_\alpha\big)\right]. \tag{2}
$$

3   Algorithm Structure

In order to solve (2), a standard constrained optimization procedure operates as follows:

Simulation. This is the inner-most loop where the SSP is simulated for several episodes and the resulting costs are aggregated.

Policy Update. This is the intermediate loop where the gradient of the Lagrangian along θ is estimated using simulated values above. The gradient estimates are then used to update policy parameter θ along a descent direction. Note that this loop is for a given value of λ; and

Lagrange Multiplier Update. This is the outer-most loop where the Lagrange multiplier λ is updated along an ascent direction, using the converged values of the inner two loops.

Using two-timescale stochastic approximation (see Chapter 6 of [7]), the policy and Lagrange multiplier update can run in parallel as follows:

$$
\theta_{n+1} = \Gamma\left(\theta_n - \gamma_n \nabla_\theta \mathcal{L}^{\theta,\lambda}(s^0)\right) \quad \text{and} \quad \lambda_{n+1} = \Gamma_\lambda\left(\lambda_n + \beta_n \nabla_\lambda \mathcal{L}^{\theta,\lambda}(s^0)\right), \tag{3}
$$

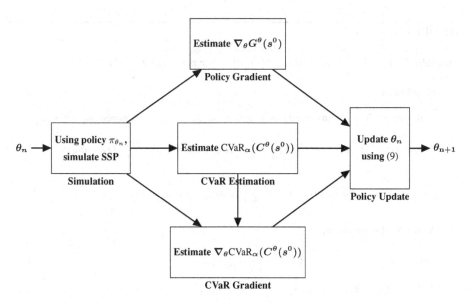

Fig. 1. Overall flow of our algorithms

where Γ and Γ_λ are projection operators that keep the iterates θ_n and λ_n bounded within the compacts sets Θ and $[0, \lambda_{\max}]$ for some $\lambda_{\max} > 0$, respectively. Further, $\gamma_n, \beta_n, n \geq 0$ are step-sizes that satisfy the following assumption:

$$\sum_{n=1}^{\infty} \beta_n = \infty, \sum_{n=1}^{\infty} \gamma_n = \infty, \sum_{n=1}^{\infty} \left(\beta_n^2 + \gamma_n^2\right) < \infty \text{ and } \frac{\beta_n}{\gamma_n} \to 0. \tag{4}$$

The last condition above ensures that θ-recursion proceeds on a faster timescale in comparison to λ-recursion.

Simulation Optimization. No closed form expression for the gradient of the Lagrangian $L^{\theta,\lambda}(s^0)$ is available and moreover, $G^\theta(s^0)$ and $C^\theta(s^0)$ are observable only via simulation. Observe that $\nabla_\theta L^{\theta,\lambda}(s^0) = \nabla_\theta G^\theta(s^0) + \lambda \nabla_\theta \text{CVaR}_\alpha(C^\theta(s^0))$ and $\nabla_\lambda L^{\theta,\lambda}(s^0) = \text{CVaR}_\alpha(C^\theta(s^0)) - K_\alpha$. Hence, in order to update according to (3), we need to estimate, for any policy parameter θ, the following quantities via simulation:
(i) $\text{CVaR}_\alpha(C^\theta(s^0))$;   (ii) $\nabla_\theta G^\theta(s^0)$;   and   (iii) $\nabla_\theta \text{CVaR}_\alpha(C^\theta(s^0))$.
In the following sections, we describe two algorithms that differ in the way they estimate each of the above quantities and subsequently establish that the estimates (and hence the overall algorithms) converge.

4   Algorithm 1: PG-CVaR-SA

Algorithm 1 describes the complete algorithm along with the update rules for the various parameters. The algorithm involves the following crucial components - simulation of the SSP, VaR/CVaR estimation and policy gradients for the objective as well as the CVaR constraint. Each of these components is described in detail in the following.

Algorithm 1. PG-CVaR-SA

Input: parameterized policy $\pi_\theta(\cdot|\cdot)$, step-sizes $\{\zeta_{n,1}, \zeta_{n,2}, \gamma_n, \beta_n\}_{n\geq1}$
Initialization: Starting state s^0, initial policy θ_0, , number of iterations $M \gg 1$.
for $n = 1, 2, \ldots, M$ **do**
    Simulation

    Simulate the SSP for an episode using actions $a_{n,0}, \ldots, a_{n,\tau_n-1}$ generated using $\pi_{\theta_{n-1}}$

    Obtain cost estimates: $G_n := \sum_{j=0}^{\tau_n-1} g(s_{n,j}, a_{n,j})$ and $C_n := \sum_{j=0}^{\tau_n-1} c(s_{n,j}, a_{n,j})$

    Obtain likelihood derivative: $z_n := \sum_{j=0}^{\tau_n-1} \nabla \log \pi_\theta(s_{n,j}, a_{n,j})$

    VaR/CVaR estimation:

    VaR: $\xi_n = \xi_{n-1} - \zeta_{n,1}\left(1 - \frac{1}{1-\alpha}\mathbf{1}_{\{C_n \geq \xi_{n-1}\}}\right),$ (5)

    CVaR: $\psi_n = \psi_{n-1} - \zeta_{n,2}(\psi_{n-1} - v(\xi_{n-1}, C_n)).$ (6)

    Policy Gradient:

    Total Cost: $\bar{G}_n = \bar{G}_{n-1} - \zeta_{n,2}(G_n - \bar{G}_n),$ Gradient: $\partial G_n = \bar{G}_n z_n.$ (7)

    CVaR Gradient:

    Total Cost: $\tilde{C}_n = \tilde{C}_{n-1} - \zeta_{n,2}(C_n - \tilde{C}_n),$ Gradient: $\partial C_n = (\tilde{C}_n - \xi_n)z_n \mathbf{1}_{\{C_n \geq \xi_n\}}.$

                                                                          (8)

    Policy and Lagrange Multiplier Update:

    $\theta_n = \theta_{n-1} - \gamma_n(\partial G_n + \lambda_{n-1}(\partial C_n)),$ $\lambda_n = \Gamma_\lambda\left(\lambda_{n-1} + \beta_n(\psi_n - K_\alpha)\right).$ (9)

end for
Output $(\theta_M, \lambda_M).$

4.1   SSP Simulation

In each iteration of PG-CVaR-SA, an episode of the underlying SSP is simulated. Each episode ends with a visit to the recurrent state 0 of the SSP. Let τ_n denote the time of this visit in episode n. The actions $a_{n,j}, j = 0, \ldots, \tau_n-1$ in episode n are chosen according to the policy $\pi_{\theta_{n-1}}$. Let $G_n := \sum_{j=0}^{\tau_n-1} g(s_{n,j}, a_{n,j})$ and $C_n := \sum_{j=0}^{\tau_n-1} c(s_{n,j}, a_{n,j})$ denote the accumulated cost values. Further, let $z_n := \sum_{j=0}^{\tau_n-1} \nabla \log \pi_\theta(s_{n,j}, a_{n,j})$ denote the likelihood derivative (see Section 4.3 below). The tuple (G_n, C_n, z_n) obtained at the end of the nth episode is used to estimate CVaR as well as policy gradients.

4.2 Estimating VaR and CVaR

A well-known result from [16] is that both VaR and CVaR can be obtained from the solution of a certain convex optimization problem and we recall this result next.

Theorem 1. *For any random variable X, let*

$$v(\xi, X) := \xi + \frac{1}{1 - \alpha}(X - \xi)_+ \text{ and } V(\xi) = \mathbb{E}\left[v(\xi, X)\right] \tag{10}$$

Then, $\text{VaR}_\alpha(X) = (\arg\min V := \{\xi \in \mathbb{R} \mid V'(\xi) = 0\})$, *where V' is the derivative of V w.r.t. ξ. Further,* $\text{CVaR}_\alpha(X) = V(\text{VaR}_\alpha(X))$.

From the above, it is clear that in order to estimate VaR/CVaR, one needs to find a ξ that satisfies $V'(\xi) = 0$. Stochastic approximation (SA) is a natural tool to use in this situation. We briefly introduce SA next and later develop a scheme for estimating CVaR along the lines of [3] on the faster timescale of PG-CVaR-SA.

Stochastic Approximation. The aim is to solve the equation $F(\theta) = 0$ when analytical form of F is not known. However, noisy measurements $F(\theta_n) + \xi_n$ can be obtained, where $\theta_n, n \geq 0$ are the input parameters and $\xi_n, n \geq 0$ are zero-mean random variables, that are not necessarily i.i.d.

The seminal Robbins Monro algorithm solved this problem by employing the following update rule:

$$\theta_{n+1} = \theta_n + \gamma_n(F(\theta_n) + \xi_n). \tag{11}$$

In the above, γ_n are step-sizes that satisfy $\sum_{n=1}^{\infty} \gamma_n = \infty$ and $\sum_{n=1}^{\infty} \gamma_n^2 < \infty$. Under a stability assumption for the iterates and bounded noise, it can be shown that θ_n governed by (11) converges to the solution of $F(\theta) = 0$ (cf. Proposition 1 in Section 6).

CVaR Estimation Using SA. Using the stochastic approximation principle and the result in Theorem 1, we have the following scheme to estimate the VaR/CVaR simultaneously from the simulated samples C_n:

$$\text{VaR:} \ \xi_n = \xi_{n-1} - \zeta_{n,1}\underbrace{(1 - \frac{1}{1 - \alpha}\mathbf{1}_{\{C_n \geq \xi\}})}_{\frac{\partial v}{\partial \xi}(\xi, C_n)}, \tag{12}$$

$$\text{CVaR:} \ \psi_n = \psi_{n-1} - \zeta_{n,2}\left(\psi_{n-1} - v(\xi_{n-1}, C_n)\right). \tag{13}$$

In the above, (12) can be seen as a gradient descent rule, while (13) can be seen as a plain averaging update. The scheme above is similar to the one proposed in [3], except that the random variable $C^\theta(s^0)$ (whose CVaR we try to estimate) is a sum of costs obtained at the end of each episode, unlike the single-shot r.v. considered in [3]. The step-sizes $\zeta_{n,1}, \zeta_{n,2}$ satisfy

$$\sum_{n=1}^{\infty} \zeta_{n,1} = \infty, \sum_{n=1}^{\infty} \zeta_{n,2} = \infty, \sum_{n=1}^{\infty} \left(\zeta_{n,1}^2 + \zeta_{n,2}^2\right) < \infty, \frac{\zeta_{n,2}}{\zeta_{n,1}} \to 0 \text{ and } \frac{\gamma_n}{\zeta_{n,2}} \to 0. \tag{14}$$

The last two conditions above ensure that VaR estimation recursion (12) proceeds on a faster timescale in comparison to CVaR estimation recursion (13) and further, the CVaR recursion itself proceeds on a faster timescale as compared to the policy parameter θ-recursion.

Using the ordinary differential equation (ODE) approach, we establish later that the tuple (ξ_n, ψ_n) converges to $\text{VaR}_\alpha(C^\theta(s^0)), \text{CVaR}_\alpha(C^\theta(s^0))$, for any fixed policy parameter θ (see Theorem 2 in Section 6).

4.3 Policy Gradient

We briefly introduce the technique of likelihood ratios for gradient estimation [9] and later provide the necessary estimate for the gradient of total cost $G^\theta(s^0)$.

Gradient Estimation Using Likelihood Ratios. Consider a Markov chain $\{X_n\}$ with a single recurrent state 0 and transient states $1, \ldots, r$. Let $P(\theta) := [[p_{X_i X_j}(\theta)]]_{i,j=0}^r$ denote the transition probability matrix of this chain. Here $p_{X_i X_j}(\theta)$ denotes the probability of going from state X_i to X_j and is parameterized by θ. Let τ denote the first passage time to the recurrent state 0.

Let $X := (X_0, \ldots, X_{\tau-1})^T$ denote the sequence of states encountered between visits to the recurrent state 0. The aim is to optimize a performance measure $F(\theta) = \mathbb{E}[f(X)]$ for this chain using simulated values of X. The likelihood estimate is obtained by first simulating the Markov chain according to $P(\theta)$ to obtain the samples $X_0, \ldots, X_{\tau-1}$ and then estimate the gradient as follows:

$$\nabla_\theta F(\theta) = \mathbb{E}\left[f(X) \sum_{m=0}^{\tau-1} \frac{\nabla_\theta p_{X_m X_{m+1}}(\theta)}{p_{X_m X_{m+1}}(\theta)}\right].$$

Policy Gradient for the Objective. For estimating the gradient of the objective $G^\theta(s^0)$, we employ the following well-known estimate (cf. [4]):

$$\nabla_\theta G^\theta(s^0) = \mathbb{E}\left[\left(\sum_{n=0}^{\tau-1} g(s_n, a_n)\right) \nabla \log P(s_0, \ldots, s_{\tau-1}) \mid s_0 = s^0\right], \tag{15}$$

where $\nabla \log P(s_0, \ldots, s_\tau)$ is the likelihood derivative for a policy parameterized by θ, defined as

$$\nabla \log P(s_0, \ldots, s_{\tau-1}) = \sum_{m=0}^{\tau-1} \nabla \log \pi_\theta(a_m \mid s_m). \tag{16}$$

The above relation holds owing to the fact that we parameterize the policies and hence, the gradient of the transition probabilities can be estimated from the policy alone. This is the well-known policy gradient technique that makes it amenable for estimating gradient of a performance measure in MDPs, since the transition probabilities are not required and one can work with policies and simulated transitions from the MDP.

4.4 Policy Gradient for the CVaR Constraint

For estimating the gradient of the CVaR of $C^\theta(s^0)$ for a given policy parameter θ, we employ the following likelihood estimate proposed in [20]:

$$\nabla_\theta \text{CVaR}_\alpha(C^\theta(s^0)) \tag{17}$$
$$= \mathbb{E}\left[\left(C^\theta(s^0) - \text{VaR}_\alpha(C^\theta(s^0))\right) \nabla \log P(s_0, \ldots, s_{\tau-1}) \mid C^\theta(s^0) \geq \text{VaR}_\alpha(C^\theta(s^0))\right],$$

where $\nabla \log P(s_0, \ldots, s_\tau)$ is as defined before in (16).

Since we do not know $\text{VaR}_\alpha(C^\theta(s^0))$, in Algorithm 1 we have an online scheme that uses ξ_n (see (12)) to approximate $\text{VaR}_\alpha(C^\theta(s^0))$, which is then used to derive an approximation to the gradient $\nabla_\theta \text{CVaR}_\alpha(C^\theta(s^0))$ (see (8)).

5 Algorithm 2: PG-CVaR-mB

As illustrated in Figure 2, in each iteration n of PG-CVaR-mB, we simulate the SSP for m_n episodes. Recall that each episode starts in the state s^0 and ends in the absorbing state 0. At the end of the simulation, we obtain the total costs and likelihood derivative estimates $\{G_{n,j}, C_{n,j}, z_{n,j}\}_{j=1}^{m_n}$. Using these, the following quantities - $\text{CVaR}_\alpha(C^\theta(s^0))$, $\nabla_\theta \text{CVaR}_\alpha(C^\theta(s^0))$ and $\nabla_\theta G^\theta(s^0)$ - are approximated as follows:

$$\text{VaR: } \xi_n = \frac{1}{m_n} \sum_{j=1}^{m_n} \left(1 - \frac{\mathbf{1}_{\{C_{n,j} \geq \xi_{n-1}\}}}{1 - \alpha}\right), \quad \text{CVaR: } \psi_n = \frac{1}{m_n} \sum_{j=1}^{m_n} v(\xi_{n-1}, C_{n,j})$$

$$\text{Total Cost: } \bar{G}_n = \frac{1}{m_n} \sum_{j=1}^{m_n} G_{n,j}, \quad \text{Policy Gradient: } \partial G_n = \bar{G}_n z_n.$$

$$\text{Total Cost: } \bar{C}_n = \frac{1}{m_n} \sum_{j=1}^{m_n} C_{n,j}, \quad \text{CVaR Gradient: } \partial C_n = (\tilde{C}_n - \xi_n) z_n \mathbf{1}_{\{\bar{C}_n \geq \xi_n\}}.$$

The above approximations can be seen as empirical means of functions of $G_{n,j}, C_{n,j}, z_{n,j}$, respectively.

The policy and Lagrange multiplier updates are as in the earlier algorithm, i.e., according to (9).

Mini-Batch Size. A simple setting for the batch-size m_n is Cn^δ for some $\delta > 0$, i.e., m_n increases as a function of n. We cannot have constant batches, i.e., $\delta = 0$, since the bias of the CVaR estimates and the gradient approximations has to vanish asymptotically.

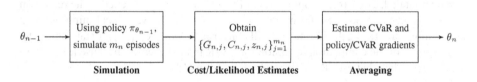

Fig. 2. Illustration of mini-batch principle in PG-CVaR-mB algorithm

6 Outline of Convergence

We analyze our algorithms using the theory of multiple time-scale stochastic approximation [7, Chapter 6]. For the analysis of our algorithms, we make the following assumptions:

(A1) For any $\theta \in \Theta$, the random variable $C^\theta(s^0)$ has a continuous distribution.

(A2) For any $\theta \in \Theta$, the policy π_θ is proper and continuously differentiable in θ.

(A3) Step-sizes β_n, γ_n satisfy the conditions in (4), while $\zeta_{n,1}, \zeta_{n,2}$ satisfy those in (14). We first provide the analysis for PG-CVaR-SA algorithm and later describe the necessary modification for the mini-batch variant[1].

Before the main proof, we recall the following well-known result (cf. Chapter 2 of [7]) related to convergence of stochastic approximation schemes under the existence of a so-called *Lyapunov function*:

Proposition 1. *Consider the following recursive scheme:*

$$\theta_{n+1} = \theta_n + \gamma_n(F(\theta_n) + \xi_{n+1}), \tag{18}$$

where $F : \mathrm{R}^d \to \mathrm{R}^d$ is a L-Lipschitz map and ξ_n a square-integrable martingale difference sequence with respect to the filtration $\mathcal{F}_n := \sigma(\theta_m, \xi_m, m \leq n)$. Moreover,
$\mathbb{E}[\|\xi_{n+1}\|_2^2 \mid \mathcal{F}_n] \leq K(1 + \|\theta_n\|_2^2)$ *for some $K > 0$. The step-sizes γ_n satisfy $\sum_{n=1}^{\infty} \gamma_n = \infty$ and $\sum_{n=1}^{\infty} \gamma_n^2 < \infty$.*

Lyapunov function. *Suppose there exists a continuously differentiable $V : \mathrm{R}^d \to [0, \infty)$ such that $\lim_{\|\theta\|_2 \to \infty} V(\theta) = \infty$. Writing $\mathbb{Z} := \{\theta \in \mathrm{R}^d \mid V(\theta) = 0\} \neq \phi$, V satisfies $\langle F(\theta), \nabla V(\theta) \rangle \leq 0$ with equality if and only if $\theta \in \mathbb{Z}$.*

Then, θ_n governed by (18) converges a.s. to an internally chain transitive set contained in $\mathbb{Z}$.

The steps involved in proving the convergence of PG-CVaR-SA are as follows:

Step 1: CVaR Estimation on Fastest Time-scale

Owing to the time-scale separation, θ and λ can be assumed to be constant while analyzing the VaR/CVaR recursions (12)–(13). The main claim is given as follows:

Theorem 2. *For any given policy parameter θ and Lagrange multiplier λ, the tuple (ξ_n, ψ_n) governed by (12)–(13) almost surely converges to the corresponding true values $(VaR_\alpha(C^\theta(s^0)), CVaR_\alpha(C^\theta(s^0)))$, as $n \to \infty$.*

The claim above regarding ξ_n can inferred by observing that V (see (10)) itself serves as the Lyapunov function and the fact that the step-sizes satisfy (A3) implies the iterates remain bounded. Thus, by an application of Proposition 1, it is evident that the recursion (12) converges to a point in the set $\{\xi \mid V(\xi) = 0\}$. Since every local minimum is a global minimum for V, the iterates ξ_n will converge to $VaR_\alpha(C^\theta(s^0))$. Establishing the convergence of the companion recursion for CVaR in (13) is easier because it is a plain averaging update that uses the VaR estimate ξ_n from (12).

---

[1] Due to space limitations, the detailed convergence proofs will be presented in a longer version of this paper.

Step 2: Policy Update on Intermediate Time-scale

We provide the main arguments to show that θ_n governed by (9) converges to asymptotically stable equilibrium points of the following ODE:

$$\dot{\theta}_t = \check{\Gamma}\left(\nabla_\theta \mathcal{L}^{\theta_t,\lambda}(s^0)\right) = \check{\Gamma}\left(\nabla_\theta G^{\theta_t}(s^0) + \lambda\nabla_\theta \mathrm{CVaR}_\alpha(C^{\theta_t}(s^0))\right), \qquad (19)$$

where $\check{\Gamma}$ is a projection operator that keeps θ_t evolving according to (19) bounded with the compact and convex set $\Theta \in \mathbb{R}^d$. Since λ is on the slowest timescale, its effect is 'quasi-static' on the θ-recursion. Further, since the CVaR estimation and necessary gradient estimates using likelihood ratios are on the faster timescale, the θ-update in (9) views these quantities as almost equilibrated. Thus, the θ-update in (9) can be seen to be asymptotically equivalent to the following in the sense that the difference between the two updates is $o(1)$:

$$\theta_{n+1} = \theta_n - \gamma_n\left(\nabla_\theta G^{\theta_n}(s^0) + \lambda\nabla_\theta \mathrm{CVaR}_\alpha(C^{\theta_n}(s^0))\right),$$

Thus, (9) can be seen to be a discretization of the ODE (19). Moreover, $\mathcal{L}^{\theta,\lambda}(s^0)$ serves as the Lyapunov function for the above recursion, since $\dfrac{d\mathcal{L}^{\theta,\lambda}(s^0)}{dt} = \nabla_\theta\mathcal{L}^{\theta,\lambda}(s^0)\dot{\theta} = \nabla_\theta\mathcal{L}^{\theta,\lambda}(s^0)\left(-\nabla_\theta\mathcal{L}^{\theta,\lambda}(s^0)\right) < 0$. Thus, by an application of Kushner-Clark lemma [10], one can claim the following:

Theorem 3. *For any given Lagrange multiplier λ, θ_n governed by (9) almost surely converges to the asymptotically stable attractor, say $\mathbb{Z}_\lambda$, for the ODE (19), as $n \to \infty$.*

Step 3: Lagrange Multiplier Update on Slowest Time-scale

This is easier in comparison to the other steps and follows using arguments similar to that used for constrained MDPs in general in [6]. The λ recursion views θ as almost equilibrated owing to time-scale separation and converges to the set of asymptotically stable equilibria of the following system of ODEs:

$$\dot{\lambda}_t = \check{\Gamma}_\lambda\left(\nabla_\lambda \mathcal{L}^{\theta^{\lambda_t},\lambda_t}(s^0)\right) = \check{\Gamma}_\lambda\left(\mathrm{CVaR}_\alpha(C^{\theta^{\lambda_t}}(s^0)) - K_\alpha\right) \qquad (20)$$

where θ^λ is the value of the converged policy parameter θ when multiplier λ is used. $\check{\Gamma}_\lambda$ is a suitably defined projection operator that keeps λ_t bounded within $[0, \lambda_{\max}]$.

Theorem 4. *Let $\mathcal{F} \triangleq \{\lambda \mid \lambda \in [0, \lambda_{\max}], \check{\Gamma}_\lambda[\mathrm{CVaR}_\alpha(C^{\theta^\lambda}(s^0)) - K_\alpha] = 0, \theta^\lambda \in \mathbb{Z}_\lambda\}$. Then, λ_n governed by (9) converges almost surely to $\mathcal{F}$ as $n \to \infty$.*

The proof of the above theorem follows using a standard stochastic approximation argument, as in [6, 14], that views λ-recursion as performing gradient ascent. By invoking the envelope theorem of mathematical economics [13], the PG-CVaR-SA algorithm can be shown to converge to a (local) saddle point of $\mathcal{L}^{\theta,\lambda}(s^0)$, i.e., to a tuple (θ^*, λ^*) that are a local minimum w.r.t. θ and a local maximum w.r.t. λ of $\mathcal{L}^{\theta,\lambda}(s^0)$.

PG-CVaR-mB. The proof for mini-batch variant differs only in the first step, i.e., estimation of VaR/CVaR and necessary gradients. Assuming that the number of mini-batch samples $m_n \to \infty$, a straightforward application of law of large numbers establishes that the empirical mean estimates for VaR, CVaR and the necessary gradients in PG-CVaR-mB converge to their corresponding true values. The rest of the proof follows in a similar manner as PG-CVaR-SA.

7   Extension to Incorporate Importance Sampling

In this section, we incorporate an importance sampling procedure in the spirit of [11, 3] to speed up the estimation procedure for VaR/CVaR in our algorithms.

Importance Sampling. Given a random variable X with density $p(\cdot)$ and a function $H(\cdot)$, the aim of an IS based scheme is to estimate the expectation $\mathbb{E}(H(X))$ in a manner that reduces the variance of the estimates. Suppose X is sampled using another distribution with density $\tilde{p}(X, \eta)$ that is parameterized by η, such that $\tilde{p}(X, \eta) = 0 \Rightarrow p(X) = 0$, i.e., satisfies an absolute continuity condition. Then,

$$\mathbb{E}(H(X)) = \mathbb{E}\left[H(X)\frac{p(X)}{\tilde{p}(X, \eta)}\right]. \tag{21}$$

The problem is to choose the parameter η of the sampling distribution so as to minimize the variance of the above estimate.

A slightly different approach based on mean-translation is taken in a recent method proposed in [11]. By translation invariance, we have

$$\mathbb{E}[H(X)] = \mathbb{E}\left[H(X + \eta)\frac{p(X + \eta)}{p(X)}\right], \tag{22}$$

and the objective is to find a η that minimizes the following variance term:

$$Q(\eta) := \mathbb{E}\left[H^2(X + \eta)\frac{p^2(X + \eta)}{p^2(X)}\right]. \tag{23}$$

If ∇Q can be written as an expectation, i.e., $\nabla Q(\eta) = \mathbb{E}[q(\eta, X)]$, then one can hope to estimate this expectation (and hence minimize Q) using a stochastic approximation recursion. However, this is not straightforward since $\|q(\eta, x)\|_2$ is required to be sub-linear to ensure convergence of the resulting scheme[2].

One can get around this problem by double translation of η as suggested first in [11] and later used in [3] for VaR/CVaR estimation. Formally, under classic log-concavity assumptions on $p(X)$, it can be shown that Q is finite, convex and differentiable, so that

$$\nabla Q(\eta) := \mathbb{E}\left[H(X - \eta)^2 \frac{p^2(X - \eta)}{p(X)p(X - 2\eta)} \frac{\nabla p(X - 2\eta)}{p(X - 2\eta)}\right]. \tag{24}$$

Writing $K(\eta, X) := \frac{p^2(X-\eta)}{p(X)p(X-2\eta)} \frac{\nabla p(X-2\eta)}{p(X-2\eta)}$, one can bound $K(\eta, X)$ by a deterministic function of η as follows: $|K(\eta, X)| \leq e^{2\rho|\eta|^b}(A|x|^{b-1} + A|\eta|^{b-1} + B)$, for some constants ρ, A and B. The last piece before present an IS scheme is related to controlling the growth of $H(X)$. We assume that $H(X)$ is controlled by another function $W(X)$ that satisfies $\forall x, |H(x)| \leq W(x), W(x+y) \leq C(1 + W(x))^c(1 + W(y))^c$ and $\mathbb{E}\left[|X|^{2(b-1)}W(X)^{4c}\right] < \infty$.

An IS scheme based on stochastic approximation updates as follows:

$$\eta_n = \eta_{n-1} - \gamma_n \tilde{q}(\eta_{n-1}, X_n), \tag{25}$$

---

[2] As illustrated in [3, Section 2.3], even for a standard Gaussian distributed X, i.e., $X \sim \mathcal{N}(0, 1)$, the function $q(\eta, x) = \exp(|\eta|^2/2 - \eta x)H^2(x)(\eta - x)$ and hence not sub-linear.

where $\tilde{q}(\eta, X) := H(X - \eta)^2 e^{-2\rho|\theta|^b} K(\eta, X)$. In lieu of the above discussion, $\|\tilde{q}(\eta, X)\|_2$ can be bounded by a linear function of $\|\eta\|_2$ and hence, the recursion (25) converges to the set $\{\eta \mid \nabla Q(\eta) = 0\}$ (See Section 2.3 in [3] for more details).

IS for VaR/CVaR Estimation. Let $D := (s_0, a_0, \ldots, s_{\tau-1}, a_{\tau-1})$ be the random variable corresponding to an SSP episode and let $D_n := (s_{n,0}, a_{n,0}, \ldots, s_{n,\tau-1}, a_{n,\tau-1})$ be the nth sample simulated using the distribution of D. Recall that the objective is to estimate the VaR/CVaR of the total cost $C^\theta(s^0)$, for a given policy parameter θ using samples from D.

Applying the IS procedure described above to our setting is not straightforward, as one requires the knowledge of the density, say $p(\cdot)$, of the random variable D. Notice that the density $p(D)$ can be written as $p(D) = \prod_{m=0}^{\tau-1} \pi_\theta(a_m \mid s_m) P(s_{m+1} \mid s_m, a_m)$. As pointed out in earlier works (cf. [17]), the ratio $\frac{p(d)}{p(d')}$ can be computed for two (independent) episodes d and d' without requiring knowledge of the transition dynamics.

In the following, we use $\tilde{p}(D_n) := \prod_{m=0}^{\tau-1} \pi_\theta(a_{n,m} \mid s_{n,m})$ as a proxy for the density $p(D_n)$ and apply the IS scheme described above to reduce the variance of the VaR/CVaR estimation scheme (12)–(13). The update rule of the resulting recursion is as follows:

$$\xi_n = \xi_{n-1} - \zeta_{n,1} e^{-\rho|\eta|^b} \left(1 - \frac{1}{1-\alpha} \mathbf{1}_{\{C_n + \eta_{n-1} \geq \xi_{n-1}\}} \frac{\tilde{p}(D_n + \eta_{n-1})}{\tilde{p}(D_n)}\right), \tag{26}$$

$$\eta_n = \eta_{n-1} - \zeta_{n,1} e^{-2\rho|\eta_{n-1}|^b} \mathbf{1}_{\{C_n - \eta_{n-1} \geq \xi_{n-1}\}} \frac{\tilde{p}^2(D_n - \eta_{n-1}) \nabla \tilde{p}(D_n - 2\eta_{n-1})}{\tilde{p}(D_n)\tilde{p}(D_n - 2\eta)\tilde{p}(D_n - 2\eta_{n-1})}, \tag{27}$$

$$\psi_n = \psi_{n-1} - \zeta_{n,2} \left(\psi_{n-1} - \xi_{n-1} - \frac{1}{1-\alpha}(C_n + \mu_{n-1} - \xi_{n-1}) \right. \tag{28}$$

$$\left. \mathbf{1}_{\{C_n + \mu_{n-1} \geq \xi_{n-1}\}} \frac{\tilde{p}(D_n + \mu_{n-1})}{\tilde{p}(D_n)} \right),$$

$$\mu_n = \mu_{n-1} - \zeta_{n,2} \frac{e^{-2\rho|\mu_{n-1}|^b}}{1 + W(-\mu_{n-1})^{2c} + \xi_{n-1}^2} (C_n - \mu_{n-1} - \xi_{n-1})^2. \tag{29}$$

$$\times \mathbf{1}_{\{C_n - \mu_{n-1} \geq \xi_{n-1}\}} \frac{\tilde{p}^2(D_n - \mu_{n-1})}{\tilde{p}(D_n)\tilde{p}(D_n - 2\mu_{n-1})} \frac{\nabla \tilde{p}(D_n - 2\mu_{n-1})}{\tilde{p}(D_n - 2\mu_{n-1})}.$$

In the above, (26) estimates the VaR, while (27) attempts to find the best variance reducer parameter for VaR estimation procedure. Similarly, (28) estimates the CVaR, while (27) attempts to find the best variance reducer parameter for CVaR estimation procedure.

Note on Convergence. Since we approximated the true density $p(D)$ above using the policy, the convergence analysis of the above scheme is challenging. The nontrivial part is to establish that one can use the approximation $\tilde{p}(\cdot)$ in place of the true density $p(\cdot)$ and this is left for future work. Assuming that this substitution holds, it can be shown that the tuple (η_n, μ_n) updated according to (27) and (29), converge to the optimal variance reducers (η^*, μ^*), using arguments similar to that in Proposition 3.1 of [3]. (η^*, μ^*) minimize the convex functions

$$Q_1(\eta, \xi_\alpha^*) := \mathbb{E}\left[\mathbf{1}_{\{C^\theta(s^0) \geq \xi_\alpha^*\}} \frac{p(D)}{p(D-\eta)} \right] \text{ and}$$

$$Q_2(\mu, \xi_\alpha^*) := \mathbb{E}\left[\left(C^\theta(s^0) - \xi_\alpha^*\right)^2 \mathbf{1}_{\{C^\theta(s^0) \geq \xi_\alpha^*\}} \frac{p(D)}{p(D-\mu)} \right], \text{ where } \xi_\alpha^* \text{ is a VaR}_\alpha(C^\theta(s^0)).$$

8 Comparison to Previous Work

In comparison to [8] and [20], which are the most closely related contributions, we would like to point out the following:

(i) The authors in [8] develop an algorithm for a (finite horizon) CVaR constrained MDP, under a separability condition for the single-stage cost. On the other hand, without a separability condition, we devise policy gradient algorithms in a SSP setting and our algorithms are shown to converge as well; and

(ii) The authors in [20] derive a likelihood estimate for the gradient of the CVaR of a random variable. However, they do not consider a risk-constrained SSP and instead optimize only CVaR. In contrast, we employ a convergent procedure for estimating CVaR that is motivated by a well-known convex optimization problem [16] and then employ policy gradients for both the objective and constraints to find a locally risk-optimal policy.

9 Conclusions

In this paper, we considered the problem of solving a risk-constrained stochastic shortest path. We used Conditional Value-at-Risk (CVaR) as a risk measure and this is motivated by applications in finance and energy markets. Using a careful synthesis of well-known techniques from stochastic approximation, likelihood ratios and importance sampling, we proposed a policy gradient algorithm that is provably convergent to a locally risk-optimal policy. We also proposed another algorithm based on the idea of mini-batches for estimating CVaR from the simulated samples. Both the algorithms incorporated a CVaR estimation procedure along the lines of [3], which in turn is based on the well-known convex optimization representation by Rockafellar-Uryasev [16]. Stochastic approximation or mini-batches are used to approximate CVaR estimates/necessary gradients in the algorithms, while the gradients themselves are obtained using the likelihood ratio technique. Further, since CVaR is an expectation that conditions on the tail probability, to speed up CVaR estimation we incorporated an importance sampling procedure along the lines of [3].

There are several future directions to be explored such as (i) obtaining finite-time bounds for our proposed algorithms , (ii) handling very large state spaces using function approximation, and (iii) applying our algorithms in practical contexts such as portfolio management in finance/energy sectors and revenue maximization in the re-insurance business.

Acknowledgments. The author would like to thank the European Community's Seventh Framework Programme (FP7/2007 − 2013) under grant agreement n° 270327 for funding the research leading to these results.

References

[1] Artzner, P., Delbaen, F., Eber, J.M., Heath, D.: Coherent measures of risk. Mathematical Finance 9(3), 203–228 (1999)

[2] Atchade, Y.F., Fort, G., Moulines, E.: On stochastic proximal gradient algorithms. arXiv preprint arXiv:1402.2365 (2014)

[3] Bardou, O., Frikha, N., Pages, G.: Computing VaR and CVaR using stochastic approximation and adaptive unconstrained importance sampling. Monte Carlo Methods and Applications 15(3), 173–210 (2009)

[4] Bartlett, P.L., Baxter, J.: Infinite-horizon policy-gradient estimation. arXiv preprint arXiv:1106.0665 (2011)

[5] Bertsekas, D.P.: Dynamic Programming and Optimal Control, 3rd edn., vol. II. Athena Scientific (2007)

[6] Borkar, V.: An actor-critic algorithm for constrained Markov decision processes. Systems & Control Letters 54(3), 207–213 (2005)

[7] Borkar, V.: Stochastic approximation: a dynamical systems viewpoint. Cambridge University Press (2008)

[8] Borkar, V., Jain, R.: Risk-constrained Markov decision processes. In: IEEE Conference on Decision and Control (CDC), pp. 2664–2669 (2010)

[9] Glynn, P.W.: Likelilood ratio gradient estimation: an overview. In: Proceedings of the 19th Conference on Winter Simulation, pp. 366–375. ACM (1987)

[10] Kushner, H., Clark, D.: Stochastic approximation methods for constrained and unconstrained systems. Springer (1978)

[11] Lemaire, V., Pages, G.: Unconstrained recursive importance sampling. The Annals of Applied Probability 20(3), 1029–1067 (2010)

[12] Mannor, S., Tsitsiklis, J.: Mean-variance optimization in Markov decision processes. arXiv preprint arXiv:1104.5601 (2011)

[13] Mas-Colell, A., Whinston, M., Green, J.: Microeconomic theory. Oxford University Press (1995)

[14] Prashanth, L.A., Ghavamzadeh, M.: Actor-critic algorithms for risk-sensitive MDPs. In: Neural Information Processing Systems 26, pp. 252–260 (2013)

[15] Robbins, H., Monro, S.: A stochastic approximation method. The Annals of Mathematical Statistics, 400–407 (1951)

[16] Rockafellar, R.T., Uryasev, S.: Optimization of conditional value-at-risk. Journal of Risk 2, 21–42 (2000)

[17] Sutton, R., Barto, A.: Reinforcement learning: An introduction. MIT Press (1998)

[18] Tamar, A., Di Castro, D., Mannor, S.: Policy gradients with variance related risk criteria. In: International Conference on Machine Learning, pp. 387–396 (2012)

[19] Tamar, A., Mannor, S.: Variance Adjusted Actor-Critic Algorithms. arXiv preprint arXiv:1310.3697 (2013)

[20] Tamar, A., Glassner, Y., Mannor, S.: Policy Gradients Beyond Expectations: Conditional Value-at-Risk. arXiv preprint arXiv:1404.3862 (2014)

Bayesian Reinforcement Learning
with Exploration

Tor Lattimore[1] and Marcus Hutter[2]

[1] University of Alberta
tor.lattimore@gmail.com
[2] Australian National University
marcus.hutter@anu.edu.au

Abstract. We consider a general reinforcement learning problem and show that carefully combining the Bayesian optimal policy and an exploring policy leads to minimax sample-complexity bounds in a very general class of (history-based) environments. We also prove lower bounds and show that the new algorithm displays adaptive behaviour when the environment is easier than worst-case.

1 Introduction

We study the question of finding the minimax sample-complexity of reinforcement learning without making the usual Markov assumption, but where the learner has access to a finite set of reinforcement learning environments to which the truth is known to belong. This problem was tackled previously by Dyagilev et al. (2008) and Lattimore et al. (2013a). The new algorithm improves on the theoretical results in both papers and is simultaneously simpler and more elegant. Unlike the latter work, in certain circumstances the new algorithm enjoys adaptive sample-complexity bounds when the true environment is benign. We show that if $\mathcal{M} = \{\mu_1, \cdots, \mu_K\}$ is a carefully chosen finite set of history-based reinforcement learning environments, then every algorithm is necessarily ε-suboptimal for $\Omega\left(\frac{K}{\varepsilon^2(1-\gamma)^3} \log \frac{K}{\delta}\right)$ time-steps with probability at least δ where γ is the discount factor. The algorithm presented has a sample-complexity bound equal to that bound except for one factor of $\log \frac{1}{\varepsilon(1-\gamma)}$, so the minimax sample-complexity of this problem is essentially known.

Aside from the previously mentioned papers, there has been little work on this problem, although sample-complexity bounds have been proven for MDPs (Lattimore and Hutter, 2012; Szita and Szepesvári, 2010; Kearns and Singh, 2002, and references there-in), as well as partially observable and factored MDPs (Chakraborty and Stone, 2011; Even-Dar et al., 2005). There is also a significant literature on the regret criterion for MDPs (Azar et al., 2013; Auer et al., 2010, and references there-in), but meaningful results cannot be obtained without a connectedness assumption that we avoid here. Regret bounds are known if the true environment is finite-state, Markov and communicating, but where the state is not observed directly (Odalric-Ambrym et al., 2013). Less restricted settings have also

P. Auer et al. (Eds.): ALT 2014, LNAI 8776, pp. 170–184, 2014.

been studied. Sunehag and Hutter (2012) proved sample-complexity bounds for the same type of reinforcement learning problems that we do, but only for deterministic environments (for the stochastic case they gave asymptotic results). Also similar is the k-meteorologist problem studied by Diuk et al. (2009), but they consider only the 1-step problem, which is equivalent to the case where the discount factor $\gamma = 0$. In that case their algorithm is comparable to the one developed by Lattimore et al. (2013a) and suffers from the same drawbacks, most notable of which is non-adaptivity. A more detailed discussion is given in the conclusion. Recently there has been a growing interesting in algorithms based on the "near-Bayesian" Thompson sampling. See, for example, the work by Osband et al. (2013) and references there-in. Note that the aforementioned paper deals with a Bayesian regret criterion for MDPs, rather than the frequentist sample-complexity results presented here.

The new algorithm is based loosely on the universal Bayesian optimal reinforcement learning algorithm studied in depth by Hutter (2005). Unfortunately, a pure Bayesian approach may not explore sufficiently to enjoy a finite sample-complexity bound (Orseau, 2010) (some exceptions by Hutter (2002)). For this reason we add exploration periods to ensure that sufficient exploration occurs for sample-complexity bounds to become possible.

2   Notation

Due to lack of space, many of the easier proofs or omitted, along with results that are periphery to the main bound on the sample-complexity. All proofs can be found in the technical report (Lattimore and Hutter, 2014).

Strings/Sequences. A finite string of length n over non-empty alphabet $\mathcal{H}$ is a finite sequence $x_1 x_2 x_3 \cdots x_n$ where $x_k \in \mathcal{H}$. An infinite sequence over $\mathcal{H}$ is a sequence $x_1 x_2 x_3 \cdots$. The set of sequences over alphabet $\mathcal{H}$ of length n is denoted by $\mathcal{H}^n$. The set of finite sequences over alphabet $\mathcal{H}$ is denoted by $\mathcal{H}^* := \bigcup_{n=0}^{\infty} \mathcal{H}^n$. The set of sequences of length at most n is $\mathcal{H}^{\leq n} := \bigcup_{k=0}^{n} \mathcal{H}^k$. The uncountable set of infinite sequences is $\mathcal{H}^\infty$. For $x \in \mathcal{H}^* \cup \mathcal{H}^\infty$, the length of x is $\ell(x)$. The empty string of length zero is denoted by ϵ, which should not be confused with small constants denoted by ε. Subsequences are $x_{1:t} := x_1 x_2 x_3 \cdots x_t$ and $x_{<t} := x_{1:t-1}$. We say x is a prefix of y and write $x \sqsubseteq y$ if $\ell(x) \leq \ell(y)$ and $x_k = y_k$ for all $k \leq \ell(x)$. The words string and sequence are used interchangeably, although the former is more likely to be finite and the latter more likely to be infinite. Strings may be concatenated in the obvious way. If $x \in \mathcal{H}^*$, then x^k is defined to be k concatenations of x. A set $A \subset \mathcal{H}^*$ is prefix free if for all $x, y \in \mathcal{H}$, $x \sqsubseteq y \implies x = y$. A prefix free set A is complete if for all infinite histories $y \in \mathcal{H}^\infty$ there exists an $x \in A$ such that $x \sqsubseteq y$.

History Sequences. Let $\mathcal{A}$, $\mathcal{O}$ and $\mathcal{R} \subset [0,1]$ be finite sets of actions, observations and rewards respectively and $\mathcal{H} := \mathcal{A} \times \mathcal{O} \times \mathcal{R}$. The set of infinite history sequences is denoted $\mathcal{H}^\infty$ while $\mathcal{H}^*$ is the set of all finite-length histories. The action/observation/reward at time-step t of history x are denoted by $a_t(x)$, $o_t(x)$, $r_t(x)$ respectively.

Environments and Policies. An environment μ is a set of conditional probability distributions $\mu(\cdot|x,a) : \mathcal{R} \times \mathcal{O} \to [0,1]$ where $x \in \mathcal{H}^*$ is a finite history and $a \in \mathcal{A}$ is an action. The value $\mu(r,o|x,a)$ is the probability of environment μ generating reward $r \in \mathcal{R}$ and observation $o \in \mathcal{O}$ given finite history $x \in \mathcal{H}^*$ has occurred and action $a \in \mathcal{A}$ has just been taken by the agent. A deterministic policy is a function $\pi : \mathcal{H}^* \to \mathcal{A}$ where $\pi(x)$ is the action taken by policy π given history x. The space of all deterministic policies is denoted by Π. A deterministic policy π is consistent with history $x \in \mathcal{H}^*$ if $\pi(x_{<t}) = a_t(x)$ for all $t \leq \ell(x)$. The set of policies consistent with history x is denoted by $\Pi(x)$.

Probability Spaces. A policy and environment interact sequentially to stochastically generate infinite histories. In order to be rigorous, it is necessary to define a (filtered) probability space on the set of infinite histories $\mathcal{H}^\infty$. Let $x \in \mathcal{H}^*$ be a finite history, then $\Gamma_x := \{y \in \mathcal{H}^\infty : x \sqsubseteq y\}$ is the set of all infinite histories starting with x and is called the cylinder set of x. Now define σ-algebras generated by the cylinders of $\mathcal{H}^*$ and $\mathcal{H}^t$ by $\mathcal{F} := \sigma(\{\Gamma_x : x \in \mathcal{H}^*\})$ and $\mathcal{F}_{<t} := \sigma(\{\Gamma_x : x \in \mathcal{H}^{t-1}\})$. Then $(\mathcal{H}^\infty, \mathcal{F}, \{\mathcal{F}_{<t}\})$ is a filtered probability space. Throughout we use the convention that time starts at 1 with the empty history. An environment and policy interact sequentially to induce a measure $\mu^\pi : \mathcal{F} \to [0,1]$ on the filtered probability space $(\mathcal{H}^\infty, \mathcal{F}, \{\mathcal{F}_{<t}\})$. If $A \in \mathcal{F} \subseteq \mathcal{H}^\infty$, then $\mu^\pi(A)$ is the probability of the event A occurring. As is common in the literature, we abuse notation and use the short-hand $\mu^\pi(x) := \mu^\pi(\Gamma_x)$. If $x,y \in \mathcal{H}^*$, then conditional probabilities are $\mu^\pi(y|x) := \mu^\pi(xy)/\mu^\pi(x)$. Expectations with respect to μ^π are denoted by $\mathbb{E}_\mu^\pi$. If ρ is any measure on $(\mathcal{H}^\infty, \mathcal{F}, \{\mathcal{F}_{<t}\})$, then we define useful random variables:

$$\rho_{<t}(x) := \rho(x_{<t}) \qquad \rho_{1:t}(x) := \rho(x_{1:t}) \qquad \rho_{t:t+d}(x) := \frac{\rho(x_{1:t+d})}{\rho(x_{<t})}.$$

Discounting and Value Functions. Let $\gamma \in [0,1)$ be the discount factor, then the discounted value of history x is the expected discounted cumulative reward.

$$V_\mu^\pi(x;d) := \mathbb{E}_\mu^\pi\left[\sum_{t=\ell(x)+1}^{\ell(x)+d} \gamma^{t-\ell(x)-1} r_t \,\Big|\, x\right] \qquad V_\mu^\pi(x) := \lim_{d\to\infty} V_\mu^\pi(x;d),$$

where d is a horizon after which reards are not counted and we assume that $0^0 = 1$ when $\gamma = 0$. The optimal policy in environment μ is $\pi_\mu^* := \arg\max_{\pi \in \Pi} V_\mu^\pi(\epsilon)$. Since rewards are bounded in $[0,1]$ and values are discounted, the value function is also bounded: $V_\mu^\pi(x) \in [0, \frac{1}{1-\gamma}]$. The value of the optimal policy in environment μ and having observed history x is $V_\mu^*(x)$. Since the discount factor does not vary within the results we omit it from the notation for the value function, but it is important to note that all values depend on this quantity.

3   Algorithm

To begin, we consider only the prediction problem where π is fixed, but μ is unknown and the task is to predict future observations and rewards given the history. We assume that π is some fixed policy and that $\mu \in \mathcal{M} = \{\nu_1 \cdots \nu_K\}$ where

$\mathcal{M}$ is known, but not μ. The Bayesian mixture measure is $\xi^\pi := \sum_{\nu \in \mathcal{M}} w_\nu \nu^\pi$ where $w : \mathcal{M} \to [0, 1]$ is a probability distribution on $\mathcal{M}$. The Bayesian optimal policy is defined by $\pi_\xi^* := \arg\max_\pi V_\xi^\pi(\epsilon) \equiv \arg\max_\pi \sum_{\nu \in \mathcal{M}} w_\nu V_\nu^\pi(\epsilon)$. It is reasonably well-known that the predictive distribution of the Bayesian mixture converges almost surely to the truth for all μ, and that it does so fast with respect to a variety of different metrics. To measure convergence we define the d-step total variation and squared Hellinger distances between predictive distributions of ξ^π and ν^π given the history at time-step t.

$$\delta_x^d(\nu^\pi, \xi^\pi) := \frac{1}{2} \sum_{y \in \mathcal{H}^d} |\nu^\pi(y|x) - \xi^\pi(y|x)| \qquad \delta_t^d(\nu^\pi, \xi^\pi)(x) := \delta_{x_{<t}}^d(\nu^\pi, \xi^\pi)$$

$$h_x^d(\nu^\pi, \xi^\pi) := \frac{1}{2} \sum_{y \in \mathcal{H}^d} \left(\sqrt{\nu^\pi(y|x)} - \sqrt{\xi^\pi(y|x)} \right)^2 \qquad h_t^d(\nu^\pi, \xi^\pi)(x) := h_{x_{<t}}^d(\nu^\pi, \xi^\pi).$$

where the distances on the right hand side are defined as random variables. The following theorem by Hutter and Muchnik (2007) will be useful.

Theorem 1. If $\mu \in \mathcal{M}$, then $\mathbb{E}_\mu^\pi \exp\left(\frac{1}{2} \sum_{t=1}^\infty h_t^1(\mu^\pi, \xi^\pi) \right) \leq \sqrt{\frac{1}{w_\mu}}$.

More usual than the Hellinger distance in the analysis of Bayesian sequence prediction is the relative entropy, but this quantity is unbounded, which somewhat surprisingly leads to weaker results (Lattimore et al., 2013b). The following theorem is a simple generalisation of Theorem 1 to the multi-step case.

Theorem 2. Let $d \geq 1$ and $\{\tau_k\}_{k=1}^\infty$ be a sequence of $(\mathcal{H}^\infty, \mathcal{F}, \{\mathcal{F}_{<t}\})$-measurable stopping times such that $\tau_k + d \leq \tau_{k+1}$ for all k. Then for all $\mu \in \mathcal{M}$

$$\mathbb{E}_\mu^\pi \exp\left(\frac{1}{2} \sum_{k=1}^\infty h_{\tau_k}^d(\mu^\pi, \xi^\pi) \right) \leq \sqrt{\frac{1}{w_\mu}}.$$

Theorem 1 is regained by choosing $\tau_k = k$ and $d = 1$. The proof of Theorem 2 can be found in the technical report. Theorem 2 shows that the predictive distribution of the Bayesian mixture converges fast to the true predictive distribution. In particular, with high probability the cumulative squared total-variation distance does not greatly exceed $\log \frac{1}{w_\mu}$.

Corollary 3. If $\delta > 0$, then $\mu^\pi \left(\sum_{k=1}^\infty \delta_{\tau_k}^d(\mu^\pi, \xi^\pi)^2 \geq \log \frac{1}{w_\mu} + \log \frac{1}{\delta^2} \right) \leq \delta$.

Proof. We combine Markov's inequality with Theorem 2.

$$\mu^\pi \left(\sum_{k=1}^\infty \delta_{\tau_k}^d(\mu^\pi, \xi^\pi)^2 \geq \log \frac{1}{w_\mu} + \log \frac{1}{\delta^2} \right) \overset{(a)}{\leq} \mu^\pi \left(\sum_{k=1}^\infty h_{\tau_k}^d(\mu^\pi, \xi^\pi) \geq \log \frac{1}{w_\mu \delta^2} \right)$$

$$\overset{(b)}{=} \mu^\pi \left(\exp\left(\frac{1}{2} \sum_{k=1}^\infty h_{\tau_k}^d(\mu^\pi, \xi^\pi) \right) \geq \frac{1}{\delta} \sqrt{\frac{1}{w_\mu}} \right) \overset{(c)}{\leq} \delta,$$

where (a) follows since the Hellinger distance upper bounds the total variation distance, (b) is trivial, and (c) by Markov's inequality. $\square$

The consequence of the above is that a Bayesian predictor quickly learns the true distribution of the rewards and observations it will receive. On first sight this might seem promising for Bayesian reinforcement learning, but there is a problem. Bayesian sequence prediction is only capable of learning to predict given a fixed policy. But in RL the agent must choose its action at each time-step, and to do this effectively it must be able to predict the consequences of *all* actions, not only the action it ultimately ends up taking. We side-step this problem in the new algorithm called BayesExp by only following the Bayesian optimal policy when it is guaranteed to be nearly optimal and exploring otherwise. The BayesExp algorithm is as follows:

Algorithm 1. BayesExp

1: **Inputs:** ε, δ and $\mathcal{M} = \{\nu_1, \nu_2, \cdots, \nu_K\}$
2: $\delta_1 \leftarrow \delta/2$ and $\varepsilon_1 \leftarrow \varepsilon(1-\gamma)/4$ and $\varepsilon_2 \leftarrow \varepsilon/12$ and $d \leftarrow \frac{\log \varepsilon_2(1-\gamma)}{\log \gamma}$
3: $x \leftarrow \epsilon$ and $t \leftarrow 1$ and $w_\nu \leftarrow 1/K$ and $D(\nu) \leftarrow 0, \forall \nu$
4: **loop**
5:      $\Pi^* \leftarrow \{\pi_\nu^* : \nu \in \mathcal{M}\} \cup \{\pi_\xi^*\}$
6:      $\pi \leftarrow \arg\max_{\pi \in \Pi^*} \max_{\nu \in \mathcal{M}} \delta_x^d(\nu^\pi, \xi^\pi)$
7:      $\Delta \leftarrow \max_{\pi \in \Pi^*, \nu \in \mathcal{M}} \delta_x^d(\nu^\pi, \xi^\pi)$
8:      **if** $\Delta > \varepsilon_1$ **then**
9:          $D(\nu) \leftarrow D(\nu) + \delta_x^d(\nu^\pi, \xi^\pi)^2, \forall \nu$
10:         **for** $j = 1 \rightarrow d$ **do**
11:             ACT(π)
12:         $\mathcal{M} \leftarrow \{\nu : D(\nu) \leq \log K/\delta_1^2\}$
13:     **else**
14:         $D(\nu) \leftarrow D(\nu) + \delta_x^1(\nu^{\pi_\xi^*}, \xi^{\pi_\xi^*})^2, \forall \nu$
15:         ACT(π_ξ^*)
16: **function** ACT(π)
17:     Take action $a = \pi(x)$ and observe $o \in \mathcal{O}$ and $r \in \mathcal{R}$ from environment
18:     $t \leftarrow t+1$ and $x \leftarrow xaor$

Indices. For the sake of readability the time indices have been omitted in the pseudo-code above. Throughout the analysis we write $\mathcal{M}_t$, $D_t(\nu)$ and Δ_t for the values of Δ, $D(\nu)$ and $\mathcal{M}$ as computed by BayesExp at time-step t. Similarly, $\mathcal{M}_z$, $D_z(\nu)$ and Δ_z are the values of $\mathcal{M}$, $D(\nu)$ and Δ respectively as they would be computed given the algorithm had reached history $z \in \mathcal{H}^*$.

Exploration Phases. The algorithm operates in phases of exploration and exploitation. If there exists an optimal policy π' with respect to some plausible environment such that the d-step total-variation distance between $\nu^{\pi'}$ and $\xi^{\pi'}$ is larger than ε_1, then the algorithm follows π' for exactly d time-steps. This period is called an exploration phase. The set of time-steps triggering exploring phases is denoted by $E \subseteq \mathbb{N}$. While the set of time-steps spent in exploration phases is denoted by $E_d := \bigcup_{t \in E} \{t, t+1, \cdots, t+d-1\}$.

Exploitation Time-Steps. If BayesExp is not exploring at time-step t, then t is an exploiting time-step where BayesExp is following the Bayes optimal policy. The set of all exploitation time-steps is denoted by $T := \mathbb{N} - E_d$.

Failure Phases. For the remainder of this section the policy π refers to the policy of BayesExp. A failure phase is a period of d time-steps triggered at time-step t provided t is not part of a previous exploration/failure phase and $\mu \in M_t$ and $V_\mu^*(x_{<t}) - V_\mu^\pi(x_{<t}) > \varepsilon$. We denote the set of time-steps triggering failure phases by $F \subset \mathbb{N}$ and the set of time-steps spent in failure phases by $F_d := \bigcup_{t \in F} \{t, t+1, \cdots, t+d-1\}$. Failure phases depend on the unknown μ, so are not known to the algorithm and are only used in the analysis.

4 Upper Bound on Sample-Complexity

Theorem 4. *Suppose π is the policy of Algorithm 1 given input $\varepsilon > 0$, $\delta > 0$ and $M = \{\nu_1, \cdots, \nu_K\}$. If $\mu \in M$, then*

$$\mu^\pi \left(\sum_{t=1}^{\infty} \mathbb{1}\{V_\mu^*(x_{<t}) - V_\mu^\pi(x_{<t}) > \varepsilon\} > \frac{416Kd}{\varepsilon^2(1-\gamma)^2} \log \frac{4K}{\delta^2} \right) \le \delta$$

where x is the infinite history sampled from μ^π and $d = \frac{\log(\varepsilon_2(1-\gamma))}{\log \gamma}$ is the effective horizon.

Noting that $d \in O(\frac{1}{1-\gamma} \log \frac{1}{\varepsilon(1-\gamma)})$, the sample-complexity is bounded by

$$O \left(\frac{K}{\varepsilon^2(1-\gamma)^3} \left(\log \frac{1}{\varepsilon(1-\gamma)} \right) \left(\log \frac{K}{\delta} \right) \right).$$

Proof Overview
(a) By definition, if $V_\mu^*(x_{<t}) - V_\mu^\pi(x_{<t}) > \varepsilon$, then either $\mu \notin M_t$ or t is part of an exploration/failure phase, $t \in E_d \cup F_d$.

(b) First we show that $\mu \in M_t$ for all t with probability at least $1 - \delta_1$.

(c) We then use the definition of the algorithm to bound

$$|E| \le \frac{K}{\varepsilon_1^2} \log \frac{K}{\delta_1^2} \implies |E_d| \le \frac{Kd}{\varepsilon_1^2} \log \frac{K}{\delta_1^2}.$$

(d) If $\mu \in M_t$ and BayesExp is exploiting, then all plausible environments are sufficiently close under all optimal policies and so

$$V_\mu^*(x_{<t}) - V_\mu^{\pi_\xi^*}(x_{<t}) \lesssim \varepsilon. \qquad (1)$$

(e) Unfortunately (1) does not imply that $V_\mu^*(x_{<t}) - V_\mu^\pi(x_{<t}) \le \varepsilon$. A careful argument is required to ensure that the number of errors in exploitation periods is also small, which essentially means bounding the number of failure phases. This eventually follows from the fact that if BayesExp is sub-optimal while exploiting, then there must be some probability of triggering an exploration phase, which cannot happen too often.

The following lemmas are required for the proof of Theorem 4 and could be skipped until they are referred to.

Lemma 5. *Suppose t is a time-step when BayesExp is exploiting given history $x_{<t}$. Then $V_\mu^{\pi_\xi^*}(x_{<t}) - V_\mu^\pi(x_{<t}) \le \sum_{y \in Y} \mu^\pi(y|x_{<t}) \left(V_\mu^{\pi_\xi^*}(x_{<t}y) - V_\mu^\pi(x_{<t}y) \right) + \varepsilon_2$, where Y is the set of finite history sequences y of length at most d such that BayesExp would explore given history $x_{<t}y$.*

$$Y = \left\{ y \in \mathcal{H}^{\le d} : BayesExp \text{ explores given history } x_{<t}y \right\}.$$

Proof. Define $\bar{Y} = Y \cup \{y \in \mathcal{H}^d : \forall z \in Y, z \not\sqsubseteq y\}$, which is complete and prefix free by definition. Since t is an exploitation time-step, BayesExp will follow policy π_ξ^* until such a time as it starts an exploration phase. Therefore by Lemma 13

$$V_\mu^{\pi_\xi^*}(x_{<t}) - V_\mu^\pi(x_{<t}) \overset{(a)}{=} \sum_{y \in \bar{Y}} \mu^\pi(y|x_{<t}) \gamma^{\ell(y)} \left(V_\mu^{\pi_\xi^*}(x_{<t}y) - V_\mu^\pi(x_{<t}y) \right)$$

$$\overset{(b)}{\le} \sum_{y \in Y} \mu^\pi(y|x_{<t}) \left| V_\mu^{\pi_\xi^*}(x_{<t}y) - V_\mu^\pi(x_{<t}y) \right| + \varepsilon_2$$

where (a) follows from Lemma 13. (b) by dropping all $y \in \bar{Y} - Y$ and using the fact that for $y \in \bar{Y} - Y$ we have $\ell(y) = d$, which by the definition of the horizon $d = \frac{\log \varepsilon_2(1-\gamma)}{\log \gamma}$ implies that the ratio $\gamma^d \le \varepsilon_2(1-\gamma)$. □

Lemma 6. *Let $x_{<t}$ be the history at an exploitation time-step $t \in T$ and assume $\mu \in \mathcal{M}_t$. Then $V_\mu^*(x_{<t}; d) - V_\mu^{\pi_\xi^*}(x_{<t}; d) \le \frac{2\varepsilon_1}{1-\gamma}$.*

Proof. Since t is an exploitation time-step we have that $\Delta_t \le \varepsilon_1$. Therefore

$$V_\mu^*(x_{<t}; d) - V_\mu^{\pi_\xi^*}(x_{<t}; d) \overset{(a)}{=} V_\mu^{\pi_\mu^*}(x_{<t}; d) - V_\mu^{\pi_\xi^*}(x_{<t}; d)$$

$$\overset{(b)}{\le} V_\mu^{\pi_\mu^*}(x_{<t}; d) - V_\xi^{\pi_\mu^*}(x_{<t}; d) + V_\xi^{\pi_\mu^*}(x_{<t}; d) - V_\xi^{\pi_\xi^*}(x_{<t}; d)$$

$$+ V_\xi^{\pi_\xi^*}(x_{<t}; d) - V_\mu^{\pi_\xi^*}(x_{<t}; d)$$

$$\overset{(c)}{\le} \frac{1}{1-\gamma} \left(\delta_{x_{<t}}^d(\mu^{\pi_\mu^*}, \xi^{\pi_\mu^*}) + \delta_{x_{<t}}^d(\mu^{\pi_\xi^*}, \xi^{\pi_\xi^*}) \right) \overset{(d)}{\le} \frac{2\Delta_t}{1-\gamma} \overset{(e)}{\le} \frac{2\varepsilon_1}{1-\gamma}$$

where (a) is the definition of $V_\mu^*(x_{<t}; d)$. (b) by adding and subtracting value functions. (c) by Lemma 11. (d) by the definition of Δ_t and $\mu \in \mathcal{M}_t$. (e) since t is an exploitation time-step. □

Lemma 7. *Let $x_{<t}$ be the history at an exploration time-step $t \in E$. Then*

$$V_\mu^*(x_{<t}; d) - V_\mu^\pi(x_{<t}; d) \le \frac{\max\{4\Delta_t, \mathbb{1}\{\mu \notin \mathcal{M}_t\}\}}{1-\gamma}.$$

Proof. If $\mu \notin \mathcal{M}_t$, then we use the trivial bound of $\frac{1}{1-\gamma}$. Now assume $\mu \in \mathcal{M}_t$ and let $\pi_\rho^* = \arg\max_{\pi \in \Pi_t^*} \max_{\nu \in \mathcal{M}_t} \delta_t^d(\nu^{\pi_\rho^*}, \xi^{\pi_\rho^*})$, which means that $\rho \in \mathcal{M}_t \cup \{\xi\}$. Therefore

$$
V_\mu^*(x_{<t}; d) - V_\mu^\pi(x_{<t}; d) \overset{(a)}{=} V_\mu^{\pi_\mu^*}(x_{<t}; d) - V_\mu^{\pi_\rho^*}(x_{<t}; d)
$$

$$
\overset{(b)}{\leq} V_\rho^{\pi_\mu^*}(x_{<t}; d) - V_\rho^{\pi_\rho^*}(x_{<t}; d) + \left(V_\mu^{\pi_\mu^*}(x_{<t}; d) - V_\rho^{\pi_\mu^*}(x_{<t}; d) \right)
$$

$$
+ \left(V_\rho^{\pi_\rho^*}(x_{<t}; d) - V_\mu^{\pi_\rho^*}(x_{<t}; d) \right)
$$

$$
\overset{(c)}{\leq} \frac{1}{1-\gamma} \left(\delta_{x_{<t}}^d(\rho^{\pi_\mu^*}, \mu^{\pi_\mu^*}) + \delta_{x_{<t}}^d(\rho^{\pi_\rho^*}, \mu^{\pi_\rho^*}) \right)
$$

$$
\overset{(d)}{\leq} \frac{1}{1-\gamma} \left(\delta_{x_{<t}}^d(\rho^{\pi_\mu^*}, \xi^{\pi_\mu^*}) + \delta_{x_{<t}}^d(\xi^{\pi_\mu^*}, \mu^{\pi_\mu^*}) + \delta_{x_{<t}}^d(\rho^{\pi_\rho^*}, \xi^{\pi_\rho^*}) + \delta_{x_{<t}}^d(\xi^{\pi_\rho^*}, \mu^{\pi_\rho^*}) \right)
$$

$$
\overset{(e)}{\leq} \frac{4\Delta_t}{1-\gamma}
$$

where (a) follows since BayesExp follows policy π_ρ^* while exploring. (b) by expanding the values. (c) by Lemma 11. (d) by the triangle inequality. (e) by the definition of Δ_t and because $\rho, \mu \in \mathcal{M} \cup \{\xi\}$. □

The proof of Theorem 4 uses a number of constants that are functions of each other. For convenience they are described in the table below.

Table 1. Constants for Theorem 4

constant	ε_1	ε_2	ε_3	ε_4	δ_1	d
constraint			$= \varepsilon_2 + \frac{2\varepsilon_1}{1-\gamma}$	$= (\varepsilon - \varepsilon_3 - 2\varepsilon_2)(1-\gamma)$		
value	$\varepsilon(1-\gamma)/4$	$\varepsilon/12$	$7\varepsilon/12$	$\varepsilon(1-\gamma)/4$	$\delta/2$	$\frac{\log \varepsilon_2}{\log \gamma}$

Proof (of Theorem 4). Following the plan, we start by bounding the probability that μ is removed from $\mathcal{M}_t$.

Step 1: Bounding Inconsistency Probability. Let A_1 be the event that $\mu \in \mathcal{M}_t$ for all time-steps t. Environment μ is removed from the model class $\mathcal{M}_t$ only once the counter $D(\mu)$ exceeds $\log K/\delta_1^2$. But $D(\mu)$ is the cumulative squared total variation distance between μ^π and ξ^π, which by Corollary 3 is bounded by $\log K/\delta_1^2$ with μ^π-probability at least $1 - \delta_1$ and so $\mu^\pi(A_1) \geq 1 - \delta_1$.

Step 2: Bounding Exploration Phases. Let t be the start of an exploration phase. Then by definition there exists a $\nu \in \mathcal{M}_t$ such that $\delta_t^d(\nu^\pi, \xi^\pi) > \varepsilon_1$ and so $D(\nu)$ is incremented by at least ε_1^2. Since an environment is removed from $\mathcal{M}$ once $D(\nu)$ exceeds $\log K/\delta_1^2$, the number of exploration phases is bounded by $E_{\max} := \frac{K}{\varepsilon_1^2} \log \frac{K}{\delta_1^2}$. Since each exploration phase is exactly d time-steps long, the number of time-steps spent in exploration phases satisfies

$$
|E_d| \leq \frac{Kd}{\varepsilon_1^2} \log \frac{K}{\delta_1^2}. \tag{2}
$$

By identical reasoning it holds that $\sum_{t \in E} \Delta_t^2 \leq K \log \frac{K}{\delta_1^2}$.   (3)

Note that both (2) and (3) hold surely over all history trajectories.

Step 3: Exploitation Success. Assume that event A_1 is true, which means that $\mu \in \mathcal{M}_t$ for all time-steps. Let $t \in T$ be a time-step when BayesExp is exploiting. Therefore

$$V_\mu^*(x_{<t}) - V_\mu^{\pi_\xi^*}(x_{<t}) \overset{(a)}{\leq} \varepsilon_2 + V_\mu^*(x_{<t}; d) - V_\mu^{\pi_\xi^*}(x_{<t}; d) \overset{(b)}{\leq} \frac{2\varepsilon_1}{1-\gamma} + \varepsilon_2 =: \varepsilon_3 < \varepsilon$$

where (a) follows by truncating the horizon (Lemma 12) and (b) by Lemma 6.

Step 4: Connecting the Policies. We now bound the number of failure phases. The intuition is that if BayesExp is exploiting at time-step t, then the Bayes-optimal policy π_ξ^* is near-optimal. Since BayesExp follows this policy until an exploration phase, $V_\mu^*(x) - V_\mu^\pi(x)$ can only be large if there is a reasonable probability of encountering an exploration phase within the next d time-steps. By some form of concentration inequality this cannot happen too often before an exploration phase actually occurs, which will lead to the correct bound on the number of time-steps when $V_\mu^*(x_{<t}) - V_\mu^\pi(x_{<t}) > \varepsilon$. Let $F = \{t_1, t_2, \cdots\}$ be the set of time-steps triggering failure phases with corresponding histories $x_{<t_k}$. For $k > |F|$ define $t_k = \infty$. At time-step t_k having observed history $x_{<t_k}$ define Y as in the statement of Lemma 5 to be the set of finite histories of length at most d such that BayesExp would explore upon reaching history $x_{<t_k} y$.

$$Y := \{y \in \mathcal{H}^{\leq d} : \text{BayesExp explores given history } x_{<t_k} y\}.$$

For $t_k < \infty$ we have that

$$\varepsilon \overset{(a)}{<} V_\mu^*(x_{<t_k}) - V_\mu^\pi(x_{<t_k}) \overset{(b)}{=} V_\mu^*(x_{<t_k}) - V_\mu^{\pi_\xi^*}(x_{<t_k}) + V_\mu^{\pi_\xi^*}(x_{<t_k}) - V_\mu^\pi(x_{<t_k})$$

$$\overset{(c)}{\leq} \varepsilon_3 + V_\mu^{\pi_\xi^*}(x_{<t_k}) - V_\mu^\pi(x_{<t_k})$$

$$\overset{(d)}{\leq} \varepsilon_3 + \varepsilon_2 + \sum_{y \in Y} \mu^\pi(y|x_{<t_k}) \left(V_\mu^{\pi_\xi^*}(x_{<t_k} y) - V_\mu^\pi(x_{<t_k} y) \right)$$

$$\overset{(e)}{\leq} \varepsilon_3 + 2\varepsilon_2 + \sum_{y \in Y} \mu^\pi(y|x_{<t_k}) \left(V_\mu^*(x_{<t_k} y; d) - V_\mu^\pi(x_{<t_k} y; d) \right)$$

$$\overset{(f)}{\leq} \varepsilon_3 + 2\varepsilon_2 + \sum_{y \in Y} \mu^\pi(y|x_{<t_k}) \left(\frac{\max\left\{ 4\Delta_{x_{<t_k} y}, \mathbb{1}\{\mu \notin \mathcal{M}_{x_{<t_k} y}\} \right\}}{1-\gamma} \right)$$ (4)

where (a) follows from the definition of t_k as a time-step when π is ε-suboptimal. (b) by splitting the difference sum. (c) by the fact that π_ξ^* is at worst ε_3-suboptimal when BayesExp is exploiting and $\mu \in \mathcal{M}_t$ (Step 3). Note that $\mu \in \mathcal{M}_{t_k}$ is assumed in the definition of a failure phase. (d) by Lemma 5.

(e) by the fact that $V_\mu^* \geq V_\mu^\pi$ for all π and by Lemma 12. (f) by Lemma 7. Define random variable X_k by

$$X_k := \sum_{t=t_k}^{t_k+d} \mathbb{1}\{t \in E\} \left(\max\{4\Delta_t, \mathbb{1}\{\mu \notin \mathcal{M}_t\}\}\right) \in [0,4].$$

By the definition of X_k and (4), if $t_k < \infty$, then

$$\mathbb{E}_\mu^\pi[X_k|x_{<t_k}] = \sum_{y \in Y} \mu^\pi(y|x_{<t_k}) \left(4\Delta_{x_{<t_k}y} + \mathbb{1}\{\mu \notin \mathcal{M}_t\}\right)$$

$$\geq (\varepsilon - \varepsilon_3 - 2\varepsilon_2)(1-\gamma) =: \varepsilon_4 \equiv \varepsilon_1. \tag{5}$$

Using the bounds on the number of exploration phases given in Step 2 we have

$$\sum_{k=1}^\infty X_k \overset{(a)}{\leq} 1 + \sum_{t \in E} 4\Delta_t \overset{(b)}{\leq} 1 + 4\sqrt{|E| \sum_{t \in E} \Delta_t^2}$$

$$\overset{(c)}{\leq} 1 + 4\sqrt{\frac{K}{\varepsilon_1^2} \log \frac{K}{\delta_1^2} \cdot K \log \frac{K}{\delta_1^2}} \leq \frac{5K}{\varepsilon_1} \log \frac{K}{\delta_1^2} \tag{6}$$

where (a) follows from the definition of X_k and the fact that $\mu \in \mathcal{M}_{t_k}$ for all $t_k < \infty$. (b) by Jensen's inequality. (c) by Equations (2) and (3). Finally we can apply concentration inequalities by noting that $\sum_{k=1}^n \mathbb{E}_\mu^\pi[X_k|x_{<t_k}] - X_k$ is a martingale with zero expectation and differences bounded by 4. Let $F_{\max} \in \mathbb{N}$ be a constant to be defined shortly and let A_2 be the event that:

$$\sum_{k=1}^{F_{\max}} \mathbb{E}_\mu^\pi[X_k|x_{<t_k}] \leq \sum_{k=1}^{F_{\max}} X_k + \sqrt{2 \cdot 4^2 \cdot F_{\max} \log \frac{1}{\delta_1}}.$$

By Azuma's inequality $\mu^\pi(A_2) \geq 1 - \delta_1$. If A_2 occurs, then

$$\frac{1}{F_{\max}} \sum_{k=1}^{F_{\max}} \mathbb{E}[X_k|X_{k-1}] \overset{(a)}{\leq} \frac{5K}{\varepsilon_1 F_{\max}} \log \frac{K}{\delta_1^2} + \sqrt{\frac{2 \cdot 4^2}{F_{\max}} \log \frac{1}{\delta_1}} \overset{(b)}{<} \varepsilon_4 \tag{7}$$

where (a) follows by substituting (6) and (b) by choosing

$$F_{\max} := \frac{25K}{\varepsilon_1 \varepsilon_4} \log \frac{K}{\delta_1^2} \equiv \frac{400K}{\varepsilon^2(1-\gamma)^2} \log \frac{4K}{\delta^2}.$$

But (7) implies that there exists a $k < F_{\max}$ such that $\mathbb{E}[X_k|x_{<t_k}] \leq \varepsilon_4$, which by (5) implies that $t_k = \infty$ and so $|F| \leq F_{\max}$ and $|F_d| \leq F_{\max}d$.

Step 5: Finishing Up. Assuming events A_1 and A_2 both occur, then it holds that both $|E_d| \leq dE_{\max}$ and $|F_d| \leq dF_{\max}$. Since $V_\mu^*(x_{<t}) - V_\mu^\pi(x_{<t}) > \varepsilon$ implies that $t \in F_d \cup E_d$ or $\mu \notin \mathcal{M}_t$ it follows that

$$\mu^\pi \left(\sum_{t=1}^\infty \mathbb{1}\{V_\mu^*(x_{<t}) - V_\mu^\pi(x_{<t}) > \varepsilon\} \leq d(E_{\max} + F_{\max})\right)$$

$$\geq \mu^\pi(A_1 \cap A_2) \geq 1 - 2\delta_1 = 1 - \delta.$$

Substituting $E_{\max} = \frac{16K}{\varepsilon^2(1-\gamma)^2} \log \frac{4K}{\delta^2}$ and $F_{\max} = \frac{400K}{\varepsilon^2(1-\gamma)^2} \log \frac{4K}{\delta^2}$ completes the proof that

$$\mu^\pi \left(\sum_{t=1}^{\infty} \mathbb{1}\{V_\mu^*(x_{<t}) - V_\mu^\pi(x_{<t}) > \varepsilon\} > \frac{416Kd}{\varepsilon^2(1-\gamma)^2} \log \frac{4K}{\delta^2} \right) \leq \delta$$

as required.   □

Remark 8. The constant can be reduced to ~ 200 by making ε_2 significantly smaller (and paying only a log cost) and increasing $\varepsilon_1 = \varepsilon_4 \approx \varepsilon(1-\gamma)/3$.

5   Lower Bound on Sample-Complexity

In the last section we showed for any finite environment class $\mathcal{M}$ of size K that the algorithm BayesExp is ε-optimal except for at most

$$O\left(\frac{K}{\varepsilon^2(1-\gamma)^3} \left(\log \frac{1}{\varepsilon(1-\gamma)} \right) \left(\log \frac{K}{\delta} \right) \right) \qquad (\star)$$

time-steps with probability at least $1 - \delta$. We now describe the counter-example leading to a nearly-matching lower-bound in the sense that there exist environment classes where no algorithm has sample-complexity much better than $(\star)$. We do not claim that BayesExp achieves the optimal sample-complexity bound in all classes (it does not), only that there exists a class where it (very nearly) does. The gap between the lower and upper bounds is only a $\log \frac{1}{\varepsilon(1-\gamma)}$ factor. The most natural approach to proving a lower bound on the sample-complexity would be to use the famous result by Mannor and Tsitsiklis (2004) on the sample-complexity of exploration for multi-armed bandits. But environment classes based on stationary bandit-like environments lead only to an $\Omega(K)$ bounds on the sample-complexity rather than the desired $\Omega(K \log K)$. The reason is that for such environments the median elimination algorithm for minimising bandit sample-complexity can be used, which achieves the $O(K)$ bound (Even-Dar et al., 2002). This highlights a distinction between the two settings. Even if $\gamma = 0$ (1 step lookahead), the non-stationary version of the problem considered here is harder than the (stationary) bandit case.

Theorem 9. *For each $K > 1$ and $\gamma > 0$ there exists an environment class $\mathcal{M}$ such that for all policies π there exists a $\mu \in \mathcal{M}$ where*

$$\mu^\pi \left(\sum_{t=1}^{\infty} \mathbb{1}\{V_\mu^*(x_{<t}) - V_\mu^\pi(x_{<t}) > \varepsilon\} > \varepsilon \right) > c \cdot \left(\frac{\log \frac{1}{2}}{\log \gamma} \right) \frac{K}{\varepsilon^2(1-\gamma)^2} \log \frac{K}{\delta} \right) > \delta.$$

for some $c > 0$ independent of K, π and γ.

The complete proof is left for the technical report, but we describe the counter-example and justify the bound.

Counter-Example. Let $\mathcal{A} = \{\rightarrow, \rightsquigarrow\}$ consist of two actions, $\mathcal{O}$ be a singleton and $\mathcal{R} = \{0, \frac{1}{2}, 1\}$. Let $K \geq 2$ and $\varepsilon, \delta > 0$ be sufficiently small. Define environment class $\mathcal{M} = \{\mu_1, \mu_2, \cdots, \mu_K\}$ as in Figure 1. The parameter $\varepsilon_{t,k}$ determines the optimal action at each time-step. Let L be some large constant, then define $\varepsilon_{t,k}$ in environment μ_k by $\varepsilon_{t,k} = \frac{\varepsilon}{2} \operatorname{sign}\{kL - t\}$. So if the learner chooses action $\rightsquigarrow$, then with probability $\frac{1}{2} + \varepsilon_{t,k}$ it receives reward 1 and otherwise no reward. For $\rightarrow$ it deterministically receives reward $\frac{1}{2}$ regardless of the time-step or environment.

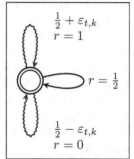

Fig. 1. Counter-example for lower bound. Environment μ_k

Explanation of the Bound. The optimal action in environment μ_k is to take action $\rightarrow$ until time-step kL and there-after take action $\rightsquigarrow$. We call each period of L time-steps a phase and consider the number of ε-errors made in the first $K-1$ phases. The difficulty arises because at the start of the ℓth phase an agent cannot distinguish between environment μ_ℓ and $\mu_{\ell+1}$. But in environment μ_ℓ the agent should take action $\rightsquigarrow$ while in environment $\mu_{\ell+1}$ the agent should take action $\rightarrow$. Now $\rightarrow$ is uninformative, so the only question is how many times a policy must sample action $\rightsquigarrow$ before switching to action $\rightarrow$. In order to guarantee that it is correct in phase ℓ with probability δ/K it should take action $\rightarrow$ approximately $\frac{1}{\varepsilon^2} \log \frac{K}{\delta}$ times. Since it must be correct in all phases, which are essentially independent, the number of times $\rightsquigarrow$ must be taken in environment μ_K in phases $\ell < K$ is $O(\frac{K}{\varepsilon^2} \log \frac{K}{\delta})$. In order to add the dependence on γ we must make two modifications.

1. Add a near-absorbing state corresponding to the times when the agent receives rewards 1, 0 and 1/2 respectively. If the agent stays in these states for $O(\frac{1}{1-\gamma})$, then the cost of a mistake becomes $\varepsilon/(1-\gamma)$ and the mistake bound will depend on $\varepsilon^{-2}(1-\gamma)^{-2}$.

2. To obtain an additional factor of the horizon we proceed in the same fashion as the lower bound given by Lattimore and Hutter (2012). Adapt the environment again so that the agent stays in the decision node for exactly $O(\frac{1}{1-\gamma})$ time-steps, regardless of its action. Only the action at the end of this period decides whether or not the agent gets reward 1/2 or 0 or 1. But if the agent is following a policy that makes an error, then this is counted for $O(\frac{1}{1-\gamma})$ time-steps before the error actually occurs, which multiplies the total number of errors by this quantity.

6   Adaptivity of BayesExp

We now show that the algorithm may learn faster when environments are easy to distinguish. Assume $\gamma = 0$, which implies that the effective horizon $d = 1$. A K-armed Bernoulli bandit is characterised by a vector $p \in [0,1]^K$. At each time-step the learner chooses arm $I_t \in \{1, \cdots, K\}$ and receives reward 1 with probability p_{I_t} and reward 0 otherwise. The value p_k is called the bias of the kth arm. There is now a huge literature on bandits, which we will not discuss, but see Bubeck and Cesa-Bianchi (2012) and references there-in for a good introduction. Choose $\mathcal{M} = \{\nu_1, \cdots, \nu_K\}$ to be the set of K-armed Bernoulli bandits where in environment ν_k the bias of the k arm is $\frac{1}{2}$ while for all other arms it is equal to $\frac{1}{2} - \Delta_k$ where $\Delta_k \geq \varepsilon$. Thus the optimal action in environment ν_k is to always choose arm k. Note that in this setting there are no observations ($\mathcal{O} \equiv$ singleton) and $\mathcal{R} = \{0, 1\}$. We show that the performance of BayesExp is substantially improved for large Δ_k where the environments are more easily distinguished.

Theorem 10. *If BayesExp is run on the environment class described above, then*

$$\mu^\pi \left(\sum_{t=1}^{\infty} \mathbb{1}\{V_\mu^*(x_{<t}) - V_\mu^\pi(x_{<t}) > \varepsilon\} > \sum_{k:\nu_k \neq \mu} \frac{4}{\Delta_k^2} \log \frac{K}{\delta^2} \right) \leq \delta.$$

The proof may be found in the associated technical report.

7   Conclusion

We adapted the Bayesian optimal agent studied by Hutter (2005) and others by adding an exploration component. The new algorithm achieves minimax finite sample-complexity bounds for finite environment classes. The theoretical results improve substantially on those given for the MERL algorithm by Lattimore et al. (2013a). In that work only two environments are compared in each exploration phase and models were discarded based on rewards alone, with observations completely ignored. Like the k-meteorologist algorithm (Diuk et al., 2009) models were only removed in discrete blocks. In contrast, the approach used here eliminates environments smoothly, which in benign environments may occur significantly faster than the worst-case bounds suggest. An example of this adaptivity is given for bandit environments in Section 6. There is another benefit of BayesExp illustrated by the example in Section 6. While the analysis in the proof of Theorem 4 leads to a largish constant, it is not used by the algorithm, which means that in simple cases the analysis can be improved substantially.

Future work could focus on proving more general problem-dependent bounds on the sample-complexity of algorithms like BayesExp, and characterising the difficulty of reinforcement learning environments and classes. This problem is now reasonably understood for bandit environments, but even for MDPs there is only limited work on problem-dependent bounds, and nothing for general RL as far as we are aware. Larger environment classes are also worth considering, including countable or separable spaces where uniform sample-complexity bounds are not possible, but problem-dependent asymptotic bounds are. We are optimistic that BayesExp can be extended to these cases.

Acknowledgements. This work was supported by the Alberta Innovates Technology Futures and NSERC.

References

Auer, P., Jaksch, T., Ortner, R.: Near-optimal regret bounds for reinforcement learning. Journal of Machine Learning Research 99, 1532–4435 (2010) ISSN 1532-4435

Azar, M.G., Lazaric, A., Brunskill, E.: Regret bounds for reinforcement learning with policy advice. In: Blockeel, H., Kersting, K., Nijssen, S., Železný, F. (eds.) ECML PKDD 2013, Part I. LNCS, vol. 8188, pp. 97–112. Springer, Heidelberg (2013)

Bubeck, S., Cesa-Bianchi, N.: Regret Analysis of Stochastic and Nonstochastic Multi-armed Bandit Problems. Foundations and Trends in Machine Learning. Now Publishers Incorporated (2012) ISBN 9781601986269

Chakraborty, D., Stone, P.: Structure learning in ergodic factored mdps without knowledge of the transition function's in-degree. In: Proceedings of the Twenty Eighth International Conference on Machine Learning (2011)

Diuk, C., Li, L., Leffler, B.: The adaptive k-meteorologists problem and its application to structure learning and feature selection in reinforcement learning. In: Danyluk, A.P., Bottou, L., Littman, M.L. (eds.) Proceedings of the 26th Annual International Conference on Machine Learning, pp. 249–256. ACM (2009)

Dyagilev, K., Mannor, S., Shimkin, N.: Efficient reinforcement learning in parameterized Models: Discrete parameter case. In: Girgin, S., Loth, M., Munos, R., Preux, P., Ryabko, D. (eds.) EWRL 2008. LNCS (LNAI), vol. 5323, pp. 41–54. Springer, Heidelberg (2008)

Even-Dar, E., Mannor, S., Mansour, Y.: PAC Bounds for Multi-armed Bandit and Markov Decision Processes. In: Kivinen, J., Sloan, R.H. (eds.) COLT 2002. LNCS (LNAI), vol. 2375, pp. 255–270. Springer, Heidelberg (2002)

Even-Dar, E., Kakade, S., Mansour, Y.: Reinforcement learning in POMDPs without resets. In: International Joint Conference on Artificial Intelligence, pp. 690–695 (2005)

Hutter, M.: Self-optimizing and Pareto-optimal policies in general environments based on Bayes-mixtures. In: Kivinen, J., Sloan, R.H. (eds.) COLT 2002. LNCS (LNAI), vol. 2375, pp. 364–379. Springer, Heidelberg (2002)

Hutter, M.: Universal Artificial Intelligence: Sequential Decisions based on Algorithmic Probability. Springer, Berlin (2005)

Hutter, M., Muchnik, A.: On semimeasures predicting Martin-Löf random sequences. Theoretical Computer Science 382(3), 247–261 (2007)

Kearns, M., Singh, S.: Near-optimal reinforcement learning in polynomial time. Machine Learning 49(2-3), 209–232 (2002)

Lattimore, T., Hutter, M.: PAC bounds for discounted MDPs. In: Bshouty, N.H., Stoltz, G., Vayatis, N., Zeugmann, T. (eds.) ALT 2012. LNCS, vol. 7568, pp. 320–334. Springer, Heidelberg (2012)

Lattimore, T., Hutter, M.: Bayesian reinforcement learning with exploration. arxiv (2014)

Lattimore, T., Hutter, M., Sunehag, P.: The sample-complexity of general reinforcement learning. In: Proceedings of the 30th International Conference on Machine Learning (2013a)

Lattimore, T., Hutter, M., Sunehag, P.: Concentration and confidence for discrete bayesian sequence predictors. In: Jain, S., Munos, R., Stephan, F., Zeugmann, T. (eds.) ALT 2013. LNCS, vol. 8139, pp. 324–338. Springer, Heidelberg (2013)

Mannor, S., Tsitsiklis, J.: The sample complexity of exploration in the multi-armed bandit problem. Journal of Machine Learning Research 5, 623–648 (2004)

Odalric-Ambrym, M., Nguyen, P., Ortner, R., Ryabko, D.: Optimal regret bounds for selecting the state representation in reinforcement learning. In: Proceedings of the Thirtieth International Conference on Machine Learning (2013)

Orseau, L.: Optimality issues of universal greedy agents with static priors. In: Hutter, M., Stephan, F., Vovk, V., Zeugmann, T. (eds.) ALT 2010. LNCS (LNAI), vol. 6331, pp. 345–359. Springer, Heidelberg (2010)

Osband, I., Russo, D., Van Roy, B.: (More) efficient reinforcement learning via posterior sampling. In: Advances in Neural Information Processing Systems, pp. 3003–3011 (2013)

Sunehag, P., Hutter, M.: Optimistic agents are asymptotically optimal. In: Thielscher, M., Zhang, D. (eds.) AI 2012. LNCS, vol. 7691, pp. 15–26. Springer, Heidelberg (2012)

Szita, I., Szepesvári, C.: Model-based reinforcement learning with nearly tight exploration complexity bounds. In: Proceedings of the 27th International Conference on Machine Learning, pp. 1031–1038. ACM, New York (2010)

A   Properties of Value Functions

Lemma 11. *Let $\pi \in \Pi$ and μ and ν be two environments. Then*

$$V_\mu^\pi(x;d) - V_\nu^\pi(x;d) \leq \frac{\delta_x^d(\mu^\pi, \nu^\pi)}{1-\gamma}.$$

Proof. The difference in value functions is a difference in expected returns with respect to μ^π and ν^π. This is bounded by the total variation distance multiplied by the maximum return, which is $1/(1-\gamma)$. □

Lemma 12. *If x is a history at time-step t and $\varepsilon > 0$ and $d \geq \left\lceil \frac{\log \varepsilon(1-\gamma)}{\log \gamma} \right\rceil$, then $V_\mu^\pi(x) \geq V_\mu^\pi(x;d)$ and $V_\mu^\pi(x) - V_\mu^\pi(x;d) \leq \varepsilon$.*

Proof. That $V_\mu^\pi(x) \geq V_\mu^\pi(x;d)$ is trivial. For the second claim:

$$V_\mu^\pi(x) - V_\mu^\pi(x;d) \stackrel{(a)}{=} \mathbb{E}_\mu^\pi\left[\sum_{k=t+d}^\infty \gamma^{k-t} r_k \Big| x\right] \stackrel{(b)}{\leq} \sum_{k=t+d}^\infty \gamma^{k-t} \stackrel{(c)}{=} \frac{\gamma^d}{1-\gamma} \stackrel{(d)}{\leq} \varepsilon$$

where (a) follows by adding and subtracting the tail sum. (b) because $r_k \in [0,1]$. (c) is trivial while (d) follows from the definition of d. □

Lemma 13. *Let μ be an environment, $x \in \mathcal{H}^*$ a history and $Y \subset \mathcal{H}^*$ be complete and prefix free. If π_1 and π_2 are policies such that $\pi_1(xz) = \pi_2(xz)$ for all y, z for which $z \sqsubset y$. Then*

$$V_\mu^{\pi_1}(x) - V_\mu^{\pi_2}(x) = \sum_{y \in Y} \mu^{\pi_1}(y|x)\gamma^{\ell(y)}\left(V_\mu^{\pi_1}(xy) - V_\mu^{\pi_2}(xy)\right).$$

Extreme State Aggregation beyond MDPs

Marcus Hutter

Research School of Computer Science
Australian National University
Canberra, ACT, 0200, Australia
http://www.hutter1.net/

Abstract. We consider a Reinforcement Learning setup without any (esp. MDP) assumptions on the environment. State aggregation and more generally feature reinforcement learning is concerned with mapping histories/raw-states to reduced/aggregated states. The idea behind both is that the resulting reduced process (approximately) forms a small stationary finite-state MDP, which can then be efficiently solved or learnt. We considerably generalize existing aggregation results by showing that even if the reduced process is not an MDP, the (q-)value functions and (optimal) policies of an associated MDP with same state-space size solve the original problem, as long as the solution can approximately be represented as a function of the reduced states. This implies an upper bound on the required state space size that holds uniformly for all RL problems. It may also explain why RL algorithms designed for MDPs sometimes perform well beyond MDPs.

Keywords: State aggregation, reinforcement learning, non-MDP.

1 Introduction

In *Reinforcement Learning* (RL) [SB98], an *agent Π* takes actions in some *environment P* and observes its consequences and is rewarded for them. A well-understood and efficiently solvable [Put94] and efficiently learnable [SLL09, LH12] case is where the environment is (modelled as) a finite-state stationary *Markov Decision Process* (MDP). Unfortunately most interesting real-world problems P are neither finite-state, nor stationary, nor Markov. One way of dealing with this mismatch is to somehow transform the real-world problem into a small MDP: *Feature Reinforcement Learning* (FRL) [Hut09b] and U-tree [McC96] deal with the case of arbitrary unknown environments, while state aggregation assumes the environment is a large known stationary MDP [GDG03, FPP04]. The former maps histories into states (Section 2), the latter groups raw states into aggregated states.

Here we follow the FRL approach and terminology, since it is arguably most general: It subsumes the cases where the original process P is an MDP, a k-order MDP, a POMDP, and others [Hut09b]. Thinking in terms of histories also naturally stifles any temptation of a naive frequency estimate of P (no history ever repeats).

P. Auer et al. (Eds.): ALT 2014, LNAI 8776, pp. 185–199, 2014.
© Springer International Publishing Switzerland 2014

More importantly, we consider maps ϕ from histories to states for which the reduced process P_ϕ is not (even approximately) an MDP (Section 3). At first this seems to defeat the original purpose, namely of reducing P to a well-understood and efficiently solvable problem class, namely small MDPs. The main novel contribution of this paper is to show that there is still an associated finite-state stationary MDP p whose solution (approximately) solves the original problem P, as long as the solution can still be represented (Section 4). Indeed, we provide an upper bound on the required state space size that holds uniformly for all P (Section 5). We also show how to learn p from experience (Section 6), and sketch an overall learning algorithm and regret/PAC analysis based on our main theorems (Section 7). We conclude with an outlook on future work and open problems (Section 8). All proofs can be found in the extended technical report [Hut14].

2   Feature Markov Decision Processes (ΦMDP)

This section formally describes the setup of [Hut09b]. It consists of the agent-environment framework and maps ϕ from observation-reward-action histories to MDP states. This arrangement is called "Feature MDP" or short ΦMDP. We use upper-case letters P, Q, V, and Π for the Probability, (Q-)Value, and Policy of the original (agent-environment interactive) Process, and lower-case letters p, q, v, and π for the probability, (q-)value, and policy of the (reduced/aggregated) MDP.

Agent-environment Setup [Hut09b]. We start with the standard agent-environment setup [RN10] in which an agent Π interacts with an environment P. The agent can choose from actions $a \in \mathcal{A}$ and the environment provides observations $o \in \mathcal{O}$ and real-valued rewards $r \in \mathcal{R} \subseteq [0; 1]$ to the agent. This happens in cycles $t = 1, 2, 3, ...$: At time t, after observing o_t and receiving reward r_t, the agent takes action a_t based on history

$$h_t := o_1 r_1 a_1 ... o_{t-1} r_{t-1} a_{t-1} o_t r_t \in \mathcal{H}_t := (\mathcal{O} \times \mathcal{R} \times \mathcal{A})^{t-1} \times \mathcal{O} \times \mathcal{R}$$

Then the next cycle $t+1$ starts. The agent's objective is to maximize its long-term reward. To avoid integrals and densities, we assume spaces $\mathcal{O}$ and $\mathcal{R}$ are finite. They may be huge, so this is not really restrictive. Indeed, the ΦMDP framework has been specifically developed for huge observation spaces. Generalization to continuous $\mathcal{O}$ and $\mathcal{R}$ is routine [Hut09a]. Furthermore we assume that $\mathcal{A}$ is finite and smallish, which is restrictive. Potential extensions to continuous $\mathcal{A}$ are discussed in Section 8.

The agent and environment may be viewed as a pair of interlocking functions of the history $\mathcal{H} := (\mathcal{O} \times \mathcal{R} \times \mathcal{A})^* \times \mathcal{O} \times \mathcal{R}$:

Env. $P : \mathcal{H} \times \mathcal{A} \rightsquigarrow \mathcal{O} \times \mathcal{R}$, $P(o_{t+1} r_{t+1} | h_t a_t)$,

Agent $\Pi : \mathcal{H} \rightsquigarrow \mathcal{A}$, $\Pi(a_t | h_t)$ or $a_t = \Pi(h_t)$, where $\rightsquigarrow$ indicates that mappings $\to$ are in general stochastic. We make no (stationarity or Markov or other) assumption on environment P. For most parts, environment P is assumed to be fixed, so dependencies on P will be suppressed. For convenience and since optimal policies can be chosen to be deterministic, we consider deterministic policies $a_t = \Pi(h_t)$ only.

Value Functions, Optimal Policies, and History Bellman Equations. We measure the performance of a policy Π in terms of the P-expected γ-discounted reward $(0 \le \gamma < 1)$, called (Q-)Value of Policy Π at history h_t (and action a_t)

$$V^{\Pi}(h_t) := \mathbb{E}^{\Pi}[R_{t+1}|h_t], \qquad Q^{\Pi}(h_t, a_t) := \mathbb{E}^{\Pi}[R_{t+1}|h_t a_t], \qquad R_t := \sum_{\tau=t}^{\infty} \gamma^{\tau-t} r_{\tau}$$

The optimal Policy and (Q-)Value functions are

$$V^*(h_t) := \max_{\Pi} V^{\Pi}(h_t), \;\; Q^*(h_t, a_t) := \max_{\Pi} Q^{\Pi}(h_t, a_t), \;\; \Pi^* :\in \arg\max_{\Pi} V^{\Pi}(\epsilon) \quad (1)$$

The maximum over all policies Π always exists [LH14] but may not be unique, in which case $\arg\max$ denotes the set of optimal policies and Π^* denotes a representative or the whole set of optimal policies. Despite being history-based we can write down (pseudo)recursive Bellman (optimality) equations for the (optimal) (Q-)Values [Hut05, Sec.4.2]:

$$Q^{\Pi}(h_t, a_t) = \sum_{o_{t+1} r_{t+1}} P(o_{t+1} r_{t+1}|h_t a_t)[r_{t+1} + \gamma V^{\Pi}(h_{t+1})], \;\; V^{\Pi}(h_t) = Q^{\Pi}(h_t, \Pi(h_t)) \, (2)$$

$$Q^*(h_t, a_t) = \sum_{o_{t+1} r_{t+1}} P(o_{t+1} r_{t+1}|h_t a_t)[r_{t+1} + \gamma V^*(h_{t+1})], \;\; V^*(h_t) = \max_{a_t \in \mathcal{A}} Q^*(h_t, a_t) \, (3)$$

$$\Pi^*(h_t) \in \arg\max_{a_t \in \mathcal{A}} Q^*(h_t, a_t) \tag{4}$$

Unlike their classical state-space cousins (see below), they are *not* self-consistency equations: The r.h.s. refers to a longer history h_{t+1} which is always different from the history h_t on the l.h.s, which precludes any learning algorithm based on estimating the frequency of state/history visits. Still the recursions will be convenient for the mathematical development.

From Histories to States (ϕ). The space of histories is huge and unwieldy and no history ever repeats. Standard ways of dealing with this are to define a similarity metric on histories [McC96] or to aggregate histories [Hut09b]. We pursue the latter via a feature map $\phi : \mathcal{H} \to \mathcal{S}$ which reduces histories $h_t \in \mathcal{H}$ to states $s_t := \phi(h_t) \in \mathcal{S}$. W.l.g. we assume that ϕ is surjective. We also assume that state space $\mathcal{S}$ is finite; indeed we are interested in small $\mathcal{S}$. This corresponds and indeed is equivalent to a partitioning of histories $\{\phi^{-1}(s) : s \in \mathcal{S}\}$. Classical state aggregation usually uses the partitioning view [GDG03, FPP04], but the map notation is a bit more convenient here.

The state s_t is supposed to summarize all relevant information in history h_t, which lower bounds the size of $\mathcal{S}$. We pass from the complete history $o_1 r_1 a_1 ... o_n r_n$ to a 'reduced' history $s_1 r_1 a_1 ... s_n r_n$. Traditionally, 'relevant' means that the future is predictable from s_t (and a_t) alone, or technically that the reduced history forms a Markov decision process. This is precisely the condition this paper intends to lift (later).

From Histories to MDPs. The probability of the successor states and rewards can be obtained by marginalization

$$P_\phi(s_{t+1}r_{t+1}|h_t a_t) := \sum_{\tilde{o}_{t+1}:\phi(h_t a_t \tilde{o}_{t+1} r_{t+1})=s_{t+1}} P(\tilde{o}_{t+1}r_{t+1}|h_t a_t) \qquad (5)$$

The reduced process P_ϕ is a Markov Decision Process, or Markov for short, if P_ϕ only depends on h_t through s_t, i.e. is the same for all histories mapped to the same state. Formally

$$P_\phi \in \mathrm{MDP} \quad :\Longleftrightarrow \quad \exists p : P_\phi(s_{t+1}r_{t+1}|\tilde{h}_t a_t) = p(s_{t+1}r_{t+1}|s_t a_t) \; \forall \phi(\tilde{h}_t) = s_t \quad (6)$$

Here and elsewhere a quantifier such as $\forall \phi(\tilde{h}_t) = s_t$ shall mean: for all values of all involved variables consistent with the constraint $\phi(\tilde{h}_t) = s_t$. The MDP P_ϕ is assumed to be stationary, i.e. independent of t; another condition to be lifted later. Condition (6) is essentially the stochastic bisimulation condition generalized to histories and being somewhat more restrictive regarding rewards [GDG03]: It is a condition on the reward distribution, while [GDG03] constrains its expectation only. This could easily be rectified but is besides the point of this paper. The bisimulation metric [FPP04] is an approximate version of (6), which measures the deviation of P_ϕ from being an MDP.

Many problems P can be reduced (approximately) to stationary MDPs [Hut09b]: Full-information *games* such as chess with static opponent are already Markov, classical *physics* is approximately 2nd-order Markov, (conditional) i.i.d. processes such as *Bandits* have counting sufficient statistics, and for a *POMDP planning* problem, the belief vector is Markov.

Markov Decision Processes (MDP). We have used and continue to use upper-case letters V, Q, Π for the general process P. We will use lower-case letters v, q, π for (stationary) MDPs p. We use s and a for the current state and action, and s' and r' for successor state and reward. Consider a stationary finite-state MDP $p : \mathcal{S} \times \mathcal{A} \rightsquigarrow \mathcal{S} \times \mathcal{R}$ and stationary deterministic policy $\pi : \mathcal{S} \to \mathcal{A}$. In this paper, p will *not* be given by (6), but *in general p will be different from* (6). In any case, the p-expected γ-discounted reward sum, called (q-)value of (optimal) policy $\pi^{(*)}$ in MDP p, are given by the Bellman (optimality) equations

$$q^\pi(s,a) = \sum_{s'r'} p(s'r'|sa)[r' + \gamma v^\pi(s')] \quad \text{and} \quad v^\pi(s) = q^\pi(s,\pi(s)) \qquad (7)$$

$$q^*(s,a) = \sum_{s'r'} p(s'r'|sa)[r' + \gamma v^*(s')] \quad \text{and} \quad v^*(s) = \max_a q^*(s,a) \qquad (8)$$

$$\pi^*(s) \in \arg\max_a q^*(s,a). \quad \text{Note: } v^\pi(s) \le v^*(s), \; q^\pi(s,a) \le q^*(s,a) \qquad (9)$$

Using $p(s'r'|sa) = p(r'|sas')p(s'|sa)$ we could also rewrite them in terms of transition matrix $p(s'|sa)$ and expected reward $\mathbb{E}[r'|sa]$ [SB98].

One can show that if P reduces via ϕ to an MDP p, the solution of these equations, yields (Q-)Values and optimal Policy of the original process P. This is not surprising and just history-based versions of classical state-aggregation results [GDG03]. Approximate versions based on bisimulation metric [FPP04] can also be historized.

More Notation. While our equations often assume or imply $s = s_t$, $a = a_t$, $s' = s_{t+1}$, $r' = r_{t+1}$, (and $h_{t+1} = h_t ao'r'$) for some t, technically s, a, s', r' are *different* variables from all variables in history $h_n = o_1 r_1 a_1 ... o_t r_t a_t o_{t+1} r_{t+1} ... a_n r_n$. Less prone to confusion are $o = o_t$, $o' = o_{t+1}$, $h = h_t$, $h' = hao'r'$. We call a function $f(h)$, piecewise constant or ϕ-uniform iff $f(h) = f(\tilde{h})$ for all $\phi(h) = \phi(\tilde{h})$. Here and elsewhere $\forall \phi(h) = \phi(\tilde{h})$ is short for $\forall h, \tilde{h} : \phi(h) = \phi(\tilde{h})$. Similarly $\forall s = \phi(h)$ is short for $\forall s, h : s = \phi(h)$. Etc. The Iverson bracket, $[\![R]\!] := 1$ if R=true and $[\![R]\!] := 0$ if R=false, denotes the indicator function. Throughout, $\varepsilon, \delta \geq 0$ denote approximation accuracy. Note that this includes the exact $= 0$ case.

3 Approximate Aggregation for General P

This section prepares for the main technical contribution of the paper in the next section. The key quantity to relate original and reduced Bellman equations is a form of stochastic inverse of ϕ, whose choice and analysis will be deferred to Section 6. Proofs can be found in [Hut14].

Dispersion Probability B. Let $B_\phi : \mathcal{S} \times \mathcal{A} \rightsquigarrow \mathcal{H}$ be a probability distribution on finite histories for each state-action pair such that $B_\phi(h|sa) = 0$ if $s \neq \phi(h)$. $B \equiv B_\phi$ may be viewed as a stochastic inverse of ϕ that assigns non-zero probability only to $h \in \phi^{-1}(s)$. The formal constraints we pose on B are

$$B(h|sa) \geq 0 \quad \text{and} \quad \sum_{h \in \mathcal{H}} B(h|sa) = \sum_{h:\phi(h)=s} B(h|sa) = 1 \ \forall s, a \qquad (10)$$

This implicitly requires ϕ to be surjective, i.e. $\phi(\mathcal{H}) = \mathcal{S}$, which can always be made true by defining $\mathcal{S} := \mathcal{S}_\phi := \phi(\mathcal{H})$. Note that the sum is taken over histories of any/mixed length. In general, B is a somewhat weird distribution, since it assigns probabilities to past and future observations given the current state and action. The interpretation and choice of B does not need to concern us, except later when we want to learn p.

The MDP requirement (6) will be replaced by the following definition:

$$p(s'r'|sa) := \sum_{h \in \mathcal{H}} P_\phi(s'r'|ha)B(h|sa) \qquad (11)$$

$$\equiv \sum_{t=1}^{\infty} \sum_{h_t \in \mathcal{H}_t} P_\phi(s_{t+1} = s', r_{t+1} = r'|h_t, a_t = a)B(h_t|sa)$$

That is, the finite-state stationary MDP p is built from feature map ϕ, dispersion probability B, and environment P: The p-probability of observing state-reward pair (s', r') from state-action pair (s, a) is defined as the B-average over all histories h consistent with (s, a) of the P_ϕ-probability of observing (s', r') (obtained from P by ϕ-marginalizing) given history h and action a. The r.h.s. of the first line is merely shorthand for the second line. Note that $sas'r'$ are fixed and do not appear in h which ranges over histories $\mathcal{H}$ of all lengths. It is easy to see that p is a probability distribution, and it is Markov by definition. If $P_\phi \in$ MDP, then definition (11) coincides with p defined in (6). In general, the MDP p, depending on arbitrary B, is *not* the state distribution induced by P (and Π), which in general is non-Markov. Note that p is a stationary MDP for any B satisfying (10) and *any* ϕ and P. We need the following lemmas:

Some Lemmas. The first lemma establishes the key relation between P and p via B used later to relate original history Bellman (optimality) equations (2–4) with reduced state Bellman (optimality) equations (7–9).

Lemma 1 (B-P-p relation). *For any function $f : \mathcal{S} \times \mathcal{R} \to \mathbb{R}$ and p defined in (11) in terms of P via (5), and $s' := \phi(h')$ and $h' := hao'r'$ it holds*

$$\sum_{h \in \mathcal{H}} B(h|sa) \underbrace{\sum_{o'r'} P(o'r'|ha) f(s', r')}_{\text{depends on } hao'r'} = \sum_{s'r'} p(s'r'|sa) f(s', r')$$

The following lemma trivially bounds $v - V$ differences in terms of $q - Q$ differences, essentially $|v - V| \le \max_a |q - Q|$.

Lemma 2 ($|v - V| \le |q - Q|$).

   (i) *If* $\Pi(h) = \pi(s) \; \forall s = \phi(h)$ *and* $\delta := \sup\limits_{s=\phi(h),a} |q^\pi(s,a) - Q^\Pi(h,a)|$

      then $|v^\pi(s) - V^\Pi(h)| \le \delta \; \forall s = \phi(h)$ (12)

   (ii) $\delta := \sup\limits_{s=\phi(h),a} |q^*(s,a) - Q^*(h,a)|$

      implies $|v^*(s) - V^*(h)| \le \delta \; \forall s = \phi(h)$ (13)

(i) follows from (2) and (7), and (ii) follows from (3) and (8) and $|\max_x f(x) - \max_x g(x)| \le \max_x |f(x) - g(x)|$.

The next lemma shows that a reverse holds in B-expectation, i.e. $|q - \langle Q \rangle_B| \le \gamma |v - V|$. The expectation can (only) be dropped if Q is constant for all $h \in \phi^{-1}(s)$. Formally define

$$\langle f(h,a) \rangle_B := \sum_{\tilde{h} \in \mathcal{H}} B(\tilde{h}|sa) f(\tilde{h}, a), \quad \text{where} \quad s := \phi(h) \qquad (14)$$

That is, $\langle f(h,a) \rangle_B$ takes a B-average over all $\tilde{h}$ that ϕ maps to the same state as h. For convenience we will drop the tilde, which we can do if we declare $s := \phi(h)$ to refer to the 'global' h in $\langle f(h,a) \rangle_B$ and not to the 'local' variable in the $h \in \mathcal{H}$ sum.

Lemma 3 ($|q - \langle Q \rangle| \le \gamma |v - V|$). *For any P, ϕ, B, define p via (11) and (5)*
(i) *If* $|v^\pi(s) - V^\Pi(h)| \le \delta \; \forall s = \phi(h)$
    then $|q^\pi(s,a) - \langle Q^\Pi(h,a) \rangle_B| \le \gamma \delta \; \forall s = \phi(h) \; \forall a$.
(ii) *If* $|v^*(s) - V^*(h)| \le \delta \; \forall s = \phi(h)$
    then $|q^*(s,a) - \langle Q^*(h,a) \rangle_B| \le \gamma \delta \; \forall s = \phi(h) \; \forall a$.

The proof uses Lemma 1 for (i) together with (2) and (7) and for (ii) together with (3) and (8). Note that in general $\Pi^* \ne \pi^*$.

4   Approximate Aggregation Results

This section contains the main technical contribution of the paper. We show that histories (or raw states) can be aggregated and modeled by an MDP even if the true aggregated process is actually not an MDP. A necessary condition for successful aggregation is of course that the quantities of interest, namely (Q-)Value functions and Policies can be represented as functions of the aggregated states. The results in this section roughly show that this necessary condition, which is significantly weaker than the MDP requirement, is also sufficient. All but one result also holds for approximate aggregation, i.e. approximate conditions lead to approximate reductions. We also lift the stationarity assumption. Proofs can be found in [Hut14].

- Theorem 4 shows how (approximately) ϕ-uniform Q^Π and Π can be obtained from the reduced Bellman equations (7).
- Theorem 5 weakens the assumptions and conclusions to (approximately) ϕ-uniform V^Π and Π.
- Theorem 6 shows that for (approximately) ϕ-uniform Q^*, the optimal policy is (approximately) ϕ-uniform, and (an approximation of it) can be obtained via the reduced Bellman optimality equations (8).
- Theorem 7 shows that for (approximately) ϕ-uniform V^* and Π^* we can obtain similar but somewhat weaker results. The proof of the latter involves extra complications not present in the other three proofs. Indeed, whether the arguably most desirable bound holds is Open Problem 8.

Note that all theorems crucially differ in their conditions and conclusions.

Theorem 4 ($\phi Q\pi$). *For any P, ϕ, and B, define p via (11) and (5). Let Π be some policy such that $\Pi(h) = \Pi(\tilde{h})$ and $|Q^\Pi(h,a) - Q^\Pi(\tilde{h},a)| \leq \varepsilon$ for all $\phi(h) = \phi(\tilde{h})$ and all a. Then for all a and h it holds:*

$$|Q^\Pi(h,a) - q^\pi(s,a)| \leq \frac{\varepsilon}{1-\gamma} \quad and \quad |V^\Pi(h) - v^\pi(s)| \leq \frac{\varepsilon}{1-\gamma},$$
$$where \quad \pi(s) := \Pi(h) \ and \ s = \phi(h)$$

Note that $\pi(s)$ is well-defined, since ϕ is surjective and $\Pi(h)$ is the same for all $h \in \phi^{-1}(s)$. The proof uses (12) and Lemma 3i.

Theorem 5 ($\phi V\pi$). *For any P, ϕ, and B, define p via (11) and (5). Let Π be some policy such that $\Pi(h) = \Pi(\tilde{h})$ and $|V^\Pi(h) - V^\Pi(\tilde{h})| \leq \varepsilon$ for all $\phi(h) = \phi(\tilde{h})$. Then for all a and h it holds:*

$$|V^\Pi(h) - v^\pi(s)| \leq \frac{\varepsilon}{1-\gamma} \quad and \quad |q^\pi(s,a) - \langle Q^\Pi(h,a)\rangle_B| \leq \frac{\varepsilon\gamma}{1-\gamma}$$
$$where \quad \pi(s) := \Pi(h) \ and \ s = \phi(h)$$

The proof also uses Lemma 3i but otherwise is different. A simple example of a P and ϕ that satisfy the conditions of Theorems 4 and 5, but violate the bisimulation condition [GDG03] and indeed have large bisimulation distance [FPP04] is given in [Hut14]. We now turn from the fixed policy case to similar theorems for optimal policies.

Theorem 6 ($\phi Q*$). *For any P, ϕ, and B, define p via (11) and (5). Assume $|Q^*(h,a) - Q^*(\tilde{h},a)| \leq \varepsilon$ for all $\phi(h) = \phi(\tilde{h})$ and all a. Then for all a and h and $s = \phi(h)$ it holds:*

(i) $|Q^*(h,a) - q^*(s,a)| \leq \dfrac{\varepsilon}{1-\gamma}$ *and* $|V^*(h) - v^*(s)| \leq \dfrac{\varepsilon}{1-\gamma}$,

(ii) $0 \leq V^*(h) - V^{\tilde{\Pi}}(h) \leq \dfrac{2\varepsilon}{(1-\gamma)^2}$, *where* $\tilde{\Pi}(h) := \pi^*(s)$

(iii) *If $\varepsilon = 0$ then $\Pi^*(h) = \pi^*(s)$*

The proof of (i) follows the same steps as the proof of Theorem 4, but uses (13) and Lemma 3ii to justify the steps. (ii) follows from (i) and (8) and an additional lemma [Hut14]. (iii) follows from (i).

Theorem 7 ($\phi V*$). *For any P, ϕ, and B, define p via (11) and (5). Assume $\Pi^*(h) = \Pi^*(\tilde{h})$ and $|V^*(h) - V^*(\tilde{h})| \leq \varepsilon$ for all $\phi(h) = \phi(\tilde{h})$. Then for all a and h and $s = \phi(h)$ it holds:*

(i) $|V^*(h) - v^*(s)| \leq \dfrac{3\varepsilon}{(1-\gamma)^2}$ *and* $|q^*(s,a) - \langle Q^*(h,a)\rangle_B| \leq \dfrac{3\varepsilon\gamma}{(1-\gamma)^2}$,

(ii) *If $\varepsilon = 0$ then $\Pi^*(h) = \pi^*(s)$*

The proof actually implies the stronger lower bound $V^*(h) - v^*(s) \geq \frac{3\varepsilon}{1-\gamma}$ and similarly for Q^*, but we do not know whether the upper bound can be improved. The proof involves (7) for $\pi^0 := \Pi^* \neq \pi^*$, (8), (4), (3), and (10), and uses Theorem 5 applied to $\Pi := \Pi^*$ (with $\pi = \pi^0$) and Lemma 3ii.

We are primarily interested in the optimal policy $\Pi^*(h)$; to correctly represent the value $V^*(h)$ is only of indirect interest. If Π^* is ϕ-uniform, it can be represented as $\Pi^*(h) = \pi^0(s)$ for some π^0, but if the ϕ-uniformity condition on V^* in Theorem 7 is dropped, the conclusion $\Pi^*(h) = \pi^*(s)$ can fail [Hut14].

Open Problem 8 ($\phi V*$) *Under the same conditions as Theorem 7, is*

$$V^*(h) - V^{\tilde{\Pi}}(h) \overset{??}{=} O\left(\dfrac{\varepsilon}{(1-\gamma)^?}\right) \quad \text{where} \quad \tilde{\Pi}(h) := \pi^*(s) \qquad (15)$$

We only know that this holds for $\varepsilon = 0$, which follows from Theorem 7. See [Hut14] for why it may be true or false for $\varepsilon > 0$.

Discussion. Open Problem 8 would be the main result if we had a proof for $\varepsilon > 0$. Absent of it we have to be content with Theorem 6ii. Both statements imply that we can aggregate histories as much as we wish, as long as the optimal value function and policy are still approximately representable as functions of aggregated states. Whether the reduced process P_ϕ is Markov or not is immaterial. We can use surrogate MDP p to find an ε-optimal policy for P.

Most RL work, including on state aggregation, is formulated in terms of MDPs, i.e. the original process P is already an MDP. Let us call this the original or raw MDP. We could interpret the whole history as a raw state, which formally makes every P an MDP, but normally only observations are identified with raw states, i.e. P is a raw MDP iff $P(o'r'|ha) = P(o'r'|oa)$. In this case, $V^*(h_t) = V^*(o_t)$ etc. depends on raw states only (which is well known). Since our results hold for all P, they clearly hold if P is a raw MDP and if $\phi(h_t) := \phi(o_t)$ maps raw states to aggregated states.

The remainder of this paper shows how much we can aggregate and how to develop RL algorithms exploiting these insights.

5   Extreme Aggregation

The results of Section 4 showed that histories can be aggregated and modeled by an MDP even if the true aggregated process is not an MDP. The only restrictions were that the (Q-)Value functions and Policies could still be (approximately) represented as functions of the aggregated states. We will see in this section that in theory this allows to represent *any* process P as a small finite-state MDP.

Extreme Aggregation Based on Theorem 6. Consider ϕ that maps each history to the vector-over-actions of optimal Q-values $Q^*(h, \cdot)$ discretized to some finite ε-grid:

$$\phi(h) := \left(\lfloor Q^*(h,a)/\varepsilon \rfloor\right)_{a \in \mathcal{A}} \in \{0, 1, ..., \lfloor \tfrac{1}{\varepsilon(1-\gamma)} \rfloor\}^{\mathcal{A}} =: \mathcal{S} \qquad (16)$$

That is, all histories with ε-close Q^*-values are mapped to the same state:

$$|Q^*(h,a) - Q^*(\tilde{h},a)| \leq \varepsilon \quad \forall \phi(h) = \phi(\tilde{h}) \ \forall a$$

Now choose some B and determine p from P via (11) and (5). Find the optimal policy π^* of MDP p of size $|\mathcal{S}|$. Define $\tilde{\Pi}(h) := \pi^*(\phi(h))$. By Theorem 6ii, $\tilde{\Pi}$ is an ε'-optimal policy of original process P in the sense that

$$|V^{\tilde{\Pi}}(h) - V^*(h)| \leq \frac{2\varepsilon}{(1-\gamma)^2} =: \varepsilon'$$

Extreme Aggregation Based on Open Problem 8. If (15) holds, we can aggregate even better: Consider ϕ that maps each history to the optimal Value $V^*(h)$ discretized to some finite ε-grid and to the optimal action $\Pi^*(h)$:

$$\phi(h) := \left(\lfloor V^*(h)/\varepsilon \rfloor, \Pi^*(h)\right) \in \{0, 1, ..., \lfloor \tfrac{1}{\varepsilon(1-\gamma)} \rfloor\} \times \mathcal{A} =: \mathcal{S} \qquad (17)$$

That is, all histories with ε-close V^*-Values and same optimal action are mapped to the same state:

$$|V^*(h) - V^*(\tilde{h})| \leq \varepsilon \quad \text{and} \quad \Pi^*(h) = \Pi^*(\tilde{h}) \qquad \forall \phi(h) = \phi(\tilde{h})$$

As before, determine p, find its optimal policy π^*, and define $\tilde{\Pi}(h) := \pi^*(\phi(h))$. If (15) holds, then $\tilde{\Pi}$ is an ε'-optimal policy of original process P in the sense that

$$|V^{\tilde{\Pi}(h)} - V^*(h)| = O\left(\frac{\varepsilon}{(1-\gamma)^?}\right) =: \varepsilon'$$

The following theorem summarizes the considerations for the two choices of ϕ above:

Theorem 9 (Extreme ϕ). *For every process P there exists a reduction ϕ ((16) or (17) will do) and MDP p defined via (11) and (5) whose optimal policy π^* is an ε'-optimal policy $\tilde{\Pi}(h) := \pi^*(\phi(h))$ for P. The size of the MDP is bounded (uniformly for any P) by*

$$|\mathcal{S}| \le \left(\frac{3}{\varepsilon'(1-\gamma)^3}\right)^{|\mathcal{A}|} \quad \text{and if (15) holds even by} \quad |\mathcal{S}| = O\left(\frac{|\mathcal{A}|}{\varepsilon'(1-\gamma)^{1+?}}\right)$$

Discussion. A valid question is of course whether Theorem 9 is just an interesting theoretical insight/curiosity or of any practical use. After all, ϕ depends on Q^* (or V^* and Π^*), but if we knew Q^*, Π^* would readily be available and the detour through p and π^* pointless.

Theorem 9 reaches relevance by the following observation: If we start with a sufficiently rich class of maps Φ that contains at least one ϕ approximately representing $Q^*(h, \cdot)$, and have a learning algorithm that favors such ϕ, then Theorems 4–7 tell us that we do not need to worry about whether P_ϕ is MDP or not; we "simply" use/learn MDP p instead. Theorem 9 tells us that this allows for extreme aggregation far beyond MDPs.

This program is in parts worked out in the next two sections, but more research is needed for its completion. Learning p from (real) P-samples is considered in Section 6 and learning ϕ in Section 7.

6   Reinforcement Learning

In RL, P and therefore p are unknown. We now show how to learn p from samples from P. For this we have to link B to the distribution over histories induced by P and to the behavior policy Π_B the agent follows. We still assume ϕ is given.

Behavior Policy Π_B. Let $\Pi_B : \mathcal{H} \rightsquigarrow \mathcal{A}$ be the behavior policy of our RL agent, which in general is non-stationary due to learning, often stochastic to ensure exploration, and (usually) different from any policy considered so far ($\Pi^*, \pi^*, \tilde{\Pi}, \Pi, \pi$).

Choice of B. The interaction of agent Π_B with environment P stochastically generates some history h_t followed by action a_t with joint probability, say $P_B(h_t a_t)$. We use subscripts B and/or ϕ to indicate dependence on Π_B and/or ϕ. We can get $P_{\phi B}(h_t|s_t a_t)$ from $P_B(h_t a_t)$ by marginalization and conditioning in the usual way, and similar for other arguments. $P_{\phi B}(h_t|s_t a_t)$ seems a natural choice for $B(h|sa)$. It nearly satisfies the required condition (10) for B but not quite. See [Hut14] for details. We can fix this mismatch by introducing weights $w_t : S \times A \rightsquigarrow [0; 1]$ and define

$$B(h_t|sa) := w_t(sa)P_{\phi B}(h_t|s_t = s, a_t = a) \; \forall t, \quad \text{where} \quad \sum_{t=1}^{\infty} w_t(sa) = 1 \; \forall s, a$$
(18)

which now satisfies (10) (due to $\sum_{h \in \mathcal{H}} = \sum_{t=1}^{\infty} \sum_{h_t \in \mathcal{H}_t}$). MDP p can now be represented as

$$p(s'r'|sa) = \sum_{t=1}^{\infty} w_t(sa) \sum_{h_t \in \mathcal{H}_t} P_\phi(s_{t+1} = s', r_{t+1} = r'|h_t, a_t = a) P_{\phi B}(h_t|s_t = s, a_t = a)$$

$$= \sum_{t=1}^{\infty} w_t(sa) P_{\phi B}^t(s'r'|sa) \tag{19}$$

That is, p is the w-weighted time-average of $P_{\phi B}^t$. The first equality follows from (11) and (18); the second one from from the definition of conditional probability. We also introduced the shorthand $P_{\phi B}^t(s'r'|sa) := P_{\phi B}(s_{t+1} = s', r_{t+1} = r'|s_t = s, a_t = a)$.

Choice of w_t. If $P_{\phi B}^t$ in (19) is stationary, i.e. independent of t, then $p(s'r'|sa) = P_{\phi B}^t(s'r'|sa)$ for all t, since the weights sum to one, and estimation is easy. Note that in general we cannot estimate non-stationary $P_{\phi B}^t$, since for each t we have only one sample available, but we will see that estimation of p is still possible. Assume we have observed h_n, and choose

$$w_t(sa) := \frac{P_{\phi B}^t(sa)}{\sum_{t=1}^{n} P_{\phi B}^t(sa)} \quad \text{for } t \le n \quad \text{and} \quad 0 \quad \text{for } t > n \tag{20}$$

Inserting this into (19) gives

$$p(s'r'|sa) = \frac{\frac{1}{n}\sum_{t=1}^{n} P_{\phi B}^t(sas'r')}{\frac{1}{n}\sum_{t=1}^{n} P_{\phi B}^t(sa)} \tag{21}$$

We estimate numerator and denominator separately.

Law of Large Numbers. For $t = 1, 2, 3, \ldots$ let $X_t \in \{0, 1\}$ be binary random variables with expectation $\mathbb{E}[X_t]$. Define $n_1 = \sum_{t=1}^{n} X_t$ be the number of sampled 1s. The strong law of large numbers says that

$$\frac{n_1}{n} - \frac{1}{n}\sum_{t=1}^{n} \mathbb{E}[X_t] \stackrel{n \to \infty}{\longrightarrow} 0 \quad \text{almost surely} \quad \text{under weak conditions} \tag{22}$$

Note that the law holds far beyond i.i.d. random variables under a variety of conditions [FK01, VGS05] which we collectively call 'weak conditions'. It is not even necessary for n_1/n to converge.

Estimation of p. Now fix some (s, a), and let $X_t := [\![s_t = s, a_t = a]\!]$. (Here we assume that variables in h_t are random variables and $sas'r'$ are realizations.) Then

$$n(sa) := n_1 = \sum_{t=1}^{n} X_t = \#\{t \le n : s_t = s, a_t = a\}$$

is the number of times action a is taken in state s, and $\mathbb{E}[X_t] = P(X_t = 1) = P^t_{\phi B}(sa)$, hence (22) implies

$$\frac{n(sa)}{n} - \frac{1}{n}\sum_{t=1}^{n} P^t_{\phi B}(sa) \overset{n\to\infty}{\longrightarrow} 0 \quad \text{a.s. under weak conditions} \tag{23}$$

Similarly for $Y_t := [\![s_t a_t s_{t+1} r_{t+1} = sas'r']\!]$ and $n(sas'r') := \sum_{t=1}^{n} Y_t$ we have

$$\frac{n(sas'r')}{n} - \frac{1}{n}\sum_{t=1}^{n} P^t_{\phi B}(sas'r') \overset{n\to\infty}{\longrightarrow} 0 \quad \text{with } P\text{-probability } 1 \tag{24}$$

under weak conditions. (23) and (24) via (21) are nearly sufficient to imply

$$\frac{n(sas'r')}{n(sa)} - p(s'r'|sa) \overset{n\to\infty}{\longrightarrow} 0 \quad \text{almost surely} \tag{25}$$

A sufficient but by far not necessary condition is

$$\liminf_{n\to\infty} \frac{n(sa)}{n} > 0 \quad \text{almost surely} \tag{26}$$

Theorem 10 (p-estimation). *For B defined in (18) and (20) we have: If (24) and (26) hold, then (25) holds. For example, if Y_t are stationary ergodic processes, then (24) and (26) hence (25) hold for all state-action pairs that matter (i.e. for those occurring with non-zero probability).*

Discussion. Limit (25) shows that standard frequency estimation for p will converge to the true p under weak conditions. If P_ϕ is MDP, samples are conditionally i.i.d. and the 'weak conditions' are satisfied. But the law of large numbers and hence (25) holds far beyond the i.i.d. case [FK01, VGS05], e.g. for stationary ergodic processes. Condition (26) that every state-action pair be visited with non-vanishing relative frequency can be significantly relaxed. Stationarity is also not necessary, and indeed often does not hold due to a non-stationary environment P or a non-stationary behavior policy Π_B (or both). Other choices for w_t are possible, e.g. we could multiply numerator and denominator of (20) by some arbitrary positive function $u_t(as)$, which leads to a weighted average estimator. We estimate p in order to estimate q^* and ultimately π^*. This is model-based RL. We can also learn π^* model-free. For instance, condition (25) should be sufficient for Q-learning to converge to Q^*. Q-learning and other RL algorithms designed for MDPs have been observed to often (but not always) perform well even if applied to non-MDP domains. Our results appear to explain why, but this calls for further investigations.

7   Feature Reinforcement Learning

The idea of FRL is to *learn* ϕ [Hut09b]. FRL starts with a class of maps Φ, compares different $\phi \in \Phi$, and *selects* the most appropriate one given the experience h_t so far. Several criteria based on how well ϕ reduces P to an MDP have been devised [Hut09b, Hut09a] and theoretically [SH10] and experimentally [NSH11] investigated [Ngu13]. Theorems 4–7 show that demanding P_ϕ to be approximately MDP is overly restrictive. Theorem 9 suggests that if we relax this condition, much more substantial aggregation is possible, provided $\varPhi$ is rich enough.

The BLB algorithm [MMR11] and its extensions IBLB [NMRO13] and improvements OMS [NOR13] can (nearly) readily be used for our purpose. The BLB family uses the same basic FRL setup from [Hut09b] used also here. The authors consider a countable class $\varPhi$ assumed to contain at least one ϕ such that P_ϕ is an MDP (6). They consider average reward, rather then discounting, and analyze regret, which (in general) requires some assumption on the mixing rate or 'diameter' of the MDP. They prove that the total regret grows with $\tilde{O}(n^{1/2\ldots2/3})$, depending on the algorithm.

Their algorithms and analyses rely on UCRL2 [JOA10], an exploration algorithm for finite-state MDPs. Going through the BLB proofs, it appears that the condition that P_ϕ is an MDP can be removed if p (11) is used instead, modulo the analysis of UCRL2 itself. The proofs for the bounds for UCRL2 exploit that s', r' conditioned on s, a are i.i.d., which is true if P_ϕ is Markov but not in general. Asymptotic versions should remain valid under the 'weak conditions' alluded to in (25). With some stronger assumptions that guarantee good convergence rates, the regret analysis of UCRL2 should remain valid too. Formally, the use of Hoeffding's inequality for i.i.d. need to be replaced by comparable bounds with weaker conditions, e.g. Azuma's inequality for martingales.

There is one serious gap in the argument above. BLB uses average reward while our theorems are for discounted reward. It is often possible to adapt algorithms and proofs which come with regret bounds for average reward to PAC bounds for discounted reward or vice versa. This would have to be done first: either a PAC version of BLB by combining MERL [LHS13] with UCRLγ [LH12], or average reward versions of the bounds derived in this paper.

See [Hut14] for a general outline of how to learn ϕ beyond MDPs by introducing partial orders on $\varPhi$ justified by our results.

8 Discussion

Summary. Our results show that RL algorithms for finite-state MDPs can be utilized even for problems P that have arbitrary history dependence and history-to-state reductions/aggregations ϕ that induce P_ϕ that are also neither stationary nor MDP. The only condition to be placed on the reduction is that the quantities of interest, (Q-)Values and (optimal) Policies, can approximately be represented. This considerably generalizes previous work on feature reinforcement learning and MDP state aggregation and allows for extreme state aggregations beyond MDPs. The obtained results may also explain why RL algorithms designed for MDPs sometimes perform well beyond MDPs.

Outlook. As usual, lots remains to be done. A list of the more interesting remaining tasks and open questions follows:

- While the approximate ϕ-uniformity condition on Q^* in Theorem 6 is very weak compared to bisimilarity, uniformity of V^* in Theorem 7 is even weaker (Theorem 9 shows how much of a difference this can make). It is an Open Problem 8 whether an analogue of Theorem 6ii also holds for Theorem 7 beyond $\varepsilon = 0$.

- An algorithm learning ϕ beyond MDPs that comes with regret or PAC guarantees has yet to be developed. This could be done by adapting the class and proofs of BLB algorithms, or by integrating MERL with UCRLγ, or by other means [Hut14].

- All bounds contain $\frac{1}{1-\gamma}$ to some power. Can the exponents be improved? For which environments/examples are the bounds tight?

- The trick to use a-dependent Q^* as a-independent map ϕ in Section 5 was to vectorize Q^* in a. Unfortunately this leads to a state-space size exponential in $\mathcal{A}$. Solution ϕ based on (V^*, Π^*) pair is only linear in $\mathcal{A}$, but rests on Open Problem 8. Are there other/better ways of dealing with actions? Other extreme aggregations ϕ, or are a-dependent ϕ possible?

- Are average-reward total-regret versions of our discounted reward results possible, under suitable mixing rate conditions?

- For small discrete action spaces typical for many board games, the exact conditions on Π are met. For continuous action spaces as in robotics, we can simply discretize the action space, introducing another ε-error, but action-continuous versions of our results would be nicer. Except for Theorem 6, any interesting generalization should replace the exact by approximate ϕ-uniformity conditions on Π.

- Our theorems and/or proof ideas should allow to extend existing convergence theorems for RL algorithms such as Q-learning and others from MDPs to beyond MDPs.

- The bisimulation conditions of classical state aggregation results are for reward and transition probabilities. It would be interesting to derive explicit weaker conditions for them that still imply our conditions on (Q-)Values.

References

[FK01]   Fazekas, I., Klesov, O.: A general approach to the strong law of large numbers. Theory of Probability & Its Applications 45(3), 436–449 (2001)

[FPP04] Ferns, N., Panangaden, P., Precup, D.: Metrics for finite Markov decision processes. In: Proc. 20th Conf. on Uncertainty in Artificial Intelligence (UAI 2004), pp. 162–169 (2004)

[GDG03] Givan, R., Dean, T., Greig, M.: Equivalence notions and model minimization in Markov decision processes. Artificial Intelligence 147(1–2), 163–223 (2003)

[Hut05] Hutter, M.: Universal Artificial Intelligence: Sequential Decisions based on Algorithmic Probability. Springer, Berlin (2005)

[Hut09a] Hutter, M.: Feature dynamic Bayesian networks. In: Proc. 2nd Conf. on Artificial General Intelligence (AGI 2009), vol. 8, pp. 67–73. Atlantis Press (2009)

[Hut09b] Hutter, M.: Feature reinforcement learning: Part I: Unstructured MDPs. Journal of Artificial General Intelligence 1, 3–24 (2009)

[Hut14]   Hutter, M.: Extreme state aggregation beyond MDPs. Technical report
          (2014), http://www.hutter1.net/publ/exsaggx.pdf
[JOA10]   Jaksch, T., Ortner, R., Auer, P.: Near-optimal regret bounds for reinforce-
          ment learning. Journal of Machine Learning Research 11, 1563–1600 (2010)
[LH12]   Lattimore, T., Hutter, M.: PAC bounds for discounted MDPs. In: Bshouty,
          N.H., Stoltz, G., Vayatis, N., Zeugmann, T. (eds.) ALT 2012. LNCS (LNAI),
          vol. 7568, pp. 320–334. Springer, Heidelberg (2012)
[LH14]   Lattimote, T., Hutter, M.: General time consistent discounting. Theoretical
          Computer Science 519, 140–154 (2014)
[LHS13]   Lattimore, T., Hutter, M., Sunehag, P.: The sample-complexity of gen-
          eral reinforcement learning. Journal of Machine Learning Research, W&CP:
          ICML 28(3), 28–36 (2013)
[McC96]   McCallum, A.K.: Reinforcement Learning with Selective Perception and
          Hidden State. PhD thesis, Department of Computer Science, University of
          Rochester (1996)
[MMR11]   Maillard, O.-A., Munos, R., Ryabko, D.: Selecting the state-representation
          in reinforcement learning. In: Advances in Neural Information Processing
          Systems (NIPS 2011), vol. 24, pp. 2627–2635 (2011)
[Ngu13]   Nguyen, P.: Feature Reinforcement Learning Agents. PhD thesis, Research
          School of Computer Science, Australian National University (2013)
[NMRO13] Nguyen, P., Maillard, O., Ryabko, D., Ortner, R.: Competing with an in-
          finite set of models in reinforcement learning. JMLR WS&CP AISTATS 31,
          463–471 (2013)
[NOR13]   Maillard, O.-A., Nguyen, P., Ortner, R., Ryabko, D.: Optimal regret bounds
          for selecting the state representation in reinforcement learning. JMLR W&CP
          ICML 28(1), 543–551 (2013)
[NSH11]   Nguyen, P., Sunehag, P., Hutter, M.: Feature reinforcement learning in prac-
          tice. In: Sanner, S., Hutter, M. (eds.) EWRL 2011. LNCS, vol. 7188, pp.
          66–77. Springer, Heidelberg (2012)
[Put94]   Puterman, M.L.: Markov Decision Processes — Discrete Stochastic Dynamic
          Programming. Wiley, New York (1994)
[RN10]   Russell, S.J., Norvig, P.: Artificial Intelligence. A Modern Approach, 3rd edn.
          Prentice-Hall, Englewood Cliffs (2010)
[SB98]   Sutton, R.S., Barto, A.G.: Reinforcement Learning: An Introduction. MIT
          Press, Cambridge (1998)
[SH10]   Sunehag, P., Hutter, M.: Consistency of feature Markov processes. In: Hutter,
          M., Stephan, F., Vovk, V., Zeugmann, T. (eds.) ALT 2010. LNCS (LNAI),
          vol. 6331, pp. 360–374. Springer, Heidelberg (2010)
[SLL09]   Strehl, A.L., Li, L., Littman, M.L.: Reinforcement learning in finite MDPs:
          PAC analysis. Journal of Machine Learning Research 10, 2413–2444 (2009)
[VGS05]   Vovk, V., Gammerman, A., Shafer, G.: Algorithmic Learning in a Random
          World. Springer, New York (2005)

On Learning the Optimal Waiting Time

Tor Lattimore[1], András György[1], and Csaba Szepesvári[1,2]

[1] Department of Computing Science, University of Alberta, Canada
[2] Microsoft Research, Redmond
tor.lattimore@gmail.com, {gyorgy,szepesva}@ualberta.ca

Abstract. Consider the problem of learning how long to wait for a bus before walking, experimenting each day and assuming that the bus arrival times are independent and identically distributed random variables with an unknown distribution. Similar uncertain optimal stopping problems arise when devising power-saving strategies, e.g., learning the optimal disk spin-down time for mobile computers, or speeding up certain types of satisficing search procedures by switching from a potentially fast search method that is unreliable, to one that is reliable, but slower. Formally, the problem can be described as a repeated game. In each round of the game an agent is waiting for an event to occur. If the event occurs while the agent is waiting, the agent suffers a loss that is the sum of the event's "arrival time" and some fixed loss. If the agents decides to give up waiting before the event occurs, he suffers a loss that is the sum of the waiting time and some other fixed loss. It is assumed that the arrival times are independent random quantities with the same distribution, which is unknown, while the agent knows the loss associated with each outcome. Two versions of the game are considered. In the full information case the agent observes the arrival times regardless of its actions, while in the partial information case the arrival time is observed only if it does not exceed the waiting time. After some general structural observations about the problem, we present a number of algorithms for both cases that learn the optimal weighting time with nearly matching minimax upper and lower bounds on their regret.

1 Introduction

Each day a student travels to school, either by bus or on foot, whichever is faster. The expected travel time for the bus is five minutes and is denoted by β while walking takes twenty minutes and is denoted by ω. Unfortunately, the bus is not always on time, so on each day t the student must decide how long he wants to wait for the bus, Y_t. The bus comes at random time X_t and if $X_t \leq Y_t$, then the student catches the bus. If $X_t > Y_t$, then they walk. The loss at time step t is the total travel time, which the student wants to keep as small as possible and is defined by

$$\ell_t(Y_t) \doteq \begin{cases} X_t + \beta, & \text{if } X_t \leq Y_t\,; \\ Y_t + \omega, & \text{otherwise}\,. \end{cases}$$

P. Auer et al. (Eds.): ALT 2014, LNAI 8776, pp. 200–214, 2014.

Table 1. Examples of losses and optimal waiting times for particular choices of arrival distributions. The first column gives the density of the arrival times.

	$p(x)$	$\ell(y)$	y^*	$\ell(y^*)$
Power	$\frac{\mathbb{1}\{x \geq 1\}}{x^2}$	$\beta + 1 + \frac{\omega - \beta}{y} + \log y$	$\omega - \beta$	$\beta + 2 + \log(\omega - \beta)$
Exponential	$\lambda e^{-\lambda x}$	$\beta + \frac{1}{\lambda} + e^{-\lambda y}\left[(\omega - \beta) - \frac{1}{\lambda}\right]$	0, if $\omega - \beta < \frac{1}{\lambda}$; ∞, otherwise	ω, if $\omega - \beta < \frac{1}{\lambda}$; $\beta + \frac{1}{\lambda}$, otherwise

We assume that $\beta < \omega$ and that the arrival times for the bus $(X_t)_t$ are identically and independently distributed according to some unknown distribution over the positive real line. We make no additional assumptions on the distribution of arrival times.[1] The expected loss of a fixed deterministic waiting time $y \geq 0$ is

$$\ell(y) = \mathbb{E}\left[\mathbb{1}\{X \leq y\}(X + \beta) + \mathbb{1}\{X > y\}(y + \omega)\right],$$

where X is identically distributed to X_t.[2] An optimal waiting time is given by $y^* \doteq \arg\min_{y \geq 0} \ell(y)$, which we will show to exist (y^* may be infinite and optimal waiting may not be unique, as we will demonstrate below).

To guide the reader's intuition, in Table 1 we tabulate the loss function, the optimal action, and the loss of the optimal action for two particular arrival time distributions. The examples show that the loss may be convex, or concave, it can be unbounded and the optimal action can also take on any value between 0 and infinity. These examples should not mislead the reader. Our methods do not need to know the form of the arrival time distributions, i.e., we consider the nonparametric setting.

Since the distribution of arrival times is unknown, the student cannot know when to stop waiting and must experiment to gain information. The regret at time step t is the difference between the actual travel time and the travel time under an optimal waiting time, $r_t \doteq \ell_t(Y_t) - \ell_t(y^*)$. Note that r_t may be negative, but has non-negative expectation, and that r_t does not depend on the choice of y^*. The cumulative regret until time step n is

$$R_n \doteq \sum_{t=1}^{n}\left(\ell_t(Y_t) - \ell_t(y^*)\right).$$

In the long run the student hopes to choose Y_t in such a way as to learn the optimal waiting time, in which case $\lim_{n \to \infty} \mathbb{E}[R_n]/n = 0$.

Two observation models will be considered. The first is a *full information setting* where X_t is always observed. This assumption is unnatural for the problem of waiting for a bus because the student would not usually observe the arrival time of the bus if they decided to walk. There are, however, waiting problems

---

[1] Note that the game is trivial if $\beta \geq \omega$, since in this case the student should always walk regardless of the expected arrival time of the bus.

[2] One can show that the expected loss is minimized by a fixed deterministic waiting time, i.e., there is no advantage to using a stopping rule. The simple reason is that when the bus arrives, due to our assumption that $\omega > \beta$ it is better to take the bus then to continue waiting and then eventually walk (since no more buses are coming).

for which the full information setting is appropriate. An example is maximising hard-disk efficiency in mobile computing, previously considered by Krishnan, Long and Vitter [12] where a hard-disk controller must decide after each inter-action how long to wait before spinning down the disk to conserve energy. This is modelled by choosing $\beta = 0$ and ω to be some value that reflects the cost (in terms of time/energy/annoyance) of spinning up the disk. The goal of the controller is to minimise the sum of energy consumption and spin-up costs.

The second setting, called the *partial information setting*, is trickier, but often more natural, e.g., for the bus-stop problem, when the student in general will not observe X_t unless $X_t \leq Y_t$. More precisely, the student observes the pair (Z_t, δ_t) where $Z_t = \min\{X_t, Y_t\}$ and $\delta_t = \mathbb{1}\{X_t \leq Y_t\}$. So δ_t is 1 if the student travelled by bus and 0 otherwise, while Z_t is the time at which the travel starts. Another appli-cation of the partial information setting is the problem of combining algorithms to solve a number of instances of a satisficing search problems.[3] We assume that the agent has access to two algorithms for a given type of search problem. The first is potentially fast, but unreliable, while the second is typically slower, but has known guarantees on its performance. For each problem instance, the agent tries to use the potentially fast solver, switching to the more consistent algorithm if the first fails to deliver a solution within a certain amount of time. The task of the agent is to learn when to switch between solvers. Formally, the unreliable solver pro-vides a solution to instance t at random time X_t, and the completion time of the slower method is deterministically ω for all instances. As for the hard-disk prob-lem, $\beta = 0$. Comparisons between stochastic satisficing search algorithms have been made before (e.g., [16] and references there-in), but to our knowledge the sequential setting combined with the regret criterion are new.

Estimating the common distribution of X_t (or other quantities depending on this distribution) in the partial information (or "censored") setting is heavily studied in the statistics literature [e.g. 6], but the focus tends to be on the natural medical applications where the censoring times are uncontrolled and independent of the arrival times. We know of no previous work on the decision problem studied here. Optimising the regret is more complex when only partial information is available because in this case the actions influence the observations.

The censored information problem is an instance of stochastic partial moni-toring, first studied by Agrawal, Teneketzis and Anantharam [1]. In recent years there has been significant progress towards understanding partial monitoring with finitely many actions, both in the stochastic and adversarial settings [4, 9, 3], but the case where the number of actions is infinite/continuous the work has been more limited and specialised [11].

Summary of Results. The full information setting is analysed in Section 3 where we present two algorithms. The first is based on discretising the action space and applying the exponential weighting algorithm (EWA), while the second is an instance of the Follow-the-Leader (FTL) algorithm. We prove that EWA suffers a regret of at most $O(\log^{3/2}(n)\sqrt{n})$ while for FTL we were able to shave

---

[3] A search problem is satisficing if the searcher can stop once a satisfactory solution has been found, with SAT being a prototypical example.

off a small amount and bound the regret by $O(\log(n)\sqrt{n})$. We also establish a lower bound of $\Omega(\sqrt{n})$.

For the partial information case we also consider two algorithms (Section 4). The first is again based on a variant of the exponential weights algorithm, which cleverly controls the exploration of actions to deal with the partial information setting [2]. We establish that this algorithm enjoys a regret of $O(\log^2(n)\sqrt{n})$. Next we propose a novel optimistic algorithm that conservatively waits for the longest time that it cannot prove to be sub-optimal with high probability. We prove that this algorithm enjoys a regret of $O(\log^{3/2}(n)\sqrt{n})$. Thus, for both algorithms, the cost of partial information is surprisingly small and of order $O(\log^{1/2}(n))$. Some proofs have been omitted or sketched, but complete versions may be found in our report [13].

The theoretical findings are complemented by computer simulations in a variety of controlled scenarios (Section 5). Results are presented for the full information setting only (similar results were observed in the partial information setting). The most interesting finding here is that for the exponential distribution, both algorithms perform better than is predicted by theory, with at least FTL achieving $O(\log n)$ regret. The EWA algorithm behaves comparably to FTL, but only when the learning rate is tuned to be much larger than is theoretically justified.

Notation. At time step t define the empirical probability measure by $\mathbb{P}_t\{A\} \doteq \frac{1}{t}\sum_{s=1}^{t} \mathbb{1}\{X_s \in A\}$ where A is any Borel-measurable subset of the real line. The cumulative distribution of the samples $X_1, \ldots, X_t$ is $F_t(x) = \mathbb{P}_t\{(-\infty, x]\}$. Expectations with respect to the empirical distribution $\mathbb{P}_t\{\cdot\}$ are denoted by $\mathbb{E}_t\{\cdot\}$. Further, by slightly abusing the notation for any measurable function $f : \mathbb{R} \to \mathbb{R}$, we define $\mathbb{E}_t[f(X)] = \int f(x)d\mathbb{P}_t(x)$ and for any Borel measurable subset A of the real line, $\mathbb{P}_t\{X \in A\} = \mathbb{E}_t[\mathbb{1}\{X \in A\}]$.

2 Structure of the Waiting Problem

Before the main theorems we present a crucial lemma that characterises the cumulative distribution of the arrival times in terms of the optimal action y^*. The result shows that the tail of X decays exponentially for times before y^*. As a consequence, if the optimal waiting time is large then the loss of choosing y much smaller than optimal cannot be too large. This latter fact should not be surprising. If it is optimal to wait for the bus for a very long time, then there must be a reasonable probability that it will arrive soon. This means that the bus is still likely to arrive if you wait for a shorter time. The critical case occurs when arrival times are exponentially distributed. As a result, it is not hard to see that to achieve a polynomially decreasing regret in n time steps, it is enough to consider waiting times below some $O(\log n)$ threshold.

Lemma 1. *Let $0 < \hat{y} \leq \tilde{y}$ such that $\inf_{y\in[0,\hat{y}]} \ell(y) \geq \ell(\tilde{y})$. Then, the following hold true for any $y \in [0, \hat{y}]$:*

1. $\mathbb{P}\{X > y\} \leq 2^{-\left\lfloor \frac{y}{2(\omega - \beta)} \right\rfloor}$.
2. $\ell(y) - \ell(\hat{y}) \leq (\omega - \beta)2^{-\left\lfloor \frac{y}{2(\omega - \beta)} \right\rfloor}$.

In particular, if $y^* \in [0, \infty]$ is optimal, then the above holds with $\tilde{y} = \hat{y} = y^*$.

The proof of the lemma utilizes the following bounds on loss differences, which will also be useful later and follows trivially from the definitions. The proof may be found in the technical report [13].

Lemma 2. *Let* $y_2 \geq y_1$, *then*
1. $\ell(y_2) - \ell(y_1) = \mathbb{E}[\mathbb{1}\{y_1 < X \leq y_2\}(X - y_1 + \beta - \omega) + \mathbb{1}\{X > y_2\}(y_2 - y_1)]$.
2. $\ell(y_2) - \ell(y_1) \geq (y_2 - y_1)(1 - F(y_2)) - (\omega - \beta)(F(y_2) - F(y_1))$.
3. $\ell(y_2) - \ell(y_1) \leq (y_2 - y_1)(1 - F(y_1)) - (\omega - \beta)(F(y_2) - F(y_1))$.

Proof (Lemma 1). Let $c \geq 0$ be some constant to be chosen later and $0 \leq y \leq \hat{y} - c$. Then we have

$$0 \overset{(a)}{\leq} \ell(y) - \ell(\tilde{y}) \overset{(b)}{=} \mathbb{E}[(y + \omega - \beta - X)\mathbb{1}\{y < X \leq \tilde{y}\} + (y - \tilde{y})\mathbb{1}\{X > \tilde{y}\}]$$

$$\overset{(c)}{\leq} \mathbb{E}[(y + \omega - \beta - X)\mathbb{1}\{y < X \leq y + c\} + (\omega - \beta - c)\mathbb{1}\{X > y + c\}]$$

$$\overset{(d)}{\leq} (\omega - \beta)\mathbb{E}[\mathbb{1}\{y < X \leq y + c\} + (\omega - \beta - c)\mathbb{1}\{X > y + c\}]$$

$$\overset{(e)}{=} (\omega - \beta)\mathbb{P}\{y < X \leq y + c\} + (\omega - \beta - c)\mathbb{P}\{X > y + c\} \tag{1}$$

$$\overset{(f)}{\leq} (\omega - \beta)\mathbb{P}\{y < X\}, \tag{2}$$

where (a) follows since $\ell(\tilde{y}) \leq \ell(y)$ by assumption, (b) follows from Part 1 of Lemma 2, (c) follows by breaking $\mathbb{1}\{y + c < X \leq \tilde{y}\}$ off from both indicators and since $y \leq \hat{y} - c \leq \tilde{y} - c$, while (d) is true by noting that $y + \omega - \beta - X \leq \omega - \beta$ for $y \leq X$. (e) and (f) are trivial. Choosing $c = 2(\omega - \beta) > 0$, (1) implies

$$\mathbb{P}\{y < X \leq y + 2(\omega - \beta)\} \geq \mathbb{P}\{X > y + 2(\omega - \beta)\}.$$

Therefore, for any $y \geq 0$ such that $y + 2(\omega - \beta) \leq \hat{y}$,

$$\mathbb{P}\{X \leq y + 2(\omega - \beta)|X > y\} \geq \frac{1}{2} \tag{3}$$

and if $2k(\omega - \beta) \leq \hat{y}$, then

$$\mathbb{P}\{X > 2k(\omega - \beta)\} \overset{(a)}{=} \prod_{i=1}^{k} \mathbb{P}\{X > 2i(\omega - \beta)|X > 2(i-1)(\omega - \beta)\}$$

$$\overset{(b)}{=} \prod_{i=1}^{k} (1 - \mathbb{P}\{X \leq 2i(\omega - \beta)|X > 2(i-1)(\omega - \beta)\}) \overset{(c)}{\leq} 2^{-k}, \tag{4}$$

where (a) follows from the chain rule for probability, (b) is just $\mathbb{P}\{A|B\} = 1 - \mathbb{P}\{A^c|B\}$ for events A and B and (c) follows by substituting (3), which is permitted thanks to $2k(\omega - \beta) \leq \hat{y}$. The above inequality immediately implies Part 1 (for $y \leq \hat{y} < 2k(\omega - \beta)$ the result holds trivially) and, combined with (2) for $c = 0$, it also yields Part 2. $\qquad \square$

That an optimal waiting time is guaranteed to exist follows from Lemma 1 by a tedious case-based analysis. See the technical report for the proof [13].

Theorem 1. *For any arrival time distribution there exists a $y^* \in [0, \infty]$ such that $\ell(y^*) = \inf_{y \in [0,\infty]} \ell(y)$.*

Part 2 of Lemma 1 also shows that to guarantee an ε-optimal action, it suffices to consider the waiting times in an interval of length $O(\log(1/\varepsilon))$ starting at zero:

Corollary 1. *Let $\varepsilon > 0$ and $\bar{y}(\varepsilon) \doteq 2(\omega - \beta) \max\left\{1 + \log_2(\frac{\omega - \beta}{\varepsilon}), 0\right\}$. Then $\inf_{y \in [0,\bar{y}(\varepsilon)]} \ell(y) - \ell(y^*) < \varepsilon$.*

Proof. The result follows immediately from Part 2 of Lemma 1.

3   Full Information Setting

We consider the case when X_t is always observed in round t. Our first algorithm discretises the set of actions and then applies the exponential weighting algorithm [e.g., 5]. The key observation is that by Corollary 1, to guarantee an ε-optimal action, it suffices to play in the interval of length $O(\log(1/\varepsilon))$. Since the exponential weights algorithm assumes a finite action set, we need to discretise the action space. The following elementary observation, which follows directly from Part 1 of Lemma 2 shows that to achieve an ε-error, it suffices to discretise the interval with an accuracy of ε.

Proposition 1. *For any $y_2 \geq y_1 \geq 0$, $\ell(y_2) - \ell(y_1) \leq y_2 - y_1$.*

The exponential weights algorithm enjoys a regret smaller than $R\sqrt{n \log(K)/2}$, where n is the number of rounds, K is the number of actions, and R is the range of losses [5, §4.2, Thm 2.2]. So we see that this method suffers a regret of at least $O(\sqrt{n})$. This suggests choosing $\varepsilon = (\omega - \beta)/\sqrt{n}$ and using the action set $A = \{k\varepsilon : 0 \leq k \leq \bar{y}(\varepsilon)/\varepsilon, k \in \mathbb{N}\}$, leading to Algorithm 1, where for tuning the learning rate η we use that the range of the loss function is $m + \omega$ when the largest waiting time is $m = \max A$. The running time of the algorithm is $O(|A|)$ per time step, which in this case is $O(\sqrt{n} \log(n))$.

Theorem 2 (EWA Regret). *Let $n > 0$ and R_n be the regret of Algorithm 1 when used for n rounds. Then $\mathbb{E}[R_n] \in O((\omega - \beta) \log^{3/2}(n)\sqrt{n})$.*

Proof. Let ε, A and R be as in the pseudo-code of the algorithm. As noted beforehand, the expected regret[4] of EWA against the best action in A is $R\sqrt{n/2 \log K}$, where $K = |A| \leq \lceil \bar{y}(\varepsilon)/\varepsilon \rceil = \lceil \bar{y}((\omega - \beta)/\sqrt{n})\sqrt{n}/(\omega - \beta) \rceil$. By Proposition 1, $\min_{y \in A} \ell(y) - \inf_{y \in [0,\bar{y}(\varepsilon)]} \ell(y) \leq \varepsilon$ and by Corollary 1, $\inf_{y \in [0,\bar{y}(\varepsilon)]} \ell(y) - \ell(y^*) \leq \varepsilon$. Hence, $\mathbb{E}[R_n] \leq R\sqrt{n/2 \log(\bar{y}((\omega - \beta)/\sqrt{n})\sqrt{n}/(\omega - \beta) + 1)} + 2(\omega - \beta)\sqrt{n} \in O((\omega - \beta) \log^{3/2}(n)\sqrt{n})$, where we used $R = \bar{y}((\omega - \beta)/\sqrt{n}) + \omega$ and that $\bar{y}((\omega - \beta)/\sqrt{n}) \in O((\omega - \beta) \log(n))$.

---

[4] Bounds for adversarial algorithms like EWA are typically proven for the regret without the expectation, but in the stochastic case this distinction is not important with bounds on the expected regret following from a straight-forward application of standard concentration inequalities.

Algorithm 1. EWA for Optimal Waiting

1: **Input:** ω, β, and n
2: $\varepsilon \leftarrow (\omega - \beta)/\sqrt{n}, A \leftarrow \{k\varepsilon : 0 \leq k \leq \bar{y}(\varepsilon)/\varepsilon, k \in \mathbb{N}\}, R \leftarrow \bar{y}(\varepsilon) + \omega$
3: $\eta \leftarrow \sqrt{8\log(|A|)/n}/R$ and $w_1(y) \leftarrow 1$ for all $y \in A$
4: **for** $t = 1, \ldots, n$ **do**
5:     $W_t \leftarrow \sum_{y \in A} w_t(y)$ and $p_t(y) \leftarrow w_t(y)/W_t$ for each y
6:     Sample waiting time Y_t from distribution p_t on A and observe X_t
7:     **for** $y \in A$ **do**   // Update the weights

$$\ell_t(y) \leftarrow \mathbb{1}\{X_t \leq y\}(X_t + \beta) + \mathbb{1}\{X_t > y\}(y + \omega)$$

$$w_{t+1}(y) \leftarrow w_t(y)\exp(-\eta\ell_t(y))$$

8:     **end for**
9: **end for**

Under the full information stochastic setting the FTL algorithm, which at each round chooses the waiting time that minimises the empirical loss so far, is also expected to do well. The next theorem shows that FTL does indeed improve slightly on EWA.

Theorem 3 (FTL Regret). *Let Y_t be defined by $Y_1 \doteq 0$ and, for all $t \geq 2$,*

$$Y_t \in \arg\min_y \sum_{s=1}^{t-1} (\mathbb{1}\{X_s \leq y\}(X_s + \beta) + \mathbb{1}\{X_s > y\}(y + \omega)) .$$

Then, $\mathbb{E}[R_n] \leq (\omega - \beta)(11.6\sqrt{n}\log n - 11\sqrt{n} + \log n + 12)$.

Remark 4. *It is easy to see that for any $t \geq 1$, $Y_t = X_s$ for some $1 \leq s \leq t-1$, hence Y_t can be computed in $O(t)$ time. Note that Y_t is not unique.*

Proof. The empirical loss of wait-time y at time step t is

$$\hat{\ell}_t(y) \doteq \frac{1}{t}\sum_{s=1}^{t} (\mathbb{1}\{X_s \leq y\}(X_s + \beta) + \mathbb{1}\{X_s > y\}(y + \omega)) .$$

The expected regret at time step t may be decomposed. Let $(s_t)_t$ be a sequence of constants to be chosen later. Then,

$$\mathbb{E}[r_t|Y_t] = \ell(Y_t) - \ell(y^*) = \ell(Y_t) - \ell(s_t) + \ell(s_t) - \ell(y^*)$$

$$= \ell(Y_t) - \hat{\ell}_{t-1}(Y_t) + \hat{\ell}_{t-1}(Y_t) - \ell(s_t) + \hat{\ell}_{t-1}(s_t) - \hat{\ell}_{t-1}(s_t) + \ell(s_t) - \ell(y^*)$$

$$\overset{(a)}{\leq} \left|\ell(Y_t) - \hat{\ell}_{t-1}(Y_t)\right| + \left|\ell(s_t) - \hat{\ell}_{t-1}(s_t)\right| + \ell(s_t) - \ell(y^*),$$

where in (a) we used the fact that $\hat{\ell}_{t-1}(Y_t) \leq \hat{\ell}_{t-1}(s_t)$. Now,

$$|\ell(y) - \hat{\ell}_t(y)| \overset{(a)}{=} |(\mathbb{E} - \mathbb{E}_t)[\mathbb{1}\{X \leq y\}(X + \beta) + \mathbb{1}\{X > y\}(\omega + y)]|$$

$$\overset{(b)}{=} |(\mathbb{E} - \mathbb{E}_t)[\mathbb{1}\{X \leq y\}(X - y) + \mathbb{1}\{X > y\}(\omega - \beta)]|$$

$$\overset{(c)}{\leq} y|F(y) - F_t(y)| + (\omega - \beta)|F(y) - F_t(y)| ,$$

where (a) is simply the definition of the losses and $(\mathbb{E} - \mathbb{E}_t)$, (b) by rearranging and using the fact that $(\mathbb{E} - \mathbb{E}_t)\alpha = 0$ for any constant α, (c) by $|X - y| \leq y$

which holds for $0 \leq X \leq y$ and the definition of the cumulative distribution. Combined with [7, Thm. 3.3], which states that $\mathbb{E}[\sup_x |F_t(x) - F(x)|] \leq 1/\sqrt{t}$, the last inequality gives

$$\mathbb{E}\left[\sup_{y \leq s} \left|\ell(y) - \hat{\ell}_t(y)\right|\right] \leq \frac{s + \omega - \beta}{\sqrt{t}}. \tag{5}$$

Next we show that $Y_{t+1} \in O(\log t)$ for any $t \geq 1$. Since Y_{t+1} is the optimal waiting time for the empirical distribution of the arrival times, we can apply Part 1 of Lemma 1 to obtain

$$\mathbb{P}_t \{X \geq Y_{t+1}\} = \inf_{\varepsilon > 0} \mathbb{P}_t \{X > Y_{t+1} - \varepsilon\} \leq \inf_{\varepsilon > 0} 2^{-\left\lfloor \frac{Y_{t+1} - \varepsilon}{2(\omega - \beta)} \right\rfloor}$$

$$\leq \inf_{\varepsilon > 0} 2^{1 - \frac{Y_{t+1} - \varepsilon}{2(\omega - \beta)}} = 2^{1 - \frac{Y_{t+1}}{2(\omega - \beta)}}.$$

Therefore, if $Y_{t+1} > m_{t+1} \doteq 2(\omega - \beta)(1 + \log_2 t)$, then $\mathbb{P}_t \{X \geq Y_{t+1}\} < 1/t$. On the other hand, $\mathbb{P}_t \{X \geq Y_{t+1}\} \geq 1/t$ since $Y_{t+1} \in \{X_1, \ldots, X_t\}$. Thus, $Y_{t+1} \leq m_{t+1}$. Choose $s_t = \min\{y^*, m_t\}$. Then, by (5),

$$\mathbb{E}\left[r_{t+1}\right] \leq \mathbb{E}\left[\left|\ell(Y_{t+1}) - \hat{\ell}_t(Y_{t+1})\right| + \left|\ell(s_{t+1}) - \hat{\ell}_t(s_{t+1})\right|\right] + \ell(s_{t+1}) - \ell(y^*)$$

$$\leq \mathbb{E}\left[\sup_{y \leq m_{t+1}} \left|\ell(y) - \hat{\ell}_t(y)\right| + \left|\ell(s_{t+1}) - \hat{\ell}_t(s_{t+1})\right|\right] + \ell(s_{t+1}) - \ell(y^*)$$

$$\leq (m_{t+1} + s_{t+1} + 2(\omega - \beta))\frac{1}{\sqrt{t}} + \ell(s_{t+1}) - \ell(y^*)$$

$$\leq (m_{t+1} + s_{t+1} + 2(\omega - \beta))\frac{1}{\sqrt{t}} + \frac{\omega - \beta}{t},$$

where in the last step we used Part 2 of Lemma 1 to bound $\ell(s_{t+1}) - \ell(y^*)$. Summing over t ultimately leads to

$$\mathbb{E}\left[R_n\right] = \mathbb{E}\left[r_1\right] + \sum_{t=2}^{n} \mathbb{E}\left[r_t\right] \leq (\omega - \beta)\left(1 + \sum_{t=1}^{n-1} \left[\frac{6 + 4\frac{\log t}{\log 2}}{\sqrt{t}} + \frac{1}{t}\right]\right)$$

$$\leq (\omega - \beta)(11.6\sqrt{n}\log n - 11\sqrt{n} + \log n + 12).$$

as required. $\square$

If the arrival time X_t is exponentially distributed, then the regret of the FTL algorithm may be shown to be at most poly-logarithmic. Experimental results suggest that the true regret is actually logarithmic in n, but so far the proof eludes us.

Theorem 5. *Assume that X_t is exponentially distributed with parameter λ such that $1/\lambda < \omega - \beta$. Then, for the algorithm of Theorem 3, we have $\mathbb{E}[R_n] \in O(\log^2 n)$.*

3.1 Lower Bound

The general upper bounds given in the previous section cannot be greatly improved in the worst-case. Note that the following theorem is proven for the easier full information setting, so translates immediately to give an identical lower bound in the partial information setting.

Theorem 6. *There exists a universal constant $c > 0$ such that for each algorithm and fixed n there exists a distribution such that $\mathbb{E}R_n \geq c(\omega - \beta)\sqrt{n}$.*

Proof. For $p \in [0,1]$ let $\mathbb{P}_p$ be a measure defined such that $\mathbb{P}_p(X = 1/2) = p$ and $\mathbb{P}_p(X = \infty) = 1 - p$. Let us denote the expected loss under measure $\mathbb{P}_p$ by ℓ_p. A simple calculation shows that

$$\ell_p(y) = \begin{cases} y + \omega, & \text{if } y < \frac{1}{2}; \\ p(\frac{1}{2} + \beta) + (1 - p)(y + \omega), & \text{otherwise.} \end{cases}$$

Thus, ℓ_p is piecewise linear, with two increasing segments. The two local minima of ℓ_p are at 0 and $1/2$ with values $\ell_p(0) = \omega$ and $\ell_p(1/2) = 1/2 + \omega - p(\omega - \beta)$. For simplicity, we set $\omega = 1$, $\beta = 0$, the full result can be obtained by scaling. Thus, $\ell_p(0) = 1$, $\ell_p(1/2) = 3/2 - p$ and the optimal waiting time y_p^* is 0 for $p < 1/2$ and $1/2$ for $p > 1/2$. If $p = 1/2$, then 0 and 1 are both optimal. It is also clear that for the "rounding function" ρ defined by $\rho(y) = \frac{1}{2}\mathbb{1}\{y \geq 1/2\}$, then for any $y \geq 0$ it holds that $\ell_p(\rho(y)) \leq \ell_p(y)$: By "rounding down" the waiting time y to either 0 or $1/2$, one can only win in terms of the expected loss. Based on $\mathbb{P}_p$, we construct three environments and will use a fairly standard technique based on the relative entropy that shows that the regret will be large in at least in one of the environments. The three environments are given by the measures $\mathbb{P}_{1/2}$, $\mathbb{P}_{1/2+\varepsilon}$ and $\mathbb{P}_{1/2-\varepsilon}$ for some $\varepsilon \in [0,1/2)$ to be chosen later. Note that $|\ell_{1/2+\sigma\varepsilon}(0) - \ell_{1/2+\sigma\varepsilon}(1/2)| = \sigma\varepsilon$. Fix $n > 0$. Now, take any algorithm A and let Y_t be the choice made by A in round $1 \leq t \leq n$. Let R_σ be the expected regret of A during the first n rounds when used on the measure $\mathbb{P}_{1/2+\sigma\varepsilon}$, $\sigma \in \{-1, 0, +1\}$. Denoting by $\mathbb{E}_\sigma$ the expectation under $\mathbb{P}_{1/2+\sigma\varepsilon}$, we thus have $R_\sigma = \mathbb{E}_\sigma[\sum_{t=1}^n \ell_{1/2+\sigma\varepsilon}(Y_t) - \ell_{1/2+\sigma\varepsilon}(y_{1/2+\sigma\varepsilon}^*)]$. Let $\hat{Y}_t = \rho(Y_t)$ be the "rounded" decision and let $N(y) = \sum_{t=1}^n \mathbb{1}\{\hat{Y}_t = y\}$, $y \in \{0, 1/2\}$. Then,

$$R_\sigma \geq \mathbb{E}_\sigma\left[\sum_{t=1}^n \ell_{1/2+\sigma\varepsilon}(\hat{Y}_t) - \ell_{1/2+\sigma\varepsilon}(y_{1/2+\sigma\varepsilon}^*)\right]$$

and thus

$$R_1 \geq \varepsilon\mathbb{E}_1[N(0)], \qquad R_{-1} \geq \varepsilon\mathbb{E}_{-1}[N(1/2)]. \tag{6}$$

Now, a standard argument shows that

$$\mathbb{E}_0[N(0)] - \mathbb{E}_1[N(0)] \leq n\sqrt{\frac{n}{2}D(\mathbb{P}_0\|\mathbb{P}_1)} \leq 2n\varepsilon\sqrt{\frac{n}{2}},$$

$$\mathbb{E}_0[N(1/2)] - \mathbb{E}_{-1}[N(1/2)] \leq n\sqrt{\frac{n}{2}D(\mathbb{P}_0\|\mathbb{P}_{-1})} \leq 2n\varepsilon\sqrt{\frac{n}{2}},$$

where $D(\mathbb{P}_0\|\mathbb{P}_{-1})$ denotes the relative entropy between $\mathbb{P}_0$ and $\mathbb{P}_{-1}$. Summing up these two inequalities and using (6), $n - (R_1/\varepsilon + R_{-1}/\varepsilon) \leq 4n\varepsilon\sqrt{n/2}$. Setting $\varepsilon = c/\sqrt{n}$ and reordering gives $\sqrt{n}c(1 - 2\sqrt{2}c) \leq R_1 + R_{-1}$. Choose $c = 1/(4\sqrt{2})$ and note that $2\max(R_1, R_{-1}) \geq R_1 + R_{-1}$ to finish the proof. $\square$

4   Partial Information

We now consider the more challenging case where X_t is not observed if $Y_t < X_t$ and so the waiting time directly influences the amount of information gained at each time step. Just like in the previous section, our first algorithm is based on a discretisation idea. As before, we first notice that it is enough to consider stopping times in an interval of length $O((\omega - \beta)\log(n))$ and also that a discretisation accuracy of $\varepsilon = (\omega - \beta)/\sqrt{n}$ will suffice to get a $\tilde{O}(\sqrt{n})$ regret, which is conjectured to hold. In this case, however, an appropriately modified version of the exponential weights algorithm is needed which works with estimated losses and adds exploration to facilitate the estimation of losses. In fact, as it turns out, after discretisation, our problem falls into the framework of prediction with expert advice with side-observations, where after the learner chooses an action $Y_t \in A$ it observes the losses for a subset $S(Y_t) \subset A$ of actions. In our case, $S(Y_t) = \{y \in A : y \leq Y_t\}$, which means that waiting for a longer time leads to more information than waiting for a shorter time. This framework was first studied by Mannor and Shamir [14]. Here, we will use the Exp3-DOM algorithm of Alon et. al. as this algorithm improves upon the results of Mannor and Shamir for our setting [2]. The general idea of Exp3-DOM is to restrict exploration to actions in a dominating set D, which is a subset of actions such that $\cup_{a \in D} S(a) = A$. In particular, exploration is restricted to a minimal dominating set. In our case, the minimal dominating set contains a single element, $y_{\max} = \max A$. This results in Algorithm 2. If the learning rate η is chosen carefully, then Theorem 7 of [2] shows that the algorithm suffers a $\tilde{O}(\sqrt{n})$ regret. Recall the definition of $\bar{y}(\varepsilon)$

Algorithm 2. Exp3-Dom

1: **Input:** ω, β, and n
2: // Recall definition of $\bar{y}(\varepsilon)$ given in Corollary 1
3: $\varepsilon \leftarrow (\omega - \beta)/\sqrt{n}$, $A \leftarrow \{k\varepsilon : 0 \leq k \leq \bar{y}(\varepsilon)/\varepsilon, k \in \mathbb{N}\}$, $R \leftarrow \bar{y}(\varepsilon) + \omega$, $\eta \leftarrow 1/(R\sqrt{n})$
4: $w_1(y) \leftarrow 1$ for all $y \in A$
5: **for** $t = 1, \ldots, n$ **do**
6:     $W_t \leftarrow \sum_{y \in A} w_t(y)$ and $p_t(y) \leftarrow \eta w_t(y)/W_t + (1 - \eta)\mathbb{1}\{y = \max A\}$ for each y
7:     Sample waiting time Y_t from distribution p_t on A and observe Z_t, δ_t
8:     $w_{t+1}(y) \leftarrow w_t(y)$ for all $y \in A$
9:     **for** $y \in A \cap [0, Y_t]$ **do**   // Update the weights

$$\ell_t(y) \leftarrow \mathbb{1}\{Z_t \leq y\}(Z_t + \beta) + \mathbb{1}\{Z_t > y\}(y + \omega)$$

$$q_t(y) \leftarrow \sum_{y' \in A: y' \geq y} p_t(y') \quad \text{and} \quad \tilde{\ell}_t(y) \leftarrow \ell_t(y)/q_t(y)$$

$$w_{t+1}(y) \leftarrow w_t(y)\exp(-\eta\tilde{\ell}_t(y))$$

10:     **end for**
11: **end for**

Theorem 7. *Pick $n > 0$ and let R_n be the regret of Algorithm 1 when used for n rounds. Then, $\mathbb{E}[R_n] \in O((\omega - \beta)\log^2(n)\sqrt{n})$.*

Proof. Let ε, A and R be as in the pseudo-code of the algorithm. Using that in our case the the observation sets $S(a)$ are fixed, Theorem 7, Eq. (2) of [2] gives that the regret of Exp3-DOM against the best waiting time in A is $O(K\log(K) + \log(K)(1/\eta + \eta R^2 \sum_{t=1}^{n}(1 + Q_t)))$, where $K = |A|$ is the number of actions and $Q_t = \sum_{y\in A} p_t(y)/q_t(y)$ and where we used that the dominant set in our case has a single element. Now, Lemma 13 of [2] gives that $Q_t = \alpha\ln(K/\eta)$, where α is the so-called independence number of the graph (A, E) underlying the observation system: $(a_1, a_2) \in E$ if $a_1 \in S(a_2)$ or $a_2 \in S(a_1)$. In our case, the graph is a clique and hence its independence number is $\alpha = 1$. Choosing $\eta = 1/(R\sqrt{n})$ thus gives that the regret of Exp3-DOM against the best waiting time in A is $O(R\log(K)\sqrt{n} + K\log(K))$. By Proposition 1, $\min_{y\in A} \ell(y) - \inf_{y\in[0,\bar{y}(\varepsilon)]} \ell(y) \leq \varepsilon$, while by Corollary 1, $\inf_{y\in[0,\bar{y}(\varepsilon)]} \ell(y) - \ell(y^*) \leq \varepsilon$. Hence, $\mathbb{E}[R_n] \in O(R\log(K)\sqrt{n} + K\log(K) + (\omega - \beta)\sqrt{n})$. Now, using the definition of $\bar{y}$, $\bar{y}((\omega - \beta)/\sqrt{n}) \in O((\omega - \beta)\log(n))$. Thus, $K = |A| \in O(\lceil \bar{y}(\varepsilon)/\varepsilon \rceil) = O(\bar{y}((\omega - \beta)/\sqrt{n})\sqrt{n}/(\omega - \beta)) = O(\log(n)\sqrt{n})$ and $R = \bar{y}((\omega - \beta)/\sqrt{n}) + \omega \in O((\omega - \beta)\log(n))$. Plugging these into the previous bound, we get $\mathbb{E}[R_n] \in O((\omega - \beta)\log^2(n)\sqrt{n})$.

Note that since the partial information setting is strictly more difficult than the full information setting, our previous lower bound shows that the regret cannot be better than $\Omega(\sqrt{n})$. However, as in the full information setting, we can expect to improve upon the performance of Exp3-DOM by using an algorithm that exploits the fact that the environment is stochastic. In particular, as is common in sequential learning algorithms we make use of an optimistic strategy, which will wait for the bus as long as reasonably possible. The algorithm maintains an estimate of the cumulative distribution and chooses a non-increasing sequence of waiting times starting from a carefully chosen upper bound. The waiting times decrease at a data-dependent rate that is chosen to ensure some nearly-optimal waiting time is always smaller than the action chosen. This results in Algorithm 3.

The following theorem bounds the expected regret of Algorithm 3. The bound is worse by a factor of $O(\sqrt{\log n})$ than that obtained in the full information setting described in Section 3, but improves the bound announced in Theorem 7.

Theorem 8. *The regret is bounded by* $\mathbb{E}[R_n] \leq (\omega - \beta)(42 + 7\log^{3/2}(n)\sqrt{n})$.

Lemma 3. *Define* $\bar{y} = \min\{y^*, y_{\max}\}$, *where* $y_{\max}$ *is given in Algorithm 3. Then* $\ell(\bar{y}) - \ell(y^*) \leq (\omega - \beta)/\sqrt{n}$.

Proof. Apply Part 2 of Lemma 1 and the definition of $\bar{y}$. □

The following lemma shows that $Y_t \geq \bar{y}$ for all $1 \leq t \leq n$ with high probability. This means that if $y^* \geq y_{\max}$, then with high probability the algorithm will always choose $Y_t = y_{\max}$ and suffer no more than $(\omega - \beta)/\sqrt{n}$ regret per time step. On the other hand, if $y^* < y_{\max}$, then the algorithm will choose $Y_t \geq y^*$, which guarantees that it is continually learning information about the loss of the optimal action.

Algorithm 3. Optimistic Waiting

1: **Input:** ω, β, and n

2: $\alpha \leftarrow 3/2$ and $y_{\max} \leftarrow 2(\omega - \beta)(1 + \log_2 \sqrt{n})$ and $Y_1 \leftarrow y_{\max}$

3: **for** $t = 1, \ldots, n$ **do**

4:     Observe Z_t, δ_t

5:     Compute the empirical distribution: $G_t(y) \doteq \dfrac{1}{t} \sum_{s=1}^{t} \mathbb{1}\{Z_s \leq y\}$

6:     Compute waiting time for next day:

$$\varepsilon_t \leftarrow \sqrt{\frac{\log 2n^\alpha}{2t}}$$

$$Y_{t+1} \leftarrow \max \left\{ 0 \leq y \leq Y_t : G_t(y) - G_t(y') + 2\varepsilon_t \geq \right.$$

$$\left. \frac{y - y'}{\omega - \beta} (1 - G_t(y) - \varepsilon_t) - \frac{1}{\sqrt{n}}, \ 0 \leq y' \leq y \right\}$$

7: **end for**

Lemma 4. *For $1 \leq t \leq n$, we have that $\mathbb{P}\{A_n^c\} \leq n^{1-\alpha}$, where event A_t is defined by $A_t = \bigcap_{s \leq t} \left\{ Y_s \geq \bar{y} \text{ and } \sup_{y \leq Y_{s-1}} |G_{s-1}(y) - F(y)| \leq \varepsilon_{t-1} \right\}$.*

Proof. Define event $B = \bigcup_{t \leq n} \left\{ \sup_{x \leq Y_{t-1}} |G_{t-1}(x) - F(x)| \geq \varepsilon_{t-1} \right\}$ and recall that $F_t(x) = \frac{1}{t} \sum_{s=1}^{t} \mathbb{1}\{X_s \leq x\}$, which is unknown to the learner. The Dvoretzky–Kiefer–Wolfowitz–Massart theorem [8, 15] gives that

$$\mathbb{P} \left\{ \sup_x |F_t(x) - F(x)| \geq \varepsilon_t \right\} \leq 2 \exp\left(-2\varepsilon_t^2 t\right) = n^{-\alpha}.$$

Therefore, by the union bound, with probability at least $1 - n^{1-\alpha}$ it holds that $|F_t(x) - F(x)| \leq \varepsilon_t$ for all $t \leq n$ and $x \in \mathbb{R}$. By the definition of the observations $(Z_s)_s$, $G_t(y) = F_t(y)$ for all $y \leq \min_{1 \leq s \leq t} Y_s$. Further, since by construction $(Y_t)_t$ is non-increasing, $\min_{1 \leq s \leq t} Y_s = Y_t$ and so $G_t(y) = F_t(y)$ for all $y \leq Y_t$. Therefore $\mathbb{P}\{B\} \leq n^{1-\alpha}$. We now show that if B does not occur then A_t holds for $1 \leq t \leq n$. We prove this by induction on t. That B^c implies A_1 is trivial. Now, assume that B^c implies that A_t holds for some $1 \leq t < n$. On B^c we have

$$\sup_{y \leq Y_t} |G_t(y) - F(y)| \leq \varepsilon_t. \tag{7}$$

Thus, it suffices to show that on B^c, $Y_{t+1} \geq \bar{y}$ also holds. By the induction hypothesis, $Y_t \geq \bar{y}$. Combining this with (7) we get

$$\sup_{y \leq \bar{y}} |G_t(y) - F(y)| \leq \varepsilon_t. \tag{8}$$

Now let $y' \leq \bar{y} \leq y^*$. Then

$$0 \overset{(a)}{\geq} \ell_t(y^*) - \ell_t(y') \overset{(b)}{\geq} \ell_t(\bar{y}) - \ell_t(y') - \frac{\omega - \beta}{\sqrt{n}}$$

$$\overset{(c)}{\geq} (\bar{y} - y')(1 - F(\bar{y})) - (\omega - \beta)(F(\bar{y}) - F(y')) - \frac{\omega - \beta}{\sqrt{n}}$$

$$\overset{(d)}{\geq} (\bar{y} - y')(1 - G_t(\bar{y}) - \varepsilon_t) - (\omega - \beta)(G_t(\bar{y}) - G_t(y') + 2\varepsilon_t) - \frac{\omega - \beta}{\sqrt{n}},$$

where (a) follows since y^* is the optimal waiting time, (b) by Lemma 3, (c) by Part 2 of Lemma 2, and (d) holds by (8). Rearranging we obtain

$$G_t(\bar{y}) - G_t(y') + 2\varepsilon_t \geq \frac{\bar{y} - y'}{\omega - \beta}(1 - G_t(\bar{y}) - \varepsilon_t) - \frac{1}{\sqrt{n}},$$

which implies, by the definition of Y_{t+1}, that $Y_{t+1} \geq \bar{y}$. Therefore A_{t+1} holds and so B^c implies that A_n holds. Therefore $\mathbb{P}\{A_n\} \geq \mathbb{P}\{B^c\} \geq 1 - n^{1-\alpha}$. $\square$

Proof (of Theorem 8). The proof follows almost immediately from Lemmas 2 to 4. Assume that A_n holds. Then

$$\ell(Y_t) - \ell(y^*) \overset{(a)}{\leq} \ell(Y_t) - \ell(\bar{y}) + \frac{\omega - \beta}{\sqrt{n}}$$

$$\overset{(b)}{\leq} (Y_t - \bar{y})(1 - F(\bar{y})) - (\omega - \beta)(F(Y_t) - F(\bar{y}))$$

$$\overset{(c)}{\leq} (Y_t - \bar{y})(1 - G_{t-1}(\bar{y}) + \varepsilon_{t-1}) - (\omega - \beta)(G_{t-1}(Y_t) - G_{t-1}(\bar{y}) - 2\varepsilon_{t-1})$$

$$\overset{(d)}{\leq} 2\varepsilon_{t-1}(Y_t - \bar{y}) + 4\varepsilon_{t-1}(\omega - \beta) + \frac{\omega - \beta}{\sqrt{n}}, \tag{9}$$

where (a) follows from Lemma 3, (b) by Part 2 of Lemma 2 and the fact that $Y_t \geq \bar{y}$, (c) follows from the definition of A_n while (d) follows from the definition of Y_t. Therefore, on A_n,

$$\sum_{t=1}^{n} \ell(Y_t) - \ell(y^*) \overset{(a)}{\leq} \ell(Y_1) - \ell(y^*) + \sum_{t=2}^{n}\left(2\varepsilon_{t-1}(Y_t - \bar{y}) + 4\varepsilon_{t-1}(\omega - \beta) + \frac{\omega - \beta}{\sqrt{n}}\right)$$

$$\overset{(b)}{\leq} y_{\max} + \sum_{t=2}^{n}\left(2\varepsilon_{t-1}(y_{\max} + 2(\omega - \beta)) + \frac{\omega - \beta}{\sqrt{n}}\right) \overset{(c)}{\leq} (\omega - \beta)(40 + 5\log^{\frac{3}{2}}(n)\sqrt{n}),$$

where (a) follows from (9), (b) follows by naively bounding $Y_t - \bar{y} \leq y_{\max}$, while (c) follows arduously from the definition of ε_t and $y_{\max}$. In case A_n does not hold, the regret may be as much as $y_{\max}$ per day, but $\mathbb{P}\{A_n^c\} \leq n^{1-\alpha} = 1/\sqrt{n}$. Combining with the previous display completes the result. $\square$

5   Experiments

We performed three experiments comparing EWA with FTL in the full-information case with $\omega = 20$ and $\beta = 5$. We used two exponential distributions with $\lambda = 1/20$ and $1/5$ respectively, as well as a power law distribution (see Table 1). The horizon was set to $n = 10,000$ and the learning rate of exponential weighting was tuned to be a factor of 100 larger than the theoretical optimum, which was observed to give a good performance across all three problems. The FTL algorithm generally out-performs the exponential weighting algorithm, but not by an enormous margin. If the theoretically optimal learning rate is used then the performance of exponential weighting deteriorates significantly.

Figures (e) and (f) suggest that FTL suffers $\sqrt{n}$ regret on the power-law distribution, but logarithmic regret for exponentially distributed arrival times with parameter $\lambda = 1/20$. Each data point is the average of 20 independent trials. Code is available at `http://downloads.tor-lattimore.com/projects/optimal_waiting`.

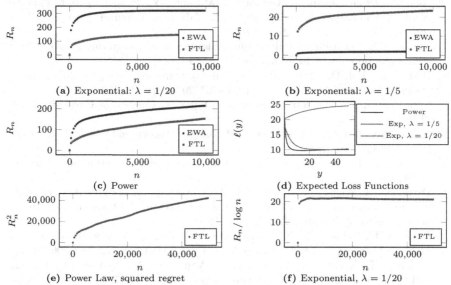

(a) Exponential: $\lambda = 1/20$

(b) Exponential: $\lambda = 1/5$

(c) Power

(d) Expected Loss Functions

(e) Power Law, squared regret

(f) Exponential, $\lambda = 1/20$

6   Conclusions

We introduced the problem of learning an optimal waiting time with two variants. In both cases, we presented two general algorithms relying on no assumptions that were shown to enjoy near-optimal worst-case regret. Interesting future work is to further analyse the problem-dependent regret bounds of FTL and other algorithms in both full and partial information settings beyond exponentially distributed arrivals. One approach for less conservative algorithms may be to use the Kaplan-Meier estimator rather than the standard empirical distribution, but the mathematical theory behind this estimator is not yet well-developed for this setting where the censoring times are known and not i.i.d. One exception is by Ganchev et. al., but unfortunately their confidence bound depends on the scale and is not suitable for obtaining optimal regret bounds in our problem [10]. Another challenge is to improve the running time of the algorithms to $O(1)$ per time step. While our results are the first in this setting, we expect various extensions to related problems, such as when one can choose between multiple options with random durations.

Acknowledgements. This work was supported by the Alberta Innovates Technology Futures and NSERC.

References

[1] Agrawal, R., Teneketzis, D., Anantharam, V.: Asymptotically efficient adaptive allocation schemes for controlled i.i.d. processes: Finite parameter space. IEEE Transaction on Automatic Control 34, 258–267 (1989)

[2] Alon, N., Cesa-Bianchi, N., Gentile, C., Mansour, Y.: From bandits to experts: A tale of domination and independence. In: Advances in Neural Information Processing Systems, pp. 1610–1618 (2013)

[3] Bartók, G.: A near-optimal algorithm for finite partial-monitoring games against adversarial opponents. In: COLT, pp. 696–710 (2013)

[4] Bartók, G., Pál, D., Szepesvári, C.: Minimax regret of finite partial-monitoring games in stochastic environments. In: COLT 2011, pp. 133–154 (2011)

[5] Cesa-Bianchi, N.: Prediction, learning, and games. Cambridge University Press (2006)

[6] Cohen, A.C.: Truncated and censored samples: theory and applications. CRC Press (1991)

[7] Devroye, L., Lugosi, G.: Combinatorial methods in density estimation. Springer (2001)

[8] Dvoretzky, A., Kiefer, J., Wolfowitz, J.: Asymptotic minimax character of the sample distribution function and of the classical multinomial estimator. The Annals of Mathematical Statistics 27, 642–669 (1956)

[9] Foster, D.P., Rakhlin, A.: No internal regret via neighborhood watch. Journal of Machine Learning Research - Proceedings Track (AISTATS) 22, 382–390 (2012)

[10] Ganchev, K., Nevmyvaka, Y., Kearns, M., Vaughan, J.W.: Censored exploration and the dark pool problem. Communications of the ACM 53(5), 99–107 (2010)

[11] Kleinberg, R., Leighton, T.: The value of knowing a demand curve: Bounds on regret for online posted-price auctions. In: Proceedings of the 44th Annual IEEE Symposium on Foundations of Computer Science, pp. 594–605. IEEE (2003)

[12] Krishnan, P., Long, P.M., Vitter, J.S.: Adaptive disk spindown via optimal rent-to-buy in probabilistic environments. Algorithmica 23(1), 31–56 (1999)

[13] Lattimore, T., György, A., Szepesvári, C.: On learning the optimal waiting time (2014), http://downloads.tor-lattimore.com/projects/optimal_waiting/

[14] Mannor, S., Shamir, O.: From bandits to experts: On the value of side-observations. In: NIPS, pp. 684–692 (2011)

[15] Massart, P.: The tight constant in the Dvoretzky-Kiefer-Wolfowitz inequality. The Annals of Probability 18, 1269–1283 (1990)

[16] Ribeiro, C.C., Rosseti, I., Vallejos, R.: Exploiting run time distributions to compare sequential and parallel stochastic local search algorithms. Journal of Global Optimization 54(2), 405–429 (2012)

Bandit Online Optimization
over the Permutahedron

Nir Ailon[1], Kohei Hatano[2], and Eiji Takimoto[2]

[1] Department of Computer Science, Technion
[2] Department of Informatics, Kyushu University
nailon@cs.technion.ac.il, {hatano,eiji}@inf.kyushu-u.ac.jp

Abstract. The permutahedron is the convex polytope with vertex set consisting of the vectors $(\pi(1), \ldots, \pi(n))$ for all permutations (bijections) π over $\{1, \ldots, n\}$. We study a bandit game in which, at each step t, an adversary chooses a hidden weight weight vector s_t, a player chooses a vertex π_t of the permutahedron and suffers an observed instantaneous loss of $\sum_{i=1}^{n} \pi_t(i) s_t(i)$.

We study the problem in two regimes. In the first regime, s_t is a point in the polytope dual to the permutahedron. Algorithm CombBand of Cesa-Bianchi et al (2009) guarantees a regret of $O(n\sqrt{T \log n})$ after T steps. Unfortunately, CombBand requires at each step an n-by-n matrix permanent computation, a #P-hard problem. Approximating the permanent is possible in the impractical running time of $O(n^{10})$, with an additional heavy inverse-polynomial dependence on the sought accuracy. We provide an algorithm of slightly worse regret $O(n^{3/2}\sqrt{T})$ but with more realistic time complexity $O(n^3)$ per step. The technical contribution is a bound on the variance of the Plackett-Luce noisy sorting process's 'pseudo loss', obtained by establishing positive semi-definiteness of a family of 3-by-3 matrices of rational functions in exponents of 3 parameters.

In the second regime, s_t is in the hypercube. For this case we present and analyze an algorithm based on Bubeck et al.'s (2012) OSMD approach with a novel projection and decomposition technique for the permutahedron. The algorithm is efficient and achieves a regret of $O(n\sqrt{T})$, but for a more restricted space of possible loss vectors.

1   Introduction

Consider a game in which, at each step, a player plays a permutation of some ground set $V = \{1, \ldots, n\}$, and then suffers (and observes) a loss. We model the loss as a sum over the items of some latent quality of the item, weighted by its position in the permutation. The game is repeated, and the items' quality can adversarially change over time. The game models many scenarios in which the player is an online system (say, a search/recommendation engine) presenting a ranked list of items (results/products) to a stream of users. A user's experience is positive if she perceives the quality of the top items on the list as higher

P. Auer et al. (Eds.): ALT 2014, LNAI 8776, pp. 215–229, 2014.

than those at the bottom. The goal of the system is to create a total positive experience for its users.

There is a myriad of methods for modelling *ranking* loss functions in the literature, especially (but not exclusively) for information retrieval. Our choice allows us to study the problem in the framework of online combinatorial optimization in the *bandit* setting, and to obtain highly nontrivial results improving on state of the art in either run time or regret bounds. More formally, we study online linear optimization over the the *n-permutahedron* action set, defined as the convex closure of all vectors in $\mathbb{R}^n$ consisting of n distinct coordinates taking values in $[n] := \{1, \ldots, n\}$ (permutations). At each step $t = 1, \ldots, T$, the player outputs an action π_t and suffers a loss $\pi_t' s_t = \sum_{i=1}^n \pi_t(i) s_t(i)$, where $s_t \in \mathbb{R}^n$ is the vector of "item qualities" chosen by some adversary who knows the player's strategy but doesn't control their random coins. The performance of the player is the difference between their total loss and that of the optimal static player, who plays the best (in hindsight) single permutation π^* throughout. This difference is known as *regret*. Note that, given $s_1, \ldots, s_T$, π^* can be computed by sorting the coordinates of $\sum_{t=1}^T s_t$ in decreasing order. This is aligned with our practical requirement that items with higher quality should be placed first, and those with lower quality should be last.

2   Results, Techniques and Contribution

Our first of two results, stated as Theorem 1, is for the setting in which at each step the loss is uniformly bounded (by 1 for simplicity) in absolute value for all possible permutations. Equivalently, the vectors s_t belong to the polytope that is dual to the permutahedron. Our algorithm, BanditRank, plays permutations from a distribution known as the Plackett-Luce model (see [13]) which is widely used in statistics and econometrics (see eg [4]). It uses an inverse covariance matrix of the distribution in order to obtain an unbiased loss vector estimator, which is a standard technique [7]. The main technical difficulty (Lemma 2) is in bounding second moment properties of Plackett-Luce, by establishing positive semidefiniteness of a certain family of 3 by 3 matrices. The lemma is interesting in its own right as a tool for studying distributions over permutations. The expected regret of our algorithm is $O(n^{3/2}\sqrt{T})$ for T steps, with running time of $O(n^3)$ per time step. This result should be compared to CombBand of [7], where a framework for playing bandit games over combinatorially structured sets was developed. Their techniques extend that of [8]. In each step, it draws a permutation from a distribution that assigns to each permutation π a probability of $e^{\eta \sum_{\tau=1}^t \pi' \tilde{s}_\tau}$, where $\tilde{s}_t$ is a *pseudo-loss* vector at time t, an unbiased estimator of the loss vector s_t. Their algorithm guarantees a regret of $O(n\sqrt{T \log n})$, which is better than ours by a factor of $\Theta(\sqrt{n/\log n})$. However, its computational requirements are much worse. In order to draw permutations, they need to compute nonnegative n by n matrix permanents. Unfortunately, nonnegative permanent computation is #P-hard, as shown by [15]. On the other hand, a groundbreaking result of [12] presents a polynomial time approximation scheme for permanent, which runs in time $O(n^{10})$ for fixed accuracy. To make things worse, the dependence in the accuracy is inverse polynomial, implying that, even if we could perform arbitrarily accurate floating point operations, the total running time would be *super linear* in T, because a regret dependence of $\sqrt{T}$ over T steps requires accuracy inverse polynomial in T. (Our algorithm does not suffer from this problem.) From a practical point of view, the runtime dependence of CombBand in both n and T is infeasible for even modest cases. For example, our algorithm can handle online ranking of $n = 100$ items in an order of few millions of operations per game iteration. In contrast, approximating the permanent of a 100-by-100 positive matrix is utterly impractical.

We note that independently of our work, Hazan et al. [10] have improved the state-of-the-art general purpose algorithm for linear bandit optimization, implying an algorithm with regret $O(n\sqrt{T})$ for our problem, but with worse running time $\tilde{O}(n^4)$.[1]

In our second result in Section 5 we further restrict s_t to have ℓ_1 norm of $1/n$. (Note that this restriction is contained in $|\pi_t' s_t| \leq 1$ by Hölder). We present and analyze an algorithm OSMDRank based on the bandit algorithm OSMD of [6] with projection and decomposition techniques over the permutahedron ([16, 14]). The projection is defined in terms of the binary relative entropy divergence. The restriction allows us to obtain an expected regret bound of $O(n\sqrt{T})$ (a $\sqrt{\log n}$ improvement over CombBand). The running time is $O(n^2 + n\tau(n))$, where $\tau(n)$ is the time complexity for some numerical procedure, which is $O(n^2)$ in a fixed precision machine.

We note previous work on playing the permutahedron online optimization game in the *full information case*, namely, when s_t is known for each t. As far as we know, Helmbold et al. [11] were the first to study a more general version of this problem, where the action set is the vertex set of the Birkhoff-von-Neumann polytope (doubly-stochastic matrices). Suehiro et al. [14] studied the problem by casting it as a submodularly constrained optimization problem, giving near optimal regret bounds, and more recently Ailon [1] both provided optimal regret bounds with improved running time and established tight regret lower bounds.

3   Definitions and Problem Statement

Let V be a ground set of n items. For simplicity, we identify V with $[n] := \{1, \ldots n\}$. Let S_n denote the set of $n!$ permutations over V, namely bijections over $[n]$. By convention, we think of $\pi(v)$ for $v \in V$ as the *position* of $v \in V$ in the ranking, where we think of lower numbered positions as *more favorable*. For distinct $u, v \in V$, we say that $u \prec_\pi v$ if $\pi(u) < \pi(v)$ (in words: u *beats* v). We use $[u, v]_\pi$ as shorthand for the indicator function of the predicate $u \prec_\pi v$.

---

[1] The running time is a product of $\tilde{O}(n^3)$ number of Markov chain steps required for drawing a random point from a convex set under a log-concave distribution, and $O(n \log n)$ time to test whether a point lies in the permutahedron. By $\tilde{O}$ we hide poly-logarithmic factors.

The convex closure of S_n is known as the permutahedron polytope. It will be more convenient for us to consider a translated version of the permutahedron, centered around the origin. More precisely, for $\pi \in S_n$ we let $\hat{\pi}$ denote

$$\hat{\pi} := (\pi(1) - (n+1)/2, \, \pi(2) - (n+1)/2, \ldots, \pi(n) - (n+1)/2) .$$

It will be convenient to define a symmetrized version of the permutation set $\hat{S}_n := \{\hat{\pi} : \pi \in S_n\}$. The symmetrized n-permutahedron, denoted $\hat{P}_n$ is the convex closure of $\hat{S}_n$. Symmetrization allows us to work with a polytope that is centered around the origin. Generalization our result to standard (un-symmetrized) permutations is a simple technicality that will be explained below. The notation $u \prec_{\hat{\pi}} v$ and $[u, v]_{\hat{\pi}}$ is defined as for $\pi \in S_n$ in an obvious manner.

At each step $t = 1, \ldots, T$, an adversary chooses and hides a nonnegative vector $s_t \in \mathbb{R}^n \equiv \mathbb{R}^V$, which assigns an elementwise quality measure $s_t(v)$ for any $v \in V$. The player-algorithm chooses a permutation $\hat{\pi}_t \in \hat{S}_n$, possibly random, and suffers an instantaneous loss

$$\ell_t := \hat{\pi}_t' s_t = \sum_{v \in V} \hat{\pi}_t(v) s_t(v) . \tag{3.1}$$

The total loss L_t is defined as $\sum_{t=1}^{T} \ell_t$. We will work with the notion of *regret*, defined as the difference $L_t - L_t^*$, where $L_T^* = \min_{\hat{\pi} \in \hat{S}_n} \sum_{t=1}^{T} \hat{\pi}' s_t$. We let $\hat{\pi}^*$ denote any minimizer achieving L_T^* in the RHS.

For any $\hat{\pi} \in \hat{S}_n$ and $s \in \mathbb{R}^n$, the dot-product $\hat{\pi}' s$ can be decomposed over pairs: $\hat{\pi}' s = \frac{1}{2} \sum_{u \neq v} [u, v]_\pi (s(v) - s(u))$. This makes the symmetrized permutahedron easier to work with. Nevertheless, our results also apply to the non-symmetrized permutahedron as well, as we shall see below.

Throughout, the notation $\sum_{u \neq v}$ means summation over distinct, ordered pairs of elements $u, v \in V$, and $\sum_{u < v}$ means summation over distinct, unordered pairs.[2] The uniform distribution over $\hat{S}_n$ will be denoted $\mathcal{U}_n$.

The smallest eigenvalue of a PSD matrix A is denoted $\lambda_{\min}(A)$. The norm $\|\cdot\|_2$ will denote spectral norm (Euclidean norm for a vector). To avoid notation such as C, C', C'', C_1 for universal constants, the expression C will denote a "general positive constant" that may change its value as necessary. For example, we may write $C = 3C + 5$.

4   Algorithm BanditRank and Its Guarantee

For this section, we will assume that the instantaneous losses are uniformly bounded by 1, in absolute value: For all t and $\hat{\pi} \in \hat{S}_n$, $|\hat{\pi}' s_t| \leq 1$. Equivalently, using geometric language, the loss vectors belong to a polytope which is *dual* to the permutahedron.

Now consider Algorithm 1. It maintains, at each time step t, a weight vector $w_t \in \mathbb{R}^n$. At each time step, it draws a random permutation $\hat{\pi}_t$ from a mixture $\mathcal{D}_t$

---

[2] We will only use expressions of the form $\sum_{u < v} f(u, v)$ for symmetric functions satisfying $f(u, v) = f(v, u)$.

Algorithm 1. Algorithm BanditRank(n, η, γ, T) (assuming $|\hat{\pi}' s_t| \leq 1$ for all t and $\hat{\pi} \in \hat{S}_n$)

1: given: ground set size n, positive parameters η, γ ($\gamma \leq 1$), time horizon T
2: set $w_0(u) = 0$ for all $u \in V = [n]$
3: **for** $t = 1..T$ **do**
4:     let distribution $\mathcal{D}_t$ over $\hat{S}_n$ denote a mixture of $\mathcal{U}_n$ (with probability γ) and $\mathcal{PL}_n(w_{t-1})$ (with probability $1 - \gamma$)
5:     draw and output $\hat{\pi}_t \sim \mathcal{D}_t$
6:     observe and suffer loss ℓ_t ($= \hat{\pi}_t' s_t$)
7:     $\tilde{s}_t = \ell_t P_t^+ \hat{\pi}_t$ where $P_t = \mathbb{E}_{\hat{\sigma} \sim \mathcal{D}_t}[\hat{\sigma}\hat{\sigma}']$
8:     set $w_t = w_{t-1} + \eta \tilde{s}$
9: **end for**

---

of the uniform distribution over $\hat{S}_n$ and a distribution $\mathcal{PL}_n(w)$ which we define shortly. The distribution mixture is determined by a parameter γ. The algorithm then plays the permutation $\hat{\pi}_t$ and thereby suffers the instantaneous loss defined in (3.1). The weights are consequently updated by adding an unbiased estimator $\tilde{s}_t$ of s_t (computed using the pseudo-inverse covariance matrix corresponding to $\mathcal{D}_t$), multiplied by another parameter $\eta > 0$.

The Plackett-Luce Random Sorting Procedure: The distribution $\mathcal{PL}_n(w)$ over $\hat{S}_n$, parametrized by $w \in \mathbb{R}^n$, is defined by the following procedure. To choose the first (most preferred) item, the procedure draws a random item, assigning probability proportional to $e^{w(u)}$ for each $u \in V$. It then removes this item from the pool of available items, and iteratively continues to choose the second item, then third and so on. As claimed in the introduction, this random permutation model is well studied in statistics. An important well known property of the distribution is that it can be equivalently defined as a *Random Utility Model (RUM)* [13, 17]: To draw a permutation, add a random iid noise variable following the Gumbel distribution to each weight, and then sort the items of V in decreasing value of noisy-weights.[3] The RUM characterization implies, in particular, that for any two *disjoint* pairs of element (u, v) and (u', v'), the events $u \prec_\pi v$ and $u' \prec_\pi v'$ are statistically independent if π is drawn from $\mathcal{PL}_n(w)$, for any w. This fact will be used later.

We are finally ready to state our main result, bounding the expected regret of the algorithm.

Theorem 1. *If algorithm BanditRank (Algorithm 1) is executed with parameters $\gamma = O(n^{3/2}/\sqrt{T})$ and $\eta = O(\gamma/n)$, then the expected regret (with respect to the game defined by the symmetrized permutahedron) is at most $O(n^{3/2}\sqrt{T})$. The running time of each iteration is $O(n^3)$. Additionally, there exists an algorithm with the same expected regret bound and running time with respect to the standard permutahedron (assuming the vectors s_t uniformly satisfy $|\pi' s_t| \leq 1, \forall \pi \in S_n$.)*

---

[3] The Gumbel distribution, also known as doubly-exponential, has a cdf of $e^{-e^{-x}}$.

The proof uses a standard technique used e.g. in Cesa-Bianchi et al.'s Comb-Band [7], which is itself an adaptation of Auer et al.'s Exp3 [3] from the finite case to the structured combinatorial case. The distribution from which the actions $\hat{\pi}_t$ are drawn in the algorithm differ from the distribution used in CombBand, and give rise to the technical difficulty of variance estimation, resolved in Lemma 2.

Proof. Let $\mathcal{T}_n$ denote the set of *tournaments* over $[n]$. More precisely, an element $A \in \mathcal{T}_n$ is a subset of $[n] \times [n]$ with either $(u, v) \in A$ or $(v, u) \in A$ (but not both) for all $u < v$. We extend our previous notation so that $u \prec_A v$ is equivalent to the predicate $(u, v) \in A$.

For any pair $\hat{\pi} \in \hat{S}_n$ and $w \in \mathbb{R}^n$, $p(\hat{\pi}|w)$ denotes the probability assigned to $\hat{\pi} \in \hat{S}_n$ by $\mathcal{PL}_n(w)$. Slightly abusing notation, we define the following shorthand:

$$p(u \prec v|w) := \sum_{\hat{\pi}:u \prec_{\hat{\pi}} v} p(\pi|w) = \frac{e^{w(u)}}{e^{w(u)} + e^{w(v)}}$$

$$p(u \prec v \prec z|w) := \sum_{\hat{\pi}:u \prec_{\hat{\pi}} v \prec_{\hat{\pi}} z} p(\hat{\pi}|w) = \frac{e^{w(u)+w(v)}}{(e^{w(u)} + e^{w(v)} + e^{w(z)})(e^{w(v)} + e^{w(z)})} .$$

The last two right hand sides are easily derived from the definition of the distribution $\mathcal{PL}_n(w)$, see also e.g. [13]. We also define the following abbreviations:

$$p(u \prec {}^v_z|w) := p(u \prec v \prec z|w) + p(u \prec z \prec v|w) = \frac{e^{w(u)}}{e^{w(u)} + e^{w(v)} + e^{w(z)}} \quad (4.1)$$

$$p({}^u_v \prec z|w) := p(u \prec v \prec z|w) + p(v \prec u \prec z|w)$$

$$= \frac{e^{w(u)+w(v)}}{e^{w(u)} + e^{w(v)} + e^{w(z)}} \left(\frac{1}{e^{w(v)} + e^{w(z)}} + \frac{1}{e^{w(u)} + e^{w(z)}} \right) \quad (4.2)$$

We will also need to define a distribution over the set of tournaments $\mathcal{T}_n$. The distribution, $\mathcal{BTL}_n(w)$ is parametrized by a weight vector $w \in \mathbb{R}^n$. Drawing $A \sim \mathcal{BTL}_n(w)$ is done by independently setting, for all $u < v$ in V,

$$(u, v) \in A \text{ with probability } p(u \prec v|w) = \frac{e^{w(u)}}{e^{w(u)} + e^{w(v)}}$$

$$(v, u) \in A \text{ with probability } p(v \prec u|w) = \frac{e^{w(v)}}{e^{w(u)} + e^{w(v)}} .$$

(Note that the distribution is equivalently defined as the product distribution, over all $u < v$ in V, of the Bradley-Terry-Luce pairwise preference model, hence the name $\mathcal{BTL}_n$. We refer to [13] for definition and history of the Bradley-Terry-Luce model.)

For $A \in \mathcal{T}_n$, we denote by $\tilde{p}(A|w)$ the probability $\prod_{u \prec_A v} p(u \prec v|w)$ of drawing A from $\mathcal{BTL}_n(w)$. The proof of the theorem proceeds roughly as the main result upper bounding the expected regret of CombBand in [7]. The following technical lemma is required in anticipation of a major hurdle (inequality (4.5)). We believe the inequality is interesting in its own right as a probabilistic statement on permutation and tournament distributions.

Lemma 2. *Let* $s, w \in \mathbb{R}^n$. *Let* $\hat{\pi} \sim \mathcal{PL}_n(w)$ *and* $A \sim \mathcal{BTL}_n(w)$ *be drawn independently. Define* $X_1 = \sum_{u,v:\ u \prec_{\hat{\pi}} v} (s(v) - s(u)) = \hat{\pi}'s$, $X_2 = \sum_{u,v:\ u \prec_A v} (s(v) - s(u))$. *Then* $\mathbb{E}[X_2^2] \leq \mathbb{E}[X_1^2]$.

(Note that clearly, $\mathbb{E}[X_2] = \mathbb{E}[X_1]$, so the lemma in fact upper bounds the variance of X_2 by that of X_1.) The proof of the lemma is deferred to Section 4.1.

Continuing the proof of Theorem 1, we let $q(\pi|w)$ denote the probability of drawing π from the mixture of the uniform distribution (with probability γ) and $\mathcal{PL}_n(w)$ (with probability $(1 - \gamma)$). Similarly to above, $q(u \prec v|w)$ denotes $\sum_{\hat{\pi}:u \prec_{\hat{\pi}} v} q(\hat{\pi}|w)$. By these definitions,

$$q(\hat{\pi}|w) = (1 - \gamma)p(\hat{\pi}|w) + \frac{\gamma}{n!} \qquad q(u \prec v|w) = (1 - \gamma)p(u \prec v|w) + \frac{\gamma}{2} . \quad (4.3)$$

The analysis proceeds by defining a potential function:

$$W_t(u, v) := e^{\frac{1}{2}\eta(w_t(u) - w_t(v))} + e^{\frac{1}{2}\eta(w_t(v) - w_t(u))} .$$

The quanatity of interest will be $\mathbb{E}\left[\sum_{u<v} \sum_t \log \frac{W_t(u,v)}{W_{t-1}(u,v)}\right]$, where the expectation is taken over all random coins used by the algorithm throughout T steps. This quantity will be bounded from above and from below, giving rise to a bound on the expected total loss, expressed using the optimal static loss. On the one hand,

$$\sum_{u<v} \log \frac{W_t(u,v)}{W_{t-1}(u,v)} = \sum_{u<v} \log \left(\frac{e^{\frac{1}{2}(w_t(u)-w_t(v))}}{W_{t-1}(u,v)} + \frac{e^{\frac{1}{2}(w_t(v)-w_t(u))}}{W_{t-1}(u,v)} \right)$$

$$= \sum_{u<v} \log \left(\frac{e^{\frac{1}{2}(w_{t-1}(u)-w_{t-1}(v))}e^{\frac{1}{2}\eta(\tilde{s}_t(u)-\tilde{s}_t(v))} + e^{\frac{1}{2}(w_t(v)-w_t(u))}e^{\frac{1}{2}\eta(\tilde{s}_t(v)-\tilde{s}_t(u))}}{W_{t-1}(u,v)} \right)$$

$$= \sum_{u<v} \log \left(p(u \prec v|w_{t-1})e^{\frac{1}{2}\eta(\tilde{s}_t(u)-\tilde{s}_t(v))} + p(v \prec u|w_{t-1})e^{\frac{1}{2}\eta(\tilde{s}_t(v)-\tilde{s}_t(u))} \right)$$

$$= \log \left(\sum_{A \in \mathcal{T}_n} \tilde{p}(A|w_{t-1})e^{\frac{1}{2}\eta\sum_{u \prec_A v}(\tilde{s}_t(u)-\tilde{s}_t(v))} \right) .$$

We will now assume that η is small enough so that for all $A \in \mathcal{T}_n$ and for all t,

$$\eta \left| \sum_{(u,v) \in A} (\tilde{s}_t(u) - \tilde{s}_t(v)) \right| \leq 1 . \quad (4.4)$$

(This will be shortly enforced.) Using $e^x \leq 1 + x + x^2 \ \forall x \in [-1/2, 1/2]$,

$$
\sum_{u,v} \log \frac{W_t(u,v)}{W_{t-1}(u,v)} \leq \log \left[\sum_{A \in \mathcal{T}_n} \tilde{p}(A|w_{t-1}) \left(1 + \frac{\eta}{2} \sum_{u \prec_A v} (\tilde{s}_t(u) - \tilde{s}_t(v)) \right. \right.
$$
$$
\left. \left. + \frac{\eta^2}{4} \left(\sum_{u \prec_A v} (\tilde{s}_t(u) - \tilde{s}_t(v)) \right)^2 \right) \right]
$$
$$
= \log \left[1 + \frac{\eta}{2} \mathbb{E}_{A \sim \mathcal{BTL}_n(w_{t-1})} \left[\sum_{u \prec_A v} (\tilde{s}_t(u) - \tilde{s}_t(v)) \right. \right.
$$
$$
\left. \left. + \frac{\eta^2}{4} \left(\sum_{u \prec_A v} (\tilde{s}_t(u) - \tilde{s}_t(v)) \right)^2 \right] \right]
$$
$$
\leq \log \left[1 + \frac{\eta}{2} \mathbb{E}_{\hat{\pi} \sim \mathcal{PL}_n(w_{t-1})} \left[\sum_{u \prec_{\hat{\pi}} v} (\tilde{s}_t(u) - \tilde{s}_t(v)) \right. \right.
$$
$$
\left. \left. + \frac{\eta^2}{4} \left(\sum_{u \prec_{\hat{\pi}} v} (\tilde{s}_t(u) - \tilde{s}_t(v)) \right)^2 \right] \right] .
$$

$$(4.5)$$

where we used Lemma 2 in the last inequality (together with the fact that the marginal probability of the event "$u \prec_Y v$" is identical for both $Y \sim \mathcal{PL}_n(w_{t-1})$ and $Y \sim \mathcal{BTL}_n(w_{t-1})$). Henceforth, for any $\hat{\pi} \in \hat{S}$, we let $\tilde{\ell}_t(\hat{\pi}) := \hat{\pi}' \tilde{s}_t = \sum_{u \prec_{\hat{\pi}} v} (\tilde{s}(v) - \tilde{s}(u))$. Using 4.3 and the fact that $\log(1 + x) \leq x$ for all x, we get

$$
\sum_{u < v} \log \frac{W_t(u,v)}{W_{t-1}(u,v)}
$$
$$
\leq \frac{\eta}{2} \sum_{u \neq v} \frac{q(u \prec v|w_{t-1}) - \frac{\gamma}{2}}{1 - \gamma} (\tilde{s}_t(u) - \tilde{s}_t(v)) + \frac{\eta^2}{4} \sum_{\hat{\pi} \in \hat{S}_n} \frac{q(\pi|w_{t-1}) - \frac{\gamma}{n!}}{1 - \gamma} \tilde{\ell}_t(\hat{\pi})^2
$$
$$
\leq \frac{-\eta}{2(1 - \gamma)} \sum_{\hat{\pi} \in \hat{S}_n} q_t(\hat{\pi}|w_{t-1}) \tilde{\ell}_t(\hat{\pi}) + \frac{\eta^2}{4(1 - \gamma)} \sum_{\hat{\pi} \in \hat{S}_n} q_t(\hat{\pi}|w_{t-1}) \tilde{\ell}_t(\hat{\pi})^2 .
$$

We now note that (1) $\sum_{\hat{\pi} \in \hat{S}} q_t(\hat{\pi}|w_{t-1}) \tilde{\ell}_t = \ell_t$ (following the properties of matrix pseudo-inverse in Line 7 in Algorithm 1), and (2) $\sum_{\hat{\pi} \in \hat{S}_n} q_t(\hat{\pi}|w_{t-1}) \tilde{\ell}_t(\pi)^2] \leq n$ (see top of page 31 together with Lemma 15 in [7]). Applying these inequalities, and then taking expectations over the algorithm's randomness and summing for $t = 1, \ldots, T$, we get

$$
\sum_{t=1}^{T} \mathbb{E} \left[\sum_{u,v} \log \frac{W_t(u,v)}{W_{t-1}(u,v)} \right] \leq -\frac{\eta}{2(1 - \gamma)} \mathbb{E}[L_T] + \frac{\eta^2}{8(1 - \gamma)} nT .
$$

On the other hand,

$$\sum_{t=1}^{T} \mathbb{E}\left[\sum_{u,v} \log \frac{W_t(u,v)}{W_{t-1}(u,v)}\right]$$

$$\geq \sum_{u,v} \mathbb{E}\left[\log\left([u,v]_{\pi^*}e^{\frac{1}{2}(w_T(u)-w_T(v))} + [v,u]_{\pi^*}e^{\frac{1}{2}(w_T(u)-w_T(v))}\right)\right)\right] - \sum_{u,v} \log 2$$

$$= \frac{1}{2}\sum_{u,v}\left(\mathbb{E}\left[[u,v]_{\pi^*}(w_T(u)-w_T(v)) + [v,u]_{\pi^*}(w_T(u)-w_T(v))\right]\right) - \binom{n}{2}\log 2$$

$$= \frac{\eta}{2}\sum_{u,v}\left(\mathbb{E}\left[[u,v]_{\pi^*}\sum_t(\tilde{s}_t(u)-\tilde{s}_t(v)) + [v,u]_{\pi^*}\sum_t(\tilde{s}_t(u)-\tilde{s}_t(v))\right]\right)$$

$$\qquad\qquad\qquad\qquad\qquad\qquad - \binom{n}{2}\log 2$$

$$= \frac{\eta}{2}\sum_{u,v}\left([u,v]_{\pi^*}\sum_t(s_t(u)-s_t(v)) + [v,u]_{\pi^*}\sum_t(s_t(u)-s_t(v))\right) - \binom{n}{2}\log 2$$

$$= -\frac{\eta}{2}L_T^* - \binom{n}{2}\log 2 \, ,$$

where L_T^* is the total loss of a player who chooses the best permutatation $\hat{\pi}^* \in \hat{S}_n$ in hindsight. Combining, we obtain $\frac{\eta}{2(1-\gamma)}\mathbb{E}[L_t] \leq \frac{\eta}{2}L_T^* + \frac{n^2}{2}\log 2 + \frac{\eta^2}{4(1-\gamma)}nT$. Multiplying both sides by $2(1-\gamma)/\eta$ yields

$$\mathbb{E}[L_T] \leq L_T^* + \gamma|L_T^*| + \frac{n^2\log 2}{\eta} + \frac{\eta}{2}nT \, . \qquad (4.6)$$

We shall now work to impose (4.4).

$$\max_t \max_{A\in\mathcal{T}(V)}\left|\sum_{(u,v)\in A}(\tilde{s}_t(u)-\tilde{s}_t(v))\right| \leq \max_t \sqrt{\sum_{v\in V}\tilde{s}_t(v)^2}\sqrt{\sum_{i=-(n-1)/2}^{(n-1)/2}i^2}$$

$$\leq C\max_t\|\tilde{s}_t\|_2 n^{3/2} \, , \qquad (4.7)$$

where the left inequality is Cauchy-Schwartz. We now note that

$$\|\tilde{s}_t\|_2 \leq |\ell_t|\|P_t^+\|_2\|\hat{\pi}_t\|_2 \, .$$

Clearly $\|\hat{\pi}\|_2$ is bounded above by $Cn^{3/2}$. Also $\|P_t^+\|_2$ equals $1/\lambda_{\min}(P_t)$. By Weyl's inequality $\lambda_{\min}(P_t) \geq \gamma\lambda_{\min}(\mathbb{E}_{\hat{\tau}\sim\mathcal{U}_n}[\hat{\tau}\hat{\tau}'])$. It is an exercise to check that $\lambda_{\min}(\mathbb{E}_{\hat{\tau}\sim\mathcal{U}_n}[\hat{\tau}\hat{\tau}']) \geq Cn^2$. We conclude (also recalling that $|\ell_t| \leq 1$) that

$$\max_t\|\tilde{s}_t\|_2 \leq C/(n^{1/2}\gamma) \, .$$

Combining, we shall satisfy (4.7) by imposing $\eta \leq \gamma/(Cn)$. Plugging in (4.6), we get

$$\mathbb{E}[L_T(\text{Alg})] \leq L_T^* + \gamma|L_T^*| + \frac{Cn^3}{\gamma} + C\gamma T . \tag{4.8}$$

Choosing $\gamma = \sqrt{\frac{Cn^3}{T}}$ gives $\mathbb{E}[L_T(\text{Alg})] \leq L_T^* + \frac{Cn^{3/2}}{\sqrt{T}}|L_T^*| + n^{3/2}\sqrt{T}$.

This concludes the required result for the symmetrized case, because $|L_T^*| \leq T$. For the standard permutahedron, notice that for any $\pi \in S_n$ and its symmetrized counterpart $\hat{\pi} \in \hat{S}_n$, $\pi's - \hat{\pi}'s = \frac{n-1}{2}\sum_{v \in V} s(v) =: f(s)$ for any $s \in \mathbb{R}^n$. Equivalently, we can write $\pi's = (\hat{\pi}', 1)(s; f(s))$, where $(\cdot, a)$ appends the scalar a to the right of a row vector and $(\cdot; a)$ appends to the bottom of a column vector. Algorithm 1 can be easily adjusted to work with action set $\hat{S}_n \times \{1\}$. For the proof, we keep the same potential function. The technical part of the proof is lower bounding the smallest eigenvalue of the expectation of $\hat{\tau}\hat{\tau}'$, where $\hat{\tau}$ is now drawn from the uniform distribution on $\hat{S}_n \times \{1\}$. We omit these simple details for lack of space.   □

4.1   Proof of Lemma 2

The expression $\mathbb{E}[X_1^2]$ can be written as

$$\mathbb{E}[X_1^2] = \sum_{u \neq v} p(u \prec v|w)((s(v) - s(u))^2$$

$$+ \sum_{|\{u,v,u',v'\}|=4} p(u \prec v \wedge u' \prec v'|w) (s(v) - s(u))(s(v') - s(u'))$$

$$+ \sum_{\substack{u \neq v, u' \neq v' \\ |\{u,v,u',v'\}|=3}} p(u \prec v \wedge u' \prec v'|w) (s(v) - s(u))(s(v') - s(u')) , \tag{4.9}$$

where $p(u \prec v \wedge u' \prec v'|w)$ is the probability that both $u \prec_{\hat{\pi}} v$ and $u' \prec_{\hat{\pi}} v'$ with $\hat{\pi} \sim \mathcal{PL}_n(w)$. Similarly,

$$\mathbb{E}[X_2^2] = \sum_{u \neq v} p(u, v|w)((s(v) - s(u))^2$$

$$+ \sum_{|\{u,v,u',v'\}|=4} p(u \prec v|w)p(u' \prec v'|w) (s(v) - s(u))(s(v') - s(u'))$$

$$+ \sum_{\substack{u \neq v, u' \neq v' \\ |\{u,v,u',v'\}|=3}} p(u \prec v|w)p(u' \prec v'|w) (s(v) - s(u))(s(v') - s(u')) . $$

$$\tag{4.10}$$

Since Plackett-Luce is a random utility model (see [13]), it is clear that whenever a pair of pairs $u \neq v, u' \neq v'$ satisfies $|\{u,v,u',v'\}| = 4$, $p(u \prec v \wedge u' \prec v'|w) = p(u \prec v|w)p(u' \prec v'|w)$. Hence, it suffices to prove that the third summand in the RHS of (4.10) is upper bounded by the third summand in the RHS of (4.9). But now notice the following identity:

$$\sum_{\substack{u \neq v, u' \neq v' \\ |\{u,v,u',v'\}|=3}} \equiv \sum_{\substack{\Delta \subseteq V \\ |\Delta|=3}} \sum_{\substack{u \neq v, u' \neq v' \\ u,v,u',v' \in \Delta \\ |\{u,v,u',v'\}|=3}} .$$

This last sum rearrangement implies that it suffices to prove that for any Δ of cardinality 3,

$$F_2(\Delta) := \sum_{\substack{u \neq v, u' \neq v' \\ u,v,u',v' \in \Delta \\ |\{u,v,u',v'\}|=3}} p(u,v|w)p(u',v'|w)\,(s(v)-s(u))(s(v')-s(u'))$$

$$\leq \sum_{\substack{u \neq v, u' \neq v' \\ u,v,u',v' \in \Delta \\ |\{u,v,u',v'\}|=3}} p(u, v \wedge u', v'|w)\,(s(v)-s(u))(s(v')-s(u')) =: F_1(\Delta) .$$

If we now denote $\Delta = \{a,b,c\}$, then both $F_1(\Delta)$ and $F_2(\Delta)$ are quadratic forms in $s(a), s(b), s(c)$ (for fixed w). It hence suffices to prove that $H(\Delta) := F_1(\Delta) - F_2(\Delta)$ is a positive semi-definite form in $s(\Delta) := (s(a), s(b), s(c))'$. We now write

$$H(\Delta) = s(\Delta)' \begin{pmatrix} H_{aa} & \frac{1}{2}H_{ab} & \frac{1}{2}H_{ac} \\ \frac{1}{2}H_{ab} & H_{bb} & \frac{1}{2}H_{bc} \\ \frac{1}{2}H_{ac} & \frac{1}{2}H_{bc} & H_{cc} \end{pmatrix} s(\Delta) .$$

The matrix is singular, because clearly $H(\Delta) = F_1(\Delta) = F_2(\Delta) = 0$ whenever $s(a) = s(b) = s(c)$. To prove positive semi-definiteness, by Sylvester's criterion it hence suffices to show that the diagonal element $H_{aa} \geq 0$ and that the principal 2-by-2 minor determinant $H_{aa}H_{bb} - \frac{1}{4}H_{ab}^2 \geq 0$. Using the definitions, together with the properties of $\mathcal{PL}_n(w)$, a technical (but quite tedious) algebraic derivation that we omit (for lack of space) gives

$$H_{aa} = \frac{4e^{s(a)+s(b)+s(c)}}{(e^{s(a)} + e^{s(b)})(e^{s(a)} + e^{s(c)})(e^{s(a)} + e^{s(b)} + e^{s(c)})} . \tag{4.11}$$

(See full version or technical report [2] for details). Similarly, by symmetry, $H_{bb} = \frac{4e^{s(a)+s(b)+s(c)}}{(e^{s(b)}+e^{s(a)})(e^{s(b)}+e^{s(c)})(e^{s(a)}+e^{s(b)}+e^{s(c)})}$. From a similar (yet more tedious) technical algebraic calculation which we omit, one gets:

$$H_{ab} = \frac{-8e^{s(a)+s(b)+2s(c)}}{(e^{s(a)} + e^{s(b)})(e^{s(a)} + e^{s(c)})(e^{s(b)} + e^{s(c)})(e^{s(a)} + e^{s(b)} + e^{s(c)})} . \tag{4.12}$$

(see full version or technical report [2] for details). One now verifies, using (4.11)-(4.12), the identity

$$H_{aa}H_{bb} - \frac{1}{4}H_{ab}^2 = \frac{16e^{2s(a)+2s(b)+2s(c)}}{Z} ,$$

where $Z := (e^{s(a)} + e^{s(b)})^2(e^{s(a)} + e^{s(c)})(e^{s(b)} + e^{s(c)})(e^{s(a)} + e^{s(b)} + e^{s(c)})^2$.

It remains to notice, trivially, that $H_{aa} \geq 0$ and $H_{aa}H_{bb} - \frac{1}{4}H_{ab}^2 \geq 0$ for all possible values of $s(a), s(b), s(c)$. The proof of the lemma is concluded.

Algorithm 2. Algorithm OSMDRank(n, η, γ, T) (assuming $\|s_t\|_1 \leq 1$ and $\hat{\pi}_t \in \hat{Q}_n$ for all t)

1: given: ground set size n, positive parameters η, γ ($\gamma \leq 1$), time horizon T
2: let $x_1 = 0 \in \hat{Q}_n$. (Note that $x_1 = \arg\min_{a \in \hat{Q}_n} F(a)$)
3: **for** $t = 1, \ldots, T$ **do**
4:     let $\tilde{x}_t = (1 - \gamma)x_t$ (Note that $\tilde{a}_t \in \hat{Q}_n$ since the origin 0 and x_t are in $\hat{Q}_n$ and $\tilde{x}_t$ is a convex combination of them).
5:     output $\pi_t = $ Decomposition($\tilde{x}_t$) (i.e., choose π_t so that $\mathbb{E}[\pi_t] = \tilde{x}_t$) and suffer loss ℓ_t ($= \pi_t' s_t$)
6:     let distribution $\mathcal{D}_t$ over $[-1,1]^n$ denote a mixture of the uniform distribution over the canonical basis with random sign (with probability γ) and a Radmacher distribution over $\{-1,1\}^n$ with parameter $(1 + x_{t,i})/2$ for each $i = 1, \ldots, n$ (with probability $1 - \gamma$)
7:     estimate the loss vector $\tilde{s}_t = \ell_t P_t^+ \pi_t$, where $P_t = \mathbb{E}_{\sigma \sim \mathcal{D}_t}[\sigma \sigma']$
8:     let $x_{t+\frac{1}{2}} = \nabla F^*(F(x_t) - \eta \tilde{s}_t)$
9:     let $x_{t+1} = $ Projection($x_{t+\frac{1}{2}}$) (that is, $x_{t+1} = \min_{x \in \hat{Q}_n} D_F(x, x_{t+\frac{1}{2}})$)
10: **end for**

---

5   Bandit Algorithm Based on Projection and Decomposition

In this section, we propose another bandit algorithm OSMDRank, described in Algorithm 2. We will be working under the more restricted assumption that $\sup \|s_t\|_1 \leq 1$ and $\sup \|\hat{\pi}_t\|_\infty \leq 1$. This in particular implies that $|\hat{\pi}_t' s_t| \leq 1$, as before. But now we shall achieve a better expected regret of $O(n\sqrt{T})$.

We prefer, for reasons clarified shortly, to require that the actions $\hat{\pi}_t$ are vertices of the rescaling $\hat{Q}_n := \frac{2}{n-1}\hat{P}_n \in [-1,1]^n$ of the symmetrized permutahedron. That is, $\sup \|\hat{\pi}_t\|_\infty \leq 1$ (and $\sup \|s_t\|_1 \leq 1$). This will allow us to work with the following standard regularizer $F : [-1,1]^n \to \mathbb{R}^+$: $F(x) = \frac{1}{2}\sum_{i=1}^n ((1+x)\ln(1+x) + (1-x)\ln(1-x))$. The regularizer $F(x)$ is the key to the OSMD (Online Stochastic Mirror Descent) algorithm of Bubeck et al. [6], on which our algorithm is based. OSMD is a bandit algorithm over the hypercube domain $[-1,1]^n$ and a variant of Follow the Regularized Leader (FTRL, e.g., [9]) for linear loss functions. To apply this algorithm, we need a new projection and decomposition technique for the polytope $\hat{Q}_n$, as well as a slightly modified perturbation step in line 4 of Algorithm 2. Our algorithm OSMDRank has the following two procedures:

1. **Projection:** Given a point $x_t \in [-1,1]^n$, return $\arg\min_{y_t \in \hat{Q}_n} \Delta_F(y_t, x_t)$, where Δ_F is the Bregman divergence defined wr.t. F, i.e., $\Delta_F(y, x) = F(y) - F(x) - \nabla F(x)'(y - x)$ (also known as *binary relative entropy*).[4]
2. **Decomposition:** Given $y_t \in \hat{Q}_n$ from the the projection step, output a random vertex $\hat{\pi}_t$ of $\hat{Q}_n$ such that $\mathbb{E}[\hat{\pi}_t] = y_t$.

---

[4] Note that the binary relative entropy is different from the relative entropy, where the relative entropy is defined as $Rel(p, q) = \sum_{i=1}^n p_i \ln \frac{p_i}{q_i}$ for probability distributions p and q over $[n]$.

The decomposition can be done using the technique of [16], which runs in $O(n \log n)$ time. (To be precise, the method there was defined for the standard permutahedron; The adjustments for the symmetrized version are trivial.) For notational purposes, we define $f := \nabla F$, and notice that $f(x)_i = \frac{1}{2} \ln \frac{1+x_i}{1-x_i}$, and its inverse function f^{-1} is given by $f^{-1}(y)_i = \frac{e^{2y_i}-1}{e^{2y_i}+1}$. Our projection procedure is presented in Algorithm 3.

Lemma 3. *(i) Given $q \in [-1,1]^n$, Algorithm 3 outputs the projection of q onto $\hat{Q}_n$, with respect to the regularizer F. (ii) The time complexity of the algorithm is $O(n\tau(n) + n^2)$, where $\tau(n)$ is the time complexity to perform step 4.*

Proof (skecth). Our projection algorithm is an extension of that in [14] and our proof follows a similar argument in [14]. For simplicity, we assume that elements in q are sorted in descending order, i.e., $q_1 \geq q_2 \geq \cdots \geq q_n$. This can be achieved in time $O(n \log n)$ by sorting q. Then, it can be shown that projection preserves the order in q by using Lemma 1 in [14]. That is, the projection p of q satisfies $p_1 \geq p_2 \geq \cdots \geq p_n$. So, if the conditions $\frac{2}{n-1}\sum_{j=1}^{i} p_j \leq \sum_{j=1}^{i}(\frac{n+1}{2} - j)$, for $i = 1, \ldots, n-1$, are satisfied, then other inequality constraints are satisfied as well since for any $S \subset [n]$ such that $|S| = i$, $\sum_{j \in S} p_j \leq \sum_{j=1}^{i} p_j$. Therefore, relevant constraints for projection onto $\hat{Q}_n$ are only linearly many.

By following a similar argument in [14], we can show that the output p indeed satisfies the KKT optimality conditions for projection, which completes the proof of the first statement. Finally, the algorithm terminates in time $O(n\tau(n) + n^2)$ since the number of iteration is at most n and each iteration takes $O(n + \tau(n))$ time, which completes the second statement of the lemma.   □

Note that with respect to other regularizers (e.g. relative entropy or Euclidean norm squared), a different projection scheme is possible in time $O(n^2)$ (see [16, 14] for the details). It is an open question whether an $O(n^2)$ algorithm can be devised with respect to the binary relative entropy we need here. In our case, we need to solve a numerical optimization problem by, say, binary search. Note that the time $\tau(n)$ is reasonably small: In fact, we can perform the binary search over the domain $[-1, 1]$ for each dimension i. Therefore, if the precision is a fixed constant, the binary search ends in time $O(n)$ for each dimension. In that case, $\tau(n)$ is $O(n^2)$. We are ready to present our main result for this section.

Theorem 4. *For $\eta = O(n\sqrt{1/T})$ and $\gamma = O(\sqrt{1/T})$, Algorithm OSMDRank has expected regret $O(n\sqrt{T})$ and running time $O(n^2 + n\tau(n))$ per step, where $\tau(n)$ is the time for a numerical optimization step depending on n. Additionally, there exists an algorithm with the same expected regret bound and running time with respect to the standard permutahedron (assuming $\|s_t\|_1 \leq 1/n$).*

Proof (sketch). The algorithm OSMDRank is a modification of OSMD for the hypercube $[-1, 1]^n$ obtained by adding (1) a projection step and (2) a decomposition step. Standard techniques show that adding the projection step does not increase the expected regret bound (see, e.g., chapters 5 and 7 on OMD and OSMD of Bubeck's lecture notes [5]). The key facts are: (i) A variant of

---

Algorithm 3. Projection onto $\hat{Q}_n$

1: given $(q_1, \ldots, q_n) \in [-1, 1]^n$ satisfying $q_1 \geq q_2 \geq \cdots \geq q_n$. *(This assumption holds by renaming the indices, and reverting to their original names at the end).*
2: set $i_0 = 0$
3: **for** $k = 1, \ldots, n$ **do**
4:     for each $i = i_{k-1} + 1, \ldots, n$, set $\delta_i^k = \min_{\delta \in \mathbb{R}} \delta$ subject to:
$$\sum_{j=i_{k-1}+1}^{i} f^{-1}(f(q_j) - \delta) \leq \frac{2}{n-1} \sum_{j=i_{k-1}+1}^{i} \left(\frac{n+1}{2} - j \right).$$
5:     $i_k = \arg\max_{i:i_{k-1}<i\leq n} \delta_i^k$. In case of multiple minimizers, choose largest as i_k.
6:     set $p_j = f^{-1}(f(q_j) - \delta_{i_k}^k)$ for $j = i_{k-1} + 1, \ldots, i_k$
7:     **if** $i_k = n$, **then** break
8: **end for**
9: **return** $(p_1, \ldots, p_n)'$

Theorem 2 of [6] (regret bound of OSMD) holds for OSMD with Projection, (ii) $E[\pi_t] = (1-\gamma)x_t$, and (iii) The estimated loss is the same one used in OSMD for the hypercube $[-1, 1]^n$. Once these three conditions are satisfied, we can prove a regret bound of OSMDRank by following the proof of Theorem 5 in Bubeck et al. [6]. In addition, the running time of OSMD per trial is $O(n)$ [6]. Combining Lemma 3 for the projection and the analysis of the decomposition from [16], the proof of the first statement is concluded. The statement related to the standard permutahedron holds based on the affine transformation between the standard permutahedron and $\hat{Q}_n$. $\qquad\square$

6　Future Work

The main open question is whether there is an algorithm of expected regret $O(n\sqrt{T})$ and time $O(n^3)$ in the setting of Section 4. Another interesting line of research is to study other ranking polytopes. For example, given any strictly monotonically increasing function $f : \mathbb{R} \mapsto \mathbb{R}$ we can consider as an action set $f^n(S_n)$, defined as $f^n(S_n) := \{(f(\pi(1)), f(\pi(2)), \ldots, f(\pi(n))) : \pi \in S_n\}$.

Acknowledgments. We thank anonymous reviewers for useful comments. Ailon acknowledges the support of a Marie Curie Reintegration Grant PIRG07-GA-2010-268403, an Israel Science Foundation (ISF) grant 127/133 and a Jacobs Technion-Cornell Innovation Institute (JTCII) grant. Hatano is grateful to the supports from JSPS KAKENHI Grant Number 25330261 and CORE Project Grant of Microsoft Research Asia. Takimoto is grateful to the supports from JSPS KAKENHI Grant Number 23300033 and MEXT KAKENHI Grant Number 24106010.

References

[1] Ailon, N.: Improved Bounds for Online Learning Over the Permutahedron and Other Ranking Polytopes. In: AISTATS (2014)

[2] Ailon, N., Hatano, K., Takimoto, E.: Bandit online optimization over the permu-
     tahedron (technical report). arXiv:1312.1530 (2013)
[3] Auer, P., Cesa-Bianchi, N., Freund, Y., Schapire, R.E.: The nonstochastic multi-
     armed bandit problem. SIAM J. Comput. 32(1), 48–77 (2003)
[4] Beggs, S., Cardell, S., Hausman, J.: Assessing the potential demand for electric
     cars. Journal of Econometrics 17(1), 1–19 (1981)
[5] Bubeck, S.: Introduction to Online Optimization (2011),
     http://www.princeton.edu/~bubeck/BubeckLectureNotes.pdf
[6] Bubeck, S., Cesa-Bianchi, N., Kakade, S.M.: Towards Minimax Policies for On-
     line Linear Optimization with Bandit Feedback. In: Proceedings of 25th Annual
     Conference on Learning Theory (COLT 2012), pp. 41.1–41.14 (2012)
[7] Cesa-Bianchi, N., Lugosi, G.: Combinatorial bandits. J. Comput. Syst. Sci. 78(5),
     1404–1422 (2012)
[8] Dani, V., Hayes, T.P., Kakade, S.: The price of bandit information for online
     optimization. In: NIPS (2007)
[9] Hazan, E.: The convex optimization approach to regret minimization. In: Sra, S.,
     Nowozin, S., Wright, S.J. (eds.) Optimization for Machine Learning, ch. 10, pp.
     287–304. MIT Press (2011)
[10] Hazan, E., Karnin, Z.S., Mehka, R.: Volumetric spanners and their applications
     to machine learning. CoRR, abs/1312.6214 (2013)
[11] Helmbold, D.P., Warmuth, M.K.: Learning Permutations with Exponential
     Weights. Journal of Machine Learning Research 10, 1705–1736 (2009)
[12] Jerrum, M., Sinclair, A., Vigoda, E.: A polynomial-time approximation algorithm
     for the permanent of a matrix with nonnegative entries. J. ACM 51(4), 671–697
     (2004)
[13] Marden, J.I.: Analyzing and Modeling Rank Data. Chapman & Hall (1995)
[14] Suehiro, D., Hatano, K., Kijima, S., Takimoto, E., Nagano, K.: Online Predic-
     tion under Submodular Constraints. In: Bshouty, N.H., Stoltz, G., Vayatis, N.,
     Zeugmann, T. (eds.) ALT 2012. LNCS, vol. 7568, pp. 260–274. Springer,
     Heidelberg (2012)
[15] Valiant, L.G.: The complexity of computing the permanent. Theor. Comput. Sci. 8,
     189–201 (1979)
[16] Yasutake, S., Hatano, K., Kijima, S., Takimoto, E., Takeda, M.: Online Linear Opti-
     mization over Permutations. In: Asano, T., Nakano, S.-i., Okamoto, Y., Watanabe,
     O. (eds.) ISAAC 2011. LNCS, vol. 7074, pp. 534–543. Springer, Heidelberg (2011)
[17] Yellott, J.: The relationship between Luce's choice axiom, Thurstone's theory of
     comparative judgment, and the double exponential distribution. Journal of Math-
     ematical Psychology 15, 109–144 (1977)

Offline to Online Conversion

Marcus Hutter

Research School of Computer Science
Australian National University
Canberra, ACT, 0200, Australia
http://www.hutter1.net/

Abstract. We consider the problem of converting offline estimators into an online predictor or estimator with small extra regret. Formally this is the problem of merging a collection of probability measures over strings of length 1,2,3,... into a single probability measure over infinite sequences. We describe various approaches and their pros and cons on various examples. As a side-result we give an elementary non-heuristic purely combinatoric derivation of Turing's famous estimator. Our main technical contribution is to determine the computational complexity of online estimators with good guarantees in general.

Keywords: Offline, online, batch, sequential, probability, estimation, prediction, time-consistency, normalization, tractable, regret, combinatorics, Bayes, Laplace, Ristad, Good-Turing.

1 Introduction

A standard problem in statistics and machine learning is to estimate or learn an in general non-i.i.d. probability distribution $q_n : \mathcal{X}^n \to [0,1]$ from a batch of data $x_1, ..., x_n$. q_n might be the Bayesian mixture over a class of distributions $\mathcal{M}$, or the (penalized) maximum likelihood (ML/MAP/MDL/MML) distribution from $\mathcal{M}$, or a combinatorial probability, or an exponentiated code length, or else. This is the batch or *offline* setting. An important problem is to predict x_{n+1} from $x_1, ..., x_n$ sequentially for $n = 0, 1, 2...$, called *online* learning if the predictor improves with n. A stochastic prediction $\tilde{q}(x_{n+1}|x_{1:n})$ can be useful in itself (e.g. weather forecasts), or be the basis for some decision, or be used for data compression via arithmetic coding, or otherwise. We use the prediction picture, but could have equally well phrased everything in terms of log-likelihoods, or perplexity, or code-lengths, or log-loss.

The naive predictor is $\tilde{q}^{\text{rat}}(x_{n+1}|x_1...x_n) := q_{n+1}(x_1...x_{n+1})/q_n(x_1...x_n)$ is not properly normalized to 1 if q_n and q_{n+1} are not compatible. We could fix the problem by normalization $\tilde{q}^{\text{nl}}(x_{n+1}|x_1...x_n) := \tilde{q}^{\text{rat}}(x_{n+1}|x_1...x_n)/\sum_{x_{n+1}} \tilde{q}^{\text{rat}}(x_{n+1}|x_1...x_n)$, but this may result in a very poor predictor. We discuss two further schemes, $\tilde{q}^{\text{lim}}$ and $\tilde{q}^{\text{mix}}$, the latter having good performance guarantees (small regret), but a direct computation of either is prohibitive. A major open problem is to find a computationally tractable online

P. Auer et al. (Eds.): ALT 2014, LNAI 8776, pp. 230–244, 2014.
© Springer International Publishing Switzerland 2014

predictor $\tilde{q}$ with provably good performance given offline probabilities (q_n). A positive answer would benefit many applications.

Applications. (i) Being able to use an offline estimator to make stochastic predictions (e.g. weather forecasts) is of course useful. The predictive probability needs to sum to 1 which $\tilde{q}^{n1}$ guarantees, but the regret should also be small, which only $\tilde{q}^{mix}$ guarantees.

(ii) Given a parameterized class of (already) online estimators $\{\tilde{q}^\theta\}$, estimating the parameter θ from data $x_1...x_n$ (e.g. maximum likelihood) for $n = 1, 2, 3, ...$ leads to a sequence of parameters $(\hat{\theta}_n)$ and a sequence of estimators $(q_n) := (\tilde{q}^{\hat{\theta}_n})$ that is usually *not* online. They need to be reconverted to become online to be useful for prediction or compression, etc.

(iii) Arithmetic coding requires an online estimator, but often is based on a class of distributions as described in (ii). The default 'trick' to get a fast and online estimator is to use $\tilde{q}^{\hat{\theta}_n}(x_{n+1}|x_{1:n})$ which is properly normalized and often very good.

(iv) Online conversions are needed even for some offline purposes. For instance, computing the cumulative distribution function $\sum_{y_{1:n} \leq x_{1:n}} q_n(y_{1:n})$ can be hard in general, but can be computed in time $O(n)$ if (q_n) is (converted to) online.

Contributions and Contents. The main purpose of this paper is to introduce and discuss the problem of converting offline estimators (q_n) to an online predictor $\tilde{q}$ (Section 2). We compare and discuss the pros and cons of the four conversion proposals (Section 3). We also define the worst-case extra regret of online $\tilde{q}$ over offline (q_n), measuring the conversion quality. We illustrate their behavior for various classical estimators (Bayes, MDL, Laplace, Good-Turing, Ristad) (Section 4). Naive normalization of the triple uniform estimator interestingly leads to the Good-Turing estimator, but induces huge extra regret, while naive normalization of Ristad's quadruple uniform estimator induces negligible extra regret. Given that $\tilde{q}^{n1}$ can fail for interesting offline estimators, natural questions to ask are: whether the excellent predictor $\tilde{q}^{mix}$ can be computed or approximated (yes), by an efficient algorithm (no), whether for every (q_n) there exists any fast $\tilde{q}$ nearly as good as $\tilde{q}^{mix}$ (no), or whether there exist (q_n) for which no fast $\tilde{q}$ can even slightly beat the trivial uniform predictor (yes) (Section 5). These results do not preclude a satisfactory positive solution in practice, in particular given the contrived nature of the constructed (q_n), but as any negative complexity result they show that a solution requires extra assumptions or to moderate our demands. This leads to some precise open problems to this effect (Section 6). Proofs for the regret bounds can be found in Appendix A and computational complexity proofs for $\tilde{q}^{mix}$ in Appendix B and for general $\tilde{q}$ in [Hut14]. As a side-result we give the arguably most convincing (simplest and least heuristic) derivation of the famous Good-Turing estimator. Other attempts at deriving the estimator Alan Turing suggested in 1941 to I.J. Good are less convincing (to us) [Goo53]. They appear more heuristic or convoluted, or are incomplete, often assuming something close to what one wants to get out [Nad85]. Our purely combinatorial derivation also feels right for 1941 and Alan Turing.

2 Problem Formulation

We now formally state the problem of offline to online conversion in three equivalent ways and the quality of a conversion. Let $x_t \in \mathcal{X}$ for $t \in \{1, ..., n\}$ and $x_{t:n} := x_t...x_n \in \mathcal{X}^{n-t+1}$, $x_{<n} := x_1...x_{n-1} \in \mathcal{X}^{n-1}$, and $x_{1:0} = x_{<1} = \epsilon$ be the empty string. ln denotes the natural logarithm and log the binary logarithm. $\tilde{q}_{|\mathcal{X}^n}$ constrains the domain $\mathcal{X}^*$ of $\tilde{q}$ to $\mathcal{X}^n$.

Formulation 1 (Measures). Given probability measures Q_n on $\mathcal{X}^n$ for $n = 1, 2, 3, ...$, find a probability measure $\tilde{Q}$ on $\mathcal{X}^\infty$ close to all Q_n in the sense of $\tilde{Q}(\mathcal{A} \times \mathcal{X}^\infty) \approx Q_n(\mathcal{A})$ for all measurable $\mathcal{A} \subseteq \mathcal{X}^n$ and all n.

For simplicity of notation, we will restrict to countable $\mathcal{X}$, and all examples will be for finite $\mathcal{X} = \{1, ..., d\}$. This allows us to reformulate the problem in terms of probability (mass) functions and predictors. A choice for $\approx$ will be given below.

Formulation 2 (Probability Mass Function). Given probability mass functions $q_n : \mathcal{X}^n \to [0; 1]$, i.e. $\sum_{x_{1:n}} q_n(x_{1:n}) = 1$, find a function $\tilde{q} : \mathcal{X}^* \to [0; 1]$ which is *time-consistent* (TC) in the sense

$$\sum_{x_n} \tilde{q}(x_{1:n}) = \tilde{q}(x_{<n}) \; \forall n, x_{<n} \quad \text{and} \quad \tilde{q}(\epsilon) = 1 \qquad \text{(TC)}$$

and is close to q_n i.e. $\tilde{q}(x_{1:n}) \approx q_n(x_{1:n})$ for all n and $x_{1:n}$.

This is equivalent to Formulation 1, via $q_n(x_{1:n}) := Q_n(\{x_{1:n}\})$, and since $\tilde{q}$ is TC iff there exists $\tilde{Q}$ with $\tilde{q}(x_{1:n}) = \tilde{Q}(\{x_{1:n}\} \times \mathcal{X}^\infty)$. We will use the following equivalent predictive formulation, discussed in the introduction, whenever convenient:

Formulation 3 (Predictors). Given q_n as before, find a predictor $\tilde{q} : \mathcal{X} \times \mathcal{X}^* \to [0; 1]$ which must be *normalized* as

$$\sum_{x_n} \tilde{q}(x_n | x_{<n}) = 1 \; \forall n, x_{<n} \qquad \text{(Norm)}$$

such that its joint probability $\tilde{q}(x_{1:n}) := \prod_{t=1}^{n} \tilde{q}(x_t | x_{<t})$ is close to q_n as before.

$\tilde{q}(x_{1:n})$ is the probability that an (infinite) sequence starts with $x_{1:n}$ and $\tilde{q}(x_n | x_{<n}) \equiv \tilde{q}(x_{1:n}) / \tilde{q}(x_{<n})$ is the probability that x_n follows given $x_{<n}$. Conditions (TC) and (Norm) are equivalent, and are the formal requirement(s) for an estimator to be *online*. We also speak of (q_n) being (not) Norm or TC.

Performance/Distance Measure. For modelling and coding we want $\tilde{q}$ as large as possible, which suggests the worst-case regret or log-loss regret

$$R_n \equiv R_n(\tilde{q}) \equiv R_n(\tilde{q}\|q_n) := \max_{x_{1:n}} \ln \frac{q_n(x_{1:n})}{\tilde{q}(x_{1:n})} \qquad (1)$$

For our qualitative considerations, other continuous $R_n \geq 0$ with $R_n = 0$ iff $\tilde{q}_{|\mathcal{X}^n} = q_n$ would also do. The R_n quantification of $\approx$ above has several convenient properties: Since an online arithmetic code of $x_{1:n}$ w.r.t. $\tilde{q}$ has code length

$|\log_2 \tilde{q}(x_{1:n})|$, and an offline Shannon-Fano or Huffman code for $x_{1:n}$ w.r.t. q_n has code length $|\log_2 q_n(x_{1:n})|$, this shows that the online coding of $x_{1:n}$ w.r.t. $\tilde{q}$ leads to codes at most $R_n \ln 2$ bits longer than offline codes w.r.t. q_n. Naturally we are interested in $\tilde{q}$ with small R_n, and indeed we will see that this is always achievable. Also, if q_n is an offline approximation of the true sampling distribution μ, then R_n upper bounds the *extra regret* of a corresponding online approximation $\tilde{q}$:

$$R_n^{\text{online}} - R_n^{\text{offline}} \equiv R_n(\tilde{q}||\mu) - R_n(q_n||\mu) \leq R_n(\tilde{q}||q_n) \equiv R_n \qquad (2)$$

Extending q_s from $\mathcal{X}^s$ to $\mathcal{X}^\infty$. Some (natural) offline $(q_n)_{n \in \mathbb{N}}$ considered later are automatically online in the sense that $\tilde{q}$ defined by $\tilde{q}(x_{1:n}) := q_n(x_{1:n})$ $\forall n, x_{1:n}$ is TC and hence $R_n = 0$ for all n. Note that it is *always* possible to choose $\tilde{q}$ such that $R_n = 0$ for *some* n: For some fixed $s \in \mathbb{N}_0$ define

$$\bar{q}_s(x_{1:n}) := \begin{cases} q_s(x_{1:s}) & \text{if } n = s, \\ \sum_{x_{n+1:s}} q_s(x_{1:s}) & \text{if } n < s, \\ q_s(x_{1:s})Q(x_{s+1:n}|x_{1:s}) & \text{if } n > s \end{cases} \qquad (3)$$

where Q can be an arbitrary measure on $\mathcal{X}^\infty$, e.g. uniform $Q(x_{s+1:n}|x_{1:s}) = |\mathcal{X}|^{n-s}$. It is easy to see that $\tilde{q} := \bar{q}_s$ is TC with $R_s(\tilde{q}) = R_s(\bar{q}_s) = R_s(q_s) = 0$, but in general $R_n(\bar{q}_s) > 0$ for $n \neq s$. Therefore naive minimization of R_n w.r.t. $\tilde{q}$ does not work. Minimizing $\lim_{n \to \infty} R_n$ can also fail for a number of reasons: the limit may not exist or is infinite, or minimizing it leads to poor finite-n performance or is not analytically possible or computationally intractable.

3 Conversion Methods

We now consider four methods of converting offline estimators to online predictors and discuss their pros and cons. They illustrate the difficulties and serve as a starting point to a more satisfactory solution.

Naive Ratio. The simplest way to define a predictor $\tilde{q}$ from q_n is via *ratio*

$$\tilde{q}^{\text{rat}}(x_t|x_{<t}) := \frac{q_t(x_{1:t})}{q_{t-1}(x_{<t})} \quad \text{or equivalently} \quad \tilde{q}^{\text{rat}}(x_{1:n}) := q_n(x_{1:n}) \qquad (4)$$

While this "solution" is tractable, it obviously only works when q_n already is TC. Otherwise $\tilde{q}^{\text{rat}}$ violates (TC). The deviation of

$$\mathcal{N}(x_{<t}) := \sum_{x_t} \tilde{q}^{\text{rat}}(x_t|x_{<t}) \equiv \frac{\sum_{x_t} q_t(x_{1:t})}{q_{t-1}(x_{<t})} \qquad (5)$$

from 1 measures the degree of violation. Note that the expectation of $\mathcal{N}(x_{<t})$ w.r.t. q_{t-1} is 1, so if $\mathcal{N}(x_{<t})$ is smaller than 1 for some $x_{<t}$ it must be larger for others, hence $\max_{x_{<t}} \mathcal{N}(x_{<t}) = 1$ iff $\mathcal{N}(x_{<t}) = 1$ for all $x_{<t} \in \mathcal{X}^{t-1}$.

Naive Normalization. Failure of $\tilde{q}^{\text{rat}}(x_t|x_{<t})$ to satisfy (Norm) is easily corrected by normalization [Sol78]:

$$\tilde{q}^{\mathrm{n1}}(x_t|x_{<t}) := \frac{q_t(x_{1:t})}{\sum_{x_t} q_t(x_{1:t})} \equiv \frac{\tilde{q}^{\mathrm{rat}}(x_t|x_{<t})}{\mathcal{N}(x_{<t})} \quad \text{and} \tag{6}$$

$$\tilde{q}^{\mathrm{n1}}(x_{1:n}) := \prod_{t=1}^{n} \tilde{q}^{\mathrm{n1}}(x_t|x_{<t}) \equiv \frac{q_n(x_{1:n})}{\prod_{t=1}^{n} \mathcal{N}(x_{<t})} \tag{7}$$

This guarantees TC and for small $\mathcal{X}$ is still tractable, but note that $\tilde{q}^{\mathrm{n1}}_{|\mathcal{X}^n} \neq q_n$ unless q_n is already TC. Unfortunately, this way of normalization can result in poor performance and very large regret R_n for finite n and asymptotically. Even if performance is good, computing R_n or finding good upper bounds can be very hard. Using (1) and (7), the regret can be represented and upper bounded as follows:

$$R_n(\tilde{q}^{\mathrm{n1}}) = \max_{x_{1:n}} \sum_{t=1}^{n} \ln \mathcal{N}(x_{<t}) \leq \sum_{t=1}^{n} \ln \max_{x_{<t}} \mathcal{N}(x_{<t}) \tag{8}$$

If q_n is TC, then $\mathcal{N} \equiv 1$, hence R_n as well as the upper bound are 0.

Let us consider here a simple but artificial example how bad things can get, following up with important practical examples in the next section. For an i.i.d. estimator $q_n(x_{1:n}) = q_n(x_1) \cdot ... \cdot q_n(x_n)$, where we slightly overloaded notation, $\tilde{q}^{\mathrm{n1}}(x_t|x_{<t}) = q_t(x_t)$ and $\tilde{q}^{\mathrm{n1}}(x_{1:n}) = q_1(x_1) \cdot ... \cdot q_n(x_n)$, therefore by definition (1)

$$R_n(\tilde{q}^{\mathrm{n1}}) = \max_{x_{1:n}} \ln \prod_{t=1}^{n} \frac{q_n(x_t)}{q_t(x_t)} = \sum_{t=1}^{n} \ln \max_{x_t} \frac{q_n(x_t)}{q_t(x_t)}$$

We now consider $\mathcal{X} = \{0,1\}$ with concrete Bernoulli($2/3$) probability $q_n(x_t = 1) = 2/3$ for even n and Bernoulli($1/3$) probability $q_n(x_t = 1) = 1/3$ for odd n. We see that for even t,

$$\tilde{q}^{\mathrm{rat}}(1_t|1_{<t}) = \frac{q_t(1_1) \cdot ... \cdot q_t(1_{t-1}) \cdot q_t(1_t)}{q_{t-1}(1_1) \cdot ... \cdot q_{t-1}(1_{t-1})} = 2^{t-1} \cdot \frac{2}{3}$$

is very badly unnormalized. Indeed $R_n(\tilde{q}^{\mathrm{n1}})$ grows linearly with n, i.e. becomes very large:

$$R_n(\tilde{q}^{\mathrm{n1}}) = \sum_{t=1}^{n} \ln \left\{ \begin{array}{ll} 1 & \text{if } n-t \text{ is even} \\ 2 & \text{if } n-t \text{ is odd} \end{array} \right\} = \left\lfloor \frac{n}{2} \right\rfloor \ln 2$$

Limit. We have seen how to make $R_s = 0$ for any fixed s using $\bar{q}_s$ (3). A somewhat natural idea is to define

$$\tilde{q}^{\mathrm{lim}}(x_{1:n}) := \lim_{s \to \infty} \bar{q}_s(x_{1:n}) = \lim_{s \to \infty} \sum_{x_{n+1:s}} q_s(x_{1:s})$$

in the hope to make $\lim_{s \to \infty} R_s = 0$. Effectively what $\tilde{q}^{\mathrm{lim}}$ does is to use q_s for very large s also for short strings of length n by marginalization. Problems are plenty: The limit may not exist, may exist but be incomputable, R_n may be hard to impossible to compute or upper bound, and even if the limit exists, $\tilde{q}^{\mathrm{lim}}$ may perform badly.

For instance, for the above Bernoulli($\frac{1}{3}|\frac{2}{3}$) example, the argument of the limit

$$\tilde{q}^{\text{lim}}(x_{1:n}) = \lim_{s \to \infty} \sum_{x_{n+1:s}} q_s(x_1) \cdot \ldots \cdot q_s(x_s) = \lim_{s \to \infty} [q_s(x_1) \cdot \ldots \cdot q_s(x_n)]$$

oscillates indefinitely (except if $x_1 + \ldots + x_n = {}^n/_2$). A template leading to a converging but badly performing $\tilde{q}^{\text{lim}}$ is $q_n(x_{1:n}) = \text{Bad}(x_{<\lfloor n/2 \rfloor}) \cdot \text{Good}(x_{\lfloor n/2 \rfloor:n})$. While offline $q_n(x_{1:n})$ is a "good" estimator on half of the data, $\tilde{q}^{\text{lim}}(x_{1:n}) = \text{Bad}(x_{1:n})$ is "bad" on all data. For example, $\text{Bad}(x_{1:n}) := |\mathcal{X}|^{-n}$ (see *Uniform* next Section) and $\text{Good}(x_{1:n}) = \binom{n+d-1}{n_1 \ldots n_d \ d-1}$ (see *Laplace* next Section) or simpler $\text{Good}(1_{1:n}) = 1$, lead to $R_n(\tilde{q}^{\text{lim}}) \propto n$.

Mixture. Another way of exploiting $\bar{q}_s$ is as follows: Rather than taking the limit $s \to \infty$ let us consider the class $\{\bar{q}_1, \bar{q}_2, \ldots\}$ of *all* $\bar{q}_s$. This corresponds to a set of measures on $\mathcal{X}^\infty$, each good in a particular circumstance, namely $\bar{q}_s$ is good and indeed perfect at time s. It is therefore natural to consider a Bayesian mixture over this class [San06]

$$\tilde{q}^{\text{mix}}(x_{1:n}) := \sum_{s=0}^{\infty} \bar{q}_s(x_{1:n}) w_s \quad \text{with prior} \quad w_s > 0, \quad \sum_{s=0}^{\infty} w_s = 1. \qquad (9)$$

$\tilde{q}^{\text{mix}}$ is TC and its regret can easily be upper bounded [San06]:

$$R_n(\tilde{q}^{\text{mix}}) = \max_{x_{1:n}} \ln \frac{q_n(x_{1:n})}{\sum_{s=0}^{\infty} \bar{q}_s(x_{1:n}) w_s} \leq \max_{x_{1:n}} \ln \frac{q_n(x_{1:n})}{\bar{q}_n(x_{1:n}) w_n} = \ln w_n^{-1} \quad (10)$$

For e.g. $w_n := \frac{1}{(n+1)(n+2)}$ we have $\ln w_n^{-1} \leq 2\ln(n+2)$ which usually can be regarded as small. This shows that *any* offline estimator can be converted into an online predictor with very small extra regret (2). Note that while $\tilde{q}^{\text{mix}}$ depends on arbitrary Q defined in (3), the upper bound (10) on R_n does not. Unfortunately it is unclear how to convert this heavy construction into an efficient algorithm.

A variation is to set $Q \equiv 0$, which makes $\tilde{q}^{\text{mix}}$ a semi-measure, which could be made TC by naive normalization (7). Bound (10) still holds since for $\tilde{q}^{\text{mix}}$ with $Q \equiv 0$ the normalizer $\mathcal{N} \leq 1$. Another variation is as follows. Often q_n violates TC only weakly, in which case a sparser prior, e.g. $w_{2^k} := \frac{1}{(k+1)(k+2)}$ and $w_n = 0$ for all other n, can lead to even smaller regret.

Further Choices for $\tilde{q}$. Of course the four presented choices for $\tilde{q}$ do not exhaust all options. Indeed, finding a tractable $\tilde{q}$ with good properties is a major open problem. Several estimation procedures do not only provide q_n on $\mathcal{X}^n$, but measures on $\mathcal{X}^\infty$ or equivalently for *each* n separately a TC $q_n : \mathcal{X}^* \to [0; 1]$ (see Bayes and crude MDL below). While this opens further options for $\tilde{q}$, e.g. $\tilde{q}(x_{n+1}|x_{1:n}) := q_n(x_{1:n+1})/q_n(x_{1:n})$ with some (weak) results for MDL [PH05], it does not solve our main problem.

Notes. Each solution attempt has its down-sides, and a solution satisfying all our criteria remains open. It is easy to verify that, if q_n is already TC, the first three definitions of $\tilde{q}$ coincide, and $R_n = 0$, which is reassuring, but $\tilde{q}_n^{\text{mix}}$ in general differs due to the arbitrary w in (9) and arbitrary Q in $\bar{q}$ in (3).

4   Examples

All examples below fall in one of two major strategies for designing estimators (the introduction mentions others we do not consider). One strategy is to start with a class $\mathcal{M}$ of probability measures ν on $\mathcal{X}^\infty$ in the hope one of them is good. For instance, $\mathcal{M}$ may contain (a subset of) i.i.d. measures $\nu_\theta(x_{1:n}) := \theta_{x_1} \cdot ... \cdot \theta_{x_n}$ with $\theta_i \geq 0$ and $\theta_1 + ... + \theta_d = 1$ and $d := |\mathcal{X}|$. One may either select a ν from $\mathcal{M}$ informed by given data $x_{1:n}$ or take an average over the class. The other strategy assigns uniform probabilities over subsets of $\mathcal{X}^n$. This combinatorial approach will be described later. Some strategies lead to TC and some examples are TC. For the others we will discuss the various online conversions $\tilde{q}$.

Bayes. The Bayesian mixture over $\mathcal{M}$ w.r.t. some prior (density) $w()$ is defined as

$$q_n(x_{1:n}) := \int_{\mathcal{M}} \nu(x_{1:n}) \, w(\nu) \, d\nu$$

Since q_n is TC, $(q_n^{\mathrm{rat}}) \equiv (q_n^{\mathrm{n1}}) \equiv (q_n^{\mathrm{lim}})$ coincide with $\tilde{q}$, $R_n = 0$, and $\tilde{q}^{\mathrm{rat}}$ is tractable if the Bayes mixture is. Note that $\tilde{q} \notin \mathcal{M}$ in general, in particular it is not i.i.d. Assume the true sampling distribution μ is in $\mathcal{M}$. For countable $\mathcal{M}$ and counting measure $d\nu$, we have $q_n(x_{1:n}) \geq \mu(x_{1:n})w(\mu)$, hence $R_n^{\mathrm{online}} = R_n^{\mathrm{offline}} \leq \ln w(\mu)^{-1}$. For continuous classes $\mathcal{M}$ we have $R_n^{\mathrm{online}} = R_n^{\mathrm{offline}} \lesssim \ln w(\mu)^{-1} + O(\ln n)$ under some mild conditions [BC91, Hut03, RH07].

MDL/NML/MAP. The MAP or MDL estimator is

$$\hat{q}_n(x_{1:n}) := \sup_{\nu \in \mathcal{M}} \{\nu(x_{1:n}) \, w(\nu)\} \quad \text{and} \quad q_n(x_{1:n}) := \frac{\hat{q}_n(x_{1:n})}{\sum_{x_{1:n}} \hat{q}_n(x_{1:n})}$$

Since $\hat{q}_n$ is not even a probability on $\mathcal{X}^n$, we have to normalize it to q_n. For uniform prior density $w()$, $\hat{q}_n$ is the maximum likelihood (ML) estimator, and q_n is known under the name normalized maximum likelihood (NML) or modern minimum description length (MDL). Unlike Bayes, q_n is *not* TC, which causes all kinds of complications [Grü07, Hut09], many of them can be traced back to our main open problem and the unsatisfactory choices for $\tilde{q}$ [PH05]. R_n^{offline} is essentially the same as for Bayes under similar conditions, but R_n^{online} depends on the choice of $\tilde{q}$. Crude MDL simply selects $q_n := \arg\max_{\nu \in \mathcal{M}} \{\nu(x_{1:n}) \, w(\nu)\}$ at time n, which is a probability measure on $\mathcal{X}^\infty$. While this opens additional options for defining $\tilde{q}$, they also can perform poorly in the worst case [PH05]. Note that most versions of MDL perform often very well in practice, comparable to Bayes; robustness and proving guarantees are the open problems.

Uniform. The uniform probability $q_n(x_{1:n}) := |\mathcal{X}|^{-n}$ is TC, hence all four $\tilde{q}$ coincide and $R_n = 0$ (only for uniform Q in case of q_n^{mix}). Unless data is uniform, this is a lousy estimator, since predictor $\tilde{q}(x_t|x_{<t}) = 1/|\mathcal{X}|$ is indifferent and ignores all evidence $x_{<t}$ to the contrary. But the basic idea of uniform probabilities is sound, if applied smartly: The general idea is to partition the sample space (here $\mathcal{X}^n$) into $\mathcal{P} = \{S_1, ..., S_{|\mathcal{P}|}\}$ and assign uniform probabilities to each partition: $q_n(x_{1:n}|S_r) = 1/|S_r|$ and a (possibly) uniform probability to

the parts themselves $q_n(S_r) = 1/|\mathcal{P}|$. For small $|\mathcal{P}|$, $q_n(x_{1:n}) = q_n(x_{1:n}|S_r)q_n(S_r)$ is never more than a small factor $|\mathcal{P}|$ smaller than uniform $|\mathcal{X}|^{-n}$ but may be a huge factor of $|\mathcal{X}|^n/|S_r||\mathcal{P}|$ larger. The Laplace rule can be derived that way, and the Good-Turing and Ristad estimators by further sub-partitioning.

Laplace. More interesting than the uniform probability is the following double uniform combinatorial probability: Let $n_i := |\{t : x_t = i\}|$ be the number of times, symbol $i \in \mathcal{X} = \{1, ..., d\}$ appears in $x_{1:n}$. We assign a uniform probability to all sequences $x_{1:n}$ with the same counts $\boldsymbol{n} := (n_1, ..., n_d)$, therefore $q_n(x_{1:n}|\boldsymbol{n}) = \binom{n}{n_1 ... n_d}^{-1}$. We also assign a uniform probability to the counts $\boldsymbol{n}$ themselves, therefore $q_n(\boldsymbol{n}) = |\{\boldsymbol{n} : n_1 + ... + n_d = n\}|^{-1} = \binom{n+d-1}{d-1}^{-1}$. Together

$$q_n(x_{1:n}) = \binom{n}{n_1 \ ... \ n_d}^{-1} \binom{n+d-1}{d-1}^{-1} = \binom{n+d-1}{n_1 \ ... \ n_d \ \ d-1}^{-1}$$

hence $\tilde{q}^{\text{rat}}(x_{n+1} = i|x_{1:n}) = \dfrac{q_{n+1}(x_{1:n}i)}{q_n(x_{1:n})} = \dfrac{n_i + 1}{n + d}$

is properly normalized (Norm), so $\tilde{q}^{\text{rat}}$ is TC, and $(q_n^{\text{rat}}) \equiv (q_n^{\text{n1}}) \equiv (q_n^{\text{lim}})$ coincide with $\tilde{q}$ and $R_n = 0$. $\tilde{q}^{\text{rat}}$ is nothing but Laplace's famous rule.

Good-Turing. Even more interesting is the following triple uniform probability: Let $M_r := \{i : n_i = r\}$ be the symbols that appear exactly $r \in \mathbb{N}_0$ times in $x_{1:n}$, and $m_r := |M_r|$ be their number. Clearly $m_r = 0$ for all $r > n$, but due to $\sum_{r=0}^{n} r \cdot m_r = n$, $m_r = 0$ also for many $r < n$. We assign uniform probabilities to $q_n(x_{1:n}|\boldsymbol{n})$ as before and to $q_n(\boldsymbol{n}|\boldsymbol{m})$ and to $q_n(\boldsymbol{m})$, where $\boldsymbol{m} := (m_0, ..., m_n)$. There are $\binom{d}{m_0 ... m_n}$ ways to distribute symbols $1, ..., d$ into sets $(M_0, ..., M_n)$ (many of them empty) of sizes $m_0, ..., m_n$. Therefore $q_n(\boldsymbol{n}|\boldsymbol{m}) = \binom{d}{m_0 ... m_n}^{-1}$. Each $\boldsymbol{m}$ constitutes a decomposition of n into natural summands with repetition but without regard to order. The number of such decompositions is a well-known function [AS74, §24.2.2] which we denote by Part(n). Therefore $q_n(\boldsymbol{m}) = $ Part$(n)^{-1}$. Together

$$q_n(x_{1:n}) = \binom{n}{n_1 \ ... \ n_d}^{-1} \binom{d}{m_0 \ ... \ m_n}^{-1} \text{Part}(n)^{-1}, \quad \text{hence} \qquad (11)$$

$$\tilde{q}^{\text{rat}}(x_{n+1} = i|x_{1:n}) = \frac{q_{n+1}(x_{1:n}i)}{q_n(x_{1:n})} = \frac{n_i + 1}{n+1} \cdot \frac{m_{r+1}+1}{m_r} \cdot \frac{\text{Part}(n)}{\text{Part}(n+1)}, \quad r = n_i \quad (12)$$

This is not TC as can be verified by example, but is a very interesting predictor: The first term is close to a frequency estimate n_i/n. The second term is close to the Good-Turing (GT) correction m_{r+1}/m_r. The intuition is that if e.g. many symbols have appeared once (m_1 large), but few twice (m_2 small), we should be skeptical of observing a symbol that has been observed only once another time, since it would move from a likely category to an unlikely one. The third term $\frac{\text{Part}(n)}{\text{Part}(n+1)} \to 1$ for $n \to \infty$. The normalized version

$$\tilde{q}^{n1}(x_{n+1} = i|x_{1:n}) = \frac{\tilde{q}^{\mathrm{rat}}(x_{n+1} = i|x_{1:n})}{\sum_{x_{n+1}} \tilde{q}^{\mathrm{rat}}(x_{n+1}|x_{1:n})} = \frac{1}{\mathcal{N}_n} \cdot \frac{r+1}{n+1} \cdot \frac{m_{r+1}+1}{m_r} \quad (13)$$

$$\text{where} \quad \mathcal{N}_n := \frac{1}{n+1} \sum_{r=0, m_r \neq 0}^{n} (r+1)(m_{r+1}+1) \quad (14)$$

is even closer to the GT estimator. We kept $\frac{1}{n+1}$ as in [Goo53, Eq.(13)], while often $\frac{1}{n}$ is seen due to [Goo53, Eq.(2)]. Anyway after normalization there is no difference. The only difference to the GT estimator is the appearance of $m_{r+1}+1$ instead of m_{r+1}. Unfortunately its regret is very large:

Theorem 1 (Naively Normalized Triple Uniform Estimator). *Naive normalization of the triple uniform combinatorial offline estimator q_n defined in (11) leads to the (non-smoothed) Good-Turing estimator $\tilde{q}^{n1}$ given in (13) with regret*

$$R_n(\tilde{q}^{n1}||q_n) = \max_{x_{1:n}} \left\{ \sum_{t=1}^{n} \ln \mathcal{N}_{t-1} \right\} - \ln(Part(n)) \begin{cases} = n \ln 2 \pm O(\sqrt{n}) \ for \ |\mathcal{X}| = \infty \\ \geq 0.43n - O(\sqrt{n}) \ for \ |\mathcal{X}| \geq 3 \end{cases} \quad (15)$$

Inserting (12) and (14) into (6) we get $\mathcal{N}(x_{1:n}) = \frac{\tilde{q}^{\mathrm{rat}}(x_{n+1}|x_{1:n})}{\tilde{q}^{n1}(x_{n+1}|x_{1:n})} = \frac{Part(n)}{Part(n+1)}\mathcal{N}_n$ which by (8) implies the first equality. We prove the last equality in Appendix A by showing that the maximizing sequence is $x_{1:\infty} = 1223334444...$ with $\mathcal{N}_n = 2 \pm O(n^{-1/2})$ which requires infinite d or at least $d \geq \sqrt{2n}$. We also show that $R_n \geq 0.43n - O(\sqrt{n})$ for every $d \geq 3$. The linearly growing R_n shows that naive normalization severely harms the offline triple uniform estimator q_n.

Indeed, raw GT performs very poorly for large r in practice, but smoothing the function $m_{()}$ leads to an excellent estimator in practice [Goo53], e.g. Kneser-Ney smoothing for text data [CG99]. Our $m_{r+1} \rightsquigarrow m_{r+1} + 1$ is a kind of albeit insufficient smoothing. $\tilde{q}^{\mathrm{mix}}$ may be regarded as an (unusual) kind of smoothing, which comes with the strong guarantee $R_n \leq 2\ln(n+2)$, but a direct computation is prohibitive. [San06] gives a low-complexity smoothing of the original GT that comes with guarantees, namely sub-linear $O(n^{2/3})$ log worst-case sequence attenuation, but this is different from R_n in various respects: Log worst-case sequence-attenuation is relative to i.i.d. coding and unlike R_n lower bounded by $O(n^{1/3})$. Still a similar construction may lead to sublinear and ideally logarithmic R_n.

Ristad. [Ris95] designed an interesting quadruple uniform probability motivated as follows: If $\mathcal{X}$ is the set of English words and $x_{1:n}$ some typical English text, then most symbols=words will not appear ($d \gg n$). In this case, Laplace assigns not enough probability ($\frac{n_i+1}{n+d} \ll \frac{n_i}{n}$) to observed words. This can be rectified by treating symbols $\mathcal{A} := \{i : n_i > 0\}$ that do appear different from symbols $\mathcal{X} \setminus \mathcal{A}$ that don't. For $n > 0$, $x_{1:n}$ may contain $m \in \{1, ..., \min\{n, d\}\}$ different symbols, so we set $q_n(m) = 1/\min\{n, d\}$. Now choose uniformly which m symbols $\mathcal{A}$ appear, $q_n(\mathcal{A}|m) = \binom{d}{m}^{-1}$ for $|\mathcal{A}| = m$. There are $\binom{n-1}{m-1}$ ways of choosing the frequency of symbols consistent with $n_1 + ... + n_d = n$ and $n_i > 0 \Leftrightarrow i \in \mathcal{A}$, hence $q_n(\boldsymbol{n}|\mathcal{A}) = \binom{n-1}{m-1}^{-1}$. Finally, $q_n(x_{1:n}|\boldsymbol{n}) = \binom{n}{n_1...n_d}^{-1}$ as before. Together

$$q_n(x_{1:n}) = \binom{n}{n_1 \ldots n_d}^{-1} \binom{n-1}{m-1}^{-1} \binom{d}{m}^{-1} \frac{1}{\min\{n,d\}}, \quad \text{which implies}$$

$$(16)$$

$$\tilde{q}^{\text{rat}}(x_{n+1} = i|x_{1:n}) = \frac{\min\{n,d\}}{\min\{n+1,d\}} \cdot \begin{cases} \frac{(n_i+1)(n-m+1)}{n(n+1)} & \text{if} \quad n_i > 0 \\ \frac{m(m+1)}{n(n+1)} \cdot \frac{1}{d-m} & \text{if} \quad n_i = 0 \end{cases}$$

This is not TC, since

$$\mathcal{N}(x_{1:n}) = \frac{\min\{n,d\}}{\min\{n+1,d\}} \cdot \begin{cases} 1 + \frac{2m}{n(n+1)} & \text{if} \quad m < d \\ 1 - \frac{m(m-1)}{n(n+1)} & \text{if} \quad m = d \end{cases}$$

is not identically 1. Normalization leads to

$$\tilde{q}^{\text{n1}}(x_{n+1} = i|x_{1:n}) = \begin{cases} \frac{(n_i+1)(n-m+1)}{n(n+1)+2m} & \text{if} \quad n_i > 0 \text{ and } m < d \\ \frac{m(m+1)}{n(n+1)+2m} \cdot \frac{1}{d-m} & \text{if} \quad n_i = 0 \\ \frac{n_i+1}{n+m} & \text{if} \quad m = d \ [\Rightarrow n_i > 0] \end{cases} \quad (17)$$

For $n = 0$ we have $\tilde{q}^{\text{rat}}(x_1) = \tilde{q}^{\text{n1}}(x_1) = q_n(x_1) = 1/d$ and $\mathcal{N}(\epsilon) = 1$. While by construction, the offline estimator should have good performance (in the intended regime), the performance of the online version depends on how much the normalizer exceeds 1. The first factor in $\mathcal{N}$ is ≤ 1 and the $m = d$ case is ≤ 1. Therefore $\mathcal{N}(x_{1:n}) \leq 1 + \frac{2m}{n(n+1)} \leq 1 + \frac{2}{n+1}$, where we have used $m \leq n$ in the second step. The regret can hence be bounded by

$$R_n(\tilde{q}^{\text{n1}}) \leq \sum_{t=1}^{n} \ln \max_{x_{<t}} \mathcal{N}(x_{<t}) \leq \sum_{t=2}^{n} \ln(1 + \tfrac{2}{t}) \leq \sum_{t=2}^{n} \tfrac{2}{t} \leq 2 \ln n$$

Theorem 2 (Quadruple Uniform Estimator). *Naive normalization of Ristad's quadruple uniform combinatorial offline estimator q_n defined in (16) leads to Ristad's natural law $\tilde{q}^{\text{n1}}$ given in (17) with regret $R_n(\tilde{q}^{\text{n1}}\|q_n) \leq 2 \ln n$.*

This shows that simple normalization does not ruin performance. Indeed, the regret bound is as good as we are able to guarantee in general via $\tilde{q}^{\text{mix}}$.

5 Computational Complexity of $\tilde{q}$

Computability and Complexity of $\tilde{q}^{\text{mix}}$. From the four discussed online estimators only q_n^{mix} guarantees small extra regret over offline (q_n) in general, but the definition of $\tilde{q}^{\text{mix}}$ is quite heavy and at first it is not even clear whether it is computable. The following theorem proven in Appendix B shows that $\tilde{q}^{\text{mix}}$ can be computed to relative accuracy ε in double-exponential time:

Theorem 3 (Computational Complexity of $\tilde{q}^{\text{mix}}$). *There is an algorithm A that computes $\tilde{q}^{\text{mix}}$ (with uniform choice for Q) to relative accuracy $|A(x_{1:n}, \varepsilon)/\tilde{q}^{\text{mix}}(x_{1:n}) - 1| < \varepsilon$ in time $O(|\mathcal{X}|^{4|\mathcal{X}|^n/\varepsilon})$ for all $\varepsilon > 0$.*

The relative accuracy ε allows us to compute the predictive distribution $\tilde{q}^{\mathrm{mix}}(x_t|x_{<t})$ to accuracy ε, ensures $A(x_{1:n}, \varepsilon) > (1 - \varepsilon)\tilde{q}_n^{\mathrm{mix}}(x_{1:n})$, hence $R_n(A(\cdot, \varepsilon)\|q_n) \leq R_n(\tilde{q}^{\mathrm{mix}}\|q_n) + \frac{\varepsilon}{1-\varepsilon}$, and approximate normalization $|1 - \sum_{x_{1:n}} A(x_{1:n}, \varepsilon)| < \varepsilon$.

Computational Complexity of General $\tilde{q}$. The existence of $\tilde{q}^{\mathrm{mix}}$ shows that any offline estimator can be converted into an online estimator with minimal extra regret $R_n \leq 2\ln(n+2)$. While encouraging and of theoretical interest, the provided algorithm for $\tilde{q}^{\mathrm{mix}}$ is prohibitive. Indeed, Theorem 4 below establishes that there exist offline (q_n) computable in polynomial time for which the fastest algorithm for *any* online (=TC) $\tilde{q}$ with $R_n \leq O(\log n)$ is at least exponential in time.

Trivially $R_n \leq n\ln|\mathcal{X}|$ can always be achieved for any (q_n) by uniform $\tilde{q}(x_{1:n}) = |\mathcal{X}|^{-n}$. So a very modest quest would be $R_n \leq (1 - \varepsilon)n\ln|\mathcal{X}|$. If we require $\tilde{q}$ to run in polynomial time but with free oracle access to (q_n), Theorem 5 below shows that this is also not possible for some exponential time (q_n).

Together this does not rule out that for every fast (q_n) there exists a fast $\tilde{q}$ with e.g. $R_n \leq \sqrt{n}$. This is our main remaining open problem to be discussed in Section 6.

The main proof idea for both results is as follows: We construct a deterministic (q_s) that is 1 on the sequence of quasi-independent quasi-random strings $\dot{x}_{1:1}^1$, $\dot{x}_{1:2}^2$, $\dot{x}_{1:3}^3$, ... The only way for $\tilde{q}(x_{1:n})$ to be not too much smaller than $\bar{q}_s(\dot{x}_{1:n}^s)$ is to know $\dot{x}_{1:s}^s$. If $s = s(n)$ is exponential in n this costs exponential time. If $\tilde{q}$ has only oracle access to (q_s), it needs exponentially many oracle calls even for linear $s(n) = (1 + \varepsilon)n$.

The general theorem is a bit unwieldy and is deferred to the end of the section and is proven in the extended technical report [Hut14]. First we present and discuss the most interesting special cases. TIME($g(n)$) is defined as the class of all algorithms that run in time $O(g(n))$ on inputs of length n. Real-valued algorithms produce for any rational $\varepsilon > 0$ given as an extra argument, an ε-approximation in this time, as did $A(x_{1:n}, \varepsilon)$ for $\tilde{q}^{\mathrm{mix}}$ above. Algorithms in $E^c := \mathrm{TIME}(2^{cn})$ run in exponential time, while $P := \bigcup_{k=1}^{\infty} \mathrm{TIME}(n^k)$ is the classical class of all algorithms that run in polynomial time (strictly speaking Function-P or FP [AB09]). The theorems don't rest on any complexity separation assumptions such as P$\neq$NP. We only state and prove the theorems for binary alphabet $\mathcal{X} = \mathbb{B} = \{0,1\}$. The generalization to arbitrary finite alphabet is trivial. 'For all large n' shall mean 'for all but finitely many n', denoted by $\forall' n$. $m > 0$ is a constant that depends on the machine model, e.g. $m = 1$ for a random access machine (RAM).

Theorem 4 (Sub-optimal Fast Online for Fast Offline). *For all $r > 0$ and $c > 0$ and $\varepsilon > 0$*

(i) $\exists(q_s) \in \mathit{TIME}(s^{b+m}) \; \forall \tilde{q} \in E^c : R_n(\tilde{q}\|q_n) \geq r\ln n \; \forall' n$, where $b := \frac{c+1+\varepsilon}{1-\varepsilon}r$

(ii) *in particular for large c and r:* $\exists(q_s) \in P \; \forall \tilde{q} \in E^c : R_n \geq r\ln n \; \forall' n$,

(iii) *in particular for small c, ε:* $\exists(q_s) \in \mathit{TIME}(s^{r+m+\varepsilon})\forall \tilde{q} \in P : R_n \geq r\ln n \; \forall' n$,

(iv) *in particular for $\tilde{q}^{\mathrm{mix}}$:* $\exists(q_s) \in P : \tilde{q}^{\mathrm{mix}} \notin E^c$

In particular (iii) implies that there is an offline estimator (q_s) computable in quartic time s^4 on a RAM for which no polynomial-time online estimator $\tilde{q}$ is as good as $\tilde{q}^{\mathrm{mix}}$. The slower (q_s) we admit (larger r), the higher the lower bound gets. (ii) says that even algorithms for $\tilde{q}$ running in exponential time 2^{cn} cannot achieve logarithmic regret for all $(q_s) \in \mathrm{P}$. In particular this implies that (iv) any algorithm for $\tilde{q}^{\mathrm{mix}}$ requires super-exponential time for some $(q_s) \in \mathrm{P}$ on some arguments.

The next theorem is much stronger in the sense that it rules out even very modest demands on R_n but is also much weaker since it only applies to online estimators for slow (q_s) used as a black box oracle. That is, $\tilde{q}^o(x_{1:n})$ can call $q_s(z_{1:s})$ for any s and $z_{1:s}$ and receives the correct answer. We define $\mathrm{TIME}^o(g(n))$ as the class of all algorithms with such oracle access that run in time $O(g(n))$, where each oracle call is counted only as one step, and similarly P^o and $\mathrm{E}^{c,o}$.

Theorem 5 (Very Poor Fast Online Using Offline Oracle). *For all $\varepsilon > 0$*

$$\exists o \equiv (q_s) \in E^1 \; \forall \tilde{q}^o \in E^{\varepsilon/2,o} : R_n(\tilde{q}^o \| q_n) \geq (1 - \varepsilon)n \ln 2 \; \forall' n$$
$$\text{or cruder: } \exists o \equiv (q_s) \; \forall \tilde{q}^o \in P^o : R_n(\tilde{q}^o \| q_n) \geq (1 - \varepsilon)n \ln 2 \; \forall' n$$

The second line states that the trivial bound $R_n \leq n \ln 2$ achieved by the uniform distribution can in general not be improved by a fast $\tilde{q}^o$ that (only) has oracle access to the offline estimator.

Usually one Does not state the complexity of the oracle, since it does not matter, but knowing that an $o \in \mathrm{E}^1$ is sufficient (first line) tells us something: First, the negative result is not an artifact of some exotic non-computable offline estimator. On the other hand, if an exponential time offline o is indeed needed to make the result true, the result wouldn't be particularly devastating. It is an open question whether an $o \in \mathrm{P}$ can cause such bad regret.

The general computational complexity result is as follows:

Theorem 6 (Fast Offline can Imply Slow Online (General)). *Let $s(n)$ and $f(n)$ and $g(n)$ be monotone increasing functions. $s(n)$ shall be injective and $\geq n$ for large n with inverse $n(s) := \max\{n : s(n) \leq s\}$ and $g(n) < \frac{1}{2}n^{-\delta}h(n)$, where $h(n) := 2^{s(n)-n}[n^{-\gamma} - 2^{f(s(n))-n}]$. $m > 0$ is a constant depending on the machine model, e.g. $m = 1$ for a RAM. Then for all $\gamma > 0$ and $\delta > 0$ it holds that*

$$\exists o \equiv (q_s) \in TIME(n(s)^{\gamma+\delta}2^{n(s)}g(n(s))s^m)$$
$$\forall \tilde{q}^o \in TIME^o(g(n)) : R_n(\tilde{q}^o \| q_n) \geq f(n) \ln 2 \; \forall' n$$

For $s = 2^{(1-\varepsilon)n/r}$ and $f(s) = r \log s = (1 - \varepsilon)n$ and $g(n) = 2^{cn}$ this implies Theorem 4, and for $s = (1 + \varepsilon)n$ and $f(s) = (1 - \varepsilon)s$ and $g(n) = 2^{\varepsilon n/2}$ this implies Theorem 5. See the extended technical report [Hut14] for the proofs of Theorems 4, 5 and 6.

6   Open Problems

We now discuss and quantify the problems that we raised earlier and are still open. For some *specific* collection (q_n) of probabilities, does there exist a polynomial-time computable time-consistent $\tilde{q}$ with $R_n(\tilde{q}||q_n) \leq 2\ln(n+2) \, \forall n$? Note that $\tilde{q}^{\mathrm{mix}}$ satisfies the bound, but a direct computation is prohibitive. So one way to a positive answer could be to find an efficient approximation of $\tilde{q}^{\mathrm{mix}}$. If the answer is negative for a specific (q_n) one could try to weaken the requirements on R_n. We have seen that for some, (non-TC) (q_n), namely Ristad's, simple normalization $\tilde{q}^{n1}$ solves the problem.

A concrete unanswered example are the triple uniform Good-Turing probabilities (q_n). Preliminary experiments indicate that they and therefore $\tilde{q}^{\mathrm{mix}}$ are more robust than current heuristic smoothing techniques, so a tractable approximation of $\tilde{q}^{\mathrm{mix}}$ would be highly desirable. It would be convenient and insightful if such a $\tilde{q}$ had a traditional GT representation but with a smarter smoothing function $m_{()}$.

The nasty (q_n) constructed in the proof of Theorem 6 is very artificial: It assigns extreme probabilities (namely 1) to quasi-random sequences. It is unknown whether there is any offline estimator of practical relevance (such as Good-Turing) for which no fast online estimator can achieve logarithmic regret.

An open problem for general (q_n) is as follows: Does there exist for every (q_n) a polynomial-time algorithm that computes a time-consistent $\tilde{q}$ with $R_n(\tilde{q}||q_n) \leq f(n) \, \forall n$. We have shown that this is not possible for $f(n) = O(\log n)$ and not even for $f(n) = (1-\varepsilon)n\ln 2$ if $\tilde{q}$ has only oracle access to (q_n). This still allows for a positive answer to the following open problem:

Open Problem 7 (Fast Online from Offline with Small Extra Regret). *Can every polynomial-time offline estimator (q_n) be converted to a polynomial-time online estimator $\tilde{q}$ with small regret $R_n(\tilde{q}||q_n) \leq \sqrt{n} \, \forall' n$? Or weaker: $\forall (q_n) \in P \exists \tilde{q} \in P : R_n = o(n)$? Or stronger: $\forall (q_n) \in P \exists \tilde{q} \in P : R_n = O(\log n)^2$?*

A positive answer would reduce once and for all the problem of finding good online estimators to the apparently easier problem of finding good offline estimators. We could also weaken our notion of worst-case regret to e.g. expected regret $\mathbb{E}[\ln(q_n/\tilde{q})]$. Expectation could be taken w.r.t. (q_n), but other choices are possible. Other losses than logarithmic also have practical interest, but I do not see how this makes the problem easier. Ignoring computational considerations, of theoretical interest is whether $O(\log n)$ is the best one can achieve in general, say $\exists q_n \forall \tilde{q} : R_n(\tilde{q}) \geq \ln n$, or whether a constant is achievable. Devising general techniques to upper bound $R_n(\tilde{q}^{n1}||q_n)$, especially if small, is of interest too.

Acknowledgements. Thanks to Jan Leike for feedback on earlier drafts.

References

[AB09]   Arora, S., Barak, B.: Computational Complexity: A Modern Approach. Cambridge University Press (2009)

[AS74]   Abramowitz, M., Stegun, I.A. (eds.): Handbook of Mathematical Functions. Dover Publications (1974)

[BC91]   Barron, A.R., Cover, T.M.: Minimum complexity density estimation. IEEE Transactions on Information Theory 37, 1034–1054 (1991)

[CG99]   Chen, S.F., Goodman, J.: An empirical study of smoothing techniques for language modeling. Computer Speech and Language 13, 359–394 (1999)

[Goo53]   Good, I.J.: The population frequencies of species and the estimation of population parameters. Biometrika 40(3/4), 237–264 (1953)

[Grü07]   Grünwald, P.D.: The Minimum Description Length Principle. The MIT Press, Cambridge (2007)

[Hut03]   Hutter, M.: Optimality of universal Bayesian prediction for general loss and alphabet. Journal of Machine Learning Research 4, 971–1000 (2003)

[Hut05]   Hutter, M.: Universal Artificial Intelligence: Sequential Decisions based on Algorithmic Probability. Springer, Berlin (2005)

[Hut09]   Hutter, M.: Discrete MDL predicts in total variation. In: Advances in Neural Information Processing Systems 22 (NIPS 2009), pp. 817–825. Curran Associates, Cambridge (2009)

[Hut14]   Hutter, M.: Offline to online conversion. Technical report (2014), http://www.hutter1.net/publ/off2onx.pdf

[Nad85]   Nadas, A.: On Turing's formula for word probabilities. IEEE Transactions on Acoustics, Speech, and Signal Processing 33(6), 1414–1416 (1985)

[PH05]   Poland, J., Hutter, M.: Asymptotics of discrete MDL for online prediction. IEEE Transactions on Information Theory 51(11), 3780–3795 (2005)

[RH07]   Ryabko, D., Hutter, M.: On sequence prediction for arbitrary measures. In: Proc. IEEE International Symposium on Information Theory (ISIT 2007), pp. 2346–2350. IEEE, Nice (2007)

[Ris95]   Ristad, E.S.: A natural law of succession. Technical Report CS-TR-495-95. Princeton University (1995)

[San06]   Santhanam, N.: Probability Estimation and Compression Involving Large Alphabets. PhD thesis, University of California, San Diego, USA (2006)

[Sol78]   Solomonoff, R.J.: Complexity-based induction systems: Comparisons and convergence theorems. IEEE Transactions on Information Theory IT-24, 422–432 (1978)

A   Proof of Theorem 1

For GT we prove $\max_{x_{1:n}} \mathcal{N}_n \to 2$, therefore $\max_{x_{1:n}} \mathcal{N}(x_{1:n}) \to 2$ due to $\frac{\text{Part}(n)}{\text{Part}(n+1)} \to 1$ for $n \to \infty$. We can upper bound (14) as

$$(n+1)\mathcal{N}_n = \sum_{r=0,m_r\neq 0}^{n} (r+1)m_{r+1} + \sum_{r=0,m_r\neq 0}^{n} r + \sum_{r=0,m_r\neq 0}^{n} 1$$

$$\leq \sum_{r'=1}^{n+1} r'm_{r'} + \sum_{r=0}^{n} rm_r + |\{r : m_r \neq 0\}|$$

$$= n + n + |\{r : m_r \neq 0\}| \leq 2n + \sqrt{2n} + 1$$

$|\{r : m_r \neq 0\}|$ under the constraint $\sum_{r=0}^{n} rm_r = n$ is maximized for $m_0 = ... = m_k = 1$ and $m_{k+1} = ... = m_n = 0$ for suitable k. We may have to set one $m_r = 2$ to meet the constraint. Therefore $n = \sum_{r=0}^{n} rm_r \geq \sum_{r=0}^{k} r = \frac{k(k+1)}{2} \geq \frac{1}{2}k^2$, hence $|\{r : m_r \neq 0\}| = k + 1 \leq \sqrt{2n} + 1$.

For the lower bound we construct a sequence that attains the upper bound. For instance, $x_{1:k(k+1)/2} = 1223334444 \ldots k\ldots k$ has $m_1 = \ldots = m_k = 1$, hence $x_{1:\infty} = 1223334444\ldots$ has $m_1 \geq 1, \ldots, m_k \geq 1$ for all $n \geq \frac{1}{2}k(k+1)$. Conversely, for any n we have $m_1 \geq 1, \ldots, m_k \geq 1$ with $k := \lfloor\sqrt{2n}\rfloor - 1$. For the chosen sequence we therefore have

$$(n+1)\mathcal{N}_n \;\geq\; \sum_{r=0}^{k-1}(r+1)(1+1) \;=\; k(k+1) \;\geq\; 2n - 3\sqrt{2n}$$

The upper and lower bounds together imply $\max_{x_{1:n}} \mathcal{N}_n = 2 \pm O(n^{-1/2})$, therefore $\max_{x_{1:n}} \mathcal{N}(x_{1:n}) = 2 \pm O(n^{-1/2})$ due to $\frac{\mathrm{Part}(n)}{\mathrm{Part}(n+1)} = 1 - O(n^{-1/2})$ [AS74]. Inserting this into (15) gives $R_n = n\ln 2 \pm O(n^{-1/2})$.

The upper bound holds for any d, but the lower bound requires $d = \infty$ or at least $d \geq \sqrt{2n}$. We now show linear growth of R_n even for finite $d \geq 3$. The lower bound is based on the same sequence as used in [San06]: For $x_{1:\infty} = 12(132)^\infty$ elementary algebra gives $\mathcal{N}_n = \frac{5}{3} + \frac{7/3}{n+1}$ and $\mathcal{N}_{n+1} = \frac{5}{3} + \frac{5/3}{n+2}$ and $\mathcal{N}_{n+2} = \frac{4}{3} + \frac{1}{n+3}$ for n a multiple of 3, hence $\mathcal{N}_n\mathcal{N}_{n+1}\mathcal{N}_{n+2} \geq \frac{100}{27}$ (except $\mathcal{N}_0\mathcal{N}_1\mathcal{N}_2 = \frac{2}{3}$). Together with asymptotics $\ln(\mathrm{Part}(n)) \sim \pi\sqrt{2n/3}$ [AS74], this implies that $R_n \geq \frac{n}{3}\ln\frac{100}{27} - O(\sqrt{n})$. ∎

B   Proof of Theorem 3

The design of an algorithm for $\tilde{q}^{\mathrm{mix}}$ and the analysis of its run-time follows standard recipes, so will only be sketched. A real-valued function $\tilde{q}^{\mathrm{mix}} : \mathcal{X}^* \to [0;1]$ is (by definition) computable (also called estimable [Hut05]), if there is an always halting algorithm $A : \mathcal{X}^* \times \mathbb{Q}^+ \to \mathbb{Q}$ with $|A(x_{1:n}, \varepsilon) - \tilde{q}^{\mathrm{mix}}(x_{1:n})| < \varepsilon$ for all rational $\varepsilon > 0$. We assume there is an oracle q_t^ε that provides q_t to ε-accuracy in time $O(1)$. We assume that real numbers can be processed in unit time. In reality we need $O(\ln 1/\varepsilon)$ bits to represent, and time to process, real numbers to accuracy ε. This leads to some logarithmic factors in run-time which are dwarfed by our exponentials, so will be ignored. To compute $\bar{q}_s(x_{1:n})$ to accuracy $\varepsilon/2$ we need to call $q_s^{\varepsilon/2N}$ oracle $N := \max\{|\mathcal{X}|^{s-n}, 1\}$ times and add up all numbers. We can compute $\tilde{q}^{\mathrm{mix}}$ to ε-accuracy by the truncated sum $\sum_{s=0}^{2/\varepsilon} \bar{q}_s^{\varepsilon/2}(x_{1:n})w_s$ with $w_s = \frac{1}{(s+1)(s+2)}$, since the tail sum is bounded by $\varepsilon/2$. Hence overall runtime is $O(|\mathcal{X}|^{2/\varepsilon - n})$. But this is not sufficient. For large n, $\tilde{q}^{\mathrm{mix}}(x_{1:n})$ is typically small, and we need a *relative* accuracy of ε, i.e. $|A(x_{1:n}, \varepsilon')/\tilde{q}^{\mathrm{mix}}(x_{1:n}) - 1| < \varepsilon$. For $Q(x_{1:n}) = |\mathcal{X}|^{-n}$, we have $\tilde{q}^{\mathrm{mix}}(x_{1:n}) \geq \frac{1}{2}Q(x_{1:n}) = \frac{1}{2}|\mathcal{X}|^{-n}$, hence $\varepsilon' = \frac{\varepsilon}{2}|\mathcal{X}|^{-n}$ suffices. Run time becomes $O(|\mathcal{X}|^{\frac{4}{\varepsilon}|\mathcal{X}|^n - n}) \leq e^{e^{O(n)}/\varepsilon}$. ∎

A Chain Rule for the Expected Suprema
of Gaussian Processes

Andreas Maurer

Adalbertstrasse 55
D-80799 München, Germany
am@andreas-maurer.eu

Abstract. The expected supremum of a Gaussian process indexed by
the image of an index set under a function class is bounded in terms of
separate properties of the index set and the function class. The bound
is relevant to the estimation of nonlinear transformations or the analysis
of learning algorithms whenever hypotheses are chosen from composite
classes, as is the case for multi-layer models.

1 Introduction

Rademacher and Gaussian averages ([1], see also [5],[11]) provide an elegant
method to demonstrate generalization for a wide variety of learning algorithms
and are particularly well suited to analyze kernel machines, where the use of
more classical methods relying on covering numbers becomes cumbersome.

To briefly describe the use of Gaussian averages (Rademacher averages will
not concern us), let $Y \subseteq \mathbb{R}^n$ and let γ be a vector $\gamma = (\gamma_1, ..., \gamma_n)$ of independent
standard normal variables. We define the (expected supremum of the) Gaussian
average of Y as

$$G(Y) = \mathbb{E} \sup_{\mathbf{y} \in Y} \langle \gamma, \mathbf{y} \rangle, \tag{1}$$

where $\langle ., . \rangle$ denotes the inner product in $\mathbb{R}^n$. Consider a loss class $\mathcal{F}$ of functions
$f : \mathcal{X} \to \mathbb{R}$, where $\mathcal{X}$ is some space of examples (such as input-output pairs), a
sample $\mathbf{x} = (x_1, ..., x_n) \in \mathcal{X}^n$ of observations and write $\mathcal{F}(\mathbf{x})$ for the subset of
$\mathbb{R}^n$ given by $\mathcal{F}(\mathbf{x}) = \{(f(x_1), ..., f(x_n)) : f \in \mathcal{F}\}$. Then we have the following
result [1].

Theorem 1. *Let the members of $\mathcal{F}$ take values in $[0, 1]$ and let $X, X_1, ..., X_n$ be
iid random variables with values in $\mathcal{X}$, $\mathbf{X} = (X_1, ..., X_n)$. Then for $\delta > 0$ with
probability at least $1 - \delta$ we have for every $f \in \mathcal{F}$ that*

$$\mathbb{E}f(X) \leq \frac{1}{n} \sum f(X_i) + \frac{\sqrt{2\pi}}{n} G(\mathcal{F}(\mathbf{X})) + \sqrt{\frac{9 \ln 2/\delta}{2n}},$$

*where the expectation in the definition (1) of $G(\mathcal{F}(\mathbf{X}))$ is conditional to the
sample $\mathbf{X}$.*

P. Auer et al. (Eds.): ALT 2014, LNAI 8776, pp. 245–259, 2014.

The utility of Gaussian averages is not limited to functions with values in $[0,1]$. For real functions ϕ with Lipschitz constant $L(\phi)$ we have $G((\phi \circ \mathcal{F})(\mathbf{x}))$ $\leq L(\phi) \ G(\mathcal{F}(\mathbf{x}))$ (see also Slepian's Lemma, [6], [4]), where $\phi \circ \mathcal{F}$ is the class $\{x \mapsto \phi(f(x)) : f \in \mathcal{F}\}$.

The inequality $G((\phi \circ \mathcal{F})(\mathbf{x})) \leq L(\phi) \ G(\mathcal{F}(\mathbf{x}))$, which in the above form holds also for Rademacher averages [10], is extremely useful and in part responsible for the success of these complexity measures. For Gaussian averages it holds in a more general sense: if $\phi : \mathbb{R}^n \to \mathbb{R}^m$ has Lipschitz constant $L(\phi)$ with respect to the Euclidean distances, then $G(\phi(Y)) \leq L(\phi)G(Y)$. This is a direct consequence of Slepian's Lemma and can be applied to the analysis of clustering or learning to learn ([9] and [8]).

But what if we also want some freedom in the choice of ϕ *after* seeing the data? If the class of Lipschitz functions considered has small cardinality, a union bound can be used. If it is very large one can try to use covering numbers, but the matter soon becomes quite complicated and destroys the elegant simplicity of the method.

These considerations lead to a more general question: given a set $Y \subset \mathbb{R}^n$ and a class $\mathcal{F}$ of functions $f : \mathbb{R}^n \to \mathbb{R}^m$, how can we bound the Gaussian average $G(\mathcal{F}(Y)) = G(\{f(y) : f \in \mathcal{F}, y \in Y\})$ in terms of separate properties of Y and $\mathcal{F}$, properties which should preferably very closely resemble Gaussian averages? If $\mathcal{H}$ is some class of functions mapping samples to $\mathbb{R}^n$ and $Y = \mathcal{H}(\mathbf{x})$, then the bound is on $G(\mathcal{F}(Y)) = G((\mathcal{F} \circ \mathcal{H})(\mathbf{x}))$, so our question is relevant to the estimation of composite functions in general. Such estimates are necessary for multitask feature-learning, where $\mathcal{H}$ is a class of feature maps and $\mathcal{F}$ is vector-valued, with components chosen independently for each task. Other potential applications are to the currently popular subject of deep learning, where we consider functional concatenations as in $\mathcal{F}_M \circ \mathcal{F}_{M-1} \circ ... \circ \mathcal{F}_1$.

The present paper gives a preliminary answer. To state it we introduce some notation. We will always take $\gamma = (\gamma_1, ...)$ to be a random vector whose components are independent standard normal variables, while $\|.\|$ and $\langle ., . \rangle$ denote norm and inner product in a Euclidean space, the dimension of which is determined by context, as is the dimension of the vector γ.

Definition 1. *If $Y \subseteq \mathbb{R}^n$ we set*

$$D(Y) = \sup_{\mathbf{y}, \mathbf{y}' \in Y} \|\mathbf{y} - \mathbf{y}'\| \ and \ G(Y) = \mathbb{E} \sup_{\mathbf{y} \in Y} \langle \gamma, \mathbf{y} \rangle.$$

If $\mathcal{F}$ is a class of functions $f : Y \to \mathbb{R}^m$ we set

$$L(\mathcal{F}, Y) = \sup_{\mathbf{y}, \mathbf{y}' \in Y, \ \mathbf{y} \neq \mathbf{y}'} \sup_{f \in \mathcal{F}} \frac{\|f(\mathbf{y}) - f(\mathbf{y}')\|}{\|\mathbf{y} - \mathbf{y}'\|} \ and$$

$$R(\mathcal{F}, Y) = \sup_{\mathbf{y}, \mathbf{y}' \in Y, \ \mathbf{y} \neq \mathbf{y}'} \mathbb{E} \sup_{f \in \mathcal{F}} \frac{\langle \gamma, f(\mathbf{y}) - f(\mathbf{y}') \rangle}{\|\mathbf{y} - \mathbf{y}'\|}.$$

We also write $\mathcal{F}(Y) = \{f(\mathbf{y}) : f \in \mathcal{F}, y \in Y\}$. When there is no ambiguity we write $L(\mathcal{F}) = L(\mathcal{F}, Y)$ and $R(\mathcal{F}) = R(\mathcal{F}, Y)$.

Then $D(Y)$ is the diameter of Y, and $G(Y)$ is the Gaussian average already introduced above. $L(\mathcal{F})$ is the smallest Lipschitz constant acceptable for all $f \in \mathcal{F}$, and the more unusual quantity $R(\mathcal{F})$ can be viewed as a Gaussian average of Lipschitz quotients. In section 3.1 we give some properties of $R(\mathcal{F})$. Our main result is the following chain rule.

Theorem 2. *Let* $Y \subset \mathbb{R}^n$ *be finite,* $\mathcal{F}$ *a finite class of functions* $f : Y \to \mathbb{R}^m$. *Then there are universal constants* C_1 *and* C_2 *such that for any* $\mathbf{y}_0 \in Y$

$$G(\mathcal{F}(Y)) \leq C_1 L(\mathcal{F}) G(Y) + C_2 D(Y) R(\mathcal{F}) + G(\mathcal{F}(\mathbf{y}_0)). \tag{2}$$

We make some general remarks on the implications of our result.

1. The requirement of finiteness for Y and $\mathcal{F}$ is a simplification to avoid issues of measurability. The cardinality of these sets plays no role.

2. The constants C_1 and C_2 as they result from the proof are rather large, because they accumulate the constants of Talagrand's majorizing measure theorem and generic chaining [6][14][15][16]. This is a major shortcoming and the reason why our result is regarded as preliminary. Is there another proof of a similar result, avoiding majorizing measures and resulting in smaller constants? This question is the subject of current research.

3. The first term on the right hand side of (2) describes the complexity inherited from the bottom layer Y (which we may think of as $\mathcal{H}(\mathbf{x})$), and it depends on the top layer $\mathcal{F}$ only through the Lipschitz constant $L(\mathcal{F})$. The other two terms represent the complexity of the top layer, depending on the bottom layer only through the diameter $D(Y)$ of Y. If Y has unit diameter and the functions in $\mathcal{F}$ are contractions, then the two layers are completely decoupled in the bound. This decoupling is the most attractive property of our result.

4. Apart from the large constants the inequality is tight in at least two situations: first, if $Y = \{\mathbf{y}_0\}$ is a singleton, then only the last term remains, and we recover the Gaussian average of $\mathcal{F}(\mathbf{y}_0)$. This also shows that the last term cannot be eliminated. On the other hand if $\mathcal{F}$ consists of a single Lipschitz function ϕ, then we recover (up to a constant) the inequality $G(\phi(Y)) \leq L(\phi) G(Y)$ above.

5. The bound can be iterated to multiple layers by re-substitution of $\mathcal{F}(Y)$ in place of Y. A corresponding formula is given in Section 3, where we also sketch applications to vector-valued function classes.

The next section gives a proof of Theorem 2, then we explain how our result can be applied to machine learning. The last section is devoted to the proof of a technical result encapsulating our use of majorizing measures.

2   Proving the Chain Rule

To prove Theorem 2 we need the theory of majorizing measures and generic chaining. Our use of these techniques is summarized in the following theorem, which is also the origin of our large constants.

Theorem 3. *Let $X_{\mathbf{y}}$ be a random process indexed by a finite set $Y \subset \mathbb{R}^n$. Suppose that there is a number $K \geq 1$ such that for any distinct members $\mathbf{y}, \mathbf{y}' \in Y$ and any $s > 0$*

$$\Pr\{X_{\mathbf{y}} - X_{\mathbf{y}'} > s\} \leq K \exp\left(\frac{-s^2}{2\|\mathbf{y} - \mathbf{y}'\|^2}\right) \tag{3}$$

Then for any $\mathbf{y}_0 \in Y$

$$\mathbb{E}\left[\sup_{\mathbf{y} \in Y} X_{\mathbf{y}} - X_{\mathbf{y}_0}\right] \leq C'G(Y) + C''D(Y)\sqrt{\ln K},$$

where C' and C'' are universal constants.

This is obtained from Talagrand's majorizing measure theorem (Theorem 6 below) combined with generic chaining [16]. An early version of a similar result is Theorem 15 in [13], where the author remarks that his method of proof (which we also use) is very indirect, and that a more direct proof would be desirable. In Section 4 we do supply a proof, largely because the dependence on K, which can often be swept under the carpet, plays a crucial role in our arguments below.

We also need the following Gaussian concentration inequality (Tsirelson-Ibragimov-Sudakov inequality, Theorem 5.6 in [4]).

Theorem 4. *Let $F : \mathbb{R}^n \to \mathbb{R}$ be L-Lipschitz. Then for any $s > 0$*

$$\Pr\{F(\gamma) > \mathbb{E}F(\gamma) + s\} \leq e^{-s^2/(2L^2)}.$$

To conclude the preparation for the proof of Theorem 2 we give a simple lemma.

Lemma 1. *Suppose a random variable X satisfies $\Pr\{X - A > s\} \leq e^{-s^2}$, for any $s > 0$. Then*

$$\forall s > 0, \quad \Pr\{X > s\} \leq e^{A^2}e^{-s^2/2}.$$

Proof. For $s \leq A$ the conclusion is trivial, so suppose that $s > A$. From $s^2 = (s - A + A)^2 \leq 2(s - A)^2 + 2A^2$ we get $(s - A)^2 \geq (s^2/2) - A^2$, so

$$\Pr\{X > s\} = \Pr\{X - A > s - A\} \leq e^{-(s-A)^2} \leq e^{A^2}e^{-s^2/2}.$$

■

Proof (of Theorem 2). The result is trivial if $\mathcal{F}$ consists only of constants, so we can assume that $L(\mathcal{F}) > 0$. For $\mathbf{y}, \mathbf{y}' \in Y$ define a function $F : \mathbb{R}^m \to \mathbb{R}$ by

$$F(\mathbf{z}) = \sup_{f \in \mathcal{F}} \langle \mathbf{z}, f(\mathbf{y}) - f(\mathbf{y}') \rangle.$$

F is Lipschitz with Lipschitz constant bounded by $\sup_{f \in \mathcal{F}} \|f(\mathbf{y}) - f(\mathbf{y}')\| \le L(\mathcal{F}) \|\mathbf{y} - \mathbf{y}'\|$. Writing $Z_{\mathbf{y},\mathbf{y}'} = F(\boldsymbol{\gamma})$, it then follows from Gaussian concentration (Theorem 4) that

$$\Pr\{Z_{\mathbf{y},\mathbf{y}'} > \mathbb{E} Z_{\mathbf{y},\mathbf{y}'} + s\} \le \exp\left(\frac{-s^2}{2L(\mathcal{F})^2 \|\mathbf{y} - \mathbf{y}'\|^2}\right).$$

Since by definition $\mathbb{E} Z_{\mathbf{y},\mathbf{y}'} \le R(\mathcal{F}) \|\mathbf{y} - \mathbf{y}'\|$, Lemma 1 gives

$$\Pr\{Z_{\mathbf{y},\mathbf{y}'} > s\} \le \exp\left(\frac{R(\mathcal{F})^2}{2L(\mathcal{F})^2}\right) \exp\left(\frac{-s^2}{4L(\mathcal{F})^2 \|\mathbf{y} - \mathbf{y}'\|^2}\right).$$

Now define a process $X_{\mathbf{y}}$, indexed by Y, as

$$X_{\mathbf{y}} = \frac{1}{\sqrt{2} L(\mathcal{F})} \sup_{f \in \mathcal{F}} \langle \boldsymbol{\gamma}, f(\mathbf{y}) \rangle.$$

Since $X_{\mathbf{y}} - X_{\mathbf{y}'} \le Z_{\mathbf{y},\mathbf{y}'} / (\sqrt{2} L(\mathcal{F}))$ we have

$$\Pr\{X_{\mathbf{y}} - X_{\mathbf{y}'} > s\} \le \Pr\left\{Z_{\mathbf{y},\mathbf{y}'} > \sqrt{2} L(\mathcal{F}) s\right\}$$

$$\le \exp\left(\frac{R(\mathcal{F})^2}{2L(\mathcal{F})^2}\right) \exp\left(\frac{-s^2}{2\|\mathbf{y} - \mathbf{y}'\|^2}\right)$$

and by Theorem 3, with $K = \exp\left(R(\mathcal{F})^2 / \left(2L(\mathcal{F})^2\right)\right) \ge 1$,

$$\mathbb{E} \sup_{\mathbf{y} \in Y} (X_{\mathbf{y}} - X_{\mathbf{y}_0}) \le C' G(Y) + C'' D(Y) \frac{R(\mathcal{F})}{\sqrt{2} L(\mathcal{F})}.$$

Multiplication by $\sqrt{2} L(\mathcal{F})$ then gives

$$\mathbb{E} \sup_{\mathbf{y} \in Y} \left(\sup_{f \in \mathcal{F}} \langle \boldsymbol{\gamma}, f(\mathbf{y}) \rangle - \sup_{f \in \mathcal{F}} \langle \boldsymbol{\gamma}, f(\mathbf{y}_0) \rangle\right) \le C_1 L(\mathcal{F}) G(Y) + C_2 D(Y) R(\mathcal{F})$$

with $C_1 = \sqrt{2} C'$ and $C_2 = C''$. ∎

3 Applications

We first give some elementary properties of the quantity $R(\mathcal{F}, Y)$ which appears in Theorem 2. Then we apply Theorem 2 to a two layer kernel machine and give a bound for multi-task learning of low-dimensional representations.

3.1   Some Properties of $R(\mathcal{F})$

Recall the definition of $R(\mathcal{F},Y)$. If $Y \subseteq \mathbb{R}^n$ and $\mathcal{F}$ consists of functions $f : Y \to \mathbb{R}^m$

$$R(\mathcal{F},Y) = \sup_{\mathbf{y},\mathbf{y}'\in Y,\ \mathbf{y}\neq\mathbf{y}'} \mathbb{E} \sup_{f\in\mathcal{F}} \frac{\langle \gamma, f(\mathbf{y}) - f(\mathbf{y}')\rangle}{\|\mathbf{y}-\mathbf{y}'\|}.$$

$R(\mathcal{F})$ is itself a supremum of Gaussian averages. For $\mathbf{y},\mathbf{y}' \in Y$ let $\Delta\mathcal{F}(\mathbf{y},\mathbf{y}') \subseteq \mathbb{R}^m$ be the set of quotients

$$\Delta\mathcal{F}(\mathbf{y},\mathbf{y}') = \left\{ \frac{f(\mathbf{y}) - f(\mathbf{y}')}{\|\mathbf{y}-\mathbf{y}'\|} : f \in \mathcal{F} \right\}.$$

It follows from the definition that $R(\mathcal{F},Y) = \sup_{\mathbf{y},\mathbf{y}'\in Y,\ \mathbf{y}\neq\mathbf{y}'} G(\Delta\mathcal{F}(\mathbf{y},\mathbf{y}'))$. We record some simple properties. Recall that for a set S in a real vector space the convex hull $Co(S)$ is defined as

$$Co(S) = \left\{ \sum_{i=1}^{n} \alpha_i z_i : n \in \mathbb{N}, z_i \in S, \alpha_i \geq 0, \sum_i \alpha_i = 1 \right\}.$$

Theorem 5. *Let $Y \subseteq \mathbb{R}^n$ and let $\mathcal{F}$ and $\mathcal{H}$ be classes of functions $f : Y \to \mathbb{R}^m$. Then*

(i) If $\mathcal{F} \subseteq \mathcal{H}$ then $R(\mathcal{F},Y) \leq R(\mathcal{H},Y)$.
(ii) If $Y \subseteq Y'$ then $R(\mathcal{F},Y) \leq R(\mathcal{F},Y')$.
(iii) If $c \geq 0$ then $R(c\mathcal{F},Y) = cR(\mathcal{F},Y)$.
(iv) $R(\mathcal{F}+\mathcal{H},Y) \leq R(\mathcal{F},Y) + R(\mathcal{H},Y)$.
(v) $R(\mathcal{F},Y) = R(Co(\mathcal{F}),Y)$.
(vi) If $Z \subseteq \mathbb{R}^K$ and $\phi : Z \to \mathbb{R}^n$ has Lipschitz constant $L(\phi)$ and the members of $\mathcal{F}$ are defined on $\phi(Z)$, then $R(\mathcal{F}\circ\phi,Z) \leq L(\phi)R(\mathcal{F},\phi(Z))$.
(vii) $R(\mathcal{F}) \leq L(\mathcal{F})\sqrt{2\ln|\mathcal{F}|}$.

Remarks:
1. From (ii) we get $R(\mathcal{F},Y) \leq R(\mathcal{F},\mathbb{R}^n)$. In applications where $Y = \mathcal{H}(\mathbf{x})$ the quantity $R(\mathcal{F},\mathcal{H}(\mathbf{x}))$ is data-dependent, but $R(\mathcal{F},\mathbb{R}^n)$ is sometimes easier to bound.
2. We see that the properties of $R(\mathcal{F})$ largely parallel the properties of the Gaussian averages themselves, except for the inequality $G(\phi(Y)) \leq L(\phi)G(Y)$, for which there doesn't seem to be an analogous property of $R(\mathcal{F})$. Instead we have a 'backwards' version of it with (vi) above, with a rather trivial proof below.
3. Of course (vii) is relevant only when $\ln|\mathcal{F}|$ is reasonably small and serves the comparison of Theorem 2 to alternative bounds.

Proof. (i)-(iii) are obvious from the definition. (iv) follows from linearity of the inner product and the triangle inequality for the supremum. To see (v) first note that $R(\mathcal{F}) \leq R(Co(\mathcal{F}))$ follows from (i), while the reverse inequality follows from

$$\sup_{\alpha_i \geq 0, \sum \alpha_i = 1} \sup_{f_1, f_2, \ldots \in \mathcal{F}} \left\langle \gamma, \sum_i \alpha_i f_i(\mathbf{y}) - \sum_i \alpha_i f_i(\mathbf{y'}) \right\rangle$$

$$= \sup_{\alpha_i \geq 0, \sum \alpha_i = 1} \sup_{f_1, f_2, \ldots \in \mathcal{F}} \sum_i \alpha_i \left\langle \gamma, f_i(\mathbf{y}) - f_i(\mathbf{y'}) \right\rangle$$

$$\leq \sup_{\alpha_i \geq 0, \sum \alpha_i = 1} \sum_i \alpha_i \sup_{f \in \mathcal{F}} \left\langle \gamma, f(\mathbf{y}) - f(\mathbf{y'}) \right\rangle$$

$$= \sup_{f \in \mathcal{F}} \left\langle \gamma, f(\mathbf{y}) - f(\mathbf{y'}) \right\rangle.$$

For (vi) we may chose $\mathbf{y}$ and $\mathbf{y'}$ such that $\phi(\mathbf{y}) \neq \phi(\mathbf{y'})$, since otherwise both sides of the inequality to be proved are zero. But then

$$\mathbb{E} \sup_{f \in \mathcal{F} \circ \phi} \frac{\langle \gamma, f(\mathbf{y}) - f(\mathbf{y'}) \rangle}{\|\mathbf{y} - \mathbf{y'}\|} = \frac{\|\phi(\mathbf{y}) - \phi(\mathbf{y'})\|}{\|\mathbf{y} - \mathbf{y'}\|} \mathbb{E} \sup_{f \in \mathcal{F}} \frac{\langle \gamma, f(\phi(\mathbf{y})) - f(\phi(\mathbf{y'})) \rangle}{\|\phi(\mathbf{y}) - \phi(\mathbf{y'})\|}$$

$$\leq L(\phi) \mathbb{E} \sup_{f \in \mathcal{F}} \frac{\langle \gamma, f(\phi(\mathbf{y})) - f(\phi(\mathbf{y'})) \rangle}{\|\phi(\mathbf{y}) - \phi(\mathbf{y'})\|}.$$

To see (vii) note that for every $\mathbf{y}$ and $\mathbf{y'}$ and every $f \in \mathcal{F}$ it follows from Gaussian concentration (Theorem 4) that

$$\Pr \left\{ \frac{\langle \gamma, f(\mathbf{y}) - f(\mathbf{y'}) \rangle}{\|\mathbf{y} - \mathbf{y'}\|} > s \right\} \leq e^{-s^2/2L^2}.$$

The conclusion then follows from standard estimates (e.g. [4], section 2.5). ∎

3.2 A Double Layer Kernel Machine

We use the chain rule to bound the complexity of a double-layer kernel machine. The corresponding optimization problem is clearly non-convex and we are not aware of an efficient optimization method. The model is chosen to illustrate the application of Theorem 2. It is defined as follows.

Assume the data to lie in $\mathbb{R}^{m_0}$ and fix two real numbers Δ_1 and B_1. On $\mathbb{R}^{m_0} \times \mathbb{R}^{m_0}$ define a (Gaussian radial-basis-function) kernel κ by

$$\kappa(z, z') = \exp\left(\frac{-\|z - z'\|^2}{2\Delta_1^2} \right), \quad z, z' \in \mathbb{R}^{m_0},$$

and let $\phi : \mathbb{R}^{m_0} \to H$ be the associated feature map, where H is the associated RKHS with inner product $\langle ., . \rangle_H$ and norm $\|.\|_H$ (for kernel methods see . Now we let $\mathcal{H}$ be the class of vector valued functions $h : \mathbb{R}^{m_0} \to \mathbb{R}^{m_1}$ defined by

$$\mathcal{H} = \left\{ z \in \mathbb{R}^{m_0} \mapsto (\langle w_1, \phi(z) \rangle_H, \ldots, \langle w_{m_1}, \phi(z) \rangle_H) : \sum_k \|w_k\|_H^2 \leq B_1^2 \right\}.$$

This can also be written as $\mathcal{H} = \{z \in \mathbb{R}^{m_0} \mapsto W\phi(z) : \|W\|_{HS} \leq B_1\}$, where $\|W\|_{HS}$ is the Hilbert-Schmidt norm of an operator $W : H \to \mathbb{R}^{m_1}$.

For the function class $\mathcal{F}$, which we wish to compose with $\mathcal{H}$, we proceed in a similar way, defining an analogous kernel of width Δ_2 on $\mathbb{R}^{m_1} \times \mathbb{R}^{m_1}$, a corresponding feature map $\psi : \mathbb{R}^{m_1} \to H$ and a class of real valued functions

$$\mathcal{F} = \{z \in \mathbb{R}^{m_1} \mapsto \langle v, \psi(z) \rangle_H : \|v_l\|_H \le B_2\}.$$

We now want high probability bounds on the estimation error for functional compositions $f \circ h$, uniform over $\mathcal{F} \circ \mathcal{H}$. To apply our result we should really restrict to finite subsets of $\mathcal{F}$ and $\mathcal{H}$ a requirement which we simply ignore. In machine learning we could of course always restrict all representations to some fixed, very high but finite precision.

Fix a sample $\mathbf{x} \in \mathbb{R}^{nm_0}$. Then $Y = \mathcal{H}(\mathbf{x}) \subset \mathbb{R}^{nm_1}$. To use Theorem 2 we define a class $\mathcal{F}'$ of functions from $\mathbb{R}^{nm_1}$ to $\mathbb{R}^n$ by

$$\mathcal{F}' = \{(y_1, ..., y_n) \in \mathbb{R}^{nm_1} \mapsto (f(y_1), ..., f(y_n)) \in \mathbb{R}^n : f \in \mathcal{F}\}.$$

Since the first feature map ϕ maps to the unit sphere of H we have

$$D(\mathcal{H}(\mathbf{x})) \le 2B_1\sqrt{n} \text{ and}$$
$$G(\mathcal{H}(\mathbf{x})) = \mathbb{E}\sup_W \sum_{ik} \gamma_{ik} \langle w_k, \phi(x_i) \rangle_H \le B_1\sqrt{nm_1}.$$

The feature map corresponding to the Gaussian kernel Δ_2 has Lipschitz constant Δ_2^{-1}. For $\mathbf{y}, \mathbf{y}' \in \mathbb{R}^{nm_1}$ we obtain

$$\sup_v \left(\sum_i \left(\langle v, \phi(y_i) \rangle_H - \langle v, \phi(y_i') \rangle_H \right)^2 \right)^{1/2} \le B_2 \left(\sum_i \|\phi(y_i) - \phi(y_i')\|_H^2 \right)^{1/2}$$
$$\le B_2\Delta_2^{-1} \|\mathbf{y} - \mathbf{y}'\|,$$

so we have $L(\mathcal{F}', \mathbb{R}^{nm_1}) \le B_2\Delta_2^{-1}$.

On the other hand

$$\mathbb{E}\sup_v \sum_i \gamma_i \left(\langle v, \phi(y_i) \rangle_H - \langle v, \phi(y_i') \rangle_H \right) \le B_2\mathbb{E}\left\| \sum_{i=1}^n \gamma_i \left(\phi(\mathbf{y}_i) - \phi(\mathbf{y}_i') \right) \right\|$$
$$\le B_2 \left(\sum_i \|\phi(y_i) - \phi(y_i')\|_H^2 \right)^{1/2}$$
$$\le B_2\Delta_2^{-1} \|\mathbf{y} - \mathbf{y}'\|,$$

so we have $R(\mathcal{F}', \mathbb{R}^{nm_1}) \le B_2\Delta_2^{-1}$. Furthermore

$$G(\mathcal{F}'(h_0(\mathbf{x}))) \le B_2\sqrt{n},$$

similar to the bound for $G(\mathcal{H}(\mathbf{x}))$.

For the composite network Theorem 2 gives us the bound

$$G(\mathcal{F}'(\mathcal{H}(\mathbf{x}))) \le C_1 B_1 B_2 \Delta_2^{-1} \sqrt{nm_1} + 2C_2 B_1 B_2 \sqrt{n}\Delta_2^{-1} + B_2\sqrt{n}.$$

Dividing by n and appealing to Theorem 1 one obtains the uniform bound: with probability at least $1 - \delta$ we have for every $h \in \mathcal{H}$ and every $f \in \mathcal{F}$ that

$$\mathbb{E}f\left(h\left(X\right)\right) \leq \frac{1}{n}\sum f\left(h\left(X_i\right)\right) +$$

$$+\sqrt{\frac{2\pi}{n}}B_2\left(B_1\Delta_2^{-1}\left(C_1\sqrt{m_1} + 2C_2\right) + 1\right) + \sqrt{\frac{9\ln 2/\delta}{2n}}.$$

Remarks.

1. One might object that the result depends heavily on the intermediate dimension m_1 so that only a very classical relationship between sample size and dimension is obtained. In this sense our result only works for intermediate representations of rather low dimension. The mapping stages of $\mathcal{H}$ and $\mathcal{F}$ however include nonlinear maps to infinite dimensional spaces.

2. Clearly the above choice of the Gaussian kernel is arbitrary. Any positive semidefinite kernel can be used for the first mapping stage, and the application of the chain rule requires only the Lipschitz property for the second kernel in the definition of $\mathcal{F}$. The Gaussian kernel was only chosen for definiteness.

3. Similarly the choice of the Hilbert-Schmidt norm as a regularizer for W in the first mapping stage is arbitrary, one could equally use another matrix norm. This would result in different bounds for $G\left(\mathcal{H}\left(\mathbf{x}\right)\right)$ and $D\left(\mathcal{H}\left(\mathbf{x}\right)\right)$, incurring a different dependency of our bound on m_1.

3.3   Multitask Learning

As a second illustration we modify the above model to accommodate multitask learning [2][3]. Here one observes a $T \times n$ sample $\mathbf{x} = (x_{ti} : 1 \leq t \leq T, 1 \leq i \leq n) \in \mathcal{X}^{nT}$, where $(x_{ti} : 1 \leq i \leq n)$ is the sample observed for the t-th task. We consider a two layer situation where the bottom-layer $\mathcal{H}$ consists of functions $h : \mathcal{X} \to \mathbb{R}^m$, and the top layer function class is of the form

$$\mathcal{F}^T = \left\{x \in \mathbb{R}^{m_1} \mapsto \mathbf{f}\left(x\right) = \left(f_1\left(x\right), ..., f_T\left(x\right)\right) \in \mathbb{R}^T : f_t \in \mathcal{F}\right\},$$

where $\mathcal{F}$ is some class of functions mapping $\mathbb{R}^{m_1}$ to $\mathbb{R}$. The functions (or representations) of the bottom layer $\mathcal{H}$ are optimized for the entire sample, in the top layer each function f_t is optimized for the represented data corresponding to the t-th task. In an approach of empirical risk minimization one selects the composed function $\hat{\mathbf{f}} \circ \hat{h}$ which minimizes the task-averaged empirical loss

$$\min_{\mathbf{f} \in \mathcal{F}^n, h \in \mathcal{H}} \frac{1}{nT} \sum_{i=1}^{n} \sum_{t=1}^{T} f_t\left(h\left(x_{it}\right)\right).$$

We wish to give a general explanation of the potential benefits of this method over the separate learning of functions from $\mathcal{F} \circ \mathcal{H}$, as studied in the previous section. Clearly we must assume that the tasks are related in the sense that the above minimum is small, so any possible benefit can only be a benefit of improved estimation.

For the multitask model a result analogous to Theorem 1 is easily obtained (see e.g. [7]). Let $\mathbf{X} = (X_{ti})$ be a vector of independent random variables with values in $\mathcal{X}$, where X_{ti} is iid to X_{tj} for all ijt, and let X_t be iid to X_{ti}. Then with probability at least $1 - \delta$ we have for every $\mathbf{f} \in \mathcal{F}^n$ and every $h \in \mathcal{H}$

$$\frac{1}{T} \sum_t \mathbb{E} f_t \left(h \left(X_t \right) \right) \leq \frac{1}{nT} \sum_{ti} f_t \left(h \left(X_{ti} \right) \right) + \frac{\sqrt{2\pi}}{nT} G \left(\mathcal{F}^T \circ \mathcal{H} \left(\mathbf{X} \right) \right) + \sqrt{\frac{9 \ln 2/\delta}{2nT}}.$$

Here the left hand side is interpreted as the task averaged risk and

$$G \left(\mathcal{F}^T \circ \mathcal{H} \left(\mathbf{x} \right) \right) = \mathbb{E} \sup_{\mathbf{f} \in \mathcal{F}^T, h \in \mathcal{H}} \sum_{ti} \gamma_{ti} f_t \left(h \left(x_{ti} \right) \right).$$

For a definite example we take $\mathcal{H}$ and $\mathcal{F}$ as in the previous section and observe that now there is an additional factor T on the sample size. This implies the modified bounds $G \left(\mathcal{H} \left(\mathbf{x} \right) \right) \leq B_1 \sqrt{Tnm_1}$ and $D \left(\mathcal{H} \left(\mathbf{x} \right) \right) \leq 2B_1 \sqrt{Tn}$. Also for $\mathbf{y}, \mathbf{y}' \in \mathbb{R}^{Tnm_1}$ with $y_{ti}, y'_{ti} \in \mathbb{R}^{m_1}$

$$\sup_{\mathbf{f} \in \mathcal{F}^T} \sum_{ti} \left(f_t \left(y_{ti} \right) - f_t \left(y'_{ti} \right) \right)^2 \leq \sum_t \sup_{\mathbf{f} \in \mathcal{F}} \sum_i \left(f_t \left(y_{ti} \right) - f_t \left(y'_{ti} \right) \right)^2$$

$$\leq L^2 \left(\mathcal{F}, \mathbb{R}^{nm_1} \right) \sum_t \sum_i \left\| y_{ti} - y'_{ti} \right\|^2,$$

so

$$L \left(\mathcal{F}^T, \mathbb{R}^{Tnm_1} \right) = L \left(\mathcal{F}, \mathbb{R}^{nm_1} \right). \tag{4}$$

Therefore $L \left(\mathcal{F}^T, \mathbb{R}^{Tnm_1} \right) \leq B_2 \Delta_2^{-1}$. Similarly

$$\mathbb{E} \sup_{\mathbf{f} \in \mathcal{F}^T} \sum_{ti} \gamma_{ti} \left(f_t \left(y_{ti} \right) - f_t \left(y'_{ti} \right) \right)$$

$$= \sum_t \mathbb{E} \sup_{\mathbf{f} \in \mathcal{F}} \sum_i \gamma_{ti} \left(f_t \left(y_{ti} \right) - f_t \left(y'_{ti} \right) \right)$$

$$\leq \sqrt{T} \left(\sum_t \left(\mathbb{E} \sup_{\mathbf{f} \in \mathcal{F}} \sum_i \gamma_{ti} \left(f_t \left(y_{ti} \right) - f_t \left(y'_{ti} \right) \right) \right)^2 \right)^{1/2}$$

$$\leq \sqrt{T} \left(\sum_t R^2 \left(\mathcal{F}, \mathbb{R}^{nm_1} \right) \sum_i \left\| y_{ti} - y'_{ti} \right\|^2 \right)^{1/2}$$

$$= \sqrt{T} R \left(\mathcal{F}, \mathbb{R}^{nm_1} \right) \left\| \mathbf{y} - \mathbf{y}' \right\|.$$

We conclude that
$$R \left(\mathcal{F}^T, \mathbb{R}^{nmT} \right) \leq \sqrt{T} R \left(\mathcal{F}, \mathbb{R}^{nm} \right), \tag{5}$$

in the given case
$$R \left(\mathcal{F}^T, \mathbb{R}^{nmT} \right) \leq \sqrt{T} B_2 \Delta_2^{-1}.$$

Also

$$G\left(\mathcal{F}^T\left(h_0\left(\mathbf{x}\right)\right)\right) = \mathbb{E} \sup_{\mathbf{f} \in \mathcal{F}^T} \sum_{ti} \gamma_{ti} f_t\left(h_0\left(x_{ti}\right)\right)$$

$$= \sum_t \mathbb{E} \sup_{f \in \mathcal{F}} \sum_i \gamma_{ti} f\left(h_0\left(x_{ti}\right)\right)$$

$$\leq T G\left(\mathcal{F}\left(h_0\left(\mathbf{x}\right)\right)\right), \tag{6}$$

so that here $G\left(\mathcal{F}^T\left(h_0\left(\mathbf{x}\right)\right)\right) \leq B_2 T \sqrt{n}$. The chain rule then gives

$$G\left(\mathcal{F} \circ \mathcal{H}\left(\mathbf{x}\right)\right) \leq C_1 B_1 B_2 \Delta_2^{-1} \sqrt{Tnm_1} + \left(2C_2 B_1 \Delta_2^{-1} + 1\right) B_2 T \sqrt{n},$$

where the first term represents the complexity of $\mathcal{H}$ and the second that of $\mathcal{F}^T$. Dividing by nT we obtain as the dominant term for the estimation error

$$C_1 B_1 B_2 \Delta_2^{-1} \sqrt{\frac{m_1}{nT}} + \frac{\left(2C_2 B_1 \Delta_2^{-1} + 1\right) B_2}{\sqrt{n}}.$$

This reproduces a general property of multitask learning [3]: in the limit $T \to \infty$ the contribution of the common representation (including the intermediate dimension m_1) to the estimation error vanishes. There remains only the cost of estimating the task specific functions in the top layer.

We have obtained this result for a very specific model. The relations (4), (5) and (6) for $L\left(\mathcal{F}^T\right)$, $R\left(\mathcal{F}^T\right)$ and $G\left(\mathcal{F}^T\left(h_0\left(\mathbf{x}\right)\right)\right)$ are nevertheless independent of the exact model, so the chain rule could be made the basis of a fairly general result about multitask feature learning.

3.4   Iteration of the Bound

We apply the chain rule to multi-layered or "deep" learning machines, a subject which appears to be of some current interest. Here we have function classes $\mathcal{F}_1, ..., \mathcal{F}_K$, where $\mathcal{F}_k$ consists of functions $f : \mathbb{R}^{n_{k-1}} \to \mathbb{R}^{n_k}$ and we are interested in the generalization properties of the composite class

$$\mathcal{F}_K \circ ... \circ \mathcal{F}_1 = \{\mathbf{x} \in \mathbb{R}^{n_0} \mapsto f_K\left(f_{K-1}\left(...\left(f_1\left(\mathbf{x}\right)\right)\right)\right) : f_k \in \mathcal{F}_k\}.$$

To state our bound we are given some sample $\mathbf{x}$ in $\mathbb{R}^{n_0}$ and introduce the notation

$$Y_0 = \mathbf{x}$$
$$Y_k = \mathcal{F}_k\left(Y_{k-1}\right) = \mathcal{F}_k \circ ... \circ \mathcal{F}_1\left(\mathbf{x}\right) \subseteq \mathbb{R}^{n_k}, \text{ for } k > 0$$
$$G_k = \min_{\mathbf{y} \in Y_{k-1}(\mathbf{x})} G\left(\mathcal{F}_k\left(\mathbf{y}\right)\right).$$

Under the convention that the product over an empty index set is 1, induction shows that

$$G\left(Y_K\right) \leq \sum_{k=1}^{K} \left(C_1^{K-k} \prod_{j=k+1}^{K} L\left(\mathcal{F}_j\right)\right) \left(C_2 D\left(Y_{k-1}\right) R\left(\mathcal{F}_k\right) + G_k\right).$$

Clearly the large constants are prohibitive for any useful quantitative prediction of generalization, but qualitative statements are possible. Observe for example that, apart from C_1 and the Lipschitz constants, each layer only makes an additive contribution to the overall complexity. More specifically, for machine learning with a sample of size n, we can make the assumptions $n_k = nm_k$, where m_k is the dimension of the k-th intermediate representations, and it is reasonable to postulate $\max\{G_k, D(Y_k) R(\mathcal{F}_k)\} \le Cn^p$, where C is some constant not depending on n and p is some exponent $p < 1$ (for multi-layered kernel machines with Lipschitz feature maps we would have $p = 1/2$ - see above). Then the above expression is of order n^p and Theorem 1 yields a uniform law of large numbers for the multi-layered class, with a uniform bound on the estimation error decreasing as n^{p-1}.

4   Proof of Theorem 3

Talagrand has proved the following result ([14]).

Theorem 6. *There are universal constants $r \ge 2$ and C such that for every finite $Y \subset \mathbb{R}^n$ there is an increasing sequence of partitions $\mathcal{A}_k$ of Y and a probability measure μ on Y, such that, whenever $A \in \mathcal{A}_k$ then $D(A) \le 2r^{-k}$ and*

$$\sup_{y \in Y} \sum_{k > k_0}^{\infty} r^{-k} \sqrt{\ln \frac{1}{\mu(A_k(y))}} \le C\, G(Y),$$

where $A_k(y)$ denotes the unique member of $\mathcal{A}_k$ which contains y, and k_0 is the largest integer k satisfying

$$2r^{-k} \ge D(Y) = \sup_{y,y' \in Y} \|y - y'\|$$

Observe that $2r^{-k_0} \ge D(Y)$, so we can assume $\mathcal{A}_{k_0} = \{Y\}$. As explained in [14], the above Theorem is equivalent to the existence of a measure μ on Y such that

$$\sup_{y \in Y} \int_0^{\infty} \sqrt{\ln \frac{1}{\mu(B(y,\epsilon))}}\, d\epsilon \le C\, G(Y),$$

where C is some other universal constant and $B(y,\epsilon)$ is the ball of radius ϵ centered at y. The latter is perhaps the more usual formulation of the majorizing measure theorem.

We will use Talagrand's theorem to prove Theorem 3, but before please note the inequality

$$D(Y) \le \sqrt{2\pi} G(Y), \tag{7}$$

which follows from

$$\sup_{y,y' \in Y} \|y - y'\| = \sqrt{\frac{\pi}{2}} \sup_{y,y'} \mathbb{E}\,|\langle \gamma, y - y'\rangle|$$

$$\le \sqrt{\frac{\pi}{2}} \mathbb{E} \sup_{y,y'} |\langle \gamma, y - y'\rangle| = \sqrt{\frac{\pi}{2}} \mathbb{E} \sup_{y,y'} \langle \gamma, y - y'\rangle.$$

In the first equality we used the fact that $\|v\| = \sqrt{\pi/2}\,\mathbb{E}\,|\langle \gamma, v\rangle|$ for any vector v.

Proof (of Theorem 3.). Let μ and A_k be as determined for Y by Theorem 6. First we claim that for any $\delta \in (0,1)$

$$\Pr\left\{\exists \mathbf{y} \in Y : X_\mathbf{y} - X_{\mathbf{y}_0} > \sum_{k>k_0} r^{-k+1}\sqrt{8\ln\left(\frac{2^{k-k_0}K}{\mu\left(A\left(\mathbf{y}\right)\right)\delta}\right)}\right\} < \delta. \qquad (8)$$

For every $k > k_0$ and every $A \in A_k$ let $\pi(A)$ be some element chosen from A. We set $\pi(Y) = \mathbf{y}_0$. We denote $\pi_k(\mathbf{y}) = \pi(A_k(\mathbf{y}))$. This implies the chaining identity:

$$X_\mathbf{y} - X_{\mathbf{y}_0} = \sum_{k>k_0}\left(X_{\pi_k(\mathbf{y})} - X_{\pi_{k-1}(\mathbf{y})}\right).$$

For $k > k_0$ and $A \in A_k$ use $\hat{A}$ to denote the unique member of A_{k-1} such that $A \subseteq \hat{A}$. Since for $A \in A_k$ both $\pi(A)$ and $\pi\left(\hat{A}\right)$ are members of $\hat{A} \in A_{k-1}$ we must have $\left\|\pi(A) - \pi\left(\hat{A}\right)\right\| \le 2r^{-k+1}$. Also note $\pi_{k-1}(\mathbf{y}) = \pi\left(\hat{A}_k(\mathbf{y})\right) = \pi((A_k(\pi_k(\mathbf{y})))\char94)$. For $k \ge k_0$ we define a function $\xi_k : A_k \to \mathbb{R}_+$ as follows:

$$\xi_k(A) = r^{-k+1}\sqrt{8\ln\left(\frac{2^{k-k_0}K}{\mu(A)\delta}\right)}.$$

To prove the claim we have to show that

$$\Pr\left\{\exists \mathbf{y} \in Y : X_\mathbf{y} - X_{\mathbf{y}_0} - \sum_{k>k_0}\xi_k(A_k(\mathbf{y})) > 0\right\} < \delta.$$

Denote the left hand side of this inequality with P. By the chaining identity

$$P \le \Pr\left\{\exists \mathbf{y} : \sum_{k>k_0}\left(X_{\pi_k(\mathbf{y})} - X_{\pi_{k-1}(\mathbf{y})} - \xi_k(A_k(\mathbf{y}))\right) > 0\right\}.$$

If the sum is positive, at least one of the terms has to be positive, so

$$P \le \Pr\left\{\exists \mathbf{y}, k > k_0 : \left(X_{\pi_k(\mathbf{y})} - X_{\pi_{k-1}(\mathbf{y})} - \xi_k(A_k(\mathbf{y}))\right) > 0\right\}.$$

The event on the right hand side can also be written as

$$\left\{\exists k > k_0, \exists A \in A_k : X_{\pi(A)} - X_{\pi(\hat{A})} > \xi_k(A)\right\},$$

and a union bound gives

$$P \le \sum_{k>k_0}\sum_{A\in A_k}\Pr\left\{X_{\pi(A)} - X_{\pi(\hat{A})} > \xi_k(A)\right\}$$

$$\le \sum_{k>k_0}\sum_{A\in A_k}K\exp\left(\frac{-\xi_k(A)^2}{2\left\|\pi(A) - \pi\left(\hat{A}\right)\right\|^2}\right)$$

$$\le \sum_{k>k_0}\sum_{A\in A_k}K\exp\left(\frac{-\xi_k(A)^2}{2\left(2r^{-k+1}\right)^2}\right),$$

where we used the bound (3) in the second and the bound on $\left\| \pi\left(A\right) - \pi\left(\hat{A}\right) \right\|$ in the third inequality. Using the definition of $\xi_k\left(A\right)$ the last expression is equal to

$$\delta \sum_{k>k_0} \frac{1}{2^{k-k_0}} \sum_{A \in \mathcal{A}_k} \mu\left(A\right) = \delta \sum_{k>k_0} \frac{1}{2^{k-k_0}} = \delta,$$

because μ is a probability measure. This establishes the claim.

Now, using $\sqrt{a+b} \le \sqrt{a} + \sqrt{b}$ for $a, b \ge 0$, with probability at least $1 - \delta$

$$\sup_{\mathbf{y}} X_{\mathbf{y}} - X_{\mathbf{y}_0} \le r \sum_{k>k_0} r^{-k} \sqrt{8\ln\left(\frac{1}{\mu\left(A_k\left(\mathbf{y}\right)\right)}\right)}$$

$$+ r^{-k_0+1} \sum_{k>0} r^{-k+1} \sqrt{8\ln\left(\frac{2^k K}{\delta}\right)}$$

$$\le \sqrt{8}rC\, G\left(Y\right) + \sqrt{8}r^{-k_0+1} \sum_{k>0} \sqrt{k}r^{-k+1} \sqrt{\ln\left(\frac{2K}{\delta}\right)},$$

where we used Talagrand's theorem and the fact that $K > 1$. By the definition of k_0 we have $r^{-k_0+1} \le r^2 D\left(Y\right)/2$, so this is bounded by

$$C'''G\left(Y\right) + C''''D\left(Y\right)\sqrt{\ln\left(\frac{2K}{\delta}\right)},$$

with $C''' = \sqrt{8}rC$ and $C'''' = \sqrt{8}\left(r^2/2\right)\sum_{k>0}\sqrt{k}r^{-k+1}$. Converting the last bound into a tail bound and integrating we obtain

$$\mathbb{E}\left[\sup_{\mathbf{y}} X_{\mathbf{y}} - X_{\mathbf{y}_0}\right] \le C'''G\left(Y\right) + C''''D\left(Y\right)\left(\sqrt{\ln 2K} + \frac{\sqrt{\pi}}{2}\right)$$

$$\le C'''G\left(Y\right) + 3C''''D\left(Y\right)\sqrt{\ln 2K}$$

$$\le \left(C''' + 3\sqrt{2\pi\ln 2}C''''\right)G\left(Y\right) + 3C''''D\left(Y\right)\sqrt{\ln K},$$

where we again used $K \ge 1$ in the second inequality and (7) in the last inequality. This gives the conclusion with $C' = C''' + 3\sqrt{2\pi\ln 2}C''''$ and $C'' = 3C''''$.   ∎

References

[1] Bartlett, P.L., Mendelson, S.: Rademacher and Gaussian Complexities: Risk Bounds and Structural Results. Journal of Machine Learning Research 3, 463–482 (2002)

[2] Baxter, J.: Theoretical Models of Learning to Learn. In: Thrun, S., Pratt, L. (eds.) Learning to Learn. Springer (1998)

[3] Baxter, J.: A Model of Inductive Bias Learning. Journal of Artificial Intelligence Research 12, 149–198 (2000)

[4] Boucheron, S., Lugosi, G., Massart, P.: Concentration Inequalities. Oxford University Press (2013)

[5] Koltchinskii, V.I., Panchenko, D.: Rademacher processes and bounding the risk of function learning. In: Gine, E., Mason, D., Wellner, J. (eds.) High Dimensional Probability II, pp. 443–459 (2000)

[6] Ledoux, M., Talagrand, M.: Probability in Banach Spaces. Springer (1991)

[7] Maurer, A.: Bounds for linear multi-task learning. Journal of Machine Learning Research 7, 117–139 (2006)

[8] Maurer, A.: Transfer bounds for linear feature learning. Machine Learning 75(3), 327–350 (2009)

[9] Maurer, A., Pontil, M.: K-dimensional coding schemes in Hilbert spaces. IEEE Transactions on Information Theory 56(11), 5839–5846 (2010)

[10] Meir, R., Zhang, T.: Generalization error bounds for Bayesian mixture algorithms. Journal of Machine Learning Research 4, 839–860 (2003)

[11] Mendelson, S.: l-norm and its application to learning theory. Positivity 5, 177–191 (2001)

[12] Shawe-Taylor, J., Cristianini, N.: Kernel Methods for Pattern Analysis. Cambridge University Press (2004)

[13] Talagrand, M.: Regularity of Gaussian processes. Acta Mathematica 159, 99–149 (1987)

[14] Talagrand, M.: A simple proof of the majorizing measure theorem. Geometric and Functional Analysis 2(1), 118–125 (1992)

[15] Talagrand, M.: Majorizing measures without measures. Ann. Probab. 29, 411–417 (2001)

[16] Talagrand, M.: The Generic Chaining. Upper and Lower Bounds for Stochastic Processes. Springer, Berlin (2005)

Generalization Bounds for Time Series Prediction with Non-stationary Processes

Vitaly Kuznetsov[1] and Mehryar Mohri[1,2]

[1] Courant Institute of Mathematical Sciences,
251 Mercer street, New York, NY 10012, USA
[2] Google Research, 111 8th Avenue, New York, NY 10012, USA
{vitaly,mohri}@cims.nyu.edu

Abstract. This paper presents the first generalization bounds for time series prediction with a non-stationary mixing stochastic process. We prove Rademacher complexity learning bounds for both average-path generalization with non-stationary β-mixing processes and path-dependent generalization with non-stationary ϕ-mixing processes. Our guarantees are expressed in terms of β- or ϕ-mixing coefficients and a natural measure of discrepancy between training and target distributions. They admit as special cases previous Rademacher complexity bounds for non-i.i.d. stationary distributions, for independent but not identically distributed random variables, or for the i.i.d. case. We show that, using a new sub-sample selection technique we introduce, our bounds can be tightened under the natural assumption of convergent stochastic processes. We also prove that fast learning rates can be achieved by extending existing local Rademacher complexity analysis to non-i.i.d. setting.

Keywords: Generalization bounds, time series, mixing, stationary processes, fast rates, local Rademacher complexity.

1 Introduction

Given a sample $((X_1, Y_1), \ldots, (X_m, Y_m))$ of pairs in $\mathcal{Z} = \mathcal{X} \times \mathcal{Y}$, the standard supervised learning task consists of selecting, out of a class of functions H, a hypothesis $h \colon \mathcal{X} \to \mathcal{Y}$ that admits a small expected loss measured using some specified loss function $L \colon \mathcal{Y} \times \mathcal{Y} \to \mathbb{R}_+$. The common assumption in the statistical learning theory and the design of algorithms is that samples are drawn i.i.d. from some unknown distribution and generalization in this scenario has been extensively studied in the past. However, for many problems such as time series prediction, the i.i.d. assumption is too restrictive and it is important to analyze generalization in the absence of that condition. A variety of relaxations of this i.i.d. setting have been proposed in the machine learning and statistics literature. In particular, the scenario in which observations are drawn from a stationary mixing distribution has become standard and has been adopted by most previous studies [1, 10, 11, 12, 18, 20]. In this work, we seek to analyze generalization

P. Auer et al. (Eds.): ALT 2014, LNAI 8776, pp. 260–274, 2014.
© Springer International Publishing Switzerland 2014

under the more realistic assumption of non-stationary data. This covers a wide spectrum of stochastic processes considered in applications, including Markov chains, which are non-stationary.

Suppose we are given a doubly infinite sequence of $\mathcal{Z}$-valued random variables $\{Z_t\}_{t=-\infty}^{\infty}$ jointly distributed according to $\mathbf{P}$. We will write $\mathbf{Z}_a^b$ to denote a vector $(Z_a, Z_{a+1}, \ldots, Z_b)$ where a and b are allowed to take values $-\infty$ and ∞. Similarly, $\mathbf{P}_a^b$ denotes the distribution of $\mathbf{Z}_a^b$. Following [4], we define β-mixing coefficients for $\mathbf{P}$ as follows. For each positive integer a, we set

$$\beta(a) = \sup_t \|\mathbf{P}_{-\infty}^t \otimes \mathbf{P}_{t+a}^\infty - \mathbf{P}_{-\infty}^t \wedge \mathbf{P}_{t+a}^\infty\|_{TV}, \tag{1}$$

where $\mathbf{P}_{-\infty}^t \wedge \mathbf{P}_{t+a}^\infty$ denotes the joint distribution of $\mathbf{Z}_{-\infty}^t$ and $\mathbf{Z}_{t+a}^\infty$. Recall that the total variation distance $\| \cdot \|_{TV}$ between two probability measures P and Q defined on the same σ-algebra of events $\mathcal{G}$ is given by $\|P - Q\|_{TV} = \sup_{A \in \mathcal{G}} |P(A) - Q(A)|$. We say that $\mathbf{P}$ is β-mixing (or absolutely regular) if $\beta(a) \to 0$ as $a \to \infty$. Roughly speaking, this means that the future has a sufficiently weak dependence on the distant past. We remark that β-mixing coefficients can be defined equivalently as follows:

$$\beta(a) = \sup_t \mathbb{E}_{\mathbf{Z}_{-\infty}^t} \left[\|\mathbf{P}_{t+a}^\infty(\cdot|\mathbf{Z}_{-\infty}^t) - \mathbf{P}_{t+a}^\infty\|_{TV} \right], \tag{2}$$

where $\mathbf{P}(\cdot|\cdot)$ denotes conditional probability measure [4]. Another standard measure of the dependence of the future on the past is the φ-mixing coefficient defined for any $a > 0$ by

$$\varphi(a) = \sup_t \sup_{B \in \mathcal{F}_t} \|\mathbf{P}_{t+a}^\infty(\cdot|B) - \mathbf{P}_{t+a}^\infty\|_{TV}, \tag{3}$$

where $\mathcal{F}_t$ is the σ-algebra generated by $\mathbf{Z}_{-\infty}^t$. A distribution $\mathbf{P}$ is said to be φ-mixing if $\varphi(a) \to 0$ as $a \to \infty$. Note that $\beta(a) \leq \varphi(a)$, so any φ-mixing distribution is necessarily β-mixing. We also recall that a sequence of random variables $\mathbf{Z}_{-\infty}^\infty$ is (strictly) stationary provided that, for any t and any non-negative integers m and k, $\mathbf{Z}_t^{t+m}$ and $\mathbf{Z}_{t+k}^{t+m+k}$ have the same distribution.

Unlike the i.i.d. case where $\mathbb{E}[L(h(X), Y)]$ is used to measure the generalization error of h, in the case of time series prediction, there is no unique measure commonly used to assess the quality of a given hypothesis h. One approach consists of seeking a hypothesis h that performs well in the near future, given the observed trajectory of the process. That is, we would like to achieve a small *path-dependent* generalization error

$$\mathcal{L}_{T+s}(h) = \mathbb{E}_{Z_{T+s}}[L(h(X_{T+s}), Y_{T+s})|\mathbf{Z}_1^T], \tag{4}$$

where $s \geq 1$ is fixed. To simplify the notation, we will often write $\ell(h, z) = L(h(x), y)$, where $z = (x, y)$. For time series prediction tasks, we often receive a sample $\mathbf{Y}_1^T$ and wish to forecast Y_{T+s}. A large class of (bounded-memory) auto-regressive models uses q past observations $\mathbf{Y}_{T-q+1}^T$ to predict Y_{T+s}. Our scenario includes this setting as a special case where we take $\mathcal{X} = \mathcal{Y}^q$ and

$Z_{t+s} = (\mathbf{Y}_{t-q+1}^{t}, Y_{t+s})$.[1] The generalization ability of stable algorithms with error defined by (4) was studied by Mohri and Rostamizadeh [12].

Alternatively, one may wish to perform well in the near future when being on some "average" trajectory. This leads to the *averaged* generalization error:

$$\bar{\mathcal{L}}_{T+s}(h) = \mathbb{E}_{\mathbf{Z}_1^T}[\mathcal{L}_{T+s}(h)] = \mathbb{E}_{Z_{T+s}}[\ell(h, Z_{T+s})]. \tag{5}$$

We note that $\bar{\mathcal{L}}_{T+s}(h) = \mathcal{L}_{T+s}(h)$ when the training and testing sets are independent. The pioneering work of Yu [20] led to VC-dimension bounds for $\bar{\mathcal{L}}_{T+s}$ under the assumption of stationarity and β-mixing. Later, Meir [10] used that to derive generalization bounds in terms of covering numbers of H. These results have been further extended by Mohri and Rostamizadeh [11] to data-dependent learning bounds in terms of the Rademacher complexity of H.

Most of the generalization bounds for non-i.i.d. scenarios that can be found in the machine learning and statistics literature assume that observations come from a (strictly) stationary distribution. The only exception that we are aware of is the work of Agarwal and Duchi [1], who present bounds for stable on-line learning algorithms under the assumptions of suitably convergent process.[2] The main contribution of our work is the first generalization bounds for both $\mathcal{L}_{T+s}$ and $\bar{\mathcal{L}}_{T+s}$ when the data is generated by a non-stationary mixing stochastic process. These results provide a sufficient condition for the predictive PAC learnability of Pestov [3, 14]. Next, we strengthen our assumptions and give generalization bounds for convergent processes. In doing so, we establish sufficient conditions for the predictive PAC learnability of Shalizi and Kontorovich [17]. These results are algorithm-agnostic analogues of the algorithm-dependent bounds of Agarwal and Duchi [1]. In [1], Agarwal and Duchi also prove fast convergence rates when a strongly convex loss is used. Similarly, Steinwart and Christmann [18] showed that regularized learning algorithms admit faster convergence rates under the assumptions of mixing and stationarity. We conclude this paper by showing that this is in fact a general phenomenon. We use local Rademacher complexity techniques [2] to establish faster convergence rates for stationary or convergent mixing processes.

A key ingredient of the bounds we present is the notion of *discrepancy* between two probability distributions that was used by Mohri and Muñoz Medina [13] to give generalization bounds for sequences of independent (but not identically distributed) random variables. In our setting, discrepancy can be defined as

$$d(t_1, t_2) = \sup_{h \in H} |\mathcal{L}_{t_1}(h) - \mathcal{L}_{t_2}(h)| \tag{6}$$

and similarly we can define $\bar{d}(t_1, t_2)$, where we replace $\mathcal{L}_t$ with $\bar{\mathcal{L}}_t$. Discrepancy is a natural measure of the non-stationarity of a stochastic process with respect to

---

[1] Observe that if $\mathbf{Y}$ is β-mixing, then so is $\mathbf{Z}$ and $\beta_{\mathbf{Z}}(a) = \beta_{\mathbf{Y}}(a - q)$. Similarly, the φ-mixing assumption is also preserved. It is an open problem (posed by Meir [10]) to derive generalization bounds for unbounded-memory models.

[2] Agarwal and Duchi [1] additionally assume that distributions are absolutely continuous and that the loss function is convex and Lipschitz.

the hypothesis class H and a loss function L. For instance, if the process is strictly stationary then $\bar{d}(t_1, t_2) = 0$ for all $t_1, t_2 \in \mathbb{Z}$. As a more interesting example, consider a weakly stationary stochastic process,[3] together with a squared loss L and a set of linear hypothesis $H = \{\mathbf{Y}_{t-q+1}^T \mapsto w \cdot \mathbf{Y}_{t-q+1}^T : w \in \mathbb{R}^q\}$. It can be shown that in this case we again have $\bar{d}(t_1, t_2) = 0$ for all $t_1, t_2 \in \mathbb{Z}$. An additional advantage of the discrepancy measure is that it can be replaced by an upper bound that, under mild conditions, can be estimated from data [8, 6].

The rest of this paper is organized as follows. In Section 2 we discuss the main technical tool used to derive our bounds. Section 3 and Section 4 present learning guarantees for averaged and path-dependent errors respectively. In Section 5 we analyze generalization with convergent processes. We conclude with fast learning rates for the non-i.i.d. setting in Section 6.

2 Independent Blocks and Sub-sample Selection

The first step towards our generalization bounds is to reduce the setting of a mixing stochastic process to a simpler scenario of a sequence of independent random variables, where we can take advantage of the known concentration results. One way to achieve this is via the independent block technique introduced by Yu [20] which we now describe.

We can divide a given sample $\mathbf{Z}_1^T$ into $2m$ blocks such that each block has size a_i and we require $T = \sum_{i=1}^{2m} a_i$. In other words, we consider a sequence of random vectors $\mathbf{Z}(i) = \mathbf{Z}_{l(i)}^{u(i)}, i = 1, \ldots, 2m$ where $l(i) = 1 + \sum_{j=1}^{i-1} a_j$ and $u(i) = \sum_{j=1}^{i} a_j$. It will be convenient to refer to even and odd blocks separately. We will write $\mathbf{Z}^o = (\mathbf{Z}(1), \mathbf{Z}(3) \ldots, \mathbf{Z}(2m-1))$ and $\mathbf{Z}^e = (\mathbf{Z}(2), \mathbf{Z}(4), \ldots, \mathbf{Z}(2m))$. In fact, we will often work with blocks that are independent.

Let $\widetilde{\mathbf{Z}}^o = (\widetilde{\mathbf{Z}}(1), \ldots, \widetilde{\mathbf{Z}}(2m-1))$ where $\widetilde{\mathbf{Z}}(i)$, $i = 1, 3, \ldots, 2m-1$, are independent and each $\widetilde{\mathbf{Z}}(i)$ has the same distribution as $\mathbf{Z}(i)$. We construct $\widetilde{\mathbf{Z}}^e$ in the same way. The following result due to Yu [20] enables us to relate sequences of dependent and independent blocks.

Proposition 1. *Let g be a real-valued Borel measurable function such that $-M_1 \leq g \leq M_2$ for some $M_1, M_2 \geq 0$. Then, the following holds:*

$$|\mathbb{E}[g(\widetilde{\mathbf{Z}}^o)] - \mathbb{E}[g(\mathbf{Z}^o)]| \leq (M_1 + M_2) \sum_{i=1}^{m-1} \beta(a_{2i}).$$

The proof of this result is given in [20], which in turn is based on [5] and [19]. We present a sketch of the main steps of the proof as these will be useful for us as stand-alone results.

---

[3] A process $\mathbf{Z}$ is weakly stationary if $\mathbb{E}[Z_t]$ is a constant function of t and $\mathbb{E}[Z_{t_1} Z_{t_2}]$ only depends on $t_1 - t_2$.

Lemma 1. *Let Q and P be probability measures on $(\Omega, \mathcal{F})$ and let $h \colon \Omega \to \mathbb{R}$ be a Borel measurable function such that $-M_1 \le h \le M_2$ for some $M_1, M_2 \ge 0$. Then*

$$|\mathbb{E}_Q[h] - \mathbb{E}_P[h]| \le (M_1 + M_2)\|P - Q\|_{TV}.$$

The proof of Lemma 1 can be found in [5, 19, 20]. Lemma 1 extended via induction yields the following result. See [5] for further details.

Lemma 2. *Let $m \ge 1$ and $(\prod_{k=1}^m \Omega_k, \prod_{k=1}^m \mathcal{F}_k)$ be a measure space with P a measure on this space and P_j the marginal on $(\prod_{k=1}^j \Omega_k, \prod_{k=1}^j \mathcal{F}_k)$. Let Q_j be a measure on $(\Omega_j, \mathcal{F}_j)$ and define*

$$\beta_j = \mathbb{E}\left[\left\|P_{j+1}\left(\cdot \mid \prod_{k=1}^j \mathcal{F}_k\right) - Q_{j+1}\right\|_{TV}\right].$$

Then, for any Borel measurable function $h \colon \prod_{k=1}^m \Omega_k \to \mathbb{R}$ such that $-M_1 \le h \le M_2$ for some $M_1, M_2 \ge 0$, the following holds

$$|\mathbb{E}_P[h] - \mathbb{E}_Q[h]| \le (M_1 + M_2) \sum_{j=1}^{m-1} \beta_j$$

where $Q = Q_1 \otimes Q_2 \otimes \ldots \otimes Q_m$.

Proposition 1 now follows from Lemma 2 by taking Q_j to be the marginal of P on $(\Omega_j, \mathcal{F}_j)$ and applying it to the case of independent blocks.

Proposition 1 is not the only way to relate mixing and independent cases. Next, we present another technique that we term *sub-sample selection*, which is particularly useful when the process is convergent. Suppose we are given a sample $\mathbf{Z}_1^T$. Fix $a \ge 1$ such that $T = ma$ for some $m \ge 1$ and define a sub-sample $\mathbf{Z}^{(j)} = (Z_{1+j}, \ldots, Z_{m-1+j})$, $j = 0, \ldots, a-1$. An application of Lemma 2 yields the following result.

Proposition 2. *Let g be a real-valued Borel measurable function such that $-M_1 \le g \le M_2$ for some $M_1, M_2 \ge 0$. Then*

$$|\mathbb{E}[g(\widetilde{\mathbf{Z}}_\Pi)] - \mathbb{E}[g(\mathbf{Z}^{(j)})]| \le (M_1 + M_2)(m-1)\beta(a),$$

where $\widetilde{\mathbf{Z}}_\Pi$ is an i.i.d. sample of size m from a distribution Π and $\beta(a) = \sup_t \mathbb{E}[\|\mathbb{P}_{t+a}(\cdot \mid \mathbf{Z}_1^t) - \Pi\|_{TV}]$.

Proposition 2 is commonly applied with Π being the stationary probability measure of a convergent process.

3   Generalization Bound for the Averaged Error

In this section, we derive a generalization bound for averaged error $\bar{\mathcal{L}}_{T+s}$. Given a sample $\mathbf{Z}_1^T$ generated by a $(\beta\text{-})$mixing process,[4] we define $\Phi(\mathbf{Z}_1^T)$ as follows:

$$\Phi(\mathbf{Z}_1^T) = \sup_{h \in H} \left(\bar{\mathcal{L}}_{T+s}(h) - \frac{1}{T} \sum_{t=1}^{T} \ell(h, Z_t) \right). \tag{7}$$

We will also use I_1 to denote the set of indices of the elements from the sample $\mathbf{Z}_1^T$ that are contained in the odd blocks. Similarly, I_2 is used for elements in the even blocks.

We establish our bounds in a series of lemmas. We start by proving a concentration result for dependent non-stationary data.

Lemma 3. *Let L be a loss function bounded by M and H an arbitrary hypothesis set. For any $a_1, \ldots, a_{2m} > 0$ such that $T = \sum_{i=1}^{2m} a_i$, partition the given sample $\mathbf{Z}_1^T$ into blocks as described in Section 2. Then, for any $\epsilon > \max(\mathbb{E}[\Phi(\widetilde{\mathbf{Z}}^o)], \mathbb{E}[\Phi(\widetilde{\mathbf{Z}}^e)])$, the following holds:*

$$\mathbb{P}(\Phi(\mathbf{Z}_1^T) > \epsilon) \leq \mathbb{P}(\Phi(\widetilde{\mathbf{Z}}^o) - \mathbb{E}[\Phi(\widetilde{\mathbf{Z}}^o)] > \epsilon_1) + \mathbb{P}(\Phi(\widetilde{\mathbf{Z}}^e) - \mathbb{E}[\Phi(\widetilde{\mathbf{Z}}^e)] > \epsilon_2) + \sum_{i=2}^{m-1} \beta(a_i),$$

where $\epsilon_1 = \epsilon - \mathbb{E}[\Phi(\widetilde{\mathbf{Z}}^o)]$ and $\epsilon_2 = \epsilon - \mathbb{E}[\Phi(\widetilde{\mathbf{Z}}^e)]$.

Proof. By convexity of the supremum $\Phi(\mathbf{Z}_1^T) \leq \frac{|I_1|}{T} \Phi(\mathbf{Z}^o) + \frac{|I_2|}{T} \Phi(\mathbf{Z}^e)$. Since $|I_1| + |I_2| = T$, for $\frac{|I_1|}{T} \Phi(\mathbf{Z}^o) + \frac{|I_2|}{T} \Phi(\mathbf{Z}^e)$ to exceed ϵ at least one element of $\{\Phi(\mathbf{Z}^o), \Phi(\mathbf{Z}^e)\}$ must be greater than ϵ. Thus, by the union bound, we can write

$$\mathbb{P}(\Phi(\mathbf{Z}_1^T) > \epsilon) \leq \mathbb{P}(\Phi(\mathbf{Z}^o) > \epsilon) + \mathbb{P}(\Phi(\mathbf{Z}^e) > \epsilon)$$
$$= \mathbb{P}(\Phi(\mathbf{Z}^o) - \mathbb{E}[\Phi(\widetilde{\mathbf{Z}}^o)] > \epsilon_1) + \mathbb{P}(\Phi(\mathbf{Z}^e) - \mathbb{E}[\Phi(\widetilde{\mathbf{Z}}^e)] > \epsilon_2).$$

We apply Proposition 1 to the indicator functions of the events $\{\Phi(\mathbf{Z}^o) - \mathbb{E}[\Phi(\widetilde{\mathbf{Z}}^o)] > \epsilon_1\}$ and $\{\Phi(\mathbf{Z}^e) - \mathbb{E}[\Phi(\widetilde{\mathbf{Z}}^e)] > \epsilon_2\}$ to complete the proof.   □

Lemma 4. *Under the same assumptions as in Lemma 3, the following holds:*

$$\mathbb{P}(\Phi(\mathbf{Z}_1^T) > \epsilon) \leq \exp\left(\frac{-2T^2 \epsilon_1^2}{\|\mathbf{a}^o\|_2^2 M^2} \right) + \exp\left(\frac{-2T^2 \epsilon_2^2}{\|\mathbf{a}^e\|_2^2 M^2} \right) + \sum_{i=2}^{m-1} \beta(a_i),$$

where $\mathbf{a}^o = (a_1, a_3, \ldots, a_{2m-1})$ and $\mathbf{a}^e = (a_2, a_4, \ldots, a_{2m})$.

Proof. We apply McDiarmid's inequality [9] to the sequence of independent blocks. We note that if $\widetilde{\mathbf{Z}}^o$ and $\widetilde{\mathbf{Z}}$ are two sequences of independent (odd) blocks

---

[4] All the results of this section hold for a slightly weaker notion of β-mixing with $\beta(a) = \sup_t \mathbb{E}\|\mathbf{P}_{t+a}(\cdot | \mathbf{Z}_{-\infty}^t) - \mathbf{P}_{t+a}\|_{TV}$.

that differ only by one block (say block i) then $\Phi(\widetilde{\mathbf{Z}}^o) - \Phi(\widetilde{\mathbf{Z}}) \le a_i \frac{M}{T}$ and it follows from McDiarmid's inequality that

$$\mathbb{P}(\Phi(\widetilde{\mathbf{Z}}^o) - \mathbb{E}[\Phi(\widetilde{\mathbf{Z}}^o)] > \epsilon_1) \le \exp\left(\frac{-2T^2\epsilon_1^2}{\|\mathbf{a}^o\|_2^2 M^2}\right).$$

Using the same argument for $\widetilde{\mathbf{Z}}^e$ finishes the proof of this lemma. $\square$

The next step is to bound $\max(\mathbb{E}[\Phi(\widetilde{\mathbf{Z}}^o)], \mathbb{E}[\Phi(\widetilde{\mathbf{Z}}^e)])$. The bound that we give is in terms of *block* Rademacher complexity defined by

$$\mathfrak{R}(\widetilde{\mathbf{Z}}^o) = \frac{1}{|I_1|}\mathbb{E}\left[\sup_{h\in H}\sum_{i=1}^{m}\sigma_i\, l\big(h, \mathbf{Z}(2i-1)\big)\right], \tag{8}$$

where σ_i is a sequence of Rademacher random variables and $l(h, \mathbf{Z}(2i-1)) = \sum_{t\in I_1\cap\mathbf{Z}(2i-1)}\ell(h, Z_t)$. Below we will show that if the block size is constant (i.e. $a_i = a$), then the block complexity can be bounded in terms of the regular Rademacher complexity.

Lemma 5. *For $j = 1, 2$, let $\Delta^j = \frac{1}{|I_j|}\sum_{t\in I_j}\bar{d}(t, T+s)$, which is an average discrepancy. Then, the following bound holds:*

$$\max(\mathbb{E}[\Phi(\widetilde{\mathbf{Z}}^o)], \mathbb{E}[\Phi(\widetilde{\mathbf{Z}}^e)]) \le 2\max(\mathfrak{R}(\widetilde{\mathbf{Z}}^o), \mathfrak{R}(\widetilde{\mathbf{Z}}^e)) + \max(\Delta^1, \Delta^2). \tag{9}$$

Proof. In the course of this proof Z_t, denotes a sample drawn according to the distribution of $\widetilde{\mathbf{Z}}^o$ (and not that of $\mathbf{Z}^o$). Using the sub-additivity of the supremum and the linearity of expectation, we can write

$$\mathbb{E}\left[\sup_{h\in H}\bar{\mathcal{L}}_{T+s}(h) - \frac{1}{|I_1|}\sum_{t\in I_1}\ell(h, Z_t)\right]$$

$$= \mathbb{E}\left[\sup_{h\in H}\bar{\mathcal{L}}_{T+s}(h) - \frac{1}{|I_1|}\sum_{t\in I_1}\bar{\mathcal{L}}_t(h) + \frac{1}{|I_1|}\sum_{t\in I_1}\bar{\mathcal{L}}_t(h) - \frac{1}{|I_1|}\sum_{t\in I_1}\ell(h, Z_t)\right]$$

$$\le \mathbb{E}\left[\sup_{h\in H}\bar{\mathcal{L}}_{T+s}(h) - \frac{1}{|I_1|}\sum_{t\in I_1}\bar{\mathcal{L}}_t(h) + \sup_{h\in H}\frac{1}{|I_1|}\sum_{t\in I_1}\bar{\mathcal{L}}_t(h) - \frac{1}{|I_1|}\sum_{t\in I_1}\ell(h, Z_t)\right]$$

$$= \frac{1}{|I_1|}\sum_{t\in I_1}\sup_{h\in H}|\bar{\mathcal{L}}_{T+s}(h) - \bar{\mathcal{L}}_t(h)| + \frac{1}{|I_1|}\mathbb{E}\left[\sup_{h\in H}\sum_{t\in I_1}\bar{\mathcal{L}}_t(h) - \sum_{t\in I_1}\ell(h, Z_t)\right]$$

$$= \Delta^1 + \frac{1}{|I_1|}\mathbb{E}\left[\sup_{h\in H}\sum_{i=1}^{m}\mathbb{E}[l(h, \widetilde{\mathbf{Z}}(2i-1))] - l(h, \widetilde{\mathbf{Z}}(2i-1))\right].$$

The second term can be written as

$$A = \frac{1}{|I_1|}\mathbb{E}\left[\sup_{h\in H}\sum_{i=1}^{m}A_i(h)\right],$$

with $A_i(h) = \mathbb{E}[l(h, \widetilde{\mathbf{Z}}(2i-1))] - l(h, \widetilde{\mathbf{Z}}(2i-1))$ for all $i \in [1, m]$. Since the terms $A_i(h)$ are all independent, the same proof as that of the standard i.i.d. symmetrization bound in terms of the Rademacher complexity applies and A can be bounded by $\mathfrak{R}(\widetilde{\mathbf{Z}}^o)$. Using the same arguments for even blocks completes the proof.   □

Combining Lemma 4 and Lemma 5 leads directly to the main result of this section.

Theorem 1. *With the assumptions of Lemma 3, for any $\delta > \sum_{i=2}^{m-1} \beta(a_i)$, with probability $1 - \delta$, the following holds for all hypotheses $h \in H$:*

$$\bar{\mathcal{L}}_{T+s}(h) \leq \frac{1}{T} \sum_{t=1}^{T} \ell(h, Z_t) + 2 \max(\mathfrak{R}(\widetilde{\mathbf{Z}}^o), \mathfrak{R}(\widetilde{\mathbf{Z}}^e)) + \max(\Delta^1, \Delta^2)$$

$$+ M \max(\|\mathbf{a}^e\|_2, \|\mathbf{a}^e\|_2) \sqrt{\frac{\log \frac{2}{\delta'}}{2T^2}},$$

where $\delta' = \delta - \sum_{i=2}^{m-1} \beta(a_i)$.

The learning bound of Theorem 1 indicates the challenges faced by the learner when presented with data drawn from a non-stationary stochastic process. In particular, the presence of the term $\max(\Delta^1, \Delta^2)$ in the bound shows that generalization in this setting depends on the "degree" of non-stationarity of the underlying process. The dependency in the training instances reduces the effective size of the sample from T to $(T/(\|\mathbf{a}^e\|_2 + \|\mathbf{a}^e\|_2))^2$. Observe that for a general non-stationary process the learning bounds presented may not converge to zero as a function of the sample size, due to the discrepancies between the training and target distributions. In Section 5 and Section 6, we will describe some natural assumptions under which this convergence does occur.

When the same size a is used for all the blocks considered in the analysis, thus $T = 2ma$, then the block Rademacher complexity terms can be replaced with standard Rademacher complexities. Indeed, in that case, we can group the summands in the definition of the block complexity according to subsamples $\mathbf{Z}^{(j)}$ and use the sub-additivity of the supremum to find that $\mathfrak{R}(\widetilde{\mathbf{Z}}^o) \leq \frac{1}{a} \sum_{j=1}^{a} \mathfrak{R}_m(\widetilde{\mathbf{Z}}^{(j)})$, where $\mathfrak{R}_m(\widetilde{\mathbf{Z}}^{(j)}) = \frac{1}{m} \mathbb{E}[\sup_{h \in H} \sum_{i=1} \sigma_i \ell(h, Z_{i,j})]$ with $(\sigma_i)_i$ a sequence of Rademacher random variables and $(Z_{i,j})_{i,j}$ a sequence of independent random variables such that $Z_{i,j}$ is distributed according to the law of $Z_{a(2i-1)+j}$ from $\mathbf{Z}_1^T$. This leads to the following perhaps more informative but somewhat less tight bound:

$$\bar{\mathcal{L}}_{T+s}(h) \leq \frac{1}{T} \sum_{t=1}^{T} \ell(h, Z_t) + \frac{2}{a} \sum_{j=1}^{2a} \mathfrak{R}_m(\mathbf{Z}^{(j)}) + \frac{2}{T} \sum_{t=1}^{T} \bar{d}(t, T+s) + M \sqrt{\frac{\log \frac{2}{\delta'}}{8m}}.$$

If the process is stationary, then we recover as a special case the generalization bound of [11]. If $\mathbf{Z}_1^T$ is a sequence of independent but not identically distributed random variables, we recover the results of [13]. In the i.i.d. case, Theorem 1 reduces to the generalization bounds of Koltchinskii and Panchenko [7].

4 Generalization Bound for the Path-Dependent Error

In this section we give generalization bounds for a path-dependent error $\mathcal{L}_{T+s}$ under the assumption that the data is generated by a $(\varphi\text{-})$mixing non-stationary process.[5] In this section, we will use $\Phi(\mathbf{Z}_1^T)$ to denote the same quantity as in (7) except that $\bar{\mathcal{L}}_{T+s}$ is replaced with $\mathcal{L}_{T+s}$.

The key technical tool that we will use is the version of McDiarmid's inequality for dependent random variables, which requires a bound on the differences of conditional expectations of Φ (see Corollary 6.10 in [9]). We start with the following adaptation of Lemma 1 to this setting.

Lemma 6. *Let $\mathbf{Z}_1^T$ be a sequence of $\mathcal{Z}$-valued random variables and suppose that $g\colon \mathcal{Z}^{k+j} \to \mathbb{R}$ is a Borel-measurable function such that $-M_1 \leq g \leq M_2$ for some $M_1, M_2 \geq 0$. Then, for any $z_1, \ldots, z_k \in \mathcal{Z}$, the following bound holds:*

$$|\mathbb{E}[g(Z_1, \ldots, Z_k, Z_{T-j+1}, \ldots, Z_T)|z_1, \ldots, z_k] - \mathbb{E}[g(z_1, \ldots, z_k, Z_{T-j+1}, \ldots, Z_T)]|$$
$$\leq (M_1 + M_2)\varphi(T + 1 - (k + j)).$$

Proof. This result follows from an application of Lemma 1:

$$|\mathbb{E}[g(Z_1, \ldots, Z_k, Z_{T-j+1}, \ldots, Z_T)|z_1, \ldots, z_k] - \mathbb{E}[g(z_1, \ldots, z_k, Z_{T-j+1}, \ldots, Z_T)]|$$
$$\leq (M_1 + M_2)\|\mathbf{P}_{T-j+1}^T(\cdot|z_1, \ldots, z_k) - \mathbf{P}_{T-j+1}^T\|_{TV}$$
$$\leq (M_1 + M_2)\varphi(T + 1 - (k + j)),$$

where the second inequality follows from the definition of φ-mixing coefficients. $\square$

Lemma 7. *For any $z_1, \ldots, z_k, z_k' \in \mathcal{Z}$ and any $0 \leq j \leq T - k$ with $k > 1$, the following holds:*

$$\left|\mathbb{E}[\Phi(\mathbf{Z}_1^T)|z_1, \ldots, z_k] - \mathbb{E}[\Phi(\mathbf{Z}_1^T)|z_1, \ldots, z_k']\right| \leq 2M\left(\tfrac{j+1}{T} + \gamma\varphi(j + 2) + \varphi(s)\right),$$

where $\gamma = 1$ iff $j + k < T$ and 0 otherwise. Moreover, if $\mathcal{L}_{T+s}(h) = \bar{\mathcal{L}}_{T+s}(h)$, then the term $\varphi(s)$ can be omitted from the bound.

Proof. First, we observe that using Lemma 6 we have $|\mathcal{L}_{T+s}(h) - \bar{\mathcal{L}}_{T+s}(h)| \leq M\varphi(s)$. Next, we use this result, the properties of conditional expectation and Lemma 6 to show that $\mathbb{E}[\Phi(\mathbf{Z}_1^T)|z_1, \ldots, z_k]$ is bounded by

$$\mathbb{E}\left[\sup_{h \in H}\left(\bar{\mathcal{L}}_{T+s}(h) - \frac{1}{T}\sum_{t=1}^T \ell(h, Z_t)\right)\Big|z_1, \ldots, z_k\right] + M\varphi(s)$$

$$\leq \mathbb{E}\left[\sup_{h \in H}\left(\bar{\mathcal{L}}_{T+s}(h) - \frac{1}{T}\sum_{t=k+j}^T \ell(h, Z_t) - \frac{1}{T}\sum_{t=1}^{k-1} \ell(h, Z_t)\right)\Big|z_1, \ldots, z_k\right] + \eta$$

$$\leq \mathbb{E}\left[\sup_{h \in H}\left(\bar{\mathcal{L}}_{T+s}(h) - \frac{1}{T}\sum_{t=k+j}^T \ell(h, Z_t) - \frac{1}{T}\sum_{t=1}^{k-1} \ell(h, z_t)\right)\right] + M\gamma\varphi(j + 2) + \eta,$$

---

[5] As in Section 3, we can weaken the notion of φ-mixing by using $\varphi(a) = \sup_t \sup_{B \in \mathcal{F}_t} \|\mathbf{P}_{t+a}(\cdot|B) - \mathbf{P}_{t+a}\|_{TV}$.

where $\eta = M(\frac{j}{T} + \varphi(s))$. Using a similar argument to bound $\mathbb{E}[\varPhi(\mathbf{Z}_1^T)|z_1, \ldots, z_k']$ from below by $-M(\gamma\varphi(j+2) + \frac{j}{T} + \varphi(s))$ and taking the difference completes the proof. □

The last ingredient that we will need to establish a generalization bound for $\mathcal{L}_{T+s}$ is a bound on $\mathbb{E}[\varPhi]$. The bound we present is in terms of a discrepancy measure and the sequential Rademacher complexity introduced in [15].

Lemma 8. *The following bound holds*

$$\mathbb{E}[\varPhi(\mathbf{Z}_1^T)] \leq \mathbb{E}[\varDelta] + 2\mathfrak{R}_{T-s}^{seq}(H_\ell) + M\frac{s-1}{T},$$

where $\mathfrak{R}_{T-s}^{seq}(H_\ell)$ *is the sequential Rademacher complexity of the function class* $H_\ell = \{z \mapsto \ell(h, z) : h \in H\}$ *and* $\varDelta = \frac{1}{T}\sum_{t=1}^{T-s} d(t+s, T+s)$.

Proof. First, we write $\mathbb{E}[\varPhi(\mathbf{Z}_1^T)] \leq \mathbb{E}\left[\sup_{h \in H}(\mathcal{L}_{T+s}(h) - \frac{1}{T}\sum_{t=s}^{T}\ell(h, Z_t))\right] + M\frac{s-1}{T}$. Using the sub-additivity of the supremum, we bound the first term by

$$\mathbb{E}\left[\sup_{h \in H}\frac{1}{T}\sum_{t=1}^{T-s}(\mathcal{L}_{t+s}(h) - \ell(h, Z_{t+s}))\right] + \mathbb{E}\left[\sup_{h \in H}\frac{1}{T}\sum_{t=1}^{T-s}(\mathcal{L}_{T+s}(h) - \mathcal{L}_{t+s}(h))\right].$$

The first summand above is bounded by $2\mathfrak{R}_{T-s}^{seq}(H_\ell)$ by Theorem 2 of [16]. Note that the result of [16] is for $s = 1$ but it can be extended to an arbitrary s. The second summand is bounded by $\mathbb{E}[\varDelta]$ by the definition of the discrepancy. □

McDiarmid's inequality (Corollary 6.10 in [9]), Lemma 7 and Lemma 8 combined yield the following generalization bound for path-dependent error $\mathcal{L}_{T+s}(h)$.

Theorem 2. *Let L be a loss function bounded by M and let H be an arbitrary hypothesis set. Let $\mathbf{d} = (d_1, \ldots, d_T)$ with $d_t = \frac{j_t+1}{T} + \gamma_t\varphi(j_t+2) + \varphi(s)$ where $0 \leq j_t \leq T-t$ and $\gamma_t = 1$ iff $j_t + t < T$ and 0 otherwise (in case training and testing sets are independent we can take $d_t = \frac{j_t+1}{T} + \gamma_t\varphi(j_t+2)$). Then, for any $\delta > 0$, with probability at least $1 - \delta$, the following holds for all $h \in H$:*

$$\mathcal{L}_{T+s}(h) \leq \frac{1}{T}\sum_{t=1}^{T}\ell(h, Z_t) + \mathbb{E}[\varDelta] + 2\mathfrak{R}_{T-s}^{seq}(H_\ell) + M\|\mathbf{d}\|_2\sqrt{2\log\frac{1}{\delta}} + M\frac{s-1}{T}.$$

Observe that for the bound of Theorem 2 to be nontrivial the mixing rate is required to be sufficiently fast. For instance, if $\varphi(\log(T)) = O(T^2)$, then taking $s = \log(T)$ and $j_t = \min\{t, \log T\}$ yields $\|\mathbf{d}\|_2 = O(\sqrt{(\log T)^3/T})$. Combining this with an observation that by Lemma 6, $\mathbb{E}[\varDelta] \leq 2\varphi(s) + \frac{1}{T}\sum_{t=1}^{T}\bar{d}(t, T+s)$ one can show that for any $\delta > 0$ with probability at least $1 - \delta$, the following holds for all $h \in H$:

$$\mathcal{L}_{T+s}(h) \leq \frac{1}{T}\sum_{t=1}^{T}\ell(h, Z_t) + 2\mathfrak{R}_{T-s}^{seq}(H_\ell) + \frac{1}{T}\sum_{t=1}^{T}\bar{d}(t, T+s) + O\left(\sqrt{\frac{(\log T)^3}{T}}\right).$$

As commented in Section 3, in general, our bounds are convergent under some natural assumptions examined in the next sections.

5   Convergent Processes

In Section 3 and Section 4 we observed that, for a general non-stationary process, our learning bounds may not converge to zero as a function of the sample size, due to the discrepancies between the training and target distributions. The bounds that we derive suggest that for that convergence to take place, training distributions should "get closer" to the target distribution. However, the issue is that as the sample size grows, the target "is moving". In light of this, we consider a stochastic process that converges to some stationary distribution Π. More precisely, we define

$$\beta(a) = \sup_t \mathbb{E}\big[\|\mathbf{P}_{t+a}(\cdot|\mathbf{Z}_{-\infty}^t) - \Pi\|_{TV}\big] \tag{10}$$

and define $\phi(a)$ in a similar way. We say that a process is β- or ϕ-mixing if $\beta(a) \to 0$ or $\phi(a) \to 0$ as $a \to \infty$ respectively. We remark that this is precisely the mixing assumption used by Agarwal and Duchi [1]. Note that the notions of β- and ϕ-mixing are strictly stronger than the necessary mixing assumptions in Section 3 and Section 4. Indeed, consider a sequence Z_t of independent Gaussian random variables with mean t and unit variance. It is immediate that this sequence is β-mixing but it is not β-mixing. On the other hand, if we use finite-dimensional mixing coefficients, then the following holds:

$$\beta(a) = \sup_t \mathbb{E}\big[\|\mathbf{P}_{t+a}(\cdot|\mathbf{Z}_{-\infty}^t) - \mathbf{P}_{t+a}\|_{TV}\big]$$

$$\leq \sup_t \mathbb{E}\big[\|\mathbf{P}_{t+a}(\cdot|\mathbf{Z}_{-\infty}^t) - \Pi\|_{TV}\big] + \sup_t \sup_A |\mathbb{E}[\mathbb{E}_{t+a}[\mathbf{1}_A|\mathbf{Z}_{-\infty}^t]] - \Pi|$$

$$\leq 2\beta(a).$$

However, note that a stationary β-mixing process is necessarily β-mixing with $\Pi = \mathbf{P}_0$. We define the *long-term* loss or error $\mathcal{L}_\Pi(h) = \mathbb{E}_\Pi[\ell(h, Z)]$ and observe that $\bar{\mathcal{L}}_T(h) \leq \mathcal{L}_\Pi(h) + M\beta(T)$ since by Lemma 1 the following inequality holds:

$$|\bar{\mathcal{L}}_T(h) - \mathcal{L}_\Pi(h)| \leq M\|\mathbf{P}_T - \Pi\|_{TV} \leq M\mathbb{E}\big[\|\mathbf{P}_T(\cdot|\mathcal{F}_0) - \Pi\|_{TV}\big]$$

$$\leq \sup_t \mathbb{E}\big[\|\mathbf{P}_{T+t}(\cdot|\mathcal{F}_t) - \Pi\|_{TV}\big] = M\beta(T).$$

Similarly, we can show that the following holds: $\mathcal{L}_{T+s}(h) \leq \mathcal{L}_\Pi(h) + M\phi(s)$. Therefore, we can use $\mathcal{L}_\Pi$ as a proxy to derive our generalization bound. With this in mind, we consider $\Phi(\mathbf{Z}_1^T)$ defined as in (7) except $\bar{\mathcal{L}}_{T+s}$ is replaced by $\mathcal{L}_\Pi$. Using the sub-sample selection technique of Proposition 2 and the same arguments as in the proof of Lemma 3, we obtain the following result.

Lemma 9. *Let L be a loss function bounded by M and H any hypothesis set. Suppose that $T = ma$ for some $m, a > 0$. Then, for any $\epsilon > \mathbb{E}[\Phi(\widetilde{\mathbf{Z}}_\Pi)]$, the following holds:*

$$\mathbb{P}(\Phi(\mathbf{Z}_1^T) > \epsilon) \leq a\mathbb{P}(\Phi(\widetilde{\mathbf{Z}}_\Pi) - \mathbb{E}[\Phi(\widetilde{\mathbf{Z}}_\Pi)] > \epsilon') + a(m-1)\beta(a), \tag{11}$$

where $\epsilon' = \epsilon - \mathbb{E}[\Phi(\widetilde{\mathbf{Z}}_\Pi)]$ and $\widetilde{\mathbf{Z}}_\Pi$ is an i.i.d. sample of size m from Π.

Using a Rademacher complexity bound [7] for $\mathbb{P}(\Phi(\widetilde{\mathbf{Z}}_{\Pi}) - \mathbb{E}[\Phi(\widetilde{\mathbf{Z}}_{\Pi})] > \epsilon')$ yields the following result.

Theorem 3. *With the assumptions of Lemma 9, for any $\delta > a(m-1)\beta(a)$, with probability $1 - \delta$, the following holds for all hypothesis $h \in H$:*

$$\mathcal{L}_{\Pi}(h) \leq \frac{1}{T}\sum_{t=1}^{T}\ell(h, Z_t) + 2\Re_m(H, \Pi) + M\sqrt{\frac{\log\frac{a}{\delta'}}{2m}},$$

where $\delta' = \delta - a(m-1)\beta(a)$ and $\Re_m(H, \Pi) = \frac{1}{m}\mathbb{E}[\sup_{h \in H}\sum_{i=1}^{m}\sigma_i\ell(h, \widetilde{Z}_{\Pi,i})]$ with σ_i a sequence of Rademacher random variables.

Note that our bound requires the confidence parameter δ to be at least $a(m-1)\beta(a)$. Therefore, for the bound to hold with high probability, we need to require $T\beta(a) \to 0$ as $T \to \infty$. This imposes restrictions on the speed of decay of β. Suppose first that our process is algebraically β-mixing, that is $\beta(a) \leq Ca^{-d}$ where $C > 0$ and $d > 0$. Then $T\beta(a) \leq C_0 Ta^{-d}$ for some $C_0 > 0$. Therefore, we would require $a = T^{\alpha}$ with $\frac{1}{d} < \alpha \leq 1$, which leads to a convergence rate of the order $\sqrt{T^{(\alpha-1)}\log T}$. Note that we must have $d > 1$. If the processes is exponentially β-mixing, i.e. $\beta(a) \leq Ce^{-da}$ for some $C, d > 0$, then setting $a = \log T^{2/d}$ leads to a convergence rate of the order $\sqrt{T^{-1}(\log T)^2}$.

Finally, we remark that, using the same arguments, it is possible to replace $\Re_m(H, \Pi)$ by its empirical counterpart $\frac{1}{m}\mathbb{E}[\sup_{h \in H}\sum_{t=1}^{T}\sigma_t\ell(h, Z_t)|\mathbf{Z}_1^T]$ leading to data-dependent bounds.

6   Fast Rates for Non-i.i.d. Data

For stationary mixing[6] processes, Steinwart and Christmann [18] have established fast convergence rates when a class of regularized learning algorithms is considered. Agarwal and Duchi [1] also show that stable on-line learning algorithms enjoy faster convergence rates if the loss function is strictly convex. In this section, we present an extension of the local Rademacher complexity results of [2] that imply that under some mild assumptions on the hypothesis set (that are typically used in i.i.d. setting as well) it is possible to have fast learning rates when the data is generated by a convergent process.

The technical assumption that we will exploit is that the Rademacher complexity $\Re_m(H_\ell)$ of the function class $H_\ell = \{z \mapsto \ell(h, z) : h \in H\}$ is bounded by some sub-root function $\psi(r)$. A non-negative non-decreasing function $\psi(r)$ is said to be sub-root if $\psi(r)/\sqrt{r}$ is non-increasing. Note that in this section $\Re_m(F)$ always denotes the standard Rademacher complexity with respect to distribution Π defined by $\Re_m(F) = \mathbb{E}[\sup_{f \in F}\frac{1}{m}\sum_{i=1}^{m}\sigma_i f(\widetilde{Z}_i)]$ where $\widetilde{Z}_i$ is an i.i.d. sample of size m drawn according to Π. Observe that one can always find

---

[6] In fact, the results of Steinwart and Christmann hold for α-mixing processes which is a weaker statistical assumption then β-mixing.

a sub-root upper bound on $\mathfrak{R}_m(\{f \in F \colon \mathbb{E}[f^2] \leq r\})$ by considering a slightly enlarged function class. More precisely,

$$\mathfrak{R}_m(\{f \in F \colon \mathbb{E}[f^2] \leq r\}) \leq \mathfrak{R}_m(\{g \colon \mathbb{E}[g^2] \leq r, g = \alpha f, \alpha \in [0,1], f \in F\}) = \psi(r)$$

and $\psi(r)$ can be shown to be sub-root (see Lemma 3.4 in [2]). The following analogue of Theorem 3.3 in [2] for the i.i.d. setting is the main result of this section.

Theorem 4. *Let $T = am$ for some $a, m > 0$. Assume that the Rademacher complexity $\mathfrak{R}_m(\{g \in H_\ell \colon \mathbb{E}[g^2] \leq r\})$ is upper bounded by a sub-root function $\psi(r)$ with a fixed point r^*.[7] Then, for any $K > 1$ and any $\delta > a(m-1)\beta(a)$, with probability at least $1 - \delta$, the following holds for all $h \in H$:*

$$\mathcal{L}_\Pi(h) \leq \left(\frac{K}{K-1}\right)\frac{1}{T}\sum_{t=1}^{T}\ell(h, Z_t) + C_1 r^* + \frac{C_2 \log\frac{a}{\delta'}}{m} \tag{12}$$

where $\delta' = \delta - a(m-1)\beta(a)$, $C_1 = 704K/M$, and $C_2 = 26MK + 11M$.

Before we prove Theorem 4, we discuss the consequences of this result. Theorem 4 tells us that with high probability, for any $h \in H$, $\mathcal{L}_\Pi(h)$ is bounded by a term proportional to the empirical loss, another term proportional to r^*, which represents the complexity of H, and a term in $O(\frac{1}{m}) = O(\frac{2a}{T})$. Here, m can be thought of as an "effective" size of the sample and a the price to pay for the dependency in the training sample. In certain situations of interest, the complexity term r^* decays at a fast rate. For example, if H_ℓ is a class of $\{0,1\}$-valued functions with finite VC-dimension d, then we can replace r^* in the statement of the Theorem with a term of order $d \log \frac{m}{d}/m$ at the price of slightly worse constants (see Corollary 2.2, Corollary 3.7, and Theorem B.7 in [2]).

Note that unlike standard high probability results, our bound requires the confidence parameter δ to be at least $a(m-1)\beta(a)$. Therefore, for our bound to hold with high probability, we need to require $T\beta(a) \to 0$ as $T \to \infty$ which depends on mixing rate. Suppose that our process is algebraically mixing, that is $\beta(a) \leq Ca^{-d}$ where $C > 0$ and $d > 0$. Then, we can write $T\beta(a) \leq CTa^{-d}$ and in order to guarantee that $T\beta(a) \to 0$ we would require $a = T^\alpha$ with $\frac{1}{d} < \alpha \leq 1$. On the other hand, this leads to a rate of convergence of the order $T^{\alpha-1}\log T$ and in order to have a fast rate, we need $\frac{1}{2} > \alpha$ which is possible only if $d > 2$. We conclude that for a high probability fast rate result, in addition to the technical assumptions on the function class H_ℓ, we may also need to require that the process generating the data be algebraically mixing with exponent $d > 2$. We remark that if the underlying stochastic process is geometrically mixing, that is $\beta(a) \leq Ce^{-da}$ for some $C, d > 0$, then a similar analysis shows that taking $a = \log T^{2/d}$ leads to a high probability fast rate of $T^{-1}(\log T)^2$.

We now present the proof of Theorem 4.

---

[7] The existence of a unique fixed point is guaranteed by Lemma 3.2 in [2].

Proof. First, we define $\Phi(\mathbf{Z}_1^T) = \sup_{h \in H} \left(\mathcal{L}_\Pi(h) - \frac{K}{K-1} \frac{1}{T} \sum_{t=1}^T \ell(h, Z_t) \right)$. Using the sub-sample selection technique of Proposition 2, we obtain that $\mathbb{P}(\Phi(\mathbf{Z}_1^T) > \epsilon) \leq a\mathbb{P}(\Phi(\tilde{\mathbf{Z}}_\Pi) > \epsilon) + a(m-1)\beta(a)$, where $\tilde{\mathbf{Z}}_\Pi$ is an i.i.d. sample of size m from Π. By Theorem 3.3 of [2], if $\epsilon = C_1 r^* + \frac{C_2 \log \frac{a}{\delta'}}{m}$, then $a\mathbb{P}(\Phi(\tilde{\mathbf{Z}}_\Pi) > \epsilon)$ is bounded above by $\delta - a(m-1)\beta(a)$, which completes the proof. Note that Theorem 3.3 requires that there exists B such that $\mathbb{E}_\Pi[g^2] \leq B\mathbb{E}_\Pi[g]$ for all $g \in H_\ell$. This condition is satisfied with $B = M$ since each $g \in H_\ell$ is a bounded non-negative function. □

We remark that, using similar arguments, most of the results of [2] can be extended to the setting of convergent processes. Of course, these results also hold for stationary β-mixing processes since, as we pointed out in Section 5, these are just a special case of convergent processes. However, we note that a slightly tighter bound can be derived for stationary β-mixing processes by using the independent block technique directly instead of relying on the sub-sample selection method.

7   Conclusion

We presented a series of generalization guarantees for learning in presence of non-stationary stochastic processes in terms of an average discrepancy measure that appears as a natural quantity in our general analysis. Our bounds can guide the design of time series prediction algorithms that would tame non-stationarity in the data by minimizing an upper bound on the discrepancy that can be computed from the data [8, 6]. The learning guarantees that we present strictly generalize previous Rademacher complexity guarantees derived for stationary stochastic processes or a drifting setting. We also presented simpler bounds under the natural assumption of convergent processes. In doing so, we have introduced a new sub-sample selection technique that can be of independent interest. Finally, we proved new fast rate learning guarantees in the non-i.i.d. setting. The fast rate guarantees presented can be further expanded by extending in a similar way several of the results of [2].

Acknowledgments. We thank Marius Kloft and Andres Muñoz Medina for discussions about topics related to this research. This work was partly funded by the NSF award IIS-1117591 and the NSERC PGS D3 award.

References

[1] Agarwal, A., Duchi, J.C.: The Generalization Ability of Online Algorithms for Dependent Data. IEEE Transactions on Information Theory 59(1), 573–587 (2013)
[2] Bartlett, P.L., Bousquet, O., Mendelson, S.: Local Rademacher complexities. The Annals of Statistics 33, 1497–1537 (2005)

[3] Berti, P., Rigo, P.: A Glivenko-Cantelli theorem for exchangeable random variables. Statistics and Probability Letters 32, 385–391 (1997)

[4] Doukhan, P.: Mixing: Properties and Examples. Lecture Notes in Statistics, vol. 85. Springer, New York (1989)

[5] Eberlein, E.: Weak convergence of partial sums of absolutely regular sequences. Statistics & Probability Letters 2, 291–293 (1994)

[6] Kifer, D., Ben-David, S., Gehrke, J.: Detecting change in data streams. In: Proceedings of the 30th International Conference on Very Large Data Bases (2004)

[7] Koltchinskii, V., Panchenko, D.: Rademacher processes and bounding the risk of function learning. In: High Dimensional Probability II, pp. 443–459. Birkhauser (1999)

[8] Mansour, Y., Mohri, M., Rostamizadeh, A.: Domain adaptation: learning bounds and algorithms. In: Proceedings of the Annual Conference on Learning Theory (COLT 2009). Omnipress (2009)

[9] McDiarmid, C.: On the method of bounded differences. In: Surveys in Combinatorics, pp. 148–188. Cambridge University Press (1989)

[10] Meir, R.: Nonparametric time series prediction through adaptive model selection. Machine Learning 39(1), 5–34 (2000)

[11] Mohri, M., Rostamizadeh, A.: Rademacher complexity bounds for non-i.i.d. processes. In: Advances in Neural Information Processing Systems (NIPS 2008), pp. 1097–1104. MIT Press (2009)

[12] Mohri, M., Rostamizadeh, A.: Stability bounds for stationary φ-mixing and β-mixing processes. Journal of Machine Learning 11 (2010)

[13] Mohri, M., Muñoz Medina, A.: New analysis and algorithm for learning with drifting distributions. In: Bshouty, N.H., Stoltz, G., Vayatis, N., Zeugmann, T. (eds.) ALT 2012. LNCS, vol. 7568, pp. 124–138. Springer, Heidelberg (2012)

[14] Pestov, V.: Predictive PAC learnability: A paradigm for learning from exchangeable input data. In: 2010 IEEE International Conference on Granular Computing (GrC 2010), Los Alamitos, California, pp. 387–391 (2010)

[15] Rakhlin, A., Sridharan, K., Tewari, A.: Online learning: random averages, combinatorial parameters, and learnability. In: Advances in Neural Information Processing Systems (NIPS 2010), pp. 1984–1992 (2010)

[16] Rakhlin, A., Sridharan, K., Tewari, A.: Sequential complexities and uniform martingale laws of large numbers. Probability Theory and Related Fields, 1–43 (2014)

[17] Shalizi, C.R., Kontorovich, A.: Predictive PAC Learning and Process Decompositions. In: Advances in Neural Information Processing Systems (NIPS 2013), pp. 1619–1627 (2013)

[18] Steinwart, I., Christmann, A.: Fast learning from non-i.i.d. observations. In: Bengio, Y., Schuurmans, D., Lafferty, J., Williams, C.K.I., Culotta, A. (eds.) Advances in Neural Information Processing Systems (NIPS 2009), pp. 1768–1776. MIT Press (2009)

[19] Volkonskii, V.A., Rozanov, Y.A.: Some limit theorems for random functions I. Theory of Probability and Its Applications 4, 178–197 (1959)

[20] Yu, B.: Rates of convergence for empirical processes of stationary mixing sequences. Annals Probability 22(1), 94–116 (1994)

Generalizing Labeled and Unlabeled Sample Compression to Multi-label Concept Classes*

Rahim Samei, Boting Yang, and Sandra Zilles

Department of Computer Science, University of Regina, Canada
{samei20r,boting,zilles}@cs.uregina.ca

Abstract. This paper studies sample compression of maximum multi-label concept classes for various notions of VC-dimension. It formulates a sufficient condition for a notion of VC-dimension to yield labeled compression schemes for maximum classes of dimension d in which the compression sets have size at most d. The same condition also yields a so-called tight sample compression scheme, which we define to generalize Kuzmin and Warmuth's unlabeled binary scheme to the multi-label case. The well-known Graph dimension satisfies our sufficient condition, while neither Pollard's pseudo-dimension nor the Natarajan dimension does.

Keywords: Multi-label class, sample compression, Graph dimension.

1 Introduction

A *sample compression scheme (SCS)* for a concept class C compresses every set S of labeled examples for some concept in C to a subset, which is decompressed to some concept that is consistent with S [7]. The size of the SCS is the cardinality of its largest compressed set. Since this size yields sample bounds for a PAC-learner for C [7], the question arises whether the smallest possible size of an SCS for C is bounded linearly in the VC-dimension (VCD) of C [3]. This question has become a long-standing open problem in computational learning theory.

Floyd and Warmuth [3] resolved this question positively for maximum C [3], i.e., any C meeting Sauer's upper bound on the size of classes with a given VCD [11], and thus implictly for all classes of VCD 1 (their SCSs have size equal to the VCD). An astonishing observation was made by Kuzmin and Warmuth, who proved that each maximum class of VCD d even has an *unlabeled* SCS of size d, i.e., an SCS in which the compression sets have no label information [6].

Recently, the study of SCSs was extended to the case of multi-label (instead of binary) concept classes [10], which is justified by the fact that Littlestone and Warmuth's PAC bounds in the size of an SCS are immediately transferred to the multi-label case. It was proven that, for a specific notion of VCD for multi-label classes, every maximum class of such dimension d has a labeled SCS of size d, and that the same is true for all classes of dimension 1 [10]. The proof for

---

* This work was supported by the Natural Sciences and Engineering Research Council of Canada (NSERC).

P. Auer et al. (Eds.): ALT 2014, LNAI 8776, pp. 275–290, 2014.
© Springer International Publishing Switzerland 2014

maximum classes extends Floyd and Warmuth's scheme to the multi-label case, and crucially relies on a specific property of the studied notion of VCD.

The main contributions of this paper are the following:

(1) We revisit the crucial property used in [10], which we henceforth call the *reduction property*. We observe that not only the very specific notion of VCD studied in [10] allows for generalizing the Floyd-Warmuth scheme to the multi-label case, but that generally every notion of VCD from a broad and natural class does so, as long as it fulfills the reduction property.

(2) We show that Kuzmin and Warmuth's result on unlabeled compression for maximum classes finds a natural extension to the multi-label case. This is not trivial, since unlabeled SCSs of size VCD cannot exist for maximum multi-label C for any known notion of VCD—simply because the size of C is larger than the number of unlabeled sets of size VCD. To generalize Kuzmin and Warmuth's unlabeled SCSs for maximum classes, we observe that they fulfill a property we call *tightness*. As opposed to the Floyd-Warmuth scheme and its extension to the multi-label case, a tight SCS uses exactly as many compression sets as there are concepts in C (trivially, it is impossible to use fewer sets, since each concept needs a different compression set—hence the term "tight"). Our main result is the following: for every notion of VCD in a broad and natural category, the reduction property is sufficient for proving that each multi-label class of VCD d has a tight SCS (i.e., an extension of the Kuzmin-Warmuth SCS) of size d.

(3) We prove that the well-known Graph-dimension [8] fulfills the reduction property and thus that each maximum class of Graph-dimension d has a tight SCS of size d. Neither Pollard's pseudo-dimension [9] nor the Natarajan-dimension [8] satisfies the reduction property.

2   Preliminaries

Let $\mathbb{N}^+$ be the set of all positive integers. For $m \in \mathbb{N}^+$, let $[m] = \{1, \dots, m\}$. For $m \in \mathbb{N}^+$, the set $X = \{X_1, \dots, X_m\}$ is called an *instance space*, where each instance X_i is associated with the value set $X_i = \{0, \dots, N_i\}$, $N_i \in \mathbb{N}^+$, for all $i \in [m]$. We call $c \in \prod_{i=1}^m X_i$ a *(multi-label) concept* on X, and a *(multi-label) concept class* C is a set of concepts on X, i.e., $C \subseteq \prod_{i=1}^m X_i$. For $c \in C$, let $c(X_i)$ denote the ith coordinate of c. We will always implicitly assume that a given concept class C is a subset of $\prod_{i=1}^m X_i$ for some $m \in \mathbb{N}^+$, where $X_i = \{0, \dots, N_i\}$, $N_i \in \mathbb{N}^+$. When $N_i = 1$ for all $i \in [m]$, C is a *binary concept class*. .

A *sample* is a set of *labeled examples*, i.e., of pairs $(X_t, \ell) \in X \times \mathbb{N}$. For a sample S, we define $X(S) = \{X_i \in X \mid (X_i, \ell) \in S \text{ for some } \ell\}$. For $t \in [m]$ and $C' \subseteq \prod_{i=1, \, i \neq t}^m X_i$, a concept $c \in C$ is an *extension* of a concept $c' \in C'$ iff $c = c' \cup \{(X_t, l)\}$, for some $l \in X_t$. Then c' is *extended* to c with (X_t, l).

For $Y = \{X_{i_1}, \dots, X_{i_k}\} \subseteq X$ with $i_1 < \dots < i_k$, we denote the *restriction* of a concept c to Y by $c|_Y$ and define it as $c|_Y = (c(X_{i_1}), \dots, c(X_{i_k}))$. Similarly, $C|_Y = \{c|_Y \mid c \in C\}$ denotes the restriction of C to Y. We also denote $c|_{X \setminus \{X_t\}}$ and $C|_{X \setminus \{X_t\}}$ by $c - X_t$ and $C - X_t$, respectively. In the binary case, the *reduction* C^{X_t} of C w.r.t. $X_t \in X$ consists of all concepts in $C - X_t$ that have both possible extensions to concepts in C, i.e., $C^{X_t} = \{c \in C - X_t \mid c \times \{0,1\} \subseteq C\}$. It is not obvious how the definition of reduction should be extended to the multi-valued case. One could consider the class of concepts in $C - X_t$ that have at least two distinct extensions, or of those that have all $N_t + 1$ extensions to concepts in C. We denote the former with $[C]_{\geq 2}^{X_t}$ and the latter with C^{X_t}.

In the binary case, $Y \subseteq X$ is *shattered* by C iff $C|_Y = \prod_{X_i \in Y} X_i = \{0,1\}^{|Y|}$. The size of the largest set shattered by C is the *VC-dimension* of C, denoted $\mathrm{VCD}(C)$. The literature offers a variety of VCD notions for the non-binary case [1, 8, 12, 9, 4]. Gurvits' framework [4] generalizes over many of these notions:

Definition 1. [4] *Let Ψ_i, $1 \leq i \leq m$, be a family of mappings $\psi_i : X_i \to \{0,1\}$. Let $\Psi = \Psi_1 \times \cdots \times \Psi_m$. We denote the VC-dimension of C w.r.t. Ψ by $\mathrm{VCD}_\Psi(C)$ and define it by $\mathrm{VCD}_\Psi(C) = \max_{\overline{\psi} \in \Psi} \mathrm{VCD}(\overline{\psi}(C))$.*

Specific families of mappings yield specific notions of dimension. The most general case is the family Ψ^* of *all* m-tuples $(\psi_1, \ldots, \psi_m)$ with $\psi_i : X_i \to \{0,1\}$.

The term *Graph-dimension* [8] refers to VCD_{Ψ_G}, where $\Psi_G = \Psi_{G_1} \times \cdots \times \Psi_{G_m}$ and for all $i \in [m]$, $\Psi_{G_i} = \{\psi_{G,k} : k \in N_i\}$ and $\psi_{G,k}(x) = 1$ if $x = k$, $\psi_{G,k}(x) = 0$ if $x \neq k$. By *Pollard's pseudo-dimension* [9] we refer to VCD_{Ψ_P}, where $\Psi_P = \Psi_{P_1} \times \cdots \times \Psi_{P_m}$ and for all $i \in [m]$, $\Psi_{P_i} = \{\psi_{P,k} : k \in N_i\}$ and $\psi_{P,k}(x) = 1$ if $x \geq k$, $\psi_{P,k}(x) = 0$ if $x < k$. The term *Natarajan-dimension* [8] refers to VCD_{Ψ_N}, where $\Psi_N = \Psi_{N_1} \times \cdots \times \Psi_{N_m}$ and for all $i \in [m]$, $\Psi_{N_i} = \{\psi_{N,k,k'} : k, k' \in N_i, k \neq k'\}$ and $\psi_{N,k,k'}(x) = 1$ if $x = k$, $\psi_{N,k,k'}(x) = 0$ if $x = k'$, $\psi_{N,k,k'}(x) = *$, otherwise. (Here technically, ψ_i maps to $\{0,1,*\}$, where $*$ is a null element to be ignored when computing the VC-dimension.)

Clearly, VCD_{Ψ^*} upper-bounds all VCD notions. Also, $\mathrm{VCD}_{\Psi_P} \geq \mathrm{VCD}_{\Psi_N}$ and $\mathrm{VCD}_{\Psi_G} \geq \mathrm{VCD}_{\Psi_N}$ [5]. However, VCD_{Ψ_P} and VCD_{Ψ_G} are incomparable [2].

Let Ψ_i be a family of mappings $\psi_i : X_i \to \{0,1\}$. Ψ_i is spanning on X_i iff any real-valued function on X_i is a linear combination of mappings from Ψ_i.

Gurvits [4] showed that the quantity $\Phi_d(N_1, \ldots, N_m) = 1 + \sum_{1 \leq i \leq m} N_i + \sum_{1 \leq i_1 < i_2 \leq m} N_{i_1} N_{i_2} + \cdots + \sum_{1 \leq i_1 < i_2 < \cdots < i_d \leq m} N_{i_1} N_{i_2} \cdots N_{i_d}$ is an upper bound on $|C|$ if $\mathrm{VCD}_\Psi(C) = d$, assuming that $\Psi = \Psi_1 \times \cdots \times \Psi_m$ where each Ψ_i, $1 \leq i \leq m$, is a spanning family of mappings. This bound is tight for all $m, d, N_1, \ldots, N_m$.

Note that, for a spanning families of mappings, the finiteness of the resulting VCD_Ψ-value of a multi-label class C guarantees that C is PAC-learnable [2].

As in the binary case [3], a *forbidden labeling* of C with $\mathrm{VCD}_\Psi(C) = d < |X|$, is a set of $d + 1$ examples that is inconsistent with all concepts in C. For $Y = \{X_{i_1}, \ldots, X_{i_{d+1}}\} \subseteq X$, $\mathrm{Forb}(C, Y) = X_{i_1} \times \cdots \times X_{i_{d+1}} \setminus C|_Y$ is the set of forbidden labelings on Y and $\mathrm{Forb}(C) = \bigcup_{Y \subseteq X, |Y| = d+1} \mathrm{Forb}(C, Y)$ is the set of forbidden labelings of size $d + 1$. For $d = |X|$, we define $\mathrm{Forb}(C, Y) = \mathrm{Forb}(C) = \emptyset$.

For binary classes, the smallest possible *size of a sample compression scheme* yields sample bounds for PAC-learning [7, 3] and an open question is whether this parameter is linear in the VC-dimension. The proof that (in the binary case) a sample compression scheme yields a successful PAC-learner with bounds expressed in terms of its size [7] immediately generalizes to the multi-label case. The notion of sample compression trivially generalizes to the multi-label case:

Definition 2. [7] *A sample compression scheme for* C *is a pair* (f, g) *of mappings as follows. Given any sample* S *that is consistent with some concept in* C, *one requires (i)* $f(S) \subseteq S$, *and (ii)* $g(f(S)) = (l_1, \ldots, l_m)$, *where* $(X_i, \ell_i) \in S$ *implies* $\ell_i = l_i$, *for all* $i \in [m]$. *The size of* (f, g) *is the maximum cardinality of a set* $f(S)$, *taken over all samples* S *consistent with some concept in* C.

Every binary *maximum* class C with $\text{VCD}(C) = d$ (i.e., $|C| = \sum_{i=0}^{d} \binom{m}{i}$), which is the largest possible size [11]) has a compression scheme of size d [3]. This result was strengthened by showing the existence of unlabeled schemes (in which the compression sets are subsets of X without label information) of size VCD [6]. Both results rely on the fact that, for $d < m$, restrictions and reductions of binary maximum classes w.r.t. a single instance are maximum of VCD d and $d - 1$, resp. [13]. We focus on multi-label maximum classes, i.e., classes with the largest size among all classes of the same VCD_Ψ, for a fixed Ψ.

Consider the following geometric example of a class that is maximum of VCD_{Ψ^*} 2 and VCD_{Ψ_G} 2. X corresponds to m lines in general position on the plane, i.e., no two lines are parallel and no three lines share a common point. Then (i) the number of regions is $1 + m + m(m - 1)/2$; (ii) the number of segments and rays is m^2; (iii) the number of intersection points is $m(m - 1)/2$. Summing these numbers yields $1 + 2m^2 = \Phi_2(2, \ldots, 2)$. All regions, segments, rays and intersection points form a natural multi-label class concept class that is VCD_{Ψ^*}-maximum and VCD_{Ψ_G}-maximum of dimension 2. Each instance takes values in $\{-1, 0, +1\}$, depending on which side of the line the concept is on (and 0 if the concept is contained within the line itself). Each region is a concept with instance values -1 or $+1$. Each segment/ray is a concept with value 0 in one particular instance and values -1 or $+1$ in all the other instances. Each intersection point is a concept with value 0 on exactly two instances. One can verify that no set of three instances is shattered using any label mapping to a binary class.

3   The Reduction Property and Tight Compression Schemes

We next define a core notion of our work, namely the reduction property. It provides a sufficient condition for maximum classes of VCD_Ψ d to have a sample compression scheme of size d, provided that Ψ is based on spanning families.

Definition 3. *Let* $m > 1$ *and* Ψ_i, $1 \leq i \leq m$, *be a family of mappings. Let* $\Psi = \Psi_1 \times \cdots \times \Psi_m$. VCD_Ψ *fulfills the* reduction property *iff for any* VCD_Ψ-*maximum class* $C \subseteq \prod_{i=1}^{m} X_i$, *for any* $t \in [m]$ *and for any concept* $\bar{c} \in C - X_t$, $|\{c \in C \mid c - X_t = \bar{c}\}| \in \{1, N_t + 1\}$ *(i.e.,* $[C]_{\geq 2}^{X_t} = C^{X_t}$).

Recently, Samei et al. [10] extended the Floyd-Warmuth sample compression scheme [3] to VCD_{Ψ^*}-maximum concept classes by showing that VCD_{Ψ^*} fulfills the reduction property. Inspecting their proofs reveals that the Floyd-Warmuth scheme can be extended to maximum classes for a broad class of VCD notions.

Theorem 1. *Let Ψ_i, $1 \leq i \leq m$, be a spanning family of mappings. Let $\Psi = \Psi_1 \times \cdots \times \Psi_m$. If VCD_Ψ fulfills the reduction property then any VCD_Ψ-maximum class C has a labeled sample compression scheme of size $\mathrm{VCD}_\Psi(C)$.*

As discussed before, for all notions of VCD_Ψ mentioned above, unlabeled compression schemes of size d for a VCD_Ψ-maximum class C of VCD_Ψ d cannot exist, as the number of concepts in C is larger than the number of subsets of the instance space of size at most $\mathrm{VCD}_\Psi(C)$. Here, we generalize the unlabeled compression scheme for VCD-maximum classes by Kuzmin and Warmuth [6] to VCD_Ψ-maximum classes, where VCD_Ψ fulfills the reduction property and is based on spanning families of mappings, by first observing its *tightness*.

Definition 4. *Let C be a VCD_Ψ-maximum class with $\mathrm{VCD}_\Psi(C) = d$. A sample compression scheme of size d for C is tight iff (i) there are exactly $|C|$ many compression sets in the scheme, (ii) each compression set represents a unique concept c in the class, which is consistent with the compression set, (iii) each sample of a concept contains exactly one compression set that represents a concept consistent with the sample.*

For illustration, consider the class C and the representatives shown in Table 1, which yield a tight scheme. As required in (i), no concept can have more than one compression set. Condition (ii) forces the one-to-one correspondence between the concepts and the compression sets. Without Condition (iii), one might map c_2 to $(X_4, 0)$ instead of $(X_3, 1)$ and the scheme would still satisfy (i) and (ii), while the sample $\{(X_2, 0), (X_4, 0)\}$ could be compressed to either $(X_4, 0)$ or $\emptyset$.

Our main result strengthens Theorem 1 and generalizes Kuzmin-Warmuth's unlabeled scheme (which is a tight labeled scheme) to the multi-label case:

Theorem 2. *Let Ψ_i, $1 \leq i \leq m$, be a spanning family of mappings. Let $\Psi = \Psi_1 \times \cdots \times \Psi_m$. If VCD_Ψ fulfills the reduction property then any VCD_Ψ-maximum class C has a tight sample compression scheme of size $\mathrm{VCD}_\Psi(C)$.*

The critical point exploited in our tight scheme is the property of *missing labelings* in the compression sets, that is, for each set of at most $\mathrm{VCD}_\Psi(C)$ instances $\{X_{i_1}, \ldots, X_{i_k}\}$, there is a tuple of labels $(l_{i_1}, \ldots, l_{i_k}) \in \prod_{1 \leq j \leq k} X_{i_j}$, such that for each compression set S with $X(S) = \{X_{i_1}, \ldots, X_{i_k}\}$ and for all $j \in \{1, \ldots, k\}$, $(X_{i_j}, l_{i_j}) \notin S$. In the binary case, our scheme exactly coincides with the Kuzmin-Warmuth scheme, which also exploits the non-trivial property of missing labelings. If one adds labels to the compression sets in the Kuzmin-Warmuth scheme, each set $S \subseteq X$ of size $k \in \{1, \ldots, \mathrm{VCD}(C)\}$ has exactly one missing labeling, and thus $2^k - 1$ assignments of 0 and 1 to the k instances in S are not used as compression sets. But then there is only one possible assignment of labels to the instances in S left, which is why the scheme is in fact unlabeled.

Our proof has the same structure as that of Kuzmin and Warmuth for the binary case. However, various technical barriers have to be overcome for the multi-label case. Because of space constraints, many proof details are omitted.

In [6] a *representation mapping* r for a VCD-maximum class $C \subseteq 2^X$ is a bijection between C and the set of all subsets of X of size at most $\mathrm{VCD}(C)$

Labeled Representatives Construction Algorithm
Input: the set $\text{Rep}_{\leq d}(X) = \{Y \subseteq X \mid 0 \leq |Y| \leq d\}$
Output: a set of labeled representatives from $\text{Rep}_{\leq d}(X)$

1. Set $\text{LRep}_{\leq d}(X) \leftarrow \{\emptyset\}$.
2. For each $Y = \{X_{i_1}, \ldots, X_{i_k}\} \in \text{Rep}_{\leq d}(X) \setminus \{\emptyset\}$ do
    Set $\text{Rep}_{\leq d}(X) \leftarrow \text{Rep}_{\leq d}(X) \setminus \{Y\}$
    Pick some $L^Y = (l_1^Y, \ldots, l_k^Y) \in \prod_{1 \leq j \leq k} X_{i_j}$
    Set $\text{LabeledRep}(Y, L^Y) \leftarrow \prod_{1 \leq j \leq k} (X_{i_j} \setminus \{l_j^Y\})$
    Set $\text{LRep}_{\leq d}(X) \leftarrow \text{LRep}_{\leq d}(X) \cup \text{LabeledRep}(Y, L^Y)$.

Algorithm 1. Constructing a set of representatives

such that for any $c, c' \in C$, $c|_{(r(c) \cup r(c'))} \neq c'|_{(r(c) \cup r(c'))}$, that is, c and c' do not *clash* w.r.t. r. The non-clashing property for a representation mapping is equivalent to having a unique representative for each sample consistent with some concept in C [6]. Kuzmin and Warmuth showed that, given a representation mapping r for a class C, for any sample S of a concept from C with $|S| \geq \text{VCD}(C)$, there is some concept $c \in C$ that is consistent with S for which, S can be mapped to $r(c) \subseteq X(S)$ and for any $c' \in C$, $c' \neq c$, consistent with S, $r(c') \nsubseteq X(S)$.

For the rest of this section, let $C \subseteq \prod_{1 \leq i \leq m} X_i$ be VCD_Ψ-maximum of dimension d, where VCD_Ψ has the reduction property and Ψ is the direct product of spanning families of mappings. Note that $|C| = \Phi_d(N_1, \ldots, N_m)$, and, if $d < m$, $C - X_t$ and C^{X_t} are maximum of VCD_Ψ d and $d - 1$, resp. [4, 10].

As we need to use labels in the compression sets, we modify the definition of representation mapping. For a set $Y = \{X_{i_1}, \ldots, X_{i_k}\} \subseteq X$, let L^Y always denote a tuple of labels $L^Y = (l_1^Y, \ldots, l_k^Y) \in \prod_{1 \leq j \leq k} X_{i_j}$. Consider the set $\text{Rep}_{\leq d}(X) = \{Y \subseteq X \mid 0 \leq |Y| \leq d\}$. We construct a set of labeled representatives $\text{LRep}_{\leq d}(X)$ from $\text{Rep}_{\leq d}(X)$ using Algorithm 1.

For each $Y = \{X_{i_1}, \ldots, X_{i_k}\}$ with $k \leq d$, $C|_Y = \prod_{1 \leq j \leq k} X_{i_j}$. So, for any output $\text{LRep}_{\leq d}(X)$ from Algorithm 1, and for any representative $S \in \text{LRep}_{\leq d}(X)$, there is a $c \in C$ with $S \subseteq c$. Further, $|\text{LRep}_{\leq d}(X)| = \Phi_d(N_1, \ldots, N_m) = |\overline{C}|$.

We say that a bijection r between C and some $\text{LRep}_{\leq d}(X)$ is *consistent*, if for each $c \in C$, $r(c) \subseteq c$. We also say that the concepts $c, c' \in C$, $c \neq c'$, clash w.r.t. a consistent bijection r, if $r(c) \subseteq c'$ and $r(c') \subseteq c$.

Definition 5. *A representation mapping for C is a consistent bijection r between C and some representative set $\text{LRep}_{\leq d}(X)$ in which no two concepts clash.*

Essentially, we want to find a representation mapping for VCD_Ψ-maximum classes with a fixed VCD_Ψ. As in the binary case [6], the following lemma shows how the non-clashing property is useful for finding unique labeled representatives for samples in the multi-label case.

Lemma 1. *Let r be a consistent bijection between C and a set of labeled representatives $\text{LRep}_{\leq d}(X)$. Then the following two statements are equivalent:*

1. *No two concepts clash w.r.t. r.*
2. *For any sample S that is consistent with at least one concept in C, there is exactly one concept $c \in C$ that is consistent with S and $r(c) \subseteq S$.*

Lemma 1 helps us to construct a compression scheme of size VCD_Ψ for a VCD_Ψ-maximum class C from a representation mapping r. For compression, a sample S is compressed to $r(c) \subseteq S$, where c is consistent with S. For reconstruction, $r(c)$ is mapped to $c \supseteq S$, as r is a consistent bijective mapping.

We showed that a representation mapping can be used as a compression-reconstruction function for the concepts in a VCD_Ψ-maximum class C. In the next corollary, we use such a mapping to derive a compression scheme of size d for $C|_Y$, for any $Y \subseteq X$ with $|Y| > d$. For any $\bar{c} \in C|_Y$, define $r_Y(\bar{c}) := r(c)$ where c is the unique concept in C with $c|_Y = \bar{c}$ and $r(c) \subseteq \bar{c}$.

Corollary 1. *Let r be a representation mapping for C. Let $Y \subseteq X$ with $|Y| > d$. Then r_Y is a representation mapping for $C|_Y$.*

At this point, the crucial notion of *tail* comes into play. As in the binary case, we define the *tail* of a concept class C on an instance $X_t \in X$ as the set of all concepts $c \in C$ such that $c - X_t \in (C - X_t) \setminus C^{X_t}$ [6]. This corresponds to the set of concepts in $C - X_t$ that do not have all extensions onto X, or equivalently (by the reduction property), that have a unique extension onto X. That is, for any $c \in \text{tail}_{X_t}(C)$, there exists only one label $l \in \{0, 1, \ldots, N_t\}$ such that $(c - X_t) \cup \{(X_t, l)\} \in C$. Note that $C = C^{X_t} \times X_t \cup \text{tail}_{X_t}(C)$.

As in the binary case, we establish a connection between tail concepts and forbidden labelings. By assumption, for $X_p \in X$, every concept in $C - X_p$ has either a unique or all possible extensions to concepts in C. So, each concept in $\text{tail}_{X_p}(C)$ corresponds to a concept in $C - X_p$ that has only one extension onto X_p. That is, $|\text{tail}_{X_p}(C)| = |\text{tail}_{X_p}(C) - X_p|$. Further, $C - X_p = C^{X_p} \cup (\text{tail}_{X_p}(C) - X_p)$ where C^{X_p} and $(\text{tail}_{X_p}(C) - X_p)$ are disjoint. Since, for $d < m$, $C - X_p$ and C^{X_p} are VCD_Ψ-maximum of dimensions d and $d - 1$ resp. [4, 10], we have $|\text{tail}_{X_p}(C)| = |\text{tail}_{X_p}(C) - X_p| = |C - X_p| - |C^{X_p}| = \sum_{1 \leq i_1 < \cdots < i_d \leq m,\ i_j \neq p} N_{i_1} \cdots N_{i_d}$.

For $Y = \{X_{i_1}, \ldots, X_{i_{d+1}}\}$, $C|_Y$ is VCD_Ψ-maximum of dimension d and thus $|\text{Forb}(C, Y)| = (N_{i_1} + 1) \cdots (N_{i_{d+1}} + 1) - \Phi_d(N_{i_1}, \ldots, N_{i_{d+1}}) = N_{i_1} \cdots N_{i_{d+1}}$.

As in the binary case, it is easy to see that every concept in $\text{tail}_{X_p}(C)$ contains some forbidden labeling of C^{X_p} of size d and each such forbidden labeling occurs in at least one tail concept. Note that C^{X_p} is a VCD_Ψ-maximum class of dimension $d - 1$ and for each set of d instances $Y = \{X_{i_1}, \ldots, X_{i_d}\} \subseteq (X \setminus \{X_p\})$, $|\text{Forb}(C^{X_p}, Y)| = N_{i_1} \cdots N_{i_d}$. So, $|\text{Forb}(C^{X_p})| = \sum_{1 \leq i_1 < \cdots < i_d \leq m, i_j \neq p} N_{i_1} \cdots N_{i_d}$ $= |\text{tail}_{X_p}(C)|$. First, adding any concept in $\text{tail}_{X_p}(C) - X_p$ to C^{X_p} increases the VCD_Ψ of C^{X_p} due to the maximum size property of C^{X_p}. So, each concept in $\text{tail}_{X_p}(C)$ contains at least one forbidden labeling of C^{X_p}. Second, $C - X_p = C^{X_p} \cup (\text{tail}_{X_p}(C) - X_p)$ where the reduction class and the tail class are disjoint. Next, for each set of d instances $Y \subseteq (X \setminus \{X_p\})$, $(C - X_p)|_Y = \prod_{X_i \in Y} X_i$, since C is VCD_Ψ-maximum class of dimension d. That is, $C^{X_p}|_Y \cup (\text{tail}_{X_p}(C) - X_p)|_Y = \prod_{X_i \in Y} X_i$ and $(\text{tail}_{X_p}(C) - X_p)|_Y = \text{Forb}(C^{X_p}, Y) = \prod_{X_i \in Y} X_i \setminus C^{X_p}|_Y$.

In other words, all forbidden labelings of C^{X_p} on Y are in $(\text{tail}_{X_p}(C) - X_p)|_Y$. Since Y was chosen arbitrarily, we conclude that all forbidden labelings of C^{X_p} appear in $\text{tail}_{X_p}(C)$.

The Kuzmin-Warmuth scheme finds representatives for C by partitioning C into $C^{X_i} \times X_i$ and $\text{tail}_{X_i}(C)$ for some $X_i \in X$. It identifies the representatives for C^{X_i} recursively, and extends them to representatives for C. That is, for any concept $c \in C^{X_i}$ with a representative $r(c)$, $r(c \cup (X_i, 0)) := r(c)$ and $r(c \cup (X_i, 1)) := r(c) \cup X_i$. Next, it finds representatives for the remaining concepts, i.e., those in $\text{tail}_{X_i}(C)$ by assigning each of them a forbidden labeling of the class C^{X_i} of size d. Since the representative for each concept in $\text{tail}_{X_i}(C)$ is a forbidden labeling of the class C^{X_i}, the non-clashing property between $\text{tail}_{X_i}(C)$ and C^{X_i} is guaranteed.

As in the Kuzmin-Warmuth scheme, we establish a recursive structure for tails. We introduce some notation, first. For $p, q \in [m]$, with $p < q$ and a concept $\bar{c} \in C|_{X \setminus \{X_p, X_q\}}$, let $i\bar{c}$, $\bar{c}j$ and $i\bar{c}j$ denote $\bar{c} \cup \{(X_p, i)\}$, $\bar{c} \cup \{(X_q, j)\}$ and $\bar{c} \cup \{(X_p, i), (X_q, j)\}$, respectively.

Lemma 2. *Let $p, q \in [m]$ with $p \neq q$. Then the following statements are true.*

1. *For each $c \in \text{tail}_{X_p}(C^{X_q})$ there are at least N_q labels $l_1, \ldots, l_{N_q} \in X_q$ such that $c \times \{l_1, \ldots, l_{N_q}\} \subseteq \text{tail}_{X_p}(C)$. If $c \in \text{tail}_{X_p}(C^{X_q}) \setminus \text{tail}_{X_p}(C - X_q)$, then there are exactly N_q such labels.*
2. *For each $c \in \text{tail}_{X_p}(C - X_q)$ there is at least one label $l \in X_q$ such that $c \times \{l\} \in \text{tail}_{X_p}(C)$. If $c \in \text{tail}_{X_p}(C - X_q) \cap \text{tail}_{X_p}(C^{X_q})$, then $c \times X_q \subseteq \text{tail}_{X_p}(C)$.*
3. *Each concept in $\text{tail}_{X_p}(C)$ is an extension of either a concept in $\text{tail}_{X_p}(C^{X_q})$ or a concept in $\text{tail}_{X_p}(C - X_q)$.*

Proof. W.l.o.g., assume $p < q$. We omit the proof of Statement 2.

Proof of Statement 1. W.l.o.g., let $c = 0\bar{c} \in \text{tail}_{X_p}(C^{X_q})$. We show that for some set $\{l_1, \ldots, l_{N_q}\} \subset X_q$, $0\bar{c}j \in \text{tail}_{X_p}(C)$, $j \in \{l_1, \ldots, l_{N_q}\}$. Clearly, $\text{tail}_{X_p}(C^{X_q}) \subseteq C^{X_q}$, so $0\bar{c} \in C^{X_q}$ and thus $0\bar{c}0, \ldots, 0\bar{c}N_q \in C$. We need to show that N_q concepts $0\bar{c}j$, $j \in \{l_1, \ldots, l_{N_q}\}$, belong to $\text{tail}_{X_p}(C)$. For purposes of contradiction, assume that $0\bar{c}0, 0\bar{c}1 \notin \text{tail}_{X_p}(C)$, that is, $\bar{c}0, \bar{c}1 \in C^{X_p}$. Since C^{X_p} is VCD$_\Psi$-maximum, $\bar{c}$ has $N_q + 1$ extensions to concepts in C^{X_p}. Therefore,

$$\bar{c}0, \bar{c}1, \ldots, \bar{c}N_q \in C^{X_p} \Rightarrow \begin{cases} 0\bar{c}0, & 1\bar{c}0, & \ldots, N_p\bar{c}0 & \in C \\ \vdots & & \\ 0\bar{c}N_q, & 1\bar{c}N_q, & \ldots, N_p\bar{c}N_q & \in C \end{cases}$$

i.e., $0\bar{c}, \ldots, N_p\bar{c} \in C^{X_q}$ and $\bar{c} \in (C^{X_q})^{X_p}$. So, $0\bar{c} \notin \text{tail}_{X_p}(C^{X_q})$—a contradiction.

We need to show that if $0\bar{c} \in \text{tail}_{X_p}(C^{X_q}) \setminus \text{tail}_{X_p}(C - X_q)$, there is an $l \in X_q$ for which $0\bar{c}l \notin \text{tail}_{X_p}(C)$. Assume that for all $j \in X_q$, $0\bar{c}j \in \text{tail}_{X_p}(C)$, i.e., $\bar{c}j \notin C^{X_p}$. That is, for all $j \in X_q$, $0\bar{c}j \in C$ and $\bar{c}j$ has only one extension on X_p to concepts in C, namely with $(X_p, 0)$. So, for all $i \in X_p \setminus \{0\}$ and all $j \in X_q$, $i\bar{c}j \notin C$, and thus $i\bar{c} \notin C - X_q$. This would imply $0\bar{c} \in \text{tail}_{X_p}(C - X_q)$.

Proof of Statement 3. First one can show that $|\mathrm{tail}_{X_p}(C)| = N_q|\mathrm{tail}_{X_p}(C^{X_q})|$ $+|\mathrm{tail}_{X_p}(C-X_q)|$. Second, from Statements 1 and 2, any concept in $\mathrm{tail}_{X_p}(C^{X_q})$ can be mapped to N_q concepts in $\mathrm{tail}_{X_p}(C)$, and any concept in $\mathrm{tail}_{X_p}(C-X_q)$ can be mapped to 1 concept in $\mathrm{tail}_{X_p}(C)$. Hence, each concept in $\mathrm{tail}_{X_p}(C)$ is an extension of either a concept in $\mathrm{tail}_{X_p}(C^{X_q})$ or a concept in $\mathrm{tail}_{X_p}(C-X_q)$. □

The next lemma states that the reduction and restriction operations are interchangeable in the order in which they are applied.

Lemma 3. *For any $p,q \in [m]$, with $p \neq q$, $C^{X_p} - X_q = (C - X_q)^{X_p}$.*

Corollary 2. $\mathrm{Forb}((C - X_q)^{X_p}) \subseteq \mathrm{Forb}(C^{X_p})$.

Lemma 4. *Any forbidden labeling for $(C^{X_p})^{X_q}$ can be extended to N_q forbidden labelings for C^{X_p}.*

Proof. Let $\mathrm{VCD}_\Psi(C) = d$. We show that for any set of d instances $Y \subseteq X \setminus \{X_p\}$ with $X_q \in Y$, there are N_q forbidden labelings $S_i = S \cup \{(X_q, l_i)\}$, $1 \leq i \leq N_q$ and $l_i \in X_q$, for C^{X_p} such that $X(S_i) = Y$, $X(S) = Y \setminus X_q$, and S is a forbidden labeling of size $d - 1$ for $(C^{X_p})^{X_q}$.

Let $Y = \{X_{i_1}, \ldots, X_{i_{d-1}}, X_q\} \subseteq X \setminus \{X_p\}$, $X(S) = \{X_{i_1}, \ldots, X_{i_{d-1}}\}$, and let $S_1 = S \cup \{(X_q, l_1)\}$ be a forbidden labeling for C^{X_p}. We first prove by contradiction that S is a forbidden labeling for $(C^{X_p})^{X_q}$. Assume that S is not a forbidden labeling for $(C^{X_p})^{X_q}$, and thus is consistent with some concept $c \in (C^{X_p})^{X_q}$. Since $c \times X_q \subseteq C^{X_p}$, we conclude that each sample $S \cup \{(X_q, j)\}$, $j \in X_q$, is consistent with some concept in C^{X_p}. Thus, $S \cup \{(X_q, l_1)\}$ is not a forbidden labeling for C^{X_p}—a contradiction.

We next show that there are $N_q - 1$ more forbidden labels $S_i = S \cup \{(X_q, l_i)\}$, $2 \leq i \leq N_q$, $l_i \in X_q$ for C^{X_p}, i.e., for any concept $\bar{c} \in C^{X_p}$ with $\bar{c}|_{\{X_{i_1}, \ldots, X_{i_{d-1}}\}} = S$, $\bar{c}(X_q) = l$ for some $l \in X_q \setminus \{l_1, \ldots, l_{N_q}\}$. Note that C^{X_p} is VCD_Ψ-maximum of dimension $d - 1$ so that $C^{X_p}|_{\{X_{i_1}, \ldots, X_{i_{d-1}}\}} = \prod_{j \in \{1, \ldots, d-1\}} X_{i_j}$, and thus $S \in C^{X_p}|_{\{X_{i_1}, \ldots, X_{i_{d-1}}\}}$. For any $\bar{c} \in C^{X_p}$ with $\bar{c}|_{\{X_{i_1}, \ldots, X_{i_{d-1}}\}} = S$, it is clear that $\bar{c}(X_q) \neq l_1$, as $S \cup \{(X_q, l_1)\}$ is a forbidden labeling for C^{X_p}. That is, for any $c' \in C^{X_p}|_Y$ with $c' - X_q = S$, $c'(X_q) \neq l_1$. So, $C^{X_p}|_Y$ does not have all extensions of S and thus, $C^{X_p}|_Y$ has a unique extension of S on X_q, as $C^{X_p}|_Y$ is a VCD_Ψ-maximum class of dimension $d - 1$ on Y. So, there is only one concept $c' \in C^{X_p}|_Y$ with $c' - X_q = S$ and $c'(X_q) = l$, for some $l \in X_q \setminus \{l_1, \ldots, l_{N_q}\}$.

Now, we need to show that C^{X_p} has a unique extension of S on X_q, namely $S \cup \{(X_q, l)\}$. Assume that there are concepts $\bar{c}_1, \bar{c}_2 \in C^{X_p}$ with $\bar{c}_1|_{\{X_{i_1}, \ldots, X_{i_{d-1}}\}} = \bar{c}_2|_{\{X_{i_1}, \ldots, X_{i_{d-1}}\}} = S$ and $\bar{c}_1(X_q) \neq \bar{c}_2(X_q)$. Let $\bar{c}_1(X_q) = l$ and $\bar{c}_2(X_q) = l'$. Since $\bar{c}_1|_Y \neq \bar{c}_2|_Y$ and $\bar{c}_1|_Y, \bar{c}_2|_Y \in C^{X_p}|_Y$, we conclude that $C^{X_p}|_Y$ has two extensions of S with (X_q, l) and (X_q, l')—a contradiction. So, for any $\bar{c} \in C^{X_p}$ with $\bar{c}|_{\{X_{i_1}, \ldots, X_{i_{d-1}}\}} = S$, $\bar{c}(X_q) = l$. In other words, each sample $S \cup \{(X_q, l_i)\}$, $1 \leq i \leq N_q$ is a forbidden labeling for C^{X_p}.

Since C is VCD_Ψ-maximum of dimension d, C^{X_p} and $(C^{X_p})^{X_q}$ are both VCD_Ψ-maximum of dimension $d - 1$ and $d - 2$, resp. [4, 10]. One can then show

Labeled Tail Matching Function (LTMF)
Input: a VCD_Ψ-maximum multi-label concept class C, X with $|X| \geq 1$
Output: a mapping r assigning representatives to all concepts in C

```
r = LTMF(C,X)
        If VCD_Ψ(C) = 0 then r(c) := ∅; (since C = {c})
        Else { pick any X_q ∈ X; r̃ = LTMF(C^{X_p}, X \ {X_p});
            For each c̄ ∈ C^{X_p} do {
                For i = 1 to N_p do
                    r(c̄ ∪ {(X_p,i)}) := r̃(c̄) ∪ {(X_p,i)};
                r(c̄ ∪ {(X_p,0)}) := r̃(c̄); }
            Set r ← r ∪ LTS(C, X, X_p);} (see Algorithm 3 for LTS)
        return r;
```

Algorithm 2. Recursively constructing labeled compression sets for concepts

$|\text{Forb}(C^{X_p})| = N_q |\text{Forb}((C^{X_p})^{X_q})| + |\text{Forb}((C - X_q)^{X_p})|$. So, $|\text{Forb}((C^{X_p})^{X_q})| = \frac{1}{N_q}|\text{Forb}(C^{X_p}, Y)|$, for all $Y \subseteq X \setminus \{X_p\}$ with $|Y| = d$ and $X_q \in Y$.

Therefore, any set of N_q forbidden labelings $S_i = S \cup \{(X_q, l_i)\}$, $1 \leq i \leq N_q$ for C^{X_p} can be mapped to one forbidden labeling S for $(C^{X_p})^{X_q}$. By counting the number of forbidden labelings for C^{X_p} that contain X_q (as done above), we conclude that any forbidden labeling for $(C^{X_p})^{X_q}$ can be extended to N_q forbidden labelings for C^{X_p}. $\qquad \square$

The next lemma now follows from Corollary 2 and Lemma 4.

Lemma 5. *Each forbidden labeling of C^{X_p} is an extension of either a forbidden labeling of $(C^{X_p})^{X_q}$ or forbidden labeling of $C^{X_p} - X_q$.*

The following lemma is crucial in connecting the set of forbidden labelings to a labeled set of representatives. While its statement is obvious in the binary case, it is not trivial in the multi-label case.

Lemma 6. *For any set $Y = \{X_{i_1}, \ldots, X_{i_d}\} \subseteq X \setminus \{X_p\}$ with $|Y| = d$, there is a tuple $(l_1, \ldots, l_d) \in \prod_{1 \leq j \leq d} X_{i_j}$ such that $\text{Forb}(C^{X_p}, Y) = \prod_{1 \leq j \leq d} (X_{i_j} \setminus \{l_j\})$.*

Proof. (sketch) Let $m = |X|$. The case $m = 1$ ($d = 0$) is obvious. One can also prove by induction on m that the claim is true for $d = m - 1$; details are omitted. For the general case, i.e. a VCD_Ψ-maximum class on m instances with $\text{VCD}_\Psi(C) = d < m$, the proof is by induction on m. The base case is $m = d+1$ or equivalently $d = m - 1$. Assume that the claim is true for any $m' < m$. Pick $X_q \in X \setminus \{X_p\}$. By Lemma 5, each forbidden labeling of C^{X_p} is an extension of a forbidden labeling of either $(C^{X_p})^{X_q}$ or $C^{X_p} - X_q$.

By Lemma 3, $C^{X_p} - X_q = (C - X_q)^{X_p}$ and thus $\text{Forb}(C^{X_p} - X_q) = \text{Forb}((C - X_q)^{X_p})$. $C - X_q$ is VCD_Ψ-maximum on $m - 1$ instances and of VCD_Ψ d. So, by induction hypothesis, for any set $Y = \{X_{i_1}, \ldots, X_{i_d}\} \subseteq X \setminus \{X_p, X_q\}$, there is a tuple $(l_1, \ldots, l_d) \in \prod_{1 \leq j \leq d} X_{i_j}$ such that $\text{Forb}((C - X_q)^{X_p}, Y) = \prod_{1 \leq j \leq d}(X_{i_j} \setminus \{l_j\})$ and hence $\text{Forb}(C^{X_p} - X_q, Y) = \prod_{1 \leq j \leq d}(X_{i_j} \setminus \{l_j\})$. Forbidden labelings of $C^{X_p} - X_q$ are exactly all forbidden labelings of C^{X_p} the do

Labeled Tail Subroutine (LTS)
Input: a VCD_Ψ-maximum multi-label concept class C over X, $X_p \in X$
Output: a mapping r assigning representatives to all concepts in $\mathrm{tail}_{X_p}(C)$

$r=\mathbf{LTS}(C,X,X_p)$
1. **If** $\mathrm{VCD}_\Psi(C) = 0$ **then** $r(c) := \emptyset$; (since $C = \mathrm{tail}_{X_p}(C) = \{c\}$)
   Else if $\mathrm{VCD}_\Psi(C) = |X|$ **then** $r := \emptyset$; (since $C = \prod_{X_i \in X} X_i$ and $\mathrm{tail}_{X_p}(C) = \emptyset$)
   (∗) **Else** {pick $q \neq p$; $r_1 =\mathbf{LTS}(C^{X_q},X \setminus \{X_q\},X_p)$; $r_2 =\mathbf{LTS}(C - X_q,X \setminus \{X_q\},X_p)$;
2.   **For each** $\bar{c} \in \mathrm{tail}_{X_p}(C^{X_q}) \setminus \mathrm{tail}_{X_p}(C - X_q)$ **do**
               For each $c \in \mathrm{tail}_{X_p}(C)$ **do**
                   For $i = 0$ **to** N_q **do**
                       If $c = \bar{c} \cup \{(X_q,i)\}$ **then** $r(c) := r_1(\bar{c}) \cup \{(X_q,i)\}$;
3.   **For each** $\bar{c} \in \mathrm{tail}_{X_p}(C - X_q) \setminus \mathrm{tail}_{X_p}(C^{X_q})$ **do**
               For each $c \in \mathrm{tail}_{X_p}(C)$ **do** { **If** $c - X_q = \bar{c}$ **then** $r(c) := r_2(\bar{c})$; }
4.   **For each** $\bar{c} \in \mathrm{tail}_{X_p}(C^{X_q}) \cap \mathrm{tail}_{X_p}(C - X_q)$ **do**
               For each $c \in \mathrm{tail}_{X_p}(C)$ **do**
                   For $i = 0$ **to** N_q **do**
                       If $c = \bar{c} \cup \{(X_q,i)\}$ **then**
                           If $r_1(\bar{c}) \cup \{(X_q,i)\}$ inconsistent with all $\hat{c} \in C^{X_p} \setminus \{c\}$ **then**
                               $r(c) := r_1(\bar{c}) \cup \{(X_q,i)\}$;
                           Else $r(c) := r_2(\bar{c})$; } (end of (∗) Else)   **return** r;

Algorithm 3. Recursively finding representatives for the tail concepts

not contain X_q. Therefore, for each $Y = \{X_{i_1}, \ldots, X_{i_d}\} \subseteq X \setminus \{X_p, X_q\}$, there is a tuple $(l_1, \ldots, l_d) \in \prod_{1 \leq j \leq d} X_{i_j}$ with

$$\mathrm{Forb}(C^{X_p}, Y) = \prod_{1 \leq j \leq d}(X_{i_j} \setminus \{l_j\}). \tag{1}$$

Moreover, $(C^{X_p})^{X_q} = (C^{X_q})^{X_p}$, and thus $\mathrm{Forb}((C^{X_p})^{X_q}) = \mathrm{Forb}((C^{X_q})^{X_p})$. C^{X_q} is VCD_Ψ-maximum on $m-1$ instances and of VCD_Ψ $d-1$. So, by induction hypothesis, for each set $Y = \{X_{i_1}, \ldots, X_{i_{d-1}}\} \subseteq X \setminus \{X_p, X_q\}$, there is a tuple $(l_1, \ldots, l_{d-1}) \in \prod_{1 \leq j \leq d-1} X_{i_j}$ such that $\mathrm{Forb}((C^{X_q})^{X_p}, Y) = \prod_{1 \leq j \leq d-1}(X_{i_j} \setminus \{l_j\})$, and hence $\mathrm{Forb}((C^{X_p})^{X_q}, Y) = \prod_{1 \leq j \leq d-1}(X_{i_j} \setminus \{l_j\})$. By Lemma 4, any forbidden labeling on Y for $(C^{X_p})^{X_q}$, is extended to N_q forbidden labelings on $Y \cup \{X_q\}$ for C^{X_p}. That is, for some $l_q \in X_q$, (X_q, l_q) never occurs in a forbidden labeling on $Y \cup \{X_q\}$. Therefore, for each $Y' = \{X_{i_1}, \ldots, X_{i_{d-1}}, X_q\} \subseteq X \setminus \{X_p\}$, there is a tuple $(l_1, \ldots, l_{d-1}, l_q) \in (\prod_{1 \leq j \leq d-1} X_{i_j}) \times X_q$ such that

$$\mathrm{Forb}(C^{X_p}, Y') = (\prod_{1 \leq j \leq d-1}(X_{i_j} \setminus \{l_j\})) \times (X_q \setminus \{l_q\}). \tag{2}$$

Now, we need to show that if the claim holds for $C^{X_p} - X_q$ and $(C^{X_q})^{X_p}$ then it also holds for C^{X_p}. Note that $\mathrm{Forb}(C^{X_p})$ can be partitioned into the set of forbidden labelings on $Y \subseteq X \setminus \{X_p, X_q\}$, and the set of forbidden labelings on $Y' \subseteq X \setminus \{X_p\}$, with $X_q \in Y'$. By combining this fact with (1) and (2), we conclude that for each $Y = \{X_{i_1}, \ldots, X_{i_d}\} \subseteq X \setminus \{X_p\}$, there is a tuple $(l_1, \ldots, l_d) \in \prod_{1 \leq j \leq d} X_{i_j}$ such that $\mathrm{Forb}(C^{X_p}, Y) = \prod_{1 \leq j \leq d}(X_{i_j} \setminus \{l_j\})$. $\qquad \square$

Theorem 3. *For any* $X_p \in X$, *there is a bipartite graph between* $\text{tail}_{X_p}(C)$ *and* $\text{Forb}(C^{X_p})$, *with an edge between a concept and a forbidden labeling if this forbidden labeling is contained in the concept. All such graphs have a unique matching.*

Proof. (sketch) Let $m = |X|$ and $\text{VCD}_\Psi(C) = d$. The proof is by double induction on m and d. For $m = d$, there is nothing to prove as $\text{tail}_{X_p}(C) = \text{Forb}(C^{X_p}) = \emptyset$, for all $p \in \{1, \ldots, m\}$. Also, for $d = 0$, C contains a single concept which is always in the tail and gets matched to the empty set.

Suppose that the claim is true for all d' and m' such that $d' \leq d$, $m' \leq m$ and $m' + d' < m + d$. Pick $X_p, X_q \in X$. First, by Lemma 5, each forbidden labeling of C^{X_p} is an extension of a forbidden labeling of either $(C^{X_p})^{X_q}$ or $C^{X_p} - X_q$. Second, by Lemma 2.(3), any concept in $\text{tail}_{X_p}(C)$ is an extension of either a concept in $\text{tail}_{X_p}(C^{X_q})$ or a concept in $\text{tail}_{X_p}(C - X_q)$. Also, $\text{tail}_{X_p}(C^{X_q})$ is a VCD_Ψ-maximum class of dimension $d - 1$ and $\text{tail}_{X_p}(C - X_q)$ is a VCD_Ψ-maximum class of dimension d; both on the instance space $X \setminus \{X_q\}$. So, by induction hypothesis there exists a unique matching between $\text{tail}_{X_p}(C - X_q)$ and $\text{Forb}((C - X_q)^{X_p})$, and also, between $\text{tail}_{X_p}(C^{X_q})$ and $\text{Forb}((C^{X_p})^{X_q})$. We combine these two matchings to form a matching for $\text{tail}_{X_p}(C)$. This is done in steps 2, 3 and 4 in Algorithm 3. We omit the details as well as the verification of the uniqueness of the thus obtained matching.   □

Let $\text{LRep}_d(X) \subset \text{LRep}_{\leq d}(X)$ denote the set of labeled representatives of size d that are constructed from Algorithm 1. The following corollary shows that there is a representation mapping between $\text{tail}_{X_p}(C)$ and $\text{LRep}_d(X \setminus \{X_p\})$. Omitted proof uses Lemma 6.

Corollary 3. *Algorithm 3 returns a representation mapping between* $\text{tail}_{X_p}(C)$ *and some* $\text{LRep}_d(X \setminus \{X_p\})$.

The following theorem can be proven by induction on d, using Corollary 3.

Theorem 4. *Algorithm 2 returns a representation mapping between the* VCD_Ψ-*maximum class* C *on* X *with* $\text{VCD}_\Psi(C) = d$ *and some* $\text{LRep}_{\leq d}(X)$.

Now we have all the pieces in place for verfying Theorem 2.

Proof of Theorem 2. By Theorem 4, there exists a representation mapping r for C, i.e., a consistent bijection between C and some $\text{LRep}_{\leq d}(X)$ in which no two concepts clash. Condition (i) of Definition 4 is then obvious as $|\text{LRep}_{\leq d}(X)| = |C|$. Condition (ii) follows from the consistency and bijection properties of r. Condition (iii) follows from the non-clashing property of r and Lemma 1.   □

Table 1 shows the representatives computed for a VCD_{Ψ_G}-maximum class (we will see later that VCD_{Ψ_G} has the reduction property). Note the missing labeling property and the tightness of the scheme. To illustrate Steps 2 and 3 of Algorithm 3, see Table 2. Assume we want representatives for $\text{tail}_{X_2}(C^{X_1})$ and we recursively found the representative $r(c_1^1) = \emptyset$ for the (only) concept in $\text{tail}_{X_2}((C^{X_1})^{X_3})$. $r(c_1^1)$ is extended to $r(c_1^3)$ for $c_1^3 \in C^{X_1}$, that is, $r(c_1^3) =$

Table 1. VCD_{Ψ_G}-maximum class and representatives resulting from Algorithm 2

c	X_1	X_2	X_3	X_4	$r(c)$
c_1	0	0	0	0	$\emptyset$
c_2	0	0	1	0	$(X_3,1)$
c_3	0	0	2	0	$(X_3,2)$
c_4	0	0	1	1	$(X_4,1)$
c_5	0	0	1	2	$(X_4,2)$
c_6	0	1	0	0	$(X_2,1)$
c_7	0	2	0	0	$(X_2,2)$
c_8	1	0	0	0	$(X_1,1)$
c_9	2	0	0	0	$(X_1,2)$
c_{10}	1	0	1	0	$(X_1,1),(X_3,1)$
c_{11}	1	0	2	0	$(X_1,1),(X_3,2)$
c_{12}	2	0	1	0	$(X_1,2),(X_3,1)$
c_{13}	2	0	2	0	$(X_1,2),(X_3,2)$
c_{14}	1	0	1	1	$(X_1,1),(X_4,1)$
c_{15}	2	0	1	1	$(X_1,2),(X_4,1)$
c_{16}	1	0	1	2	$(X_1,1),(X_4,2)$
c_{17}	2	0	1	2	$(X_1,2),(X_4,2)$

c	X_1	X_2	X_3	X_4	$r(c)$
c_{18}	1	1	0	0	$(X_1,1),(X_2,1)$
c_{19}	1	2	0	0	$(X_1,1),(X_2,2)$
c_{20}	2	1	0	0	$(X_1,2),(X_2,1)$
c_{21}	2	2	0	0	$(X_1,2),(X_2,2)$
c_{22}	0	1	0	1	$(X_3,0),(X_4,1)$
c_{23}	0	1	0	2	$(X_3,0),(X_4,2)$
c_{24}	0	1	1	0	$(X_2,1),(X_3,1)$
c_{25}	0	1	2	0	$(X_2,1),(X_3,2)$
c_{26}	0	2	1	0	$(X_2,2),(X_3,1)$
c_{27}	0	2	2	0	$(X_2,2),(X_3,2)$
c_{28}	0	1	1	1	$(X_2,1),(X_4,1)$
c_{29}	0	2	1	1	$(X_2,2),(X_4,1)$
c_{30}	0	1	2	1	$(X_3,2),(X_4,1)$
c_{31}	0	1	1	2	$(X_2,1),(X_4,2)$
c_{32}	0	2	1	2	$(X_2,2),(X_4,2)$
c_{33}	0	1	2	2	$(X_3,2),(X_4,2)$

Table 2. Illustration of Steps 2 and 3 of Algorithm 3

$c \in \mathrm{tail}_{X_2}((C^{X_1})^{X_3})$	X_2	X_4	$r(c)$
c_1^1	0	0	$\emptyset$

$c \in \mathrm{tail}_{X_2}(C^{X_1} - X_3)$	X_2	X_4	$r(c)$
c_1^2	0	1	$\{(X_4,1)\}$
c_2^2	0	2	$\{(X_4,2)\}$

$c \in \mathrm{tail}_{X_2}(C^{X_1})$	X_2	X_3	X_4	$r(c)$
c_1^3	0	1	0	$r(c_1^1) \cup \{(X_3,1)\} = \{(X_3,1)\}$
c_2^3	0	2	0	$r(c_1^1) \cup \{(X_3,2)\} = \{(X_3,2)\}$
c_3^3	0	1	1	$r(c_1^2) = \{(X_4,1)\}$
c_4^3	0	1	2	$r(c_2^2) = \{(X_4,2)\}$

$r(c_1^1) \cup \{(X_3,1)\} = \{(X_3,1)\}$ because $c_1^3 = c_1^1 \cup \{(X_3,1)\}$. Similarly, $r(c_1^1)$ is extended to $r(c_2^3)$ for $c_2^3 \in C^{X_1}$, that is, $r(c_2^3) = r(c_1^1) \cup \{(X_3,2)\} = \{(X_3,2)\}$ because $c_2^3 = c_1^1 \cup \{(X_3,2)\}$. Next assume we want representatives for $\mathrm{tail}_{X_2}(C^{X_1})$ and we recursively found the representatives for $\mathrm{tail}_{X_2}(C^{X_1} - X_3)$. $r(c_1^2)$ for $c_1^2 \in C^{X_1} - X_3$ is extended to $r(c_3^3)$ for $c_3^3 \in C^{X_1}$ because $c_3^3 - X_3 = c_1^2$. Similarly, $r(c_2^2)$ for $c_2^2 \in C^{X_1} - X_3$ is extended to $r(c_4^3)$ for $c_4^3 \in C^{X_1}$ because $c_4^3 - X_3 = c_2^2$.

4   Which Notions of VCD Fulfill the Reduction Property?

Theorems 1 and 2 raise the question which notions of VCD fulfill the reduction property. We know that VCD_{Ψ^*} has the reduction property [10]. Here, we show that the same is true for the Graph-dimension, but not for Pollard's pseudo-dimension or the Natarajan-dimension. Since Ψ_G is the direct product over spanning families of mappings, Theorems 1 and 2 then apply to VCD_{Ψ_G}.

Theorem 5. VCD_{Ψ_G} fulfills the reduction property. In particular, each VCD_{Ψ_G}-maximum class C has a tight sample compression scheme of size $VCD_{\Psi_G}(C)$.

A challenging part of the proof of Theorem 5 is to establish the following crucial lemma. The rest of the proof is analogous to that in [10].

Lemma 7. *Let* $X_i = \{0,1\}$, *for* $i \in [m-1]$, $X_m = \{0,\ldots,N_m\}$, $N_m \geq 2$. *Let* $\Psi = \mathrm{id}_1 \times \cdots \times \mathrm{id}_{m-1} \times \Psi_G$ *and* $C \subseteq \prod_{i=1}^{m} X_i$ *be* VCD_Ψ-*maximum of* VCD_Ψ $m-1$. *Then for all* $\bar{c} \in C - X_m$, $|\{c \in C \mid c - X_m = \bar{c}\}| \in \{1, N_m + 1\}$.

Proof. We show that if some $\bar{c} \in C - X_m$ has more than one but fewer than $N_m + 1$ extensions in C, then $\mathrm{VCD}_\Psi(C) = m$. To do this, we first partition C into $N_m + 1$ classes $C_i = \{c \in C \mid c(X_m) = i\}$, for $0 \leq i \leq N_m$. We claim that

$$2^{m-1} - 1 \leq |C_i| \leq 2^{m-1}, \quad \text{for all } i \in \{0,\ldots,N_m\}. \tag{3}$$

$|C_i| \leq |C - X_m| = 2^{m-1}$ yields the upper bound. For the lower bound, assume $|C_t| = 2^{m-1} - k$, $k \geq 2$, for some $t \in X_m$. Then one can show $|C \setminus C_t| \geq (2^{m-1} - 1)N_m + 2$. So, by the pigeonhole principle and by $|C_i| \leq 2^{m-1}$, at least two $C_l, C_{l'} \subseteq (C \setminus C_t)$ satisfy $|C_l| = |C_{l'}| = 2^{m-1}$ and $C_l - X_m = C_{l'} - X_m = \{0,1\}^{m-1}$. Thus, any tuple in Ψ_G that maps l and l' to different values makes C shatter X—a contradiction.[1] Hence, for all $i \in \{0,\ldots,N_m\}$, $|C_i| \geq 2^{m-1} - 1$. We claim

(a) There exists some $t \in X_m$, such that $|C_t| = 2^{m-1}$.
(b) $|C_i| = 2^{m-1} - 1$ for all $i \in X_m \setminus \{t\}$.

Assume that for all $i \in X_m$, $|C_i| = 2^{m-1} - 1$. Then $|C| = \sum_{i=0}^{N_m} |C_i| = (N_m + 1)(2^{m-1} - 1) = 2^{m-1} + 2^{m-1}N_m - N_m - 1 < 2^{m-1} + 2^{m-1}N_m - N_m = \Phi_{m-1}(1,\ldots,1,N_m)$. So, there is at least one concept class $C_t \subseteq C$ such that $|C_t| > 2^{m-1} - 1$, that is, $|C_t| = 2^{m-1}$ from (3), which proves (a). Consequently, $\sum_{i=0, i \neq t}^{N_m} |C_i| = |C| - |C_t| = 2^{m-1} + 2^{m-1}N_m - N_m - 2^{m-1} = 2^{m-1}N_m - N_m = (2^{m-1} - 1)N_m$. Since $|C_i| \geq 2^{m-1} - 1$, for all $0 \leq i \leq N_m$, we conclude that $|C_i| = 2^{m-1} - 1$, for all $i \in X_m \setminus \{t\}$, i.e., we have proven (b).

Now let $1 \leq k < N_m$. Suppose there is a $\bar{c} \in C - X_m$ with $|\{c \in C \mid c - X_m = \bar{c}\}| = k + 1$. Let $c_0,\ldots,c_k \in C$ with $c_i \neq c_j$ and $c_i - X_m = c_j - X_m = \bar{c}$, for all $i, j \in \{0,\ldots,k\}$, $i \neq j$. W.l.o.g., $c_i(X_m) = i$ for $i \in \{0,\ldots,k\}$. On the one hand,

$$c_i = \bar{c} \times \{i\} \in C_i \quad \text{for each } i \in \{0,\ldots,k\}. \tag{4}$$

On the other hand, for $c \in C$ with $c - X_m = \bar{c}$, $c(X_m) \neq l$, for all $l \in \{k+1,\ldots,N_m\}$. Thus, for all $l \in \{k+1,\ldots,N_m\}$, $\bar{c} \times \{l\} \notin C$ and $\bar{c} \times \{l\} \notin C_l$. So, $C_l \subseteq (\{0,1\}^{m-1} \times \{l\}) \setminus \{\bar{c} \times \{l\}\}$, for $l \in \{k+1,\ldots,N_m\}$ and thus, from (3), $|C_l| = 2^{m-1} - 1$ and $C_l = (\{0,1\}^{m-1} \times \{l\}) \setminus \{\bar{c} \times \{l\}\}$, for $l \in \{k+1,\ldots,N_m\}$. Consequently, from (a), for some $t \in \{0,\ldots,k\}$, $|C_t| = 2^{m-1}$.

We show $\mathrm{VCD}_\Psi(C) = m$. Let $\bar{\psi} = (\mathrm{id}_1,\ldots,\mathrm{id}_{m-1},\psi_m)$, where $\psi_m(x) = 1$ if $x = t$, else $\psi_m(x) = 0$. First, $\bar{\psi}(C_t) = \{0,1\}^{m-1} \times \{1\}$. Second, $\bar{c} \times \{k+1\} \notin C_{k+1}$, so $\bar{\psi}(C_{k+1}) = (\{0,1\}^{m-1} \times \{0\}) \setminus \{\bar{c} \times \{0\}\}$. Hence, $\{0,1\}^m \setminus \{\bar{c} \times \{0\}\} \subseteq \bar{\psi}(C)$. By (4), $\bar{c} \times \{0\} \in \bar{\psi}(C_i)$, for all $i \in \{0,\ldots,k\} \setminus \{t\}$, so $\bar{\psi}(C) = \{0,1\}^m$. $\square$

For VCD_{Ψ_P} and VCD_{Ψ_N}, we give counterexamples to the reduction property. First, Table 3 witnesses the following claim (proof details are omitted).

Proposition 1. *There is a* VCD_{Ψ_P}-*maximum class* C *of* VCD_{Ψ_P} 2 *such that, for some* $X_t \in X$ *and some* $\bar{c} \in C - X_t$, $|\{c \in C \mid c - X_t = \bar{c}\}| = 2 \leq N_t$.

Table 3. Maximum class C of VCD_{Ψ_P} 2 used in the proof of Proposition 1

c	X_1	X_2	X_3	c	X_1	X_2	X_3	c	X_1	X_2	X_3	c	X_1	X_2	X_3	c	X_1	X_2	X_3
c_1	0	0	0	c_5	1	0	0	c_9	0	2	1	c_{13}	2	0	2	c_{17}	2	2	2
c_2	0	0	1	c_6	1	0	1	c_{10}	0	2	2	c_{14}	2	1	1	c_{18}	1	2	1
c_3	0	1	0	c_7	1	1	1	c_{11}	2	0	0	c_{15}	2	1	2	c_{19}	1	2	2
c_4	0	1	1	c_8	0	2	0	c_{12}	2	0	1	c_{16}	2	2	1				

Table 4. Maximum class C of VCD_{Ψ_N} 1 used in the proof of Proposition 2

c	X_1	X_2	X_3	c	X_1	X_2	X_3	c	X_1	X_2	X_3	c	X_1	X_2	X_3	c	X_1	X_2	X_3
c_1	0	0	0	c_3	0	1	0	c_5	1	2	2	c_7	2	2	1	c_9	2	0	0
c_2	0	0	1	c_4	1	0	0	c_6	2	1	2	c_8	2	2	2	c_{10}	2	0	2

The class in Table 3 does not stay VCD_{Ψ_P}-maximum when applying either definition of reduction w.r.t. X_3, yet it *does* have a tight compression scheme. Details are omitted. For VCD_{Ψ_N}, the class in Table 4 witnesses Proposition 2.

Proposition 2. *There is a VCD_{Ψ_N}-maximum class C of VCD_{Ψ_N} 1 such that, for some $X_t \in X$ and some $\bar{c} \in C - X_t$, $|\{c \in C \mid c - X_t = \bar{c}\}| = 2 \le N_t$.*

The reduction of the class in Table 4 is not VCD_{Ψ_N}-maximum under either definition of reduction. Interestingly, this class has *no* tight compression scheme. Note that the Natarajan-dimension violates both premises of Theorems 1 and 2—it violates the reduction property, and it is not based on a spanning family.

References

[1] Alon, N.: On the density of sets of vectors. Discrete Math. 46, 199–202 (1983)

[2] Ben-David, S., Cesa-Bianchi, N., Haussler, D., Long, P.: Characterizations of learnability for classes of $\{0, ..., n\}$-valued functions. J. Comp. Syst. Sci. 50, 74–86 (1995)

[3] Floyd, S., Warmuth, M.K.: Sample compression, learnability, and the Vapnik-Chervonenkis dimension. Machine Learning 21(3), 269–304 (1995)

[4] Gurvits, L.: Linear algebraic proofs of VC-dimension based inequalities. In: Ben-David, S. (ed.) EuroCOLT 1997. LNCS, vol. 1208, pp. 238–250. Springer, Heidelberg (1997)

[5] Haussler, D., Long, P.M.: A generalization of Sauer's lemma. Journal of Combinatorial Theory, Series A 71(2), 219–240 (1995)

[6] Kuzmin, D., Warmuth, M.K.: Unlabeled compression schemes for maximum classes. Journal of Machine Learning Research 8, 2047–2081 (2007)

[7] Littlestone, N., Warmuth, M.: Relating data compression and learnability (1986) (unpublished notes)

[8] Natarajan, B.K.: On learning sets and functions. Mach. Learn. 4, 67–97 (1989)

[9] Pollard, D.: Empirical Processes: Theory and Applications. In: NSF-CBMS Regional Conference Series in Probability and Statistics 2, pp. i, iii, v, vii, viii, 1–86 (1990)

---

[1] Note that any spanning family of mappings should contain at least one such mapping.

[10] Samei, R., Semukhin, P., Yang, B., Zilles, S.: Sample compression for multi-label concept classes. In: Proc. 27th COLT, pp. 371–393 (2014)

[11] Sauer, N.: On the density of families of sets. J. Comb. Theory, Ser. A 13, 145–147 (1972)

[12] Vapnik, V.N.: Inductive principles of the search for empirical dependences (methods based on weak convergence of probability measures). In: Proc. 2nd COLT, pp. 3–21 (1989)

[13] Welzl, E.: Complete range spaces (1987) (unpublished notes)

Robust and Private Bayesian Inference

Christos Dimitrakakis[1], Blaine Nelson[2,*],
Aikaterini Mitrokotsa[1], and Benjamin I. P. Rubinstein[3]

[1] Chalmers University of Technology, Sweden
[2] University of Potsdam, Germany
[3] The University of Melbourne, Australia

Abstract. We examine the robustness and privacy of Bayesian inference, under assumptions on the prior, and with no modifications to the Bayesian framework. First, we generalise the concept of differential privacy to arbitrary dataset distances, outcome spaces and distribution families. We then prove bounds on the robustness of the posterior, introduce a posterior sampling mechanism, show that it is differentially private and provide finite sample bounds for distinguishability-based privacy under a strong adversarial model. Finally, we give examples satisfying our assumptions.

1 Introduction

Significant research challenges for statistical learning include efficiency, robustness to noise (stochasticity) and adversarial manipulation, and preserving training data privacy. In this paper we study techniques for meeting these challenges simultaneously, through a simple unification of Bayesian inference, differential privacy and distinguishability. In particular, we examine the following problem.

Summary of Setting. A Bayesian statistician ($\mathscr{B}$) wants to communicate results about some data x to a third party ($\mathscr{A}$), but without revealing the data x itself. (x could be a single datum, or a sample of data.) More specifically:

(i) $\mathscr{B}$ selects a model family ($\mathcal{F}_\Theta$) and a prior (ξ).
(ii) $\mathscr{A}$ is allowed to see $\mathcal{F}_\Theta$ and ξ and is computationally unbounded.
(iii) $\mathscr{B}$ observes data x and calculates the posterior $\xi(\theta|x)$ but does not reveal it. Instead, $\mathscr{B}$ responds to queries at times $t = 1, \ldots$ as follows.
(iv) $\mathscr{A}$ sends a query q_t to $\mathscr{B}$.
(v) $\mathscr{B}$ responds $q_t(\theta_t)$ where θ_t is drawn from the posterior: $\theta_t \sim \xi(\theta|x)$.

We show that if $\mathcal{F}_\Theta$ or ξ are chosen appropriately, the resulting posterior-sampling mechanism satisfies generalized differential privacy and indistinguishability properties. The intuition is that robustness and privacy are linked via smoothness. Learning algorithms that are smooth mappings—their output (*e.g.*, a spam filter) varies little with perturbations to input (*e.g.*, similar training corpora)—are robust: outliers have reduced influence, and adversaries cannot

---

* Blaine Nelson is now at Google, Mountain View.

P. Auer et al. (Eds.): ALT 2014, LNAI 8776, pp. 291–305, 2014.

easily discover unknown information about the data. This suggests that robustness and privacy can be simultaneously achieved and perhaps are deeply linked. We show that under mild assumptions this is indeed true for the posterior distribution, suggesting a differentially-private mechanism for Bayesian inference.

Our contributions. (i) We generalise differential privacy to arbitrary dataset distances, outcome spaces, and distribution families. (ii) Under certain regularity conditions on the prior distribution ξ or likelihood family $\mathcal{F}_\Theta$, we show that the posterior distribution is *robust*: small changes in the dataset result in small posterior changes; (iii) We introduce a novel *posterior sampling mechanism* that is private. Unlike other common mechanisms, our approach sits squarely in the non-private (Bayesian) learning framework without modification; (iv) We introduce the notion of *dataset distinguishability* for which we provide finite-sample bounds for our mechanism (v) We provide examples of conjugate-pair distributions where our assumptions hold.

Paper organisation. Section 1.1 discusses related work. Section 2 specifies the setting and our assumptions. Section 3 proves results on robustness of Bayesian learning. Section 4 proves privacy results. Examples where our assumptions hold are given in Section 5. We present a discussion of our results in Section 6. Appendix A contains proofs of the main theorems. Proofs of the examples and a discussion on matching lower bounds are given in a technical report [8].

1.1   Related Work

In Bayesian statistical decision theory [1, 2, 7], learning is cast as a statistical inference problem and decision-theoretic criteria are used as a basis for assessing, selecting and designing procedures. In particular, for a given cost function, the Bayes-optimal procedure minimises the *Bayes risk* under a particular prior distribution.

In an adversarial setting, this is extended to a minimax risk, by assuming that the prior distribution is selected arbitrarily by nature. In the field of *robust statistics*, the minimax asymptotic bias of a procedure incurred within an ϵ-contamination neighbourhood is used as a robustness criterion giving rise to the notion of a procedure's *influence function* and *breakdown point* to characterise robustness [17, 18]. In a Bayesian context, robustness appears in several guises including minimax risk, robustness of the posterior within ϵ-contamination neighbourhoods, and robust priors [1]. In this context Grünwald and Dawid [15] demonstrated the link between robustness in terms of the minimax expected score of the likelihood function and the (generalized) maximum entropy principle, whereby nature is allowed to select a worst-case prior.

Differential privacy, first proposed by Dwork et al. [12], has achieved prominence in the theory of computer science, databases, and more recently learning communities. Its success is largely due to the semantic guarantee of privacy it formalises. Differential privacy is normally defined with respect to a randomised mechanism for responding to queries. Informally, a mechanism preserves differential privacy if perturbing one training instance results in a small change to the mechanism's response distribution. Differential privacy is detailed in Section 2.

A popular approach for differential privacy is the *exponential mechanism* [19] which generalises the *Laplace mechanism* of adding Laplace noise to released statistics [12]. This mechanism releases a response with probability exponential in a score function measuring distance to the non-private response. An alternate approach, employed for privatising regularised ERM [6], is to alter the inferential procedure itself, in that case by adding a random term to the primal objective. Further results on the accuracy of the exponential mechanism with respect to the Kolmogorov-Smirnov distance are given in [23]. Unlike previous studies, our mechanisms do not require modification to the underlying learning framework.

In a different direction, Duchi et al. [9] provided information-theoretic bounds for private learning, by modelling the protocol for interacting with an adversary as an arbitrary conditional distribution, rather than restricting it to specific mechanisms. In a similar vein Chaudhuri and Hsu [5] drew a quantitative connection between robust statistics and differential privacy by providing finite sample convergence rates for differentially private plug-in statistical estimators in terms of the *gross error sensitivity*, a common measure of robustness. These bounds can be seen as complementary to ours because our Bayesian estimators do not have private views of the data but use a suitably-defined prior instead.

Little research in differential privacy focuses on the Bayesian paradigm, and to our knowledge, none has established differentially-private Bayesian inference. Williams and McSherry [25] applied Bayesian inference to improve the utility of differentially private releases by computing posteriors in a noisy measurement model. In a similar vein, Xiao and Xiong [26] used Bayesian credible intervals to respond to queries with as high utility as possible, subject to a privacy budget. In the PAC-Bayesian setting, Mir [20] showed that the Gibbs estimator [19] is differentially private. While their algorithm corresponds to a posterior sampling mechanism, it is a posterior found by minimising risk bounds; by contrast, our results are purely Bayesian and come from conditions on the prior.

Smoothness of the learning map, achieved here for Bayesian inference by appropriate concentration of the prior, is related to *algorithmic stability* which is used in statistical learning theory to establish error rates [3]. Rubinstein et al. [22] used the γ-uniform stability of the SVM to calibrate the level of noise for using the Laplace mechanism to achieve differential privacy for the SVM. Hall et al. [16] extended this technique to adding Gaussian process noise for differentially private release of infinite-dimensional functions lying in an RKHS.

Finally, Dwork and Lei [11] made the first connection between (frequentist) robust statistics and differential privacy, developing mechanisms for the interquartile, median and B-robust regression. While robust statistics are designed to operate near an ideal distribution, they can have prohibitively high global, worst-case sensitivity. In this case privacy was still achieved by performing a differentially-private test on local sensitivity before release [13]. Little further work has explored robustness and privacy, and no general connection is known.

2 Problem Setting

We consider the problem of a Bayesian statistician ($\mathscr{B}$) communicating with an untrusted third party ($\mathscr{A}$). $\mathscr{B}$ wants to convey useful information to the queries of $\mathscr{A}$ (*e.g.*, how many people suffer from a disease or vote for a particular party) without revealing private information about the original data (*e.g.*, whether a particular person has cancer). This requires communicating information in a way that strikes a good balance between utility and privacy. In this paper, we study the inherent privacy and robustness properties of Bayesian inference and explore the question of whether $\mathscr{B}$ can select a prior distribution so that a computationally unbounded $\mathscr{A}$ cannot obtain private information from queries.

2.1 Definitions

We begin with our notation. Let $\mathcal{S}$ be the set of all possible datasets. For example, if $\mathcal{X}$ is a finite alphabet, then we might have $\mathcal{S} = \bigcup_{n=0}^{\infty} \mathcal{X}^n$, *i.e.*, the set of all possible observation sequences over $\mathcal{X}$.

Comparing datasets. Central to notions of privacy and robustness, is the concept of distance between datasets. Firstly, the effect of dataset perturbation on learning depends on the amount of noise as quantified by some distance. Secondly, the amount that an attacker can learn from queries can be quantified in terms of the distance of his guesses to the true dataset. To model these situations, we equip $\mathcal{S}$ with a pseudo-metric[1] $\rho : \mathcal{S} \times \mathcal{S} \to \mathbb{R}_+$. Using pseudo-metrics, we considerably generalise previous work on differential privacy, which considers only the special case of Hamming distance. We note that a similar generalisation has been developed in parallel and independently by Chatzikokolakis et al. [4].

Bayesian inference. This paper focuses on the *Bayesian inference* setting, where the statistician $\mathscr{B}$ constructs a posterior distribution from a prior distribution ξ and a training dataset x. More precisely, we assume that data $x \in \mathcal{S}$ have been drawn from some distribution P_{θ^*} on $\mathcal{S}$, parametrised by θ^*, from a family of distributions $\mathcal{F}_{\Theta}$. $\mathscr{B}$ defines a parameter set Θ indexing a family of distributions $\mathcal{F}_{\Theta}$ on $(\mathcal{S}, \mathfrak{S}_{\mathcal{S}})$, where $\mathfrak{S}_{\mathcal{S}}$ is an appropriate σ-algebra on $\mathcal{S}$:

$$\mathcal{F}_{\Theta} \triangleq \{ P_\theta : \theta \in \Theta \}, \tag{1}$$

and where we use p_θ to denote the corresponding densities[2] when necessary. To perform inference in the Bayesian setting, $\mathscr{B}$ selects a prior measure ξ on $(\Theta, \mathfrak{S}_{\Theta})$ reflecting $\mathscr{B}$'s subjective beliefs about which θ is more likely to be true, a priori; *i.e.*, for any measurable set $B \in \mathfrak{S}_{\Theta}$, $\xi(B)$ represents $\mathscr{B}$'s prior belief that $\theta^* \in B$. In general, the posterior distribution after observing $x \in \mathcal{S}$ is:

$$\xi(B \mid x) = \frac{\int_B p_\theta(x) \, \mathrm{d}\xi(\theta)}{\phi(x)}, \tag{2}$$

---

[1] Meaning that $\rho(x, y) = 0$ does not necessarily imply $x = y$.

[2] *I.e.*, the Radon-Nikodym derivative of P_θ relative to some dominating measure ν.

where ϕ is the corresponding marginal density given by:

$$\phi(x) \triangleq \int_{\Theta} p_{\theta}(x) \, d\xi(\theta) \ . \tag{3}$$

While the choice of the prior is generally arbitrary, this paper shows that its careful selection can yield good privacy guarantees.

Privacy. We first recall the idea of differential privacy [10]. This states that on similar datasets, a randomised query response mechanism yields (pointwise) similar distributions. We adopt the view of mechanisms as conditional distributions under which differential privacy can be seen as a measure of smoothness. In our setting, conditional distributions conveniently correspond to posterior distributions. These can also be interpreted as the distribution of a mechanism that uses posterior sampling, to be introduced in Section 4.2.

Definition 1 $((\epsilon, \delta)$-differential privacy). *A conditional distribution $P(\cdot \mid x)$ on $(\Theta, \mathfrak{S}_{\Theta})$ is (ϵ, δ)-differentially private if, for all $B \in \mathfrak{S}_{\Theta}$ and for any $x \in \mathcal{S} = \mathcal{X}^n$*

$$P(B \mid x) \leq e^{\epsilon} P(B \mid y) + \delta,$$

for all y in the hamming-1 *neighbourhood of x. That is, there is at most one $i \in \{1, \ldots, n\}$ such that $x_i \neq y_i$.*

As a first step, we generalise this definition to arbitrary dataset spaces $\mathcal{S}$ that are not necessarily product spaces. To do so, we introduce the notion of differential privacy under a pseudo-metric ρ on the space of all datasets.

Definition 2 $((\epsilon, \delta)$-differential privacy under ρ.). *A conditional distribution $P(\cdot \mid x)$ on $(\Theta, \mathfrak{S}_{\Theta})$ is (ϵ, δ)-differentially private under a pseudo-metric $\rho : \mathcal{S} \times \mathcal{S} \to \mathbb{R}_{+}$ if, for all $B \in \mathfrak{S}_{\Theta}$ and for any $x \in \mathcal{S}$, then:*

$$P(B \mid x) \leq e^{\epsilon \rho(x,y)} P(B \mid y) + \delta \rho(x,y) \quad \forall y \ .$$

Remark 1. If $\mathcal{S} = \mathcal{X}^n$ and $\rho(x,y) = \sum_{i=1}^{n} \mathbb{I}\{x_i \neq y_i\}$ is the Hamming distance, this definition is analogous to standard (ϵ, δ)-differential privacy. When considering only $(\epsilon, 0)$- differential privacy or $(0, \delta)$-privacy, it is an equivalent notion.[3]

Proof. For $(\epsilon, 0)$-DP, let $\rho(x, z) = \rho(z, y) = 1$; *i.e.*, they only differ in one element. Then, from standard DP, we have $P(B \mid x) \leq e^{\epsilon} P(B \mid z)$ and so obtain $P(B \mid x) \leq e^{2\epsilon} P(B \mid y) = e^{\rho(x,y)\epsilon} P(B \mid y)$. By induction, this holds for any x, y pair. Similarly, for $(0, \delta)$-DP, by induction we obtain $P(B \mid x) \leq P(B \mid x) + \delta \rho(x,y)$.

Definition 1 allows for privacy against a very strong attacker $\mathscr{A}$, who attempts to match the empirical distribution induced by the true dataset by querying

---

[3] Making the definition wholly equivalent is possible, but results in an unnecessarily complex definition.

the learned mechanism and comparing its responses to those given by distributions simulated using knowledge of the mechanism and knowledge of all but one datum—narrowing the dataset down to a hamming-1 ball. Indeed the requirement of differential privacy is sometimes *too strong* since it may come at the price of utility. Our Definition 2 allows for a much broader encoding of the attacker's knowledge via the selected pseudo-metric.

2.2 Our Main Assumptions

In the sequel, we show that if the distribution family $\mathcal{F}_\Theta$ or prior ξ is such that close datasets $x, y \in \mathcal{S}$, result in posterior distributions that are close. In that case, it is difficult for a third party to use such a posterior to distinguish the true dataset x from similar datasets.

To formalise these notions, we introduce two possible assumptions one could make on the smoothness of the family $\mathcal{F}_\Theta$ with respect to some metric d on $\mathbb{R}_+$. The first assumption states that the likelihood is smooth for all parameterizations of the family:

Assumption 1 (Lipschitz continuity). *Let $d(\cdot, \cdot)$ be a metric on $\mathbb{R}$. There exists $L > 0$ such that, for any $\theta \in \Theta$:*

$$d(p_\theta(x), p_\theta(y)) \leq L\rho(x, y), \qquad \forall x, y \in \mathcal{S} \ . \tag{4}$$

However, it may be difficult for this assumption to hold uniformly over Θ. This can be seen by a counterexample for the Bernoulli family of distributions. Consequently, we relax it by only requiring that $\mathcal{B}$'s *prior* probability ξ is concentrated in the parts of the family for which the likelihood is smoothest:

Assumption 2 (Stochastic Lipschitz continuity[21]). *Let $d(\cdot, \cdot)$ be a metric on $\mathbb{R}$ and let*

$$\Theta_L \triangleq \left\{ \theta \in \Theta : \sup_{x, y \in \mathcal{S}} \left\{ d(p_\theta(x), p_\theta(y)) - L\rho(x, y) \right\} \leq 0 \right\} \tag{5}$$

be the set of parameters for which Lipschitz continuity holds with Lipschitz constant L. Then there is some constant $c > 0$ such that, for all $L \geq 0$:

$$\xi(\Theta_L) \geq 1 - \exp(-cL) \ . \tag{6}$$

By not requiring uniform smoothness, this weaker assumption is easier to meet but still yields useful guarantees. In fact, in Section 5, we demonstrate that this assumption is satisfied by many important example distribution families.

To make our assumptions concrete, we now fix the distance function d to be the absolute log-ratio,

$$d(a, b) \triangleq \begin{cases} 0 & \text{if } a = b = 0 \\ \left| \ln \frac{a}{b} \right| & \text{otherwise} \end{cases}, \tag{7}$$

which is a proper metric on $\mathbb{R}_+ \times \mathbb{R}_+$. This particular choice of distance yields guarantees on differential privacy and indistinguishability.

We next show that verifying our assumptions for a distribution of a single random variable lifts to a corresponding property for the product distribution on i.i.d. samples.

Lemma 1. *If p_Θ satisfies Assumption 1 (resp. Assumption 2) with respect to pseudo-metric ρ and constant L (or c), then, for any fixed $n \in \mathbb{N}$, $p_\Theta^n(\{x_i\}) = \prod_{i=1}^n p_\Theta(x_i)$ satisfies the same assumption with respect to:*

$$\rho^n(\{x_i\}, \{y_i\}) = \sum_{i=1}^n \rho(x_i, y_i)$$

and constant $L \cdot n$ (or $\frac{c}{n}$). Further, if $\{x_i\}$ and $\{y_i\}$ differ in at most k items, the assumption holds with the same pseudo-metric but with constant $L \cdot k$ (or $\frac{c}{k}$) instead.

3   Robustness of the Posterior Distribution

We now show that the above assumptions provide guarantees on the robustness of the posterior. That is, if the distance between two datasets x, y is small, then so too is the distance between the two resulting posteriors, $\xi(\cdot \mid x)$ and $\xi(\cdot \mid y)$. We prove this result for the case where we measure the distance between the posteriors in terms of the well-known KL-divergence:

$$D\left(P \parallel Q\right) = \int_S \ln \frac{\mathrm{d}P}{\mathrm{d}Q} \, \mathrm{d}P \ . \tag{8}$$

The following theorem shows that any distribution family $\mathcal{F}_\Theta$ and prior ξ satisfying one of our assumptions is robust, in the sense that the posterior does not change significantly with small changes to the dataset. It is notable that our mechanisms are simply tuned through the choice of prior.

Theorem 1. *When $d : \mathbb{R}_+ \times \mathbb{R}_+ \to \mathbb{R}_+$ is the absolute log-ratio distance (7), ξ is a prior distribution on Θ and $\xi(\cdot \mid x)$ and $\xi(\cdot \mid y)$ are the respective posterior distributions for datasets $x, y \in S$, the following results hold:*

(i) Under a metric ρ and $L > 0$ satisfying Assumption 1,

$$D\left(\xi(\cdot \mid x) \parallel \xi(\cdot \mid y)\right) \le 2L\rho(x, y) \tag{9}$$

(ii) Under a metric ρ and $c > 0$ satisfying Assumption 2,

$$D\left(\xi(\cdot \mid x) \parallel \xi(\cdot \mid y)\right) \le \frac{\kappa}{c} \cdot \rho(x, y) \tag{10}$$

where κ is constant (see Appendix A); $\kappa \approx 4.91081$.

Note that the second claim bounds the KL divergence in terms of $\mathscr{B}$'s prior belief that L is small, which is expressed via the constant c. The larger c is, the less prior mass is placed in large L and so the more robust inference becomes. Of course, choosing c to be too large may decrease efficiency.

4   Privacy Properties of the Posterior Distribution

We next examine the differential privacy of the posterior distribution. We show in Section 4.1 that this can be achieved under either of our assumptions. The result can also be interpreted as the differential privacy of a *posterior sampling mechanism* for responding to queries, which is described in Section 4.2. Finally, Section 4.3 introduces an alternative notion of privacy: *dataset distinguishability*. We prove a high-probability bound on the sample complexity of distinguishability under our assumptions.

4.1   Differential Privacy of Posterior Distributions

We consider our generalised notion of differential privacy for posterior distributions (Definition 2); and show that the type of privacy exhibited by the posterior depends on which assumption holds.

Theorem 2. *Using the log-ratio distance (as in Theorem 1),*

(i) Under Assumption 1, for all $x, y \in S$, $B \in \mathfrak{S}_\Theta$:

$$\xi(B \mid x) \leq \exp\{2L\rho(x,y)\}\xi(B \mid y) \tag{11}$$

i.e., the posterior ξ is $(2L, 0)$-differentially private under pseudo-metric ρ.
(ii) Under Assumption 2, for all $x, y \in S$, $B \in \mathfrak{S}_\Theta$:

$$|\xi(B \mid x) - \xi(B \mid y)| \leq \sqrt{\frac{\kappa}{2c}\rho(x,y)}$$

i.e., the posterior ξ is $\left(0, \sqrt{\frac{\kappa}{2c}}\right)$-differentially private under pseudo-metric $\sqrt{\rho}$.

4.2   Posterior Sampling Query Model

Given that we have a full posterior distribution, we use it to define an algorithm achieving privacy. In this framework, we allow the adversary to submit a set of queries $\{q_k\}$ which are mappings from parameter space Θ to some arbitrary answer set Ψ; *i.e.*, $q_k : \Theta \to \Psi$. If we know the true parameter θ, then we would reply to any query with $q_k(\theta)$. However, since θ is unknown, we must select a method for conveying the required information. There are three main approaches that we are aware of. The first is to marginalise θ out. The second is to use the *maximum a posteriori* value of θ. The final, which we employ here, is to use sampling; *i.e.*, to reply to each query q_k using a θ_k sampled from the posterior.

This sample-based interactive query model is presented in Algorithm 1. First, the algorithm calculates the posterior distribution $\xi(\cdot \mid x)$. Then, for the kth received query q_k, the algorithm draws a sample θ_k from the posterior distribution and responds with $q_k(\theta_k)$.

Algorithm 1. Posterior sampling query model

---
1. **Input** prior ξ, data $x \in \mathcal{S}$
2. Calculate posterior $\xi(\cdot \mid x)$.
3. **for** $k = 1, \ldots$ **do**
4.   Observe query $q_k : \Theta \to \Psi$.
5.   Sample $\theta_k \sim \xi(\cdot \mid x)$.
6.   Return $q_k(\theta_k)$.
7. **end for**

In this context, Theorem 2 can be interpreted as proving differential privacy for the posterior sampling mechanism for the case when the response set is the parameter set; *i.e.*, $\Psi = \Theta$ and $q_k(\theta) = \theta$. Due to the data-processing inequality, this also holds for all query functions. As an example, consider querying conditional expectations:

Example 1. Let each model P_θ in the family define a distribution on the product space $\mathcal{S} = \bigcup_{n=1}^{\infty} \mathcal{X}^n$, such for any $x = (x_1, \ldots, x_n) \in \mathcal{X}^n$, $P_\theta(x) = \prod_i P_\theta(x_i)$. In addition, let $\mathcal{X} = \mathcal{Y} \times \mathcal{Z}$ (with appropriate algebras $\mathfrak{S}_{\mathcal{X}}, \mathfrak{S}_{\mathcal{Y}}, \mathfrak{S}_{\mathcal{Z}}$) and write $x_i = (x_{i,y}, x_{i,z})$ for point x_i and its two components. A conditional expectation query would require an answer to the question:

$$\mathbb{E}_\theta(x_{|\mathcal{Y}} \mid x_{|\mathcal{Z}}),$$

where the parameter θ is unknown to the questioner. In this case, the answer set Ψ would be identical to $\mathcal{Y}$, while k would index the values in $\mathcal{Z}$.

4.3   Distinguishability of Datasets

A limitation of the differential privacy framework is that it does not give us insight on the amount of effort required by an adversary to obtain private information. In fact, an adversary wishing to breach privacy, needs to distinguish x from alternative datasets y. Within the posterior sampling query model, $\mathscr{A}$ has to decide whether $\mathscr{B}$'s posterior is $\xi(\cdot \mid x)$ or $\xi(\cdot \mid y)$. However, he can only do so within some neighbourhood ϵ of the original data. In this section, we bound his error in determining the posterior in terms of the number of queries he performs. This is analogous to the dataset-size bounds on queries in interactive models of differential privacy [12].

Let us consider an adversary querying to sample $\theta_k \sim \xi(\cdot \mid x)$. This is the most powerful query possible under the model shown in Algorithm 1. Then, the adversary needs only to construct the empirical distribution to approximate the posterior up to some sample error. By bounds on the KL divergence between the empirical and actual distributions we can bound his power in terms of how many samples he needs in order to distinguish between x and y.

Due to the sampling model, we first require a finite sample bound on the quality of the empirical distribution. The adversary could attempt to distinguish different posteriors by forming the empirical distribution on any sub-algebra $\mathfrak{S}$.

Lemma 2. *For any $\delta \in (0, 1)$, let $\mathcal{M}$ be a finite partition of the sample space $\mathcal{S}$, of size $m \leq \log_2 \sqrt{1/\delta}$, generating the σ-algebra $\mathfrak{S} = \sigma(\mathcal{M})$. Let $x_1, \ldots, x_n \sim P$ be i.i.d. samples from a probability measure P on $\mathcal{S}$, let $P_{|\mathfrak{S}}$ be the restriction of P on $\mathfrak{S}$ and let $\hat{P}^n_{|\mathfrak{S}}$ be the empirical measure on $\mathfrak{S}$. Then, with probability at least $1 - \delta$:*

$$\left\| \hat{P}^n_{|\mathfrak{S}} - P_{|\mathfrak{S}} \right\|_1 \leq \sqrt{\frac{3}{n} \ln \frac{1}{\delta}}. \tag{12}$$

Of course, the adversary could choose any arbitrary estimator ψ to guess x. The accompanying technical report [8] describes how to apply Le Cam's method to obtain matching lower bound rates in this case, by defining *dataset estimators*. This is however is not essential for the remainder of the paper.

We can combine this bound on the adversary's estimation error with Theorem 1's bound on the KL divergence between posteriors resulting from similar data to obtain a measure of how fine a distinction between datasets the adversary can make after a finite number of draws from the posterior:

Theorem 3. *Under Assumption 1, the adversary can distinguish between data x, y with probability $1 - \delta$ if:*

$$\rho(x, y) \geq \frac{3}{4Ln} \ln \frac{1}{\delta}. \tag{13}$$

Under Assumption 2, this becomes:

$$\rho(x, y) \geq \frac{3c}{2\kappa n} \ln \frac{1}{\delta}. \tag{14}$$

Consequently, either smoother likelihoods (*i.e.*, decreasing L), or a larger concentration on smoother likelihoods (*i.e.*, increasing c), both increases the effort required by the adversary and reduces the sensitivity of the posterior. Note that, unlike the results obtained for differential privacy of the posterior sampling mechanism, these results have the same algebraic form under both assumptions.

5   Examples Satisfying Our Assumptions

In what follows we study, for different choices of likelihood and corresponding conjugate prior, what constraints must be placed on the prior's concentration to guarantee a desired level of privacy. These case studies closely follow the pattern in differential privacy research where the main theorem for a new mechanism are sufficient conditions on (*e.g.*, Laplace) noise levels to be introduced to a response in order to guarantee a level ϵ of ϵ-differential privacy.

For exponential families, we have $p_\theta(x) = h(x) \exp \left\{ \eta_\theta^\top T(x) - A(\eta_\theta) \right\}$, where $h(x)$ is the base measure, η_θ is the distribution's natural parameter corresponding to θ, $T(x)$ is the distribution's sufficient statistic, and $A(\eta_\theta)$ is its log-partition function. For distributions in this family, under the absolute log-ratio distance, the family of parameters Θ_L of Assumption 2 must satisfy, for all $x, y \in \mathcal{S}$: $\left| \ln \frac{h(x)}{h(y)} + \eta_\theta^\top (T(x) - T(y)) \right| \leq L\rho(x, y)$. If the left-hand side has an amenable form, then we can quantify the set Θ_L for which this requirement holds. Particularly, for distributions where $h(x)$ is constant and $T(x)$ is scalar (e.g., Bernoulli, exponential, and Laplace), this requirement simplifies to $\frac{|T(x) - T(y)|}{\rho(x, y)} \leq \frac{L}{\eta_\theta}$. One can then find the supremum of the left-hand side independent from θ, yielding a simple formula for the feasible L for any θ. Here are some examples, whose proofs can be found in [8].

Lemma 3 (Exponential conjugate prior). *For the case of an exponential distribution $\mathcal{E}xp(\theta)$ with exponential conjugate prior $\theta \sim \mathcal{E}xp(\lambda)$, $\lambda > 0$ satisfies Assumption 2 with parameter $c = \lambda$ and metric $\rho(x, y) = |x - y|$.*

Lemma 4 (Laplace conjugate prior). *The Laplace distribution $Laplace(\theta)$ and Laplace conjugate prior $\theta \sim Laplace(\mu, s, \lambda)$, $\mu \in \mathbb{R}$, $s \geq L$, $\lambda > 0$ satisfies Assumption 2 with parameters $c = \lambda$ and metric $\rho(x, y) = |x - y|$*

Lemma 5 (Beta-Binomial conjugate prior). *The Binomial distribution $\mathcal{B}inom(\theta, n)$, with Binomial prior $\theta \sim Beta(\alpha, \beta)$, $\alpha = \beta > 1$ satisfies Assumption 2 for $c = O(\alpha)$ and metric $\rho(x, y) = |x - y|$.*

Lemma 6 (Normal distribution). *The normal distribution $N(\mu, \sigma^2)$ with an exponential prior $\sigma^2 \sim \mathcal{E}xp(\lambda)$ satisfies Assumption 2 with parameter $c = \lambda$ and metric $\rho(x, y) = |x^2 - y^2| + 2|x - y|$.*

Lemma 7 (Discrete Bayesian networks). *Consider a family of discrete Bayesian networks on K variables, $\mathcal{F}_\Theta = \{ P_\theta : \theta \in \Theta \}$. More specifically, each member P_θ, is a distribution on a finite space $\mathcal{S} = \prod_{k=1}^K \mathcal{S}_k$ and we write $P_\theta(x)$ for the probability of any outcome $x = (x_1, \ldots, x_K)$ in $\mathcal{S}$. We also let $\rho(x, y) \triangleq \sum_{k=1}^K \mathbb{I}\{x_k \neq y_k\}$ be the distance between x and y. If ϵ is the smallest probability assigned to any one sub-event, then Assumption 1 is satisfied with $L = \ln 1/\epsilon$.*

The above examples demonstrate that our assumptions are reasonable. In fact, for several of them we recover standard choices of prior distributions.

6   Conclusion

We have presented a unifying framework for private and secure inference in a Bayesian setting. Under simple but general assumptions, we have shown that Bayesian inference is both robust and private in a certain formal sense. In particular, our results establish that generalised differential privacy can be achieved while using only existing constructs in Bayesian inference. Our results merely place concentration conditions on the prior. This allows us to use a general posterior sampling mechanism for responding to queries.

Due to its relative simplicity on top of non-private inference, our framework may thus serve as a fundamental building block for more sophisticated, general-purpose Bayesian inference. As an additional step towards this goal, we have demonstrated the application of our framework to deriving analytical expressions for well-known distribution families, and for discrete Bayesian networks. Finally, we bounded the amount of effort required of an attacker to breach privacy when observing samples from the posterior. This serves as a principled guide for how much access can be granted to querying the posterior, while still guaranteeing privacy.

We have not examined how privacy concerns relate to learning. While larger c improves privacy, it also concentrates the prior so much that learning would be inhibited. Thus, c should be chosen to optimise the trade-off between privacy and learning. However, we leave this issue for future work.

Acknowledgments. We gratefully thank Aaron Roth, Kamalika Chaudhuri, and Matthias Bussas for their discussion and insights as well as the anonymous reviewers for their comments on the paper. This work was partially supported by the Marie Curie Project "Efficient Sequential Decision Making Under Uncertainty", Grant Number 237816.

A   Proofs of Main Theorems

Proof (Proof of Lemma 1). For Assumption 1, the proof follows directly from the definition of the absolute log-ratio distance; namely,

$$d(p_\Theta^n(\{\,x_i\,\}), p_\Theta^n(\{\,y_i\,\})) = n\sum_{i=1}^n d(p_\Theta(x_i), p_\Theta(y_i))$$
$$\leq L \cdot n \sum_{i=1}^n d(x_i, y_i) \ .$$

This can be reduced from n to k if only k items differ since $d(p_\Theta(x_i), p_\Theta(y_i)) = 0$ if $x_i = y_i$.

For Assumption 2, the same argument shows that the Θ_L from Eq. (5) becomes $\Theta_{L \cdot n}$ (or $\Theta_{L \cdot k}$ for the k differing items case) for the product distribution. Hence, the same prior can be used to give the bound required by Eq. (6) if parameter $\frac{c}{n}$ (or $\frac{c}{k}$) is used.

Proof (Proof of Theorem 1). Let us now tackle claim (1.i). First, we can decompose the KL-divergence $D\left(\xi(\cdot \mid x) \parallel \xi(\cdot \mid y)\right)$ into two parts:

$$\int_\Theta \ln \frac{d\xi(\theta \mid x)}{d\xi(\theta \mid y)} \, d\xi(\theta) = \int_\Theta \ln \frac{p_\theta(x)}{p_\theta(y)} \, d\xi(\theta) + \int_\Theta \ln \frac{\phi(y)}{\phi(x)} \, d\xi(\theta)$$
$$\leq \int_\Theta \left| \ln \frac{p_\theta(x)}{p_\theta(y)} \right| \, d\xi(\theta) + \int_\Theta \ln \frac{\phi(y)}{\phi(x)} \, d\xi(\theta) \leq L\rho(x,y) + \left| \ln \frac{\phi(y)}{\phi(x)} \right|. \qquad (15)$$

From Ass. 1, $p_\theta(y) \leq \exp(L\rho(x,y))p_\theta(x)$ for all θ so:

$$\phi(y) = \int_\Theta p_\theta(y) \, d\xi(\theta) \leq \exp(L\rho(x,y)) \int_\Theta p_\theta(x) \, d\xi(\theta) = \exp(L\rho(x,y))\phi(x).$$
$$(16)$$

Combining this with (15) we obtain $D\left(\xi(\cdot \mid x) \parallel \xi(\cdot \mid y)\right) \le 2L\rho(x,y)$.

Claim (1.ii) is dealt with similarly. Once more, we can break down the distance in parts. Let $\Theta_{[a,b]} \triangleq \Theta_b \setminus \Theta_a$. Then $\xi(\Theta_{[a,b]}) = \xi(\Theta_b) - \xi(\Theta_a) \le e^{-ca}$, as $\Theta_b \supset \Theta_a$, while $\xi(\Theta_b) \le 1$ and $\xi(\Theta_a) \ge 1 - e^{-ca}$ from Ass 2. We can partition Θ into uniform intervals $[(L-1)\alpha, L\alpha)$ of size $\alpha > 0$ indexed by L. We bound the divergence on each partition and sum over L.

$$
D\left(\xi(\cdot \mid x) \parallel \xi(\cdot \mid y)\right)
$$

$$
\le \sum_{L=1}^{\infty} \left\{ \int_{\Theta_{[(L-1)\alpha, L\alpha)}} \left| \ln \frac{p_\theta(x)}{p_\theta(y)} \right| \, d\xi(\theta) + \int_{\Theta_{[(L-1)\alpha, L\alpha]}} \ln \frac{\phi(y)}{\phi(x)} \, d\xi(\theta) \right\}
$$

$$
\le 2\rho(x,y)\alpha \sum_{L=1}^{\infty} Le^{-c(L-1)\alpha} = 2\rho(x,y)\alpha \left(1 - e^{-c\alpha}\right)^{-2}, \tag{17}
$$

via the geometric series. This holds for any size parameter $\alpha > 0$ and is convex for $\alpha > 0$, $c > 0$. Thus, there is an optimal choice for α that minimizes this bound. Differentiating w.r.t α and setting the result to 0 yields $\alpha^* = \frac{\omega}{c}$ where ω is the unique non-zero solution to $e^\omega = 2\omega + 1$. The optimal bound is then $D\left(\xi(\cdot \mid x) \parallel \xi(\cdot \mid y)\right) \le \frac{2\omega}{(1-e^{-\omega})^2} \cdot \frac{\rho(x,y)}{c}$ As the $\omega \approx 1.25643$ is the unique positive solution to $e^\omega = 2\omega + 1$, and we define $\kappa = \frac{2\omega}{(1-e^{-\omega})^2} \approx 4.91081$.

Proof (Proof of Theorem 2). For part (2.i), we assumed that there is an $L > 0$ such that $\forall x, y \in \mathcal{S}$, $\left| \log \frac{p_\theta(x)}{p_\theta(y)} \right| \le L\rho(x,y)$, thus implying $\frac{p_\theta(x)}{p_\theta(y)} \le \exp\{L\rho(x,y)\}$. Further, in the proof of Theorem 1, we showed that $\phi(y) \le \exp\{L\rho(x,y)\}\phi(x)$ for all $x, y \in \mathcal{S}$. From Eq. 2, we can then combine these to bound the posterior of any $B \in \mathfrak{S}_\Theta$ as follows for all $x, y \in \mathcal{S}$:

$$
\xi(B \mid x) = \frac{\int_B \frac{p_\theta(x)}{p_\theta(y)} p_\theta(y) \, d\xi(\theta)}{\phi(y)} \cdot \frac{\phi(y)}{\phi(x)} \le \exp\{2L\rho(x,y)\}\xi(B \mid y) .
$$

For part (2.ii), note that from Theorem (1.ii) that the KL divergence of the posteriors under assumption is bounded by $\kappa\rho(x,y)/c$. Now, recall Pinsker's inequality [cf. 14]:

$$
D\left(Q\|P\right) \ge \frac{1}{2} \|Q - P\|_1^2. \tag{18}
$$

Using it, this bound yields: $|\xi(B \mid x) - \xi(B \mid y)| \le \sqrt{\frac{1}{2}D\left(\xi(\cdot \mid x) \parallel \xi(\cdot \mid y)\right)} \le \sqrt{\kappa\rho(x,y)/2c}$

Proof (Proof of Lemma 2). We use the inequality due to Weissman et al. [24] on the ℓ_1 norm, which states that for any multinomial distribution p with m outcomes, the ℓ_1 deviation of the empirical distribution $\hat{p}_n$ satisfies: $\mathbb{P}(\|\hat{p}_n - p\|_1 \ge \epsilon) \le (2^m - 2)e^{-\frac{1}{2}n\epsilon^2}$. The right hand side is bounded by $e^{m \ln 2 - \frac{1}{2}n\epsilon^2}$. Substituting $\epsilon = \sqrt{\frac{3}{n} \ln \frac{1}{\delta}}$:

$$\mathbb{P}(\|\hat{p}_n - p\|_1 \geq \sqrt{\frac{3}{n}\ln\frac{1}{\delta}}) \leq e^{m\ln 2 - \frac{3}{2}\ln\frac{1}{\delta}}$$

$$\leq e^{\log_2\sqrt{\frac{1}{\delta}}\ln 2 - \frac{3}{2}\ln\frac{1}{\delta}} = e^{\frac{1}{2}\ln\frac{1}{\delta} - \frac{3}{2}\ln\frac{1}{\delta}} = \delta. \tag{19}$$

where the second inequality follows from $m \leq \log_2\sqrt{1/\delta}$.

Proof (Proof of Theorem 3). Recall that the data processing inequality states that, for any sub-algebra $\mathfrak{S}$:

$$\|Q_{|\mathfrak{S}} - P_{|\mathfrak{S}}\|_1 \leq \|Q - P\|_1. \tag{20}$$

Using this and Pinsker's inequality (18) we get:

$$2L\rho(x, y) \geq 2L\epsilon \geq D\left(\xi(\cdot \mid x)\|\xi(\cdot \mid y)\right)$$

$$\geq \frac{1}{2}\|\xi(\cdot \mid x) - \xi(\cdot \mid y)\|_1^2 \geq \frac{1}{2}\|\xi_{|\mathfrak{S}}(\cdot \mid x) - \xi_{|\mathfrak{S}}(\cdot \mid y)\|_1^2. \tag{21}$$

On the other hand, due to (12) the adversary's ℓ_1 error in the posterior distribution is bounded by $\sqrt{\frac{3}{n}\ln\frac{1}{\delta}}$ with probability $1 - \delta$. Using the above inequalities, we can bound the error in terms of the distinguishability of the real dataset x from an arbitrary set y as: $4L\rho(x, y) \geq \frac{3}{n}\ln\frac{1}{\delta}$. Rearranging, we obtain the required result. The second case is treated similarly to obtain: $2\kappa\rho(x, y)/c \geq \frac{3}{n}\ln\frac{1}{\delta}$.

References

[1] Berger, J.O.: Statistical Decision Theory and Bayesian Analysis. Springer (1985)
[2] Bickel, P.J., Doksum, K.A.: Mathematical Statistics: Basic Ideas and Selected Topics, vol. 1. Holden-Day Company (2001)
[3] Bousquet, O., Elisseeff, A.: Stability and generalization. Journal of Machine Learning Research 2, 499–526 (2002)
[4] Chatzikokolakis, K., Andrés, M.E., Bordenabe, N.E., Palamidessi, C.: Broadening the scope of differential privacy using metrics. In: De Cristofaro, E., Wright, M. (eds.) PETS 2013. LNCS, vol. 7981, pp. 82–102. Springer, Heidelberg (2013)
[5] Chaudhuri, K., Hsu, D.: Convergence rates for differentially private statistical estimation. In: ICML (2012)
[6] Chaudhuri, K., Monteleoni, C., Sarwate, A.D.: Differentially private empirical risk minimization. Journal of Machine Learning Research 12, 1069–1109 (2011)
[7] DeGroot, M.H.: Optimal Statistical Decisions. John Wiley & Sons (1970)
[8] Dimitrakakis, C., Nelson, B., Mitrokotsa, A., Rubinstein, B.: Robust and private Bayesian inference. Technical report, arXiv:1306.1066 (2014)
[9] Duchi, J.C., Jordan, M.I., Wainwright, M.J.: Local privacy and statistical minimax rates. Technical report, arXiv:1302.3203 (2013)
[10] Dwork, C.: Differential privacy. In: Bugliesi, M., Preneel, B., Sassone, V., Wegener, I. (eds.) ICALP 2006. LNCS, vol. 4052, pp. 1–12. Springer, Heidelberg (2006)
[11] Dwork, C., Lei, J.: Differential privacy and robust statistics. In: STOC, pp. 371–380 (2009)

[12] Dwork, C., McSherry, F., Nissim, K., Smith, A.: Calibrating noise to sensitivity in private data analysis. In: Halevi, S., Rabin, T. (eds.) TCC 2006. LNCS, vol. 3876, pp. 265–284. Springer, Heidelberg (2006)

[13] Dwork, C., Smith, A.: Differential privacy for statistics: What we know and what we want to learn. Journal of Privacy and Confidentiality 1(2), 135–154 (2009)

[14] Fedotov, A.A., Harremoës, P., Topsoe, F.: Refinements of Pinsker's inequality. IEEE Transactions on Information Theory 49(6), 1491–1498 (2003)

[15] Grünwald, P.D., Dawid, A.P.: Game theory, maximum entropy, minimum discrepancy, and robust bayesian decision theory. The Annals of Statistics 32(4), 1367–1433 (2004)

[16] Hall, R., Rinaldo, A., Wasserman, L.: Differential privacy for functions and functional data. Journal of Machine Learning Research 14, 703–727 (2013)

[17] Hampel, F.R., Ronchetti, E.M., Rousseeuw, P.J., Stahel, W.A.: Robust Statistics: The Approach Based on Influence Functions. John Wiley and Sons (1986)

[18] Huber, P.J.: Robust Statistics. John Wiley and Sons (1981)

[19] McSherry, F., Talwar, K.: Mechanism design via differential privacy. In: FOCS, pp. 94–103 (2007)

[20] Mir, D.: Differentially-private learning and information theory. In: Proceedings of the 2012 Joint EDBT/ICDT Workshops, pp. 206–210. ACM (2012)

[21] Norkin, V.: Stochastic Lipschitz functions. Cybernetics and Systems Analysis 22(2), 226–233 (1986)

[22] Rubinstein, B.I.P., Bartlett, P.L., Huang, L., Taft, N.: Learning in a large function space: Privacy-preserving mechanisms for SVM learning. Journal of Privacy and Confidentiality 4(1) (2012)

[23] Wasserman, L., Zhou, S.: A statistical framework for differential privacy. Journal of the American Statistical Association 105(489), 375–389 (2010)

[24] Weissman, T., Ordentlich, E., Seroussi, G., Verdu, S., Weinberger, M.J.: Inequalities for the L1 deviation of the empirical distribution. Technical report, Hewlett-Packard Labs (2003)

[25] Williams, O., McSherry, F.: Probabilistic inference and differential privacy. In: NIPS, pp. 2451–2459 (2010)

[26] Xiao, Y., Xiong, L.: Bayesian inference under differential privacy. Technical report, arXiv:1203.0617 (2012)

Clustering, Hamming Embedding, Generalized LSH and the Max Norm

Behnam Neyshabur, Yury Makarychev, and Nathan Srebro

Toyota Technological Institute at Chicago
{bneyshabur,yury,nati}@ttic.edu

Abstract. We study the convex relaxation of clustering and hamming embedding, focusing on the asymmetric case (co-clustering and asymmetric hamming embedding), understanding their relationship to LSH as studied by Charikar (2002) and to the max-norm ball, and the differences between their symmetric and asymmetric versions.

Keywords: Clustering, Hamming Embedding, LSH, Max Norm.

1 Introduction

Convex relaxations play an important role in designing efficient learning and recovery algorithms, as well as in statistical learning and online optimization. It is thus desirable to understand the convex hull of hypothesis sets, to obtain tractable relaxation to these convex hulls, and to understand the tightness of such relaxations.

In this paper we consider convex relaxations of two important problems, namely *clustering* and *hamming embedding*, and of their asymmetric variants: *co-clustering* (e.g. Dhillon et al. (2003); Banerjee et al. (2004)) and *asymmetric hamming embedding* Neyshabur et al. (2013). We show how these two problems (clustering and hamming embedding) are highly related, how hamming embedding can be viewed as a generalization of clustering, and how the convex hull of both corresponds to a generalization of Locality Sensitive Hashing (LSH).

Our main conclusion is that the convex hull of co-clustering and asymmetric hamming embedding is tightly captured by a shift-invariant modification of the max-norm—a tractable SDP-representable relaxation (Theorem 2 in Section 5). We contrast this with the symmetric clustering and hamming embedding, in which the corresponding SDP relaxation is not tight, highlighting an important distinction between symmetric and asymmetric clustering, embedding and LSH.

To set the stage, we begin by formally introducing clustering and hamming embeddings and the relationship between them (Section 2). We then relate these concepts to Locality Section Hashing (LSH) as studied by Charikar (2002), as well as to more generalized notions of LSH (Section 3). Next, in Section 4, we turn to the asymmetric variants of all three notions, introducing also the appropriate generalization of asymmetric LSH. Finally, in Section 5 we formalize the notion of tightness of a convex relaxation and state our main results about the tightness of SDP relaxations in the symmetric and asymmetric cases.

P. Auer et al. (Eds.): ALT 2014, LNAI 8776, pp. 306–320, 2014.

2   Clustering and Hamming Embedding

In this Section we introduce the problems of clustering and hamming embedding. We provide a unified view of both problems, viewing hamming embedding as a direct generalization of clustering. Our starting point for both problems is an input similarity function sim : $S \times S \to [-1, +1]$ over a (possibly infinite) set of objects S. "Clustering", as we think of it here, is the problem of partitioning the elements of S into disjoint clusters so that items in the same cluster are similar while items in different clusters are not similar. "Hamming Embedding" is the problem of embedding S into a space of short strings such that similarity is approximated by the hamming distance between strings.

2.1   Clustering

We represent a clustering of S as a mapping $h : S \to \Gamma$, where Γ is a discrete alphabet representing the different clusters. We can think of h as a function that assigns a cluster identity to each element, where the meaning of the different identities is arbitrary. The alphabet Γ might have a fixed finite cardinality $|\Gamma| = k$, if we would like to have a clustering with a specific number of clusters. E.g., a binary alphabet corresponds to standard graph partitioning into two clusters. If $|\Gamma| = k$, we can assume that $\Gamma = [k]$. The alphabet Γ might be infinitely countable (e.g. $\Gamma = \mathbb{N}$), in which case we are not constraining the number of clusters.

The *cluster incidence function* $\kappa_h : S \times S \to \{\pm 1\}$ associated with a clustering h is defined as $\kappa_h(x, y) = 1$ if $h(x) = h(y)$ and $\kappa_h(x, y) = -1$ otherwise. For a finite space S of cardinality $n = |S|$ we can think of $\kappa_h \in \{\pm 1\}^{n \times n}$ as a permuted block-diagonal matrix, with $+1$s on the diagonal blocks, and -1s outside these blocks. We denote the set of all valid cluster incidence functions over S with an alphabet of size k (i.e. with at most k clusters) as $M_{S,k} = \{\kappa_h \mid h : S \to [k]\}$, where $k = \infty$ is allowed.

With this notion in hand, we can think of clustering as a problem of finding a cluster incidence function κ_h that approximates a given similarity sim, as quantified by objectives such as $\min \mathbb{E}_{x,y}[|\kappa_h(x, y) - sim(x, y)|]$ or $\max \mathbb{E}_{x,y}[sim(x, y)\kappa_h(x, y)]$ (this is essentially the correlation clustering objective). Both of these objectives are convex in κ, but the computational difficulty of clustering arises from the non-convex constraint that κ is a valid cluster incidence function. A possible approach is therefore to relax the constraint that κ is a valid cluster incidence function, or in the finite case, a cluster incidence matrix. This is the approach taken by, e.g. Jalali et al. (2011); Jalali and Srebro (2012), who relax the constraint to a trace-norm and max-norm constraint respectively. One of the questions we will be exploring here is whether this is the tightest relaxation possible, or whether there is a significantly tighter relaxation.

2.2   Hamming Embedding and Binary Matrix Factorization

In the problem of binary hamming embedding (also known as binary hashing), we want to find a mapping from each object $x \in S$ to a binary string $b(x) \in$

$\{\pm 1\}^d$ such that the similarity between strings is approximated by the hamming distance between their images:

$$\text{sim}(x, y) \approx 1 - \frac{2\delta_{\text{Ham}}(b(x), b(y))}{d} \tag{1}$$

Calculating the hamming distance of two binary hashes is an extremely fast operation, and so such a hash is useful for very fast computation of similarities between massive collections of objects. Furthermore, hash tables can be used to further speed up retrieval of similar objects Gionis et al. (1999); Indyk and Motwani (1998); Andoni and Indyk (2006).

Binary hamming embedding can be seen as a generalization of clustering as follows: For each position $i = 1, \ldots, d$ in the hash, we can think of $b_i(x)$ as a clustering into two clusters (i.e. with $\Gamma = \{\pm 1\}$). The hamming distance is then an average of the d cluster incidence functions:

$$1 - \frac{2\delta_{\text{Ham}}(b(x), b(y))}{d} = \frac{1}{d} \sum_{i=1}^{d} \kappa_{b_i}(x, y).$$

Our goal then is to approximate a similarity function by an average of d binary clusterings. For $d = 1$ this is exactly a binary clustering. For $d > 1$, we are averaging multiple binary clusterings.

Since we have $\langle b(x), b(y) \rangle = d - 2\delta_{\text{Ham}}(b(x), b(y))$, we can formulate the binary hashing problem as a binary matrix factorization where the goal is to approximate the similarity matrix by a matrix of the form $RR^\top$, where R is a d-dimensional binary matrix:

$$\begin{aligned}
\min_{R} \quad & \sum_{ij} \text{err}(\text{sim}(i, j), X(i, j)) \\
\text{s.t} \quad & X = RR^\top \\
& R \in \{\pm 1\}^{n \times d}
\end{aligned} \tag{2}$$

where $\text{err}(x, y)$ is some error function such as $\text{err}(x, y) = |x - y|$.

Going beyond binary clustering and binary embedding, we can consider hamming embeddings over larger alphabets. That is, we can consider mappings $b : S \to \Gamma^d$, where we aim to approximate the similarity as in (1), recalling that hamming distance always counts the number of positions in which the strings disagree. Again, we have that the length d hamming embeddings over a (finite or infinitely countable) alphabet Γ correspond to averages of d cluster incidence matrices over the same alphabet Γ.

3   Locality Sensitive Hashing Schemes

Moving on from a finite average of clusterings, with a fixed number of components, as in hamming embedding, to an infinite average, we arrive at the notion of LSH as studied by Charikar (2002).

Given a collection S of objects, an alphabet Γ and a similarity function sim : $S \times S \rightarrow [-1,1]$ such that for any $x \in S$ we have $\text{sim}(x,x) = 1$, we define a *locality sensitive hashing scheme* (**LSH**) as a probability distribution on the family of clustering functions (hash functions) $\mathcal{H} = \{h : S \rightarrow \Gamma\}$ such that[1]

$$\mathbb{E}_{h \in \mathcal{H}}[\kappa_h(x,y)] = \text{sim}(x,y). \tag{3}$$

The set of all locality sensitive hashing schemes with an alphabet of size k is nothing but the convex hull of the set $M_{S,k}$ of cluster incidence matrices.

The importance of an LSH, as an object in its own right as studied by Charikar (2002), is that a hamming embedding can be obtained from an LSH by randomly generating a finite number of hash functions from the distribution over the family $\mathcal{H}$. In particular, if we draw $h_1, \ldots, h_d$ i.i.d. from an LSH, then the length-d hamming embedding $b(x) = [h_1(x), \ldots, h_d(x)]$ has expected square error

$$E[(\text{sim}(x,y) - \frac{1}{d}\sum \kappa_{h_d}(x,y))^2] \le \frac{1}{d},$$

where the expectation is w.r.t. the sampling, and this holds for all x, y, and so also for any average over them.

3.1 α-LSH

If the goal is to obtain a low-error embedding, the requirement (3) might be too stringent. We can tolerate an affine relationship between the embedding and the target similarity, and instead require that

$$\alpha \mathbb{E}_{h \in \mathcal{H}}[\kappa_h(x,y)] - \theta = \text{sim}(x,y). \tag{4}$$

where $\alpha, \theta \in \mathbb{R}$, $\alpha > 0$. Note that α and θ go hand-in-hand: if $\text{sim}(x,x) = 1$, then we must have $\theta = \alpha - 1$.

A distribution over h that obeys (4) is called an α-**LSH**. We can now verify that, for $h_1, \ldots, h_d$ drawn i.i.d. from an α-LSH, and any $x, y \in S$:

$$E\left[\left(\text{sim}(x,y) - (\frac{\alpha}{d}\sum \kappa_{h_d}(x,y) - \theta)\right)^2\right] \le \frac{\alpha^2}{d}.$$

The length of the LSH required to achieve accurate approximation of a similarity function thus scales quadratically with α, and it is therefor desirable to obtain an α-LSH with as low an α as possible.

Unfortunately, the requirement (4) might be too difficult to attain, even with a very large α. As we formalize in the following Claim, which is based on Lemmas 2 and 3 of Charikar (2002), having an α-LSH is equivalent to being embeddable to hamming space with no distortion:

---

[1] Charikar (2002) discuss similarity functions with a range of $[0, 1]$, rather than $[-1, 1]$, and so require $\mathbb{P}_{h \in \mathcal{H}}[h(x) = h(y)] = \text{sim}(x,y)$. The definition (3) is equivalent, when applied to the transformed similarity function $2\text{sim}(x,y) - 1$.

Claim 1. *For any finite or countable alphabet* Γ, $k = |\Gamma| \geq 2$, *a similarity function* $sim(x, y)$ *has an* α-*LSH over* Γ *for some* α *if and only if* $\delta_{sim}(x, y) = \frac{1 - sim(x,y)}{2}$ *is embeddable to hamming space with no distortion.*

Proof. We first show that if there exist an α-LSH for function $sim(x, y)$ then $\frac{1 - sim(x,y)}{2}$ is embeddable to hamming space with no distortion. An α-LSH for function $sim(x, y)$ corresponds to an LSH for function $1 - \frac{1 - sim(x,y)}{\alpha}$. Using lemma 3 in Charikar (2002), we can say that $\frac{1 - sim(x,y)}{\alpha}$ is isometrically embeddable in the Hamming cube which means $1 - sim(x, y)$ can be embedded in Hamming cube with no distortion.

We now prove that existence of a Hamming embedding with no distortion is a sufficient condition for existence of α-LSH. Let f be the map function from set S to the Hamming space with no distortion is, i.e. there exist $\beta > 0$ such that for any $x, y \in S$, $\delta_{sim}(x, y) = \beta \delta_{\text{Ham}}(x, y)$. We have that:

$$
\begin{aligned}
\mathbb{E}_{h \sim \mathcal{H}}[\kappa_h(x, y)] &= 2\mathbb{P}_{h \sim \mathcal{H}}[h(x) = h(y)] - 1 \\
&= 1 - 2\mathbb{P}_{h \sim \mathcal{H}}[h(x) \neq h(y)] \\
&= 1 - 2\delta_{\text{Ham}}(x, y) \\
&= 1 - \frac{2\delta_{sim}(x, y)}{\beta} \\
&= 1 - \frac{1 - sim(x, y)}{\beta} \\
&= \frac{sim(x, y) + (\beta - 1)}{\beta}
\end{aligned}
$$

which gives us an α-LSH based on equation (4) by setting $\alpha = \beta$ and $\theta = \beta - 1$. $\square$

As a result of Claim 1, it can be shown that given *any* large enough set of low dimensional unit vectors, there is no α-LSH for the Euclidian inner product.

Claim 2. *Let* $\{x^{(1)}, \ldots, x^{(n)}\}$ *be an arbitrary set of distinct unit vectors in the unit sphere. Let Z be a matrix whose entries are $Z_{ij} = \langle x^{(i)}, x^{(j)} \rangle$ for $1 \leq i, j \leq n$. If $d < \log_2 n$, then there is no α-LSH for Z.*

Proof. According to Danzer and Grünbaum (1962) (see also Buchok (2010)), if $d < \log_2 n$ then in any set of n points in d-dimensional Euclidian space, there exist at least three points that form an obtuse triangle. Equivalently, there exist three vectors x, y and z in any set of n different d-dimensional unit vectors such that:

$$\langle z - x, z - y \rangle < 0$$

We rewrite the above inequality as:

$$(1 - \langle z, x \rangle) + (1 - \langle z, y \rangle) < (1 - \langle x, y \rangle)$$

The above inequality implies that the distance measure $\Delta_{ij} = (1 - Z_{ij})/2$ is not a metric. Consequently, according to Claim 1 since $\Delta_{ij} = (1 - Z_{ij})/2$ is not a metric, there is no α-LSH for the matrix Z.   □

We can unfortunately conclude that there is no α-LSH for several important similarity measures such as the Euclidian inner product, Overlap coefficient and Dice's coefficient. In fact, we see that we might not have an α-LSH even for a similarity matrix based on a finite positive semidefinite matrix, such as the matrix Z in Claim 2. It might therefore be desirable to consider even more relaxed notions of locality sensitive hashing.

3.2   Generalized α-LSH

In the following section, we will see how to break the barrier imposed by Claim 1 by allowing asymmetry, highlighting the extra power asymmetry affords us. But before doing so, let us consider a different attempt at relaxing the definition of an α-LSH, motivated by to the work of Charikar and Wirth (2004) and Alon and Naor (2006): in order to uncouple the shift θ from the scaling α, we will allow for a different, arbitrary, shift on the self-similarities $\text{sim}(x, x)$ (i.e. on the diagonal of sim).

We say that a probability distribution over $\mathcal{H} = \{h : S \to \Gamma\}$ is a **Generalized α-LSH**, for $\alpha > 0$ if there exist $\theta, \gamma \in \mathbb{R}$ such that for all x, y:

$$\alpha \mathbb{E}_{h \in \mathcal{H}}[\kappa_h(x, y))] = \text{sim}(x, y) + \theta + \gamma 1_{x=y}$$

With this definition, then any symmetric similarity function, at least over a finite domain, admits a Generalized α-LSH, with a sufficiently large α:

Claim 3. *For a finite set S, $|S| = n$, for any symmetric $\text{sim} : S \times S \to [-1, 1]$ with $\text{sim}(x, x) = 1$, there exists a Generalized α-LSH over a binary alphabet Γ ($|\Gamma| = 2$) where $\alpha = O((1 - \lambda_{\min}) \log n)$-LSH, and $\lambda_{\min}$ is the smallest eigenvalue of the matrix sim.*

Proof. We observe that $\text{sim} - \lambda_{\min} I$ is a positive semidefinite matrix. According to Charikar and Wirth (2004), if a matrix Z with unit diagonal is positive semidefinite, then there is a probability distribution over a family $\mathcal{H}$ of hash functions such that for any $x \neq y$:

$$\mathbb{E}_{h \in \mathcal{H}}[h(x)h(y)] = \frac{Z(x, y)}{C \log n}.$$

We let $Z(x, y) = (\text{sim}(x, y) - \lambda_{\min} 1_{x=y})/(1 - \lambda_{\min})$. Matrix Z is positive semidefinite and has unit diagonal. Hence, there is a probability distribution over a family $\mathcal{H}$ of hash functions such that

$$\mathbb{E}_{h \in \mathcal{H}}[h(x)h(y)] = \frac{\text{sim}(x, y) - \lambda_{\min} 1_{x=y}}{C(1 - \lambda_{\min}) \log n},$$

equivalently

$$C(1 - \lambda_{\min}) \log n \mathbb{E}_{h \in \mathcal{H}}[\kappa_h(x, y))] = \text{sim}(x, y) - \lambda_{\min} 1_{x=y}.$$

□

It is important to note that $\lambda_{\min}$ could be negative, and as low as $\lambda_{\min} = -\Omega(n)$. The required α might therefore be as large as $\Omega(n)$, yielding a terrible LSH.

4 Asymmetry

In order to allow for greater power, we now turn to *Asymmetric* variants of clustering, hamming embedding, and LSH.

Given two collections of objects S, T, which might or might not be identical, and an alphabet Γ, an *asymmetric clustering* (or co-clustering Dhillon et al. (2003)) is specified by pair of mappings $f : S \to \Gamma$ and $g : T \to \Gamma$ and is captured by the asymmetric cluster incidence matrix $\kappa_{f,g}(x,y)$ where $\kappa_{f,g}(x,y) = 1$ if $f(x) = g(y)$ and $\kappa_{f,g}(x,y) = -1$ otherwise. We denote the set of all valid asymmetric cluster incidence functions over S, T with an alphabet of size k as $M_{(S,T),k} = \{\kappa_{f,g} \mid f : S \to [k], g : T \to [k]\}$, where we again also allow $k = \infty$ to correspond to a countable alphabet $\Gamma = \mathbb{N}$.

Likewise, an asymmetric binary embedding of S, T with alphabet Γ consists of a pair of functions $f : S \to \Gamma^d, g : T \to \Gamma^d$, where we approximate a similarity as:

$$\operatorname{sim}(x,y) \approx 1 - \frac{2\delta_{\mathrm{Ham}}(f(x), g(y))}{d} = \frac{1}{d}\sum_{i=1}^{d} \kappa_{f_i,g_i}(x,y). \tag{5}$$

That is, in asymmetric hamming embedding, we approximate a similarity as an average of d asymmetric cluster incidence matrices from $M_{(S,T),k}$.

In a recent work, Neyshabur et al. (2013) showed that even when $S = T$ and the similarity function sim is a well-behaved symmetric similarity function, asymmetric binary embedding could be much more powerful in approximating the similarity, using shorter lengths d, both theoretically and empirically on datasets of interest. That is, these concepts are relevant and useful not only in an a-priori asymmetric case where $S \neq T$ or sim is not symmetric, but also when the target similarity *is* symmetric, but we allow an asymmetric embedding. We will soon see such gaps also when considering the convex hulls of $M_{S,k}$ and $M_{(S,T),k}$, i.e. when considering LSHs. Let us first formally define an asymmetric α-LSH.

Given two collections of objects S and T, an alphabet Γ, a similarity function $\operatorname{sim} : S \times T \to [-1,1]$, and $\alpha > 0$, we say that an α-**ALSH** is a distribution over pairs of functions $f : S \to \Gamma$, $g : T \to \Gamma$, or equivalently over $M_{(S,T),|\Gamma|}$, such that for some $\theta \in \mathbb{R}$ and all $x \in S, y \in T$:

$$\alpha \mathbb{E}_{(f,g)\in\mathcal{F}\times\mathcal{G}}[\kappa_{f,g}(x,y))] - \theta = \operatorname{sim}(x,y). \tag{6}$$

To understand the power of asymmetric LSH, recall that many symmetric similarity functions do not have an α-LSH for any α. On the other hand, any similarity function over finite domains necessarily has an α-ALSH:

Claim 4. *For any similarity function* $\operatorname{sim} : S \times T \to [-1,1]$ *over finite* S, T, *there exists an* α-*ALSH with* $\alpha \leq \min\{|S|, |T|\}$.

This is corollary of Theorem 2 that will be proved later in section 5. The proof follows from Theorem 2 and the following upper bound on the max-norm:

$$\|Z\|_{\max} \leq \text{rank}(Z).\|Z\|_\infty^2$$

where $\|Z\|_\infty^2 = \max_{x,y} |Z(x,y)|$.

In section 3, we saw that similarity functions that do not admit an α-LSH, still admit Generalized α-LSH. However, the gap between the α required for a Generalized α-LSH and that required for an α-ALSH might be as large as $\Omega(|S|)$:

Theorem 1. *For any even n, there exists a set S of n objects and a similarity $Z : S \times S \to \mathbb{R}$ such that*

- *there is a binary $3K_R$-ALSH for Z, where $K_R \approx 1.79$ is Krivine's constant;*
- *there is no Generalized α-LSH for any $\alpha < n - 1$.*

Proof. Let $S = [n]$ and Z be the following similarity matrix:

$$Z = 2I_{n \times n} + \begin{bmatrix} -1_{\frac{n}{2} \times \frac{n}{2}} & 1_{\frac{n}{2} \times \frac{n}{2}} \\ 1_{\frac{n}{2} \times \frac{n}{2}} & -1_{\frac{n}{2} \times \frac{n}{2}} \end{bmatrix}$$

Now we use Theorem 2, which we will prove later (our proof of Theorem 2 does not rely on the proof of this theorem). Using triangle inequality property of the norm, we have $\|Z\|_{\max} \leq \|Z - 2I_{n \times n}\|_{\max} + \|2I_{n \times n}\|_{\max} = 3$; and by Theorem 2 there is a $3K_R$-ALSH for Z. Looking at the decomposition of Z, it is not difficult to see that the smallest eigenvalue of Z is $2 - n$. So in order to have a positive semidefinite similarity matrix, we need γ to be at least $n - 2$ and θ to be at least -1 (otherwise the sum of elements of $Z + \theta + (n-2)I$ will be less than zero and so $Z + \theta + (n-2)I$ will not be positive semidefinite). So $\alpha = \theta + \gamma$ is at least $n - 1$. ☐

5   Convex Relaxations, α-LSH and Max-Norm

After setting the stage we will now turn to our two main questions, which we will see are essentially the same question: can we get a tight convex relaxation of the set $M_{(S,T),k}$ of (asymmetric) clustering incidence functions? And can we characterize the values of α for which we can get an α-ALSH for a particular similarity measure? We just stated the questions for the asymmetric case, which will be the main focus of this Section, but in Sub-Section 5.3 we will also return to the symmetric case and ask the same questions there.

For notational simplicity, we will now fix S and T and use M_k to denote $M_{(S,T),k}$.

5.1   The Ratio Function

The tightest possible convex relaxation of M_k is simply its convex hull conv M_k. Assuming P $\neq$ NP, it seems that conv M_k is not polynomially tractable[2].

---

[2]   conv M_2 is not polynomially tractable Alon and Naor (2006).

What we ask here is whether we have a tractable tight relaxation of conv M_k. To measure tightness of some convex $B \supseteq M_k$, for each $Z \in B$, we will bound its *cluster ratio*:

$$\rho_k(Z) = \min\{r | Z \in r \operatorname{conv} M_k\} = \min\{r | Z/r \in \operatorname{conv} M_k\}.$$

In fact, the function ρ_k is Minkowski gauge of conv M_k Thompson (1996). That is, by how much do we have to inflate M_k so that includes $Z \in B$. The supremum $\rho_k(B) = \sup_{Z \in B} \rho_k(Z)$ is then the maximal inflation ratio between conv M_k and B, i.e. such that conv $M_k \subseteq B \subseteq \rho_k \operatorname{conv} M_k$. Similarly, we define the *centralized cluster ratio* as:

$$\hat{\rho}_k(Z) = \min_{\theta \in \mathbb{R}} \min\{r | Z - \theta \in r \operatorname{conv} M_k\}.$$

This is nothing but the lowest α for which we have an α-ALSH:

Claim 5. *For any similarity function $sim(x, y)$, $\hat{\rho}_k(sim)$ is equal to the smallest α s.t. there exists an α-ALSH for sim over alphabet of cardinality k.*

Proof. We write the problem of minimizing α in α-ALSH as:

$$\min_{\theta \in \mathbb{R}, \alpha \in \mathbb{R}^+} \quad \alpha \tag{7}$$
$$\text{s.t.} \quad sim(x, y) = \alpha \mathbb{E}_{(f,g) \in \mathcal{F} \times \mathcal{G}}[\kappa_{f,g}(x, y)] - \theta$$

We know that:

$$\mathbb{E}_{(f,g) \in \mathcal{F} \times \mathcal{G}}[\kappa_{f,g}(x, y)] = \sum_{f \in M_{S,k}} \sum_{g \in M_{T,k}} \kappa_{f,g}(x, y) p(f, g)$$

where $p(f, g)$ is the joint probability of hash functions f and g. Define $\mu(f, g) = \alpha p(f, g)$ and write:

$$\alpha = \alpha \sum_{f \in M_{S,k}} \sum_{g \in M_{T,k}} p(f, g) = \sum_{f \in M_{S,k}} \sum_{g \in M_{T,k}} \alpha p(f, g) = \sum_{f \in M_{S,k}} \sum_{g \in M_{T,k}} \mu(f, g)$$

We have:

$$\alpha \sum_{f \in M_{S,k}} \sum_{g \in M_{T,k}} \kappa_{f,g}(x, y) p(f, g) - \theta = \sum_{f \in M_{S,k}} \sum_{g \in M_{T,k}} \kappa_{f,g}(x, y) \mu(f, g) - \theta$$

Substituting the last two equalities into formulation (7) gives us the formulation for centralized cluster ratio. $\qquad \square$

Our main goal in this section is to obtain tight bounds on $\rho_k(Z)$ and $\hat{\rho}_k(Z)$.

The Ratio Function and Cluster Norm The convex hull conv M_k is related to the cut-norm, and its generalization the cluster-norm. Although the two are not identical, it is worth understanding the relationship.

For $k = 2$, the ratio function is a norm, and is in fact the dual of a modified cut-norm:

$$\rho_2^*(W) = \|W\|_{C,2} = \max_{u:S\to\{\pm 1\},v:T\to\{\pm 1\}} \sum_{x\in S,y\in T} W(x,y)u(x)v(y)$$

The norm $\|W\|_{C,2}$ is a variant of the cut-norm, and is always within a factor of four from the cut-norm as defined by Alon and Naor (2006). The set conv M_2 in this case is the unit ball of the modified cut-norm.

For $k > 2$, the ratio function is *not* a norm, since M_k, for $k > 2$, is not symmetric about the origin: we might have $Z \in M_k$ but $-Z \notin M_k$ and so $\rho_k(Z) \neq \rho_k(-Z)$. A ratio function defined with respect to the symmetric convex hull conv$(M_k \cup -M_k)$, is a norm, and is dual to the following *cluster norm*, which is a generalization of the modified cut-norm:

$$\|W\|_{C,k} = \max_{u:S\to\Gamma,v:T\to\Gamma} \sum_{x\in S,y\in T} W(x,y)\kappa_{u,v}(x,y)$$

5.2   A Tight Convex Relaxation Using the Max-Norm

Recall that the max-norm (also known as the $\gamma_2 : \ell_1 \to \ell_\infty$ norm) of a matrix is defined as Srebro and Shraibman (2005):

$$\|Z\|_{\max} = \min_{UV^\top = Z} \max(\|U\|_{2,\infty}^2, \|V\|_{2,\infty}^2)$$

where $\|U\|_{2,\infty}$ is the maximum ℓ_2 norm of rows of the matrix U. The max-norm is SDP representable and thus tractable Srebro et al. (2005). Even when S and T are not finite, and thus sim is not a finite matrix, the max-norm can be defined as above, where now U and V can be thought of as mappings from S and T respectively into a Hilbert space, with $\text{sim}(x,y) = (UV^\top = Z)(x,y) = \langle U(x), V(y)\rangle$ and $\|U\|_{2,\infty} = \sup_x \|U(x)\|$.

We also define the *centralized max-norm*, which, even though it is *not* a norm, we denote as:

$$\|Z\|_{\widehat{\max}} = \min_{\theta\in\mathbb{R}} \|Z - \theta\|_{\max}$$

The centralized max-norm is also SDP-representable.

Our main result is that the max-norm provides a tight bound on the ratio function:

Theorem 2. *For any similarity function* sim $: S \times T \to \mathbb{R}$ *we have that:*

$$\frac{1}{2}\|sim\|_{\widehat{\max}} \leq \frac{1}{2}\hat{\rho}_2(sim) \leq \hat{\rho}(sim) \leq \hat{\rho}_k(sim) \leq \hat{\rho}_2(sim) \leq K\|sim\|_{\widehat{\max}}$$

and also

$$\frac{1}{3}\|sim\|_{\max} \leq \rho(sim) \leq \rho_k(sim) \leq \rho_2(sim) \leq K\|sim\|_{\max}$$

where all inequalities are tight and we have $1.67 \leq K_G \leq K \leq K_R \leq 1.79$ *(K_G is Grothendieck's constant and K_R is Krivine's constant).*

We can interpret Theorem 2 in two ways: a "primal" interpretation in terms of tightness of an SDP approximation to co-clustering and asymmetric hamming embedding, and a "dual" view in terms of α-ALSH.

Taking the primal view, we see that the SDP relaxation given by the max-norm provides for a very tight relaxation for co-clustering and asymmetric hamming embeddings: it is only a factor of less than 6 from the best possible convex relaxation (namely the convex hull) of M_k:

$$\{Z \mid \|Z\|_{\max} \leq 1/K\} \subseteq \operatorname{conv} M_k \subseteq \{Z \mid \|Z\|_{\max} \leq 3\} \qquad (8)$$

where recall $K < 1.79$. Allowing an additive shift, we obtain an even tighter tractable relaxation, with an inflation ratio of less than 4.

Considering the dual view of Theorem 2, and recalling Claim 5, we can also interpret the first line of inequalities as providing a tight characterization of the smallest α for which we can obtain an α-ALSH: the SDP-representable centralized max-norm gives us the smallest such α up to a factor of less than 4. In particular, since the centralized max-norm is always defined and finite for every finite matrix (and bounded by the dimensionality of the matrix), we see that for any pair S, T of finite domains, we always have an α-ALSH for some finite α, bounded by the cardinality of the domains—as claimed Theorem 1.

Interestingly, we see that the effect of the alphabet size k on the existence of α-ALSH is very small—the difference between an unbounded alphabet size and a binary alphabet size is at most a factor of two difference in α. Back to the "primal" view, this also means that the number of clusters allowed does not significantly affect the convex hull: taking a convex relaxation of correlation clustering with an unbounded number of clusters and of graph partitioning is separated by only a constant multiplicative factor (the quality of rounding might of course be different).

5.3   The Symmetric Case

It is not difficult to show that the lower bounds for α-LSH are the same as for α-ALSH and the inequalities are tight. However, there are no upper bounds for α-LSH similar to those for α-ALSH. Specifically, let $\hat{\alpha}$ and $\hat{\alpha}_g$ be the smallest values of α such that there is a (symmetric) α-LSH for sim and there is a generalized (symmetric) α-LSH for sim, respectively. Note that for some similarity functions sim there is no α-LSH at all; that is, $\hat{\alpha} = \infty$ and $\|\text{sim}\|_{\max} < \infty$. Also, as Theorem 1 shows, there is a similarity function sim such that

$$\|\text{sim}\|_{\max} = O(1) \quad \text{but} \quad \hat{\alpha}_g \geq n - 1.$$

Moreover, it follows from the result of Arora et al. (2005) that there is no efficiently computable upper bound β for $\hat{\alpha}_g$ such that

$$\frac{\beta}{\log^c n} \leq \hat{\alpha}_g \leq \beta$$

(under a standard complexity assumption that $NP \not\subseteq DTIME(n^{\log^3 n})$). That is, neither the max-norm nor any other efficiently computable norm of sim gives a constant factor approximation for $\hat{\alpha}_g$.

In the remainder of this section we prove a series of lemmas corresponding to the inequalities in Theorem 2.

5.4 Proofs

Lemma 1. *For any two sets S and T of objects and any function* $\text{sim} : S \times T \to \mathbb{R}$, *we have that* $\hat{\rho}_2(\text{sim}) \leq 2\hat{\rho}(\text{sim})$ *and the inequality is tight.*

Proof. Using Claim 5, all we need to do is to prove that given the function sim, if there exist an α-ALSH with arbitrary cardinality, then we can find a binary $2\alpha - ALSH$. In order to do so, we assume that there exists an α-ALSH for family $\mathcal{F}$ and $\mathcal{G}$ of hash functions such that:

$$\alpha \mathbb{E}_{(f,g) \in \mathcal{F} \times \mathcal{G}}[\kappa_{f,g}(x,y)] = \text{sim}(x,y) + \theta$$

where $f : S \to \Gamma$ and $g : T \to \Gamma$ are hash functions. Now let $\mathcal{H}$ be a family of pairwise independent hash functions of the form $\Gamma \to \{\pm 1\}$ such that each element $\gamma \in \Gamma$, has the equal chance of being mapped into -1 or 1. Now, we have that:

$$\begin{aligned}
2\alpha \mathbb{E}_{h \in \mathcal{H}, (f,g) \in \mathcal{F} \times \mathcal{G}}[\kappa_{hof,hog}(x,y)] &= 2\alpha \mathbb{E}_{h \in \mathcal{H}, (f,g) \in \mathcal{F} \times \mathcal{G}}[\kappa_{hof,hog}(x,y)] \\
&= 2\alpha \mathbb{E}_{h \in \mathcal{H}, (f,g) \in \mathcal{F} \times \mathcal{G}}[h(f(x))h(g(y))] \\
&= 2\alpha(2P_{h \in \mathcal{H}, (f,g) \in \mathcal{F} \times \mathcal{G}}[h(f(x)) = h(g(y))] - 1) \\
&= 2\alpha P_{(f,g) \in \mathcal{F} \times \mathcal{G}}[f(x) = g(y)] \\
&= \text{sim}(x,y) + \theta + \alpha \\
&= \text{sim}(x,y) + \tilde{\theta}
\end{aligned}$$

The tightness can be demonstrated by the example $\text{sim}(x,y) = 2_{x=y} - 1$ when S is not finite. $\qquad\square$

Lemma 2. *For any two sets S and T of objects and any function* $\text{sim} : S \times T \to \mathbb{R}$, *we have that* $\|\text{sim}\|_{\max} \leq \rho_2(\text{sim})$ *and the inequality is tight.*

Proof. Without loss of generality, we assume that $\Gamma = \{\pm 1\}$. We want to solve the following optimization problem:

$$\rho_2(\text{sim}) = \min_{\mu : M_{S,2} \times M_{T,2} \to \mathbb{R}^+} \sum_{f \in M_{S,2}} \sum_{g \in M_{T,2}} \mu(f,g)$$

$$\text{s.t. } \text{sim}(x,y) = \sum_{f \in M_{S,2}} \sum_{g \in M_{T,2}} \kappa_{f,g}(x,y)\mu(f,g)$$

For any $x \in S$ and $y \in T$, we define two new function variables $\ell_x : M_{S,2} \times M_{T,2} \to \mathbb{R}$ and $r_y : M_{S,2} \times M_{T,2} \to \mathbb{R}$:

$$\ell_x(f,g) = \sqrt{\mu(f,g)}f(x)$$
$$r_y(f,g) = \sqrt{\mu(f,g)}g(y)$$

Since cluster incidence matrix can be written as $\kappa_{f,g}(x,y) = f(x)g(y)$, we have $\text{sim}(x,y) = \langle \ell_x, r_y \rangle$ and $\|\ell_x\|_2^2 = \sum_{f \in M_{S,2}} \sum_{g \in M_{T,2}} \mu(f,g)$. Therefore, we rewrite the optimization problem as:

$$\rho_2(\text{sim}) = \min_{t,\ell,r,\mu: M_{S,2} \times M_{T,2} \to \mathbb{R}^+} t$$

$$\text{s.t. } \langle l_x, r_y \rangle = \text{sim}(x,y)$$
$$\|\ell_x\|_2^2 \leq t$$
$$\|r_y\|_2^2 \leq t$$
$$\ell_x(f,g) = \sqrt{\mu(f,g)}f(x)$$
$$r_y(f,g) = \sqrt{\mu(f,g)}g(y)$$

Finally, we relax the above problem by removing the last two constraints:

$$\|\text{sim}\|_{\max} = \min_{t,\ell,r} t$$

$$\text{s.t. } \langle l_x, r_y \rangle = \text{sim}(x,y) \qquad (9)$$
$$\|\ell_x\|_2^2 \leq t$$
$$\|r_x\|_2^2 \leq t$$

The above problem is a max-norm problem and the solution is $\|\text{sim}\|_{\max}$. Therefore, $\|\text{sim}\|_{\max} \leq \rho_2(\text{sim})$. Taking the function $\text{sim}(x,y)$ to be a binary cluster incidence function will indicate the tightness of the inequality. $\qquad \square$

Lemma 3. *(Krivine's lemma Krivine (1977)) For any two sets of unit vectors $\{u_i\}$ and $\{v_j\}$ in a Hilbert space H, there are two sets of unit vectors $\{u_i'\}$ and $\{v_j'\}$ in a Hilbert space H' such that for any u_i and v_j, $\sin(c\langle u_i, v_j \rangle) = \langle u_i', v_j' \rangle$ where $c = \sinh^{-1}(1)$.*

Lemma 4. *For any two sets S and T of objects and any function $\text{sim} : S \times T \to R$, we have that $\rho_2(\text{sim}) \leq K\|\text{sim}\|_{\max}$ where $1.67 \leq K_G \leq K \leq K_R \leq 1.79$ (K_G is Grothendieck's constant and K_R is Krivine's constant).*

Proof. A part of the proof is similar to Alon and Naor (2006). Let ℓ_x and r_y be the solution to the max-norm formulation (9). If we use Lemma 3 on the normalized $\ell_x/\|\ell_x\|_2$ and $r_y/\|r_y\|_2$ in Hilbert space H and we call the new vectors ℓ_x' and r_y' in Hilbert space H', we have that:

$$\sin\left(\frac{cZ(x,y)}{\|\ell_x\|_2\|r_x\|_2}\right) = \langle \ell_x', r_y' \rangle$$

If z is a random vector chosen uniformly from H', by Lemma 3, we have:

$$\mathbb{E}\left[\text{sign}(\langle \ell'_x, z\rangle)\text{sign}(\langle r'_y, z\rangle)\right] = \frac{2}{\pi}\arcsin(\langle \ell'_x, r'_y\rangle)) = \frac{2c}{\pi\|\ell_x\|_2\|r_y\|_2}\text{sim}(x, y)$$

Now if we set the hashing function $f(x) = s(x)\text{sign}(\langle \ell'_x, z\rangle)$ where $s(x) = 1$ with probability $\frac{1}{2} + \frac{\|\ell_x\|_2}{2\sqrt{t}}$ and $s(x) = -1$ with probability $\frac{1}{2} - \frac{\|\ell_x\|_2}{2\sqrt{t}}$ we have that:

$$\begin{aligned}
\mathbb{E}[f(x)\text{sign}(\langle r'_y, z\rangle)] &= \left(\frac{1}{2} + \frac{\|\ell_x\|_2}{2\sqrt{t}}\right)\frac{2c}{\pi\|\ell_x\|_2\|r_y\|_2}\text{sim}(x, y) \\
&\quad - \left(\frac{1}{2} - \frac{\|\ell_x\|_2}{2\sqrt{t}}\right)\frac{2c}{\pi\|\ell_x\|_2\|r_y\|_2}\text{sim}(x, y) \\
&= \frac{2c}{\pi\sqrt{t}\|r_y\|_2}\text{sim}(x, y)
\end{aligned}$$

If we do the same procedure on $g(y) = s'(x)\text{sign}(\langle r'_y, z\rangle)$, we will have:

$$\mathbb{E}[f(x)g(y)] = \frac{2c}{t\pi}\text{sim}(x, y)$$

By setting $\mu(f, g) = \frac{\pi\|sim\|_{\max}}{2c}p(f, g)$ where $p(f, g)$ is the probability distribution over the defined f and g, we can see that such $\mu(f, g)$ is a feasible solution for the formulation of cluster ratio and we have:

$$\rho_2(\text{sim}) \leq \sum_{f \in M_{s,2}}\sum_{g \in M_{T,2}}\mu(f, g) = \frac{\pi}{2c}\|\text{sim}\|_{\max} = K_R\|\text{sim}\|_{\max}$$

The inequality $K_G \leq K$ is known due to Alon and Naor (2006).   □

Acknowledgments. This research was partially supported by NSF CAREER award CCF-1150062, NSF grant IIS-1302662 and an Intel ICRI-CI award.

References

Alon, N., Naor, A.: Approximating the cut-norm via grothendieck's inequality. SIAM Journal on Computing 35(4), 787–803 (2006)

Andoni, A., Indyk, P.: Near-optimal hashing algorithms for approximate nearest neighbor in high dimensions. In: FOCS, pp. 459–468 (2006)

Arora, S., Berger, E., Hazan, E., Kindler, G., Safra, M.: On non-approximability for quadratic programs. In: FOCS, pp. 206–215 (2005)

Banerjee, A., Dhillon, I., Ghosh, J., Merugu, S., Modha, D.S.: A generalized maximum entropy approach to bregman co-clustering and matrix approximation. In: SIGKDD, pp. 509–514 (2004)

Buchok, L.V.: Two new approaches to obtaining estimates in the danzer-grunbaum problem. Mathematical Notes 87(4), 489–496 (2010)

Charikar, M., Wirth, A.: Maximizing quadratic programs: Extending grothendieck's inequality. In: FOCS, pp. 54–60 (2004)

Charikar, M.S.: Similarity estimation techniques from rounding algorithms. In: STOC (2002)

Danzer, L., Grünbaum, B.: Über zwei probleme bezüglich konvexer körper von p. erdös und von vl klee. Mathematische Zeitschrift 79(1), 95–99 (1962)

Dhillon, I., Subramanyam, M., Dharmendra, S.M.: Information-theoretic co-clustering. In: SIGKDD (2003)

Gionis, A., Indyk, P., Motwani, R.: Similarity search in high dimensions via hashing. VLDB 99, 518–529 (1999)

Indyk, P., Motwani, R.: Approximate nearest neighbors: towards removing the curse of dimensionality. In: STOC, pp. 604–613 (1998)

Jalali, A., Chen, Y., Sanghavi, S., Xuo, H.: Clustering partially observed graphs via convex optimization. In: ICML (2011)

Jalali, A., Srebro, N.: Clustering using max-norm constrained optimization. In: ICML (2012)

Krivine, J.L.: Sur la constante de grothendieck. C. R. Acad. Sci. Paris Ser. A-B 284, 445–446 (1977)

Neyshabur, B., Yadollahpour, P., Makarychev, Y., Salakhutdinov, R., Srebro, N.: The power of asymmetry in binary hashing. In: NIPS (2013)

Srebro, N., Rennie, J., Jaakkola, T.: Maximum margin matrix factorization. In: NIPS (2005)

Srebro, N., Shraibman, A.: Rank, trace-norm and max-norm. In: Auer, P., Meir, R. (eds.) COLT 2005. LNCS (LNAI), vol. 3559, pp. 545–560. Springer, Heidelberg (2005)

Thompson, A.C.: Minkowski Geometry. Cambridge University Press (1996)

Indefinitely Oscillating Martingales

Jan Leike and Marcus Hutter

The Australian National University
{jan.leike,marcus.hutter}@anu.edu.au

Abstract. We construct a class of nonnegative martingale processes that oscillate indefinitely with high probability. For these processes, we state a uniform rate of the number of oscillations for a given magnitude and show that this rate is asymptotically close to the theoretical upper bound. These bounds on probability and expectation of the number of upcrossings are compared to classical bounds from the martingale literature. We discuss two applications. First, our results imply that the limit of the minimum description length operator may not exist. Second, we give bounds on how often one can change one's belief in a given hypothesis when observing a stream of data.[1]

Keywords: Martingales, infinite oscillations, bounds, convergence rates, minimum description length, mind changes.

1 Introduction

Martingale processes model fair gambles where knowledge of the past or choice of betting strategy have no impact on future winnings. But their application is not restricted to gambles and stock markets. Here we exploit the connection between nonnegative martingales and probabilistic data streams, i.e., probability measures on infinite strings. For two probability measures P and Q on infinite strings, the quotient Q/P is a nonnegative P-martingale. Conversely, every nonnegative P-martingale is a multiple of Q/P P-almost everywhere for some probability measure Q.

One of the famous results of martingale theory is Doob's Upcrossing Inequality [Doo53]. The inequality states that in expectation, every nonnegative martingale has only finitely many oscillations (called *upcrossings* in the martingale literature). Moreover, the bound on the expected number of oscillations is inversely proportional to their magnitude. Closely related is Dubins' Inequality [Dur10] which asserts that the probability of having many oscillations decreases exponentially with their number. These bounds are given with respect to oscillations of fixed magnitude.

In Section 4 we construct a class of nonnegative martingale processes that have infinitely many oscillations of (by Doob necessarily) decreasing magnitude.

---

[1] In Theorem 4, Q needs to be absolutely continuous with respect to P on cylinder sets. In Theorem 6, Corollary 7, Corollary 8, and Corollary 13, P needs to have perpetual entropy. See technical report [LH14].

P. Auer et al. (Eds.): ALT 2014, LNAI 8776, pp. 321–335, 2014.

These martingales satisfy uniform lower bounds on the probability and the expectation of the number of upcrossings. We prove corresponding upper bounds in Section 5 showing that these lower bounds are asymptotically tight. Moreover, the construction of the martingales is agnostic regarding the underlying probability measure, assuming only mild restrictions on it. We compare these results to the statements of Dubins' Inequality and Doob's Upcrossing Inequality and demonstrate that our process makes those inequalities (in Doob's case asymptotically) tight. If we drop the uniformity requirement, asymptotics arbitrarily close to Doob and Dubins' bounds are achievable. We discuss two direct applications of these bounds.

The Minimum Description Length (MDL) principle [Ris78] and the closely related Minimal Message Length (MML) principle [WB68] recommend to select among a class of models the one that has the shortest code length for the data plus code length for the model. There are many variations, so the following statements are generic: for a variety of problem classes MDL's predictions have been shown to converge asymptotically (predictive convergence). For continuous independently identically distributed data the MDL estimator usually converges to the true distribution [Grü07, Wal05] (inductive consistency). For arbitrary (non-i.i.d.) countable classes, the MDL estimator's predictions converge to those of the true distribution for single-step predictions [PH05] and ∞-step predictions [Hut09]. Inductive consistency implies predictive convergence, but not the other way around. In Section 6 we show that indeed, the MDL estimator for countable classes is *inductively inconsistent*. This can be a major obstacle for using MDL for prediction, since the model used for prediction has to be changed over and over again, incurring the corresponding computational cost.

Another application of martingales is in the theory of mind changes [LS05]. How likely is it that your belief in some hypothesis changes by at least $\alpha > 0$ several times while observing some evidence? Davis recently showed [Dav13] using elementary mathematics that this probability decreases exponentially. In Section 7 we rephrase this problem in our setting: the stochastic process

$$P(\text{ hypothesis } | \text{ evidence up to time } t)$$

is a martingale bounded between 0 and 1. The upper bound on the probability of many changes can thus be derived from Dubins' Inequality. This yields a simpler alternative proof for Davis' result. However, because we consider nonnegative but unbounded martingales, we get a weaker bound than Davis.

Omitted proofs can be found in the extended technical report [LH14].

2 Strings, Measures, and Martingales

We presuppose basic measure and probability theory [Dur10, Chp.1]. Let Σ be a finite set, called *alphabet*. We assume Σ contains at least two distinct elements. For every $u \in \Sigma^*$, the *cylinder set*

$$\Gamma_u := \{uv \mid v \in \Sigma^\omega\}$$

is the set of all infinite strings of which u is a prefix. Furthermore, fix the σ-algebras

$$\mathcal{F}_t := \sigma\left(\{\Gamma_u \mid u \in \Sigma^t\}\right) \quad \text{and} \quad \mathcal{F}_\omega := \sigma\left(\bigcup_{t=1}^\infty \mathcal{F}_t\right).$$

$(\mathcal{F}_t)_{t\in\mathbb{N}}$ is a *filtration*: since $\Gamma_u = \bigcup_{a\in\Sigma} \Gamma_{ua}$, it follows that $\mathcal{F}_t \subseteq \mathcal{F}_{t+1}$ for every $t \in \mathbb{N}$, and all $\mathcal{F}_t \subseteq \mathcal{F}_\omega$ by the definition of $\mathcal{F}_\omega$. An *event* is a measurable set $E \subseteq \Sigma^\omega$. The event $E^c := \Sigma^\omega \setminus E$ denotes the complement of E.

Definition 1 (Stochastic Process). $(X_t)_{t\in\mathbb{N}}$ *is called ($\mathbb{R}$-valued) stochastic process iff each X_t is an $\mathbb{R}$-valued random variable.*

Definition 2 (Martingale). *Let P be a probability measure over $(\Sigma^\omega, \mathcal{F}_\omega)$. An $\mathbb{R}$-valued stochastic process $(X_t)_{t\in\mathbb{N}}$ is called a P-supermartingale (P-submartingale) iff*

(a) each X_t is $\mathcal{F}_t$-measurable, and
(b) $\mathbb{E}[X_t \mid \mathcal{F}_s] \le X_s$ ($\mathbb{E}[X_t \mid \mathcal{F}_s] \ge X_s$) almost surely for all $s,t \in \mathbb{N}$ with $s < t$.

A process that is both P-supermartingale and P-submartingale is called P-martingale.

We call a supermartingale (submartingale) process $(X_t)_{t\in\mathbb{N}}$ *nonnegative* iff $X_t \ge 0$ for all $t \in \mathbb{N}$.

A *stopping time* is an $(\mathbb{N} \cup \{\omega\})$-valued random variable T such that $\{v \in \Sigma^\omega \mid T(v) = t\} \in \mathcal{F}_t$ for all $t \in \mathbb{N}$. Given a supermartingale $(X_t)_{t\in\mathbb{N}}$, the *stopped process* $(X_{\min\{t,T\}})_{t\in\mathbb{N}}$ is a supermartingale [Dur10, Thm. 5.2.6]. If $(X_t)_{t\in\mathbb{N}}$ is bounded, the limit of the stopped process, X_T, exists almost surely even if $T = \omega$ (Martingale Convergence Theorem [Dur10, Thm. 5.2.8]). We use the following variant on Doob's Optional Stopping Theorem for supermartingales.

Theorem 3 (Optional Stopping Theorem [Dur10, Thm. 5.7.6]). *Let $(X_t)_{t\in\mathbb{N}}$ be a nonnegative supermartingale and let T be a stopping time. The random variable X_T is almost surely well defined and $\mathbb{E}[X_T] \le \mathbb{E}[X_0]$.*

We exploit the following two theorems that state the connection between probability measures on infinite strings and martingales. For any two probability measures P and Q on $(\Sigma^\omega, \mathcal{F}_\omega)$, the quotient Q/P is a nonnegative P-martingale. Conversely, for every nonnegative P-martingale there is a probability measure Q on $(\Sigma^\omega, \mathcal{F}_\omega)$ such that the martingale is P-almost surely a multiple of Q/P.

Theorem 4 (Measures → Martingales [Doo53, II§7 Ex. 3]). *Let Q and P be two probability measures on $(\Sigma^\omega, \mathcal{F}_\omega)$. The stochastic process $(X_t)_{t\in\mathbb{N}}$, $X_t(v) := Q(\Gamma_{v_{1:t}})/P(\Gamma_{v_{1:t}})$ is a nonnegative P-martingale with $\mathbb{E}[X_t] = 1$.*

Theorem 5 (Martingales → Measures [LH14]). *Let P be a probability measure on $(\Sigma^\omega, \mathcal{F}_\omega)$ and let $(X_t)_{t\in\mathbb{N}}$ be a nonnegative P-martingale with $\mathbb{E}[X_t] = 1$. There is a probability measure Q on $(\Sigma^\omega, \mathcal{F}_\omega)$ such that for all $v \in \Sigma^\omega$ and all $t \in \mathbb{N}$ with $P(\Gamma_{v_{1:t}}) > 0$, $X_t(v) = Q(\Gamma_{v_{1:t}})/P(\Gamma_{v_{1:t}})$.*

3   Martingale Upcrossings

Fix $c \in \mathbb{R}$, and let $(X_t)_{t \in \mathbb{N}}$ be a martingale over the probability space $(\Sigma^\omega, \mathcal{F}_\omega, P)$. Let $t_1 < t_2$. We say the process $(X_t)_{t \in \mathbb{N}}$ does an ε-*upcrossing* between t_1 and t_2 iff $X_{t_1} \leq c - \varepsilon$ and $X_{t_2} \geq c + \varepsilon$. Similarly, we say $(X_t)_{t \in \mathbb{N}}$ does an ε-*downcrossing* between t_1 and t_2 iff $X_{t_1} \geq c + \varepsilon$ and $X_{t_2} \leq c - \varepsilon$. Except for the first upcrossing, consecutive upcrossings always involve intermediate downcrossings. Formally, we define the stopping times

$$T_0(v) := 0,$$
$$T_{2k+1}(v) := \inf\{t > T_{2k}(v) \mid X_t(v) \leq c - \varepsilon\}, \text{ and}$$
$$T_{2k+2}(v) := \inf\{t > T_{2k+1}(v) \mid X_t(v) \geq c + \varepsilon\}.$$

The $T_{2k}(v)$ denote the indices of upcrossings. We count the number of upcrossings by the random variable $U_t^X(c - \varepsilon, c + \varepsilon)$, where

$$U_t^X(c - \varepsilon, c + \varepsilon)(v) := \sup\{k \geq 0 \mid T_{2k}(v) \leq t\}$$

and $U^X(c - \varepsilon, c + \varepsilon) := \sup_{t \in \mathbb{N}} U_t^X(c - \varepsilon, c + \varepsilon)$ denotes the total number of upcrossings. We omit the superscript X if the martingale $(X_t)_{t \in \mathbb{N}}$ is clear from context.

The following notation is used in the proofs. Given a monotone decreasing function $f : \mathbb{N} \to [0, 1)$ and $m, k \in \mathbb{N}$, we define the events $E_{m,k}^{X,f}$ that denote that there are at least k-many $f(m)$-upcrossings:

$$E_{m,k}^{X,f} := \left\{v \in \Sigma^\omega \mid U^X(1 - f(m), 1 + f(m))(v) \geq k\right\}.$$

For all $m, k \in \mathbb{N}$ we have $E_{m,k}^{X,f} \supseteq E_{m,k+1}^{X,f}$ and $E_{m,k}^{X,f} \subseteq E_{m+1,k}^{X,f}$. Again, we omit X and f in the superscript if they are clear from context.

4   Indefinitely Oscillating Martingales

In this section we construct a class of martingales that has a high probability of doing an infinite number of upcrossings. The magnitude of the upcrossings decreases at a rate of a given summable function f (a function f is called *summable* iff it has finite L_1-norm, i.e., $\sum_{i=1}^{\infty} f(i) < \infty$), and the value of the martingale X_t oscillates back and forth between $1 - f(M_t)$ and $1 + f(M_t)$, where M_t denotes the num-

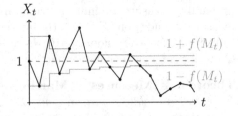

Fig. 1. An example evaluation of the martingale defined in the proof of Theorem 6

ber of upcrossings so far. The process has a monotone decreasing chance of escaping the oscillation.

Theorem 6 (An Indefinitely Oscillating Martingale). *Let $0 < \delta < \frac{2}{3}$ and let $f : \mathbb{N} \to [0, 1)$ be any monotone decreasing function such that $\sum_{i=1}^{\infty} f(i) \leq \frac{\delta}{2}$. For every probability measure P with $P(\Gamma_u) > 0$ for all $u \in \Sigma^*$ there is a nonnegative martingale $(X_t)_{t \in \mathbb{N}}$ with $\mathbb{E}[X_t] = 1$ and*

$$P[\forall m.\ U(1 - f(m), 1 + f(m)) \geq m] \geq 1 - \delta.$$

Proof. We assume $\Sigma = \{0, 1\}$ by grouping symbols into two groups. Since $P(\Gamma_{u0} \mid \Gamma_u) + P(\Gamma_{u1} \mid \Gamma_u) = 1$, we can define a function $a : \Sigma^* \to \Sigma$ that assigns to every string $u \in \Sigma^*$ a symbol $a_u := a(u)$ such that $p_u := P(\Gamma_{ua_u} \mid \Gamma_u) \leq \frac{1}{2}$. By assumption, we have $p_u > 0$.

We define the following stochastic process $(X_t)_{t \in \mathbb{N}}$. Let $v \in \Sigma^\omega$ and $t \in \mathbb{N}$ be given and define $u := v_{1:t}$. Let

$$M_t(v) := 1 + \underset{m \in \mathbb{N}}{\arg\max} \left\{ \forall k \leq m.\ U_t^X(1 - f(k), 1 + f(k)) \geq k \right\},$$

i.e., M_t is 1 plus the number of upcrossings completed up to time t. Define

$$\gamma_t(v) := \tfrac{p_u}{1 - p_u}(1 + f(M_t(v)) - X_t(v)).$$

For $t = 0$, we set $X_0(v) := 1$, otherwise we distinguish the following three cases.

(i) For $X_t(v) \geq 1$:

$$X_{t+1}(v) := \begin{cases} 1 - f(M_t(v)) & \text{if } v_{t+1} \neq a_u, \\ X_t(v) + \tfrac{1 - p_u}{p_u}(X_t(v) - (1 - f(M_t(v)))) & \text{if } v_{t+1} = a_u. \end{cases}$$

(ii) For $1 > X_t(v) \geq \gamma_t(v)$:

$$X_{t+1}(v) := \begin{cases} X_t(v) - \gamma_t(v) & \text{if } v_{t+1} \neq a_u, \\ 1 + f(M_t(v)) & \text{if } v_{t+1} = a_u. \end{cases}$$

(iii) For $X_t(v) < \gamma_t(v)$ and $X_t(v) < 1$:
let $d_t(v) := \max\{0, \min\{\tfrac{p_u}{1 - p_u}X_t(v), \tfrac{1 - p_u}{p_u}\gamma_t(v) - 2f(M_t(v))\}\};$

$$X_{t+1}(v) := \begin{cases} X_t(v) + d_t(v) & \text{if } v_{t+1} \neq a_u, \\ X_t(v) - \tfrac{1 - p_u}{p_u}d_t(v) & \text{if } v_{t+1} = a_u. \end{cases}$$

We give an intuition for the behavior of the process $(X_t)_{t \in \mathbb{N}}$. For all m, the following repeats. First X_t increases while reading a_u's until it reads one symbol that is not a_u and then jumps down to $1 - f(m)$. Subsequently, X_t decreases while not reading a_u's until it falls below γ_t or reads an a_u and then jumps up to $1 + f(m)$. If it falls below 1 and γ_t, then at every step, it can either jump up to $1 - f(m)$ or jump down to 0, whichever one is closest (the distance to the closest of the two is given by d_t). See Figure 1 for a visualization.

For notational convenience, in the following we omit writing the argument v to the random variables X_t, γ_t, M_t, and d_t.

Claim 1: $(X_t)_{t\in\mathbb{N}}$ is a martingale. Each X_{t+1} is $\mathcal{F}_{t+1}$-measurable, since it uses only the first $t+1$ symbols of v. Writing out cases (i), (ii), and (iii), we get

$$\mathbb{E}[X_{t+1} \mid \mathcal{F}_t] \overset{(i)}{=} (1 - f(M_t))(1 - p_u) + \left(X_t + \tfrac{1-p_u}{p_u}(X_t - (1 - f(M_t)))\right)p_u = X_t,$$

$$\mathbb{E}[X_{t+1} \mid \mathcal{F}_t] \overset{(ii)}{=} \left(X_t - \tfrac{p_u}{1-p_u}((1 + f(M_t)) - X_t)\right)(1 - p_u) + (1 + f(M_t))p_u = X_t,$$

$$\mathbb{E}[X_{t+1} \mid \mathcal{F}_t] \overset{(iii)}{=} (X_t + d_t)(1 - p_u) + (X_t - \tfrac{1-p_u}{p_u}d_t)p_u = X_t.$$

Claim 2: $X_t \geq 0$ and $\mathbb{E}[X_t] = 1$. The latter follows from $X_0 = 1$. Regarding the former, we use $0 \leq f(M_t) < 1$ to conclude

(i$\neq$) $1 - f(M_t) \geq 0$,

(i$=$) $\tfrac{1-p_u}{p_u}(X_t - (1 - f(M_t))) \geq 0$ for $X_t \geq 1$,

(ii$\neq$) $X_t - \gamma_t \geq 0$ for $X_t \geq \gamma_t$,

(ii$=$) $1 + f(M_t) \geq 0$,

(iii$\neq$) $X_t + d_t \geq 0$ since $d_t \geq 0$, and

(iii$=$) $X_t - \tfrac{1-p_u}{p_u}d_t \geq 0$ since $d_t \leq \tfrac{p_u}{1-p_u}X_t$.

Claim 3: $X_t \leq 1 - f(M_t)$ or $X_t \geq 1 + f(M_t)$ for all $t \geq T_1$. We use induction on t. The induction start holds with $X_{T_1} \leq 1 - f(M_t)$. The induction step is clear for (i) $X_t \geq 1$ and (ii) $1 > X_t \geq \gamma_t$ since $\gamma_t \geq 0$. In case (iii) we have either $d_t = 0$ or $d_t \leq \tfrac{1-p_u}{p_u}\gamma_t - 2f(M_t)$ and since $X_t < \gamma_t$,

$$X_{t+1} \leq X_t + d_t \leq X_t + (1 + f(M_t) - X_t) - 2f(M_t) = 1 - f(M_t).$$

Claim 4: If $X_t \geq 1 - f(M_t)$ then $X_t > \gamma_t$. In this case

$$\gamma_t = \tfrac{p_u}{1-p_u}(1 + f(M_t) - X_t) \leq 2\tfrac{p_u}{1-p_u}f(M_t),$$

and thus with $p_u \leq \tfrac{1}{2}$ and $f(M_t) \leq \sum_{k=1}^{\infty} f(k) \leq \tfrac{\delta}{2} < \tfrac{1}{3}$,

$$X_t - \gamma_t \geq 1 - f(M_t) - 2\tfrac{p_u}{1-p_u}f(M_t) = 1 - \tfrac{1+p_u}{1-p_u}f(M_t) \geq 1 - 3f(M_t) > 0.$$

Claim 5: If $X_t > 0$ and $(f(M_t) > 0$ or $X_t < 1)$ then $X_{t+1} \neq X_t$.

(i) Assume $X_t \geq 1$. Then either $X_{t+1} = 1 - f(M_t) < 1$, or $\tfrac{1-p_u}{p_u}(X_t - (1 - f(M_t))) > 0$ since $X_t > 1 - f(M_t)$.

(ii) Assume $1 > X_t \geq \gamma_t$. Then either $X_{t+1} = 1 + f(M_t) \geq 1 > X_t$, or $1 + f(M_t) - X_t \geq 1 - X_t > 0$, hence $\gamma_t > 0$ and thus $X_{t+1} = X_t - \gamma_t < X_t$.

(iii) Assume $0 < X_t < \gamma_t$ and $X_t < 1$. From Claim 4 follows $X_t < 1 - f(M_t)$, thus $\tfrac{1-p_u}{p_u}\gamma_t - 2f(M_t) = 1 - f(M_t) - X_t > 0$. By assumption, $\tfrac{p_u}{1-p_u}X_t > 0$ and therefore $d_t > 0$. Hence $X_t + d_t > X_t$ and $X_t - \tfrac{1-p_u}{p_u}d_t < X_t$.

Claim 6: For all $m \in \mathbb{N}$, if $E_{m,m-1} \neq \emptyset$ then $P(E_{m,m} \mid E_{m,m-1}) \geq 1 - 2f(m)$. Let $v \in E_{m,m-1}$ and let $t_0 \in \mathbb{N}$ be a time step such that exactly $m - 1$ upcrossings have been completed up to time t_0, i.e., $M_{t_0}(v) = m$. The subsequent downcrossing is completed eventually with probablity 1: we are in case (i) and in every step there is a chance of $1 - p_u \geq \frac{1}{2}$ of completing the downcrossing. Therefore we assume without loss of generality that the downcrossing has been completed, i.e., that t_0 is such that $X_{t_0}(v) = 1 - f(m)$. We will bound the probability $p := P(E_{m,m} \mid E_{m,m-1})$ that X_t rises above $1 + f(m)$ after t_0 to complete the m-th upcrossing.

Define the stopping time $T : \Sigma^\omega \to \mathbb{N} \cup \{\omega\}$,

$$T(v) := \inf\{t \geq t_0 \mid X_t(v) \geq 1 + f(m) \vee X_t(v) = 0\},$$

and define the stochastic process $Y_t = 1 + f(m) - X_{\min\{t_0+t,T\}}$. Because $(X_{\min\{t_0+t,T\}})_{t\in\mathbb{N}}$ is martingale, $(Y_t)_{t\in\mathbb{N}}$ is martingale. By definition, X_t always stops at $1 + f(m)$ before exceeding it, thus $X_T \leq 1 + f(m)$, and hence $(Y_t)_{t\in\mathbb{N}}$ is nonnegative. The Optional Stopping Theorem yields $\mathbb{E}[Y_{T-t_0} \mid \mathcal{F}_{t_0}] \leq \mathbb{E}[Y_0 \mid \mathcal{F}_{t_0}]$ and thus $\mathbb{E}[X_T \mid \mathcal{F}_{t_0}] \geq \mathbb{E}[X_{t_0} \mid \mathcal{F}_{t_0}] = 1 - f(m)$. By Claim 5, X_t does not converge unless it reaches either 0 or $1 + f(m)$, and thus

$$1 - f(m) \leq \mathbb{E}[X_T \mid \mathcal{F}_{t_0}] = (1 + f(m)) \cdot p + 0 \cdot (1 - p),$$

hence $P(E_{m,m} \mid E_{m,m-1}) = p \geq 1 - f(m)(1 + p) \geq 1 - 2f(m)$.

Claim 7: $E_{m+1,m} = E_{m,m}$ *and* $E_{m+1,m+1} \subseteq E_{m,m}$. By definition of M_t, the i-th upcrossings of the process $(X_t)_{t\in\mathbb{N}}$ is between $1 - f(i)$ and $1 + f(i)$. The function f is monotone decreasing, and by Claim 3 the process $(X_t)_{t\in\mathbb{N}}$ does not assume values between $1 - f(i)$ and $1 + f(i)$. Therefore the first m $f(m+1)$-upcrossings are also $f(m)$-upcrossings, i.e., $E_{m+1,m} \subseteq E_{m,m}$. By definition of $E_{m,k}$ we have $E_{m+1,m} \supseteq E_{m,m}$ and $E_{m+1,m+1} \subseteq E_{m+1,m}$.

Claim 8: $P(E_{m,m}) \geq 1 - \sum_{i=1}^{m} 2f(i)$. For $P(E_{0,0}) = 1$ this holds trivially. Using Claim 6 and Claim 7 we conclude inductively

$$P(E_{m,m}) = P(E_{m,m} \cap E_{m,m-1}) = P(E_{m,m} \mid E_{m,m-1})P(E_{m,m-1})$$
$$= P(E_{m,m} \mid E_{m,m-1})P(E_{m-1,m-1})$$
$$\geq (1 - 2f(m))\left(1 - \sum_{i=1}^{m-1} 2f(i)\right) \geq 1 - \sum_{i=1}^{m} 2f(i).$$

From Claim 7 follows $\bigcap_{i=1}^{m} E_{i,i} = E_{m,m}$ and therefore $P(\bigcap_{i=1}^{\infty} E_{i,i}) = \lim_{m\to\infty} P(E_{m,m}) \geq 1 - \sum_{i=1}^{\infty} 2f(i) \geq 1 - \delta$.   □

Theorem 6 gives a *uniform* lower bound on the probability for many upcrossings: it states the probability of the event that *for all* $m \in \mathbb{N}$, $U(1 - f(m), 1 + f(m)) \geq m$ holds. This is a lot stronger than the nonuniform bound $P[U(1 - f(m), 1 + f(m)) \geq m] \geq 1 - \delta$ for all $m \in \mathbb{N}$: the quantifier is inside the probability statement.

As an immediate consequence of Theorem 6, we get the following uniform lower bound on the *expected* number of upcrossings.

Corollary 7 (Expected Upcrossings). *Under the same conditions as in Theorem 6, for all $m \in \mathbb{N}$,*

$$\mathbb{E}[U(1 - f(m), 1 + f(m))] \geq m(1 - \delta).$$

Proof. From Theorem 6 and Markov's inequality.   □

By choosing the slowly decreasing but summable function f by setting $f^{-1}(\varepsilon) := 2\delta(\frac{1}{\varepsilon(\ln \varepsilon)^2} - \frac{e^2}{4})$, we get the following concrete results.

Corollary 8 (Concrete Lower Bound). *Let $0 < \delta < 1$. For every probability measure P with $P(\Gamma_u) > 0$ for all $u \in \Sigma^*$, there is a nonnegative martingale $(X_t)_{t \in \mathbb{N}}$ with $\mathbb{E}[X_t] = 1$ such that*

$$P\left[\forall \varepsilon > 0.\, U(1 - \varepsilon, 1 + \varepsilon) \in \Omega\left(\frac{\delta}{\varepsilon(\ln \frac{1}{\varepsilon})^2}\right)\right] \geq 1 - \delta \text{ and}$$

$$\mathbb{E}[U(1 - \varepsilon, 1 + \varepsilon)] \in \Omega\left(\frac{1}{\varepsilon(\ln \frac{1}{\varepsilon})^2}\right).$$

Moreover, for all $\varepsilon < 0.015$ we get $\mathbb{E}[U(1 - \varepsilon, 1 + \varepsilon)] > \frac{\delta(1-\delta)}{\varepsilon(\ln \frac{1}{\varepsilon})^2}$ and

$$P\left[\forall \varepsilon < 0.015.\, U(1 - \varepsilon, 1 + \varepsilon) > \frac{\delta}{\varepsilon(\ln \frac{1}{\varepsilon})^2}\right] \geq 1 - \delta.$$

The concrete bounds given in Theorem 8 are *not* the asymptotically optimal ones: there are summable functions that decrease even more slowly. For example, we could multiply f^{-1} with the factor $\sqrt{\ln(1/\varepsilon)}$ (which still is not optimal).

5   Martingale Upper Bounds

In this section we state upper bounds on the probability and expectations of many upcrossings (Dubins' Inequality and Doob's Upcrossing Inequality). We use the construction from the previous section to show that these bounds are tight. Moreover, with the following theorem we show that the uniform lower bound on the probability of many upcrossings guaranteed in Theorem 6 is asymptotically tight.

Every function f is either summable or not. If f is summable, then we can scale it with a constant factor such that its sum is smaller than $\frac{\delta}{2}$, and then apply the construction of Theorem 6. If f is not summable, the following theorem implies that there is no *uniform* lower bound on the probability of having at least m-many $f(m)$-upcrossings.

Theorem 9 (Upper Bound on Upcrossing Rate). *Let $f : \mathbb{N} \to [0, 1)$ be a monotone decreasing function such that $\sum_{t=1}^{\infty} f(t) = \infty$. For every probability measure P and for every nonnegative P-martingale $(X_t)_{t \in \mathbb{N}}$ with $\mathbb{E}[X_t] = 1$,*

$$P[\forall m.\, U(1 - f(m), 1 + f(m)) \geq m] = 0.$$

Proof. Define the events $D_m := \bigcup_{i=1}^{m} E_{i,i}^c = \{\forall i \leq m.\, U(1 - f(i), 1 + f(i)) \geq i\}$. Then $D_m \subseteq D_{m+1}$. Assume there is a constant $c > 0$ such that $c \leq P(D_m^c) = P(\bigcap_{i=1}^{m} E_{i,i})$ for all m. Let $m \in \mathbb{N}$, $v \in D_m^c$, and pick $t_0 \in \mathbb{N}$ such that the process $X_0(v), \ldots, X_{t_0}(v)$ has completed i-many $f(i)$-upcrossings for all $i \leq m$

and $X_{t_0}(v) \leq 1 - f(m+1)$. If $X_t(v) \geq 1 + f(m+1)$ for some $t \geq t_0$, the $(m+1)$-st upcrossing for $f(m+1)$ is completed and thus $v \in E_{m+1,m+1}$. Define the stopping time $T : \Sigma^\omega \to (\mathbb{N} \cup \{\omega\})$,

$$T(v) := \inf\{t \geq t_0 \mid X_t(v) \geq 1 + f(m+1)\}.$$

According to the Optional Stopping Theorem applied to the process $(X_t)_{t \geq t_0}$, the random variable X_T is almost surely well-defined and $\mathbb{E}[X_T \mid \mathcal{F}_{t_0}] \leq \mathbb{E}[X_{t_0} \mid \mathcal{F}_{t_0}] = X_{t_0}$. This yields $1 - f(m+1) \geq X_{t_0} \geq \mathbb{E}[X_T \mid \mathcal{F}_{t_0}]$ and by taking the expectation $\mathbb{E}[\cdot \mid X_{t_0} \leq 1 - f(m+1)]$ on both sides,

$$1 - f(m+1) \geq \mathbb{E}[X_T \mid X_{t_0} \leq 1 - f(m+1)]$$
$$\geq (1 + f(m+1))P[X_T \geq 1 + f(m+1) \mid X_{t_0} \leq 1 - f(m+1)]$$

by Markov's inequality. Therefore

$$P(E_{m+1,m+1} \mid D_m^c) = P[X_T \geq 1 + f(m+1) \mid X_{t_0} \leq 1 - f(m+1)]$$
$$\cdot P[X_{t_0} \leq 1 - f(m+1) \mid D_m^c]$$
$$\leq P[X_T \geq 1 + f(m+1) \mid X_{t_0} \leq 1 - f(m+1)]$$
$$\leq \tfrac{1-f(m+1)}{1+f(m+1)} \leq 1 - f(m+1).$$

Together with $c \leq P(D_m^c)$ we get

$$P(D_{m+1} \setminus D_m) = P\left(E_{m+1,m+1}^c \cap D_m^c\right)$$
$$= P\left(E_{m+1,m+1}^c \mid D_m^c\right) P(D_m^c) \geq f(m+1)c.$$

This is a contradiction because $\sum_{i=1}^\infty f(i) = \infty$:

$$1 \geq P(D_{m+1}) = P\left(\biguplus_{i=1}^m (D_{i+1} \setminus D_i)\right) = \sum_{i=1}^m P(D_{i+1} \setminus D_i) \geq \sum_{i=1}^m f(i+1)c \to \infty.$$

Therefore the assumption $P(D_m^c) \geq c$ for all m is false, and hence we get $P[\forall m. \, U(1 - f(m), 1 + f(m)) \geq m] = P(\bigcap_{i=1}^\infty E_{i,i}) = \lim_{m\to\infty} P(D_m^c) = 0.$ □

By choosing the decreasing non-summable function f by setting $f^{-1}(\varepsilon) := \frac{-a}{\varepsilon(\ln \varepsilon)} - b$ for Theorem 9, we get that $U(1 - \varepsilon, 1 + \varepsilon) \notin \Omega(\frac{1}{\varepsilon \log(1/\varepsilon)})$ P-almost surely.

Corollary 10 (Concrete Upper Bound). *Let P be a probability measure and let $(X_t)_{t \in \mathbb{N}}$ be a nonnegative martingale with $\mathbb{E}[X_t] = 1$. Then for all $a, b > 0$,*

$$P\left[\forall \varepsilon > 0. \, U(1 - \varepsilon, 1 + \varepsilon) \geq \frac{a}{\varepsilon \log(1/\varepsilon)} - b\right] = 0.$$

Theorem 11 (Dubins' Inequality [Dur10, Ex.5.2.14]). *For every nonnegative P-martingale $(X_t)_{t \in \mathbb{N}}$ and for every $c > 0$ and every $\varepsilon > 0$,*

$$P[U(c - \varepsilon, c + \varepsilon) \geq k] \leq \left(\tfrac{c-\varepsilon}{c+\varepsilon}\right)^k \mathbb{E}\left[\min\left\{\tfrac{X_0}{c-\varepsilon}, 1\right\}\right].$$

Dubins' Inequality immediately yields the following bound on the probability of the number of upcrossings.

$$P[U(1 - f(m), 1 + f(m)) \geq k] \leq \left(\frac{1-f(m)}{1+f(m)}\right)^k.$$

The construction from Theorem 6 shows that this bound is asymptotically tight for $m = k \to \infty$ and $\delta \to 0$: define the monotone decreasing function $f : \mathbb{N} \to [0, 1)$,

$$f(t) := \begin{cases} \frac{\delta}{2k}, & \text{if } t \leq k, \text{ and} \\ 0, & \text{otherwise.} \end{cases}$$

Then the martingale from Theorem 6 yields the lower bound

$$P[U(1 - \tfrac{\delta}{2k}, 1 + \tfrac{\delta}{2k}) \geq k] \geq 1 - \delta,$$

while Dubins' Inequality gives the upper bound

$$P[U(1 - \tfrac{\delta}{2k}, 1 + \tfrac{\delta}{2k}) \geq k] \leq \left(\frac{1 - \frac{\delta}{2k}}{1 + \frac{\delta}{2k}}\right)^k = \left(1 - \frac{2\delta}{2k + \delta}\right)^k \xrightarrow{k \to \infty} \exp(-\delta).$$

As δ approaches 0, the value of $\exp(-\delta)$ approaches $1 - \delta$ (but exceeds it since exp is convex). For $\delta = 0.2$ and $m = k = 3$, the difference between the two bounds is already lower than 0.021.

The following theorem places an upper bound on the rate of *expected* upcrossings.

Theorem 12 (Doob's Upcrossing Inequality [Xu12]). *Let $(X_t)_{t \in \mathbb{N}}$ be a submartingale. For every $c \in \mathbb{R}$ and $\varepsilon > 0$,*

$$\mathbb{E}[U_t(c - \varepsilon, c + \varepsilon)] \leq \tfrac{1}{2\varepsilon}\mathbb{E}[\max\{c - \varepsilon - X_t, 0\}].$$

Asymptotically, Doob's Upcrossing Inequality states that with $\varepsilon \to 0$,

$$\mathbb{E}[U(1 - \varepsilon, 1 + \varepsilon)] \in O\left(\tfrac{1}{\varepsilon}\right).$$

Again, we can use the construction of Theorem 6 to show that these asymptotics are tight: define the monotone decreasing function $f : \mathbb{N} \to [0, 1)$,

$$f(t) := \begin{cases} \frac{\delta}{2m}, & \text{if } t \leq m, \text{ and} \\ 0, & \text{otherwise.} \end{cases}$$

Then for $\delta = \tfrac{1}{2}$, Theorem 7 yields a martingale fulfilling the lower bound

$$\mathbb{E}[U(1 - \tfrac{1}{4m}, 1 + \tfrac{1}{4m})] \geq \frac{m}{2}$$

and Doob's Upcrossing Inequality gives the upper bound

$$\mathbb{E}[U(1 - \tfrac{1}{4m}, 1 + \tfrac{1}{4m})] \leq 2m,$$

which differs by a factor of 4.

The lower bound for the expected number of upcrossings given in Theorem 8 is a little looser than the upper bound given in Doob's Upcrossing Inequality. Closing this gap remains an open problem. We know by Theorem 9 that given a non-summable function f, the uniform probability for many $f(m)$-upcrossings goes to 0. However, this does not necessarily imply that expectation also tends to 0; low probability might be compensated for by high value. So for expectation there might be a lower bound larger than Theorem 7, an upper bound smaller than Doob's Upcrossing Inequality, or both.

If we drop the requirement that the rate of upcrossings to be uniform, Doob's Upcrossing Inequality is the best upper bound we can give [LH14].

6   Application to the MDL Principle

Let $\mathcal{M}$ be a countable set of probability measures on $(\Sigma^\omega, \mathcal{F}_\omega)$, called *environment class*. Let $K : \mathcal{M} \to [0,1]$ be a function such that $\sum_{Q \in \mathcal{M}} 2^{-K(Q)} \leq 1$, called *complexity function on* $\mathcal{M}$. Following notation in [Hut09], we define for $u \in \Sigma^*$ the *minimal description length* model as

$$\mathrm{MDL}^u := \underset{Q \in \mathcal{M}}{\arg\min} \left\{ -\log Q(\Gamma_u) + K(Q) \right\}.$$

That is, $-\log Q(\Gamma_u)$ is the (arithmetic) code length of u given model Q, and $K(Q)$ is a complexity penalty for Q, also called *regularizer*. Given data $u \in \Sigma^*$, MDL^u is the measure $Q \in \mathcal{M}$ that minimizes the total code length of data and model.

The following corollary of Theorem 6 states that in some cases the limit $\lim_{t \to \infty} \mathrm{MDL}^{v_{1:t}}$ does not exist with high probability.

Corollary 13 (MDL May not Converge). *Let P be a probability measure on the measurable space $(\Sigma^\omega, \mathcal{F}_\omega)$. For any $\delta > 0$, there is a set of probability measures $\mathcal{M}$ containing P, a complexity function $K : \mathcal{M} \to [0,1]$, and a measurable set $Z \in \mathcal{F}_\omega$ with $P(Z) \geq 1 - \delta$ such that for all $v \in Z$, the limit $\lim_{t \to \infty} \mathrm{MDL}^{v_{1:t}}$ does not exist.*

Proof. Fix some positive monotone decreasing summable function f (e.g., the one given in Theorem 8). Let $(X_t)_{t \in \mathbb{N}}$ be the P-martingale process from Theorem 6. By Theorem 5 there is a probability measure Q on $(\Sigma^\omega, \mathcal{F}_\omega)$ such that

$$X_t(v) = \frac{Q(\Gamma_{v_{1:t}})}{P(\Gamma_{v_{1:t}})}.$$

Choose $\mathcal{M} := \{P, Q\}$ with $K(P) := K(Q) := 1$. From the definition of MDL and Q it follows that

$$X_t(u) < 1 \iff Q(\Gamma_u) < P(\Gamma_u) \implies \mathrm{MDL}^u = P, \text{ and}$$
$$X_t(u) > 1 \iff Q(\Gamma_u) > P(\Gamma_u) \implies \mathrm{MDL}^u = Q.$$

For $Z := \bigcap_{m=1}^{\infty} E_{m,m}$ Theorem 6 yields

$$P(Z) = P[\forall m.\ U(1 - f(m), 1 + f(m)) \geq m] \geq 1 - \delta.$$

For each $v \in Z$, the measure $\mathrm{MDL}^{v_{1:t}}$ alternates between P and Q indefinitely, and thus its limit does not exist.   □

Crucial to the proof of Theorem 13 is that not only does the process Q/P oscillate indefinitely, it oscillates around the constant $\exp(K(Q) - K(P)) = 1$. This implies that the MDL estimator may keep changing indefinitely, and thus it is inductively inconsistent.

7   Bounds on Mind Changes

Suppose we are testing a hypothesis $H \subseteq \Sigma^{\omega}$ on a stream of data $v \in \Sigma^{\omega}$. Let $P(H \mid \Gamma_{v_{1:t}})$ denote our belief in H at time $t \in \mathbb{N}$ after seeing the evidence $v_{1:t}$. By Bayes' rule,

$$P(H \mid \Gamma_{v_{1:t}}) = P(H)\frac{P(\Gamma_{v_{1:t}} \mid H)}{P(\Gamma_{v_{1:t}})} =: X_t(v).$$

Since X_t is a constant multiple of $P(\ \cdot\ \mid H)/P$ and $P(\ \cdot\ \mid H)$ is a probability measure on $(\Sigma^{\omega}, \mathcal{F}_{\omega})$, the process $(X_t)_{t\in\mathbb{N}}$ is a P-martingale with respect to the filtration $(\mathcal{F}_t)_{t\in\mathbb{N}}$ by Theorem 4. By definition, $(X_t)_{t\in\mathbb{N}}$ is bounded between 0 and 1. Let $\alpha > 0$. We are interested in the question how likely it is to often change one's mind about H by at least α, i.e., what is the probability for $X_t = P(H \mid \Gamma_{v_{1:t}})$ to decrease and subsequently increase m times by at least α. Formally, we define the stopping times $T'_{0,\nu}(v) := 0$,

$$T'_{2k+1,\nu}(v) := \inf\{t > T'_{2k,\nu}(v) \mid X_t(v) \leq X_{T'_{2k,\nu}(v)}(v) - \nu\alpha\},$$
$$T'_{2k+2,\nu}(v) := \inf\{t > T'_{2k+1,\nu}(v) \mid X_t(v) \geq X_{T'_{2k+1,\nu}(v)}(v) + \nu\alpha\},$$

and $T'_k := \min\{T'_{k,\nu} \mid \nu \in \{-1, +1\}\}$. (In Davis' notation, $X_{T'_{0,\nu}}, X_{T'_{1,\nu}}, \ldots$ is an α-alternating W-sequence for $\nu = 1$ and an α-alternating M-sequence for $\nu = -1$ [Dav13, Def. 4].) For any $t \in \mathbb{N}$, the random variable

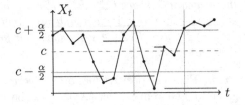

$$A_t^X(\alpha)(v) := \sup\{k \geq 0 \mid T'_k(v) \leq t\},$$

is defined as the number of α-alternations up to time t. Let $A^X(\alpha) := \sup_{t\in\mathbb{N}} A_t^X(\alpha)$ denote the total number of α-alternations.

Setting $\alpha = 2\varepsilon$, the α-alternations

Fig. 2. This example process has two up-crossings between $c - \alpha/2$ and $c + \alpha/2$ (completed at the time steps of the vertical orange bars) and four α-alternations (completed when crossing the horizontal blue bars)

differ from ε-upcrossings in three ways: first, for upcrossings, the process decreases below $c - \varepsilon$, then increases above $c + \varepsilon$, and then repeats. For alternations, the process may overshoot $c - \varepsilon$ or $c + \varepsilon$ and thus change the bar for the subsequent alternations, causing a 'drift' in the target bars over time. Second, for α-alternations the initial value of the martingale is relevant. Third, one upcrossing corresponds to two alternations, since one upcrossing always involves a preceding downcrossing. See Figure 2.

To apply our bounds for upcrossings on α-alternations, we use the following lemma by Davis. We reinterpret it as stating that every bounded martingale process $(X_t)_{t \in \mathbb{N}}$ can be modified into a martingale $(Y_t)_{t \in \mathbb{N}}$ such that the probability for many α-alternations is not decreased and the number of alternations equals the number of upcrossings plus the number of downcrossings [LH14].

Lemma 14 (Upcrossings and Alternations [Dav13, Lem. 9]). *Let $(X_t)_{t \in \mathbb{N}}$ be a martingale with $0 \leq X_t \leq 1$. There exists a martingale $(Y_t)_{t \in \mathbb{N}}$ with $0 \leq Y_t \leq 1$ and a constant $c \in (\alpha/2, 1 - \alpha/2)$ such that for all $t \in \mathbb{N}$ and for all $k \in \mathbb{N}$,*

$$P[A_t^X(\alpha) \geq 2k] \leq P[A_t^Y(\alpha) \geq 2k] = P[U_t^Y(c - \alpha/2, c + \alpha/2) \geq k].$$

Theorem 15 (Upper Bound on Alternations). *For every martingale process $(X_t)_{t \in \mathbb{N}}$ with $0 \leq X_t \leq 1$,*

$$P[A(\alpha) \geq 2k] \leq \left(\frac{1 - \alpha}{1 + \alpha}\right)^k.$$

Proof. We apply Theorem 14 to $(X_t)_{t \in \mathbb{N}}$ and $(1 - X_t)_{t \in \mathbb{N}}$ to get the processes $(Y_t)_{t \in \mathbb{N}}$ and $(Z_t)_{t \in \mathbb{N}}$. Dubins' Inequality yields

$$P[A_t^X(\alpha) \geq 2k] \leq P[U_t^Y(c_+ - \tfrac{\alpha}{2}, c_+ - \tfrac{\alpha}{2}) \geq k] \leq \left(\frac{c_+ - \frac{\alpha}{2}}{c_+ + \frac{\alpha}{2}}\right)^k =: g(c_+) \text{ and}$$

$$P[A_t^{1-X}(\alpha) \geq 2k] \leq P[U_t^Z(c_- - \tfrac{\alpha}{2}, c_- - \tfrac{\alpha}{2}) \geq k] \leq \left(\frac{c_- - \frac{\alpha}{2}}{c_- + \frac{\alpha}{2}}\right)^k = g(c_-)$$

for some $c_+, c_- \in (\alpha/2, 1 - \alpha/2)$. Because Theorem 14 is symmetric for $(X_t)_{t \in \mathbb{N}}$ and $(1 - X_t)_{t \in \mathbb{N}}$, we have $c_+ = 1 - c_-$. Since $P[A_t^X(\alpha) \geq 2k] = P[A_t^{1-X}(\alpha) \geq 2k]$ by the definition of $A_t^X(\alpha)$, we have that both are less than $\min\{g(c_+), g(c_-)\} = \min\{g(c_+), g(1 - c_+)\}$. This is maximized for $c_+ = c_- = 1/2$ because g is strictly monotone increasing for $c > \alpha/2$. Therefore

$$P[A_t^X(\alpha) \geq 2k] \leq \left(\frac{\frac{1}{2} - \frac{\alpha}{2}}{\frac{1}{2} + \frac{\alpha}{2}}\right)^k = \left(\frac{1 - \alpha}{1 + \alpha}\right)^k.$$

Since this bound is independent of t, it also holds for $P[A^X(\alpha) \geq 2k]$.   □

The bound of Theorem 15 is the square root of the bound derived by Davis [Dav13, Thm. 10 & Thm. 11]:

$$P[A(\alpha) \geq 2k] \leq \left(\frac{1 - \alpha}{1 + \alpha}\right)^{2k} \tag{1}$$

This bound is tight [Dav13, Cor. 13].

Because $0 \leq X_t \leq 1$, the process $(1 - X_t)_{t \in \mathbb{N}}$ is also a nonnegative martingale, hence the same upper bounds apply to it. This explains why the result in Theorem 15 is worse than Davis' bound (1): Dubins' bound applies to all nonnegative martingales, while Davis' bound uses the fact that the process is bounded from below *and* above. For unbounded nonnegative martingales, downcrossings are 'free' in the sense that one can make a downcrossing almost surely successful (as done in the proof of Theorem 6). If we apply Dubins' bound to the process $(1 - X_t)_{t \in \mathbb{N}}$, we get the same probability bound for the downcrossings of $(X_t)_{t \in \mathbb{N}}$ (which are upcrossings of $(1 - X_t)_{t \in \mathbb{N}}$). Multiplying both bounds yields Davis' bound (1); however, we still require a formal argument why the upcrossing and downcrossing bounds are independent.

The following corollary to Theorem 15 derives an upper bound on the *expected* number of α-alternations.

Theorem 16 (Upper Bound on Expected Alternations). *For every martingale* $(X_t)_{t \in \mathbb{N}}$ *with* $0 \leq X_t \leq 1$, *the expectation* $\mathbb{E}[A(\alpha)] \leq \frac{1}{\alpha}$.

Proof. By Theorem 15 we have $P[A(\alpha) \geq 2k] \leq \left(\frac{1-\alpha}{1+\alpha}\right)^k$, and thus

$$\mathbb{E}[A(\alpha)] = \sum_{k=1}^{\infty} P[A(\alpha) \geq k]$$

$$= P[A(\alpha) \geq 1] + \sum_{k=1}^{\infty} \left(P[A(\alpha) \geq 2k] + P[A(\alpha) \geq 2k+1]\right)$$

$$\leq 1 + \sum_{k=1}^{\infty} 2P[A(\alpha) \geq 2k] \leq 1 + 2\sum_{k=1}^{\infty} \left(\frac{1-\alpha}{1+\alpha}\right)^k = \frac{1}{\alpha}. \qquad \square$$

We now apply the technical results of this section to the martingale process $X_t = P(\cdot \mid H)/P$, our belief in the hypothesis H as we observe data. The probability of changing our mind k times by at least α decreases exponentially with k (Theorem 15). Furthermore, the expected number of times we change our mind by at least α is bounded by $1/\alpha$ (Theorem 16). In other words, having to change one's mind a lot often is unlikely.

Because in this section we consider martingales that are bounded between 0 and 1, the lower bounds from Section 4 do not apply here. While for the martingales constructed in Theorem 6, the number of 2α-alternations and the number of α-up- and downcrossings coincide, these processes are not bounded. However, we can give a similar construction that is bounded between 0 and 1 and makes Davis' bound asymptotically tight.

8   Conclusion

We constructed an indefinitely oscillating martingale process from a summable function f. Theorem 6 and Theorem 7 give uniform lower bounds on the probability and expectation of the number of upcrossings of decreasing magnitude.

In Theorem 9 we proved the corresponding upper bound if the function f is not summable. In comparison, Doob's Upcrossing Inequality and Dubins' Inequality give upper bounds that are not uniform. In Section 5 we showed that for a certain summable function f, our martingale makes these bounds asymptotically tight as well.

Our investigation of indefinitely oscillating martingales was motivated by two applications. First, in Theorem 13 we showed that the minimum description length operator may not exist in the limit: for any probability measure P we can construct a probability measure Q such that Q/P oscillates forever around the specific constant that causes $\lim_{t \to \infty} \mathrm{MDL}^{v_{1:t}}$ to not converge.

Second, we derived bounds for the probability of changing one's mind about a hypothesis H when observing a stream of data $v \in \Sigma^\omega$. The probability $P(H \mid \Gamma_{v_{1:t}})$ is a martingale and in Theorem 15 we proved that the probability of changing the belief in H often by at least α decreases exponentially.

A question that remains open is whether there is a *uniform* upper bound on the *expected* number of upcrossings tighter than Doob's Upcrossing Inequality.

References

[Dav13] Davis, E.: Bounding changes in probability over time: It is unlikely that you will change your mind very much very often. Technical report (2013), https://cs.nyu.edu/davise/papers/dither.pdf

[Doo53] Doob, J.L.: Stochastic Processes. Wiley, New York (1953)

[Dur10] Durrett, R.: Probability: Theory and Examples, 4th edn. Cambridge University Press (2010)

[Grü07] Grünwald, P.D.: The Minimum Description Length Principle. The MIT Press, Cambridge (2007)

[Hut09] Hutter, M.: Discrete MDL predicts in total variation. In: Advances in Neural Information Processing Systems 22 (NIPS 2009), pp. 817–825. Curran Associates, Cambridge (2009)

[LH14]   Leike, J., Hutter, M.: Indefinitely oscillating martingales. Technical report, The Australian National University (2014), http://www.hutter1.net/publ/martoscx.pdf

[LS05]   Luo, W., Schulte, O.: Mind change efficient learning. In: Auer, P., Meir, R. (eds.) COLT 2005. LNCS (LNAI), vol. 3559, pp. 398–412. Springer, Heidelberg (2005)

[PH05]   Poland, J., Hutter, M.: Asymptotics of discrete MDL for online prediction. IEEE Transactions on Information Theory 51(11), 3780–3795 (2005)

[Ris78] Rissanen, J.: Modeling by shortest data description. Automatica 14(5), 465–471 (1978)

[Wal05] Wallace, C.S.: Statistical and Inductive Inference by Minimum Message Length. Springer, Berlin (2005)

[WB68]   Wallace, C.S., Boulton, D.M.: An information measure for classification. Computer Journal 11(2), 185–194 (1968)

[Xu12]   Xu, W.: Martingale convergence theorems. Technical report (2012), http://people.maths.ox.ac.uk/xu/Martingale_convergence.pdf

A Safe Approximation
for Kolmogorov Complexity

Peter Bloem[1], Francisco Mota[2], Steven de Rooij[1],
Luís Antunes[2], and Pieter Adriaans[1]

[1] System and Network Engineering Group
University of Amsterdam, Amsterdam, The Netherlands
uva@peterbloem.nl, steven.de.rooij@gmail.com, p.w.adriaans@uva.nl
[2] CRACS & INESC-Porto LA and Institute for Telecommunications
University of Porto, Porto, Portugal
fmota@fmota.eu, lfa@dcc.fc.up.pt

Abstract. Kolmogorov complexity (K) is an incomputable function. It can be approximated from above but not to arbitrary given precision and it cannot be approximated from below. By restricting the source of the data to a specific model class, we can construct a computable function $\bar{\kappa}$ to approximate K in a probabilistic sense: the probability that the error is greater than k decays exponentially with k. We apply the same method to the normalized information distance (NID) and discuss conditions that affect the safety of the approximation.

The Kolmogorov complexity of an object is its shortest description, considering all computable descriptions. It has been described as "the accepted absolute measure of information content of an individual object" [1], and its investigation has spawned a slew of derived functions and analytical tools. Most of these tend to separate neatly into one of two categories: the platonic and the practical.

On the platonic side, we find such tools as the normalized information distance [2], algorithmic statistics [1] and sophistication [3, 4]. These subjects all deal with incomputable "ideal" functions: they optimize over all computable functions, but they cannot be computed themselves.

To construct practical applications (ie. runnable computer programs), the most common approach is to take one of these platonic, incomputable functions, derived from Kolmogorov complexity (K), and to approximate it by swapping K out for a computable compressor like GZIP [5]. This approach has proved effective in the case of normalized information distance (NID) [2] and its approximation, the normalized compression distance (NCD) [6]. Unfortunately, the switch to a general-purpose compressor leaves an analytical gap. We know that the compressor serves as an upper bound to K—up to a constant—but we do not know the difference between the two, and how this error affects the error of derived functions like the NCD. This can cause serious contradictions. For instance, the normalized information distance has been shown to be non-approximable [7], yet the NCD has proved its merit empirically [6]. Why this

P. Auer et al. (Eds.): ALT 2014, LNAI 8776, pp. 336–350, 2014.

should be the case, and when this approach may fail has, to our knowledge, not yet been investigated.

We aim to provide the first tools to bridge this gap. We will define a computable function which can be said to approximate Kolmogorov complexity, with some practical limit to the error. To this end, we introduce two concepts:

- We generalize resource-bounded Kolmogorov complexity (K^t) to *model-bounded Kolmogorov complexity*, which minimizes an object's description length over any given enumerable subset of Turing machines (a *model class*). We explicitly assume that the source of the data is contained in the model class.
- We introduce a probabilistic notion of approximation. A function approximates another *safely*, under a given distribution, if the probability of them differing by more than k bits, decays at least exponentially in k.[1]

While the resource-bounded Kolmogorov complexity is computable in a technical sense, it is never computed practically. The generalization to model bounded Kolmogorov complexity creates a connection to *minimum description length* (MDL) [8, 9, 10], which does produce algorithms and methods that are used in a practical manner. Kolmogorov complexity has long been seen as a kind of platonic ideal which MDL approximates. Our results show that MDL is not just an upper bound to K, it also approximates it in a probabilistic sense.

Interestingly, the model-bounded Kolmogorov complexity itself—the smallest description using a single element from the model class—is not a safe approximation. We can, however, construct a computable, safe approximation by taking into account all descriptions the model class provides for the data.

The main result of this paper is a computable function $\overline{\kappa}$ which, under a model assumption, safely approximates K (Theorem 3). We also investigate whether a $\overline{\kappa}$-based approximation of NID is safe, for different properties of the model class from which the data originated (Theorems 5, 6 and 7).

1   Turing Machines and Probability

Turing Machines

Let $\mathbb{B} = \{0,1\}^*$. We assume that our data is encoded as a finite binary string. Specifically, the natural numbers can be associated to binary strings, for instance by the bijection: $(0, \epsilon), (1, 0), (2, 1), (3, 00), (4, 01)$, etc, where ϵ is the empty string. To simplify notation, we will sometimes conflate natural numbers and binary strings, implicitly using this ordering.

We fix a canonical prefix-free coding, denoted by $\overline{x}$, such that $|\overline{x}| \leq |x| + 2\log|x|$. See [11, Example 1.11.13] for an example. Among other things, this gives us a canonical pairing function to encode two strings x and y into one: $\overline{x}y$.

---

[1] This consideration is subject to all the normal drawbacks of asymptotic approaches. For this reason, we have foregone the use of big-O notation as much as possible, in order to make the constants and their meaning explicit.

We use the Turing machine model from [11, Example 3.1.1]. The following properties are important: the machine has a read-only, right-moving input tape, an auxiliary tape which is read-only and two-way, two read-write two-way worktapes and a read-write two-way output tape.[2] All tapes are one-way infinite. If a tape head moves off the tape or the machine reads beyond the length of the input, it enters an infinite loop. For the function computed by TM i on input p with auxiliary input y, we write $T_i(p \mid y)$ and $T_i(p) = T_i(p \mid \epsilon)$. The most important consequence of this construction is that the programs for which a machine with a given auxiliary input y halts, form a prefix-free set [11, Example 3.1.1]. This allows us to interpret the machine as a probability distribution (as described in the next subsection).

We fix an effective ordering $\{T_i\}$. We call the set of all Turing machines $\mathscr{C}$. There exists a universal Turing machine, which we will call U, that has the property that $U(\bar{i}p \mid y) = T_i(p \mid y)$ [11, Theorem 3.1.1].

Probability

We want to formalize the idea of a probability distribution that is *computable*: it can be simulated or computed by a computational process. For this purpose, we will interpret a given Turing machine T_q as a probability distribution p_q: each time the machine reads from the input tape, we provide it with a random bit. The Turing machine will either halt, read a finite number of bits without halting, or read an unbounded number of bits. $p_q(x)$ is the probability that this process halts and produces x: $p_q(x) = \sum_{p:T_q(p)=x} 2^{-|p|}$. We say that T_q *samples* p_q. Note that if p is a semimeasure, $1 - \sum_x p(x)$ corresponds to the probability that this sampling process will not halt.

We model the probability of x conditional on y by a Turing machine with y on its auxiliary tape: $p_q(x \mid y) = \sum_{p:T_q(p|y)=x} 2^{-|p|}$.

The *lower semicomputable semimeasures* [11, Chapter 4] are an alternative formalization. We show that it is equivalent to ours:

Lemma 1. ** *The set of probability distributions sampled by Turing machines in $\mathscr{C}$ is equivalent to the set of lower semicomputable semimeasures.*

The distribution corresponding to the universal Turing machine U is called m: $m(x) = \sum_{p:U(p)=x} 2^{-|p|}$. This is known as a universal distribution. K and m dominate each other, ie. $\exists c \forall x : |K(x) - \log m(x)| < c$ [11, Theorem 4.3.3].

2   Model-Bounded Kolmogorov Complexity

In this section we present a generalization of the notion of resource-bounded Kolmogorov complexity. We first review the unbounded version:

---

[2] Multiple work tapes are only required for proofs involving resource bounds.
** Proof in the appendix.

Definition 1. *Let* $k(x \mid y) = \arg\min_{p:U(p|y)=x} |p|$. *The prefix-free, conditional* Kolmogorov complexity *is*

$$K(x \mid y) = |k(x \mid y)|$$

with $K(x) = K(x \mid \epsilon)$.

Due to the halting problem, K is not computable. By limiting the set of Turing machines under consideration, we can create a computable approximation.

Definition 2. *A* model class $C \subseteq \mathscr{C}$ *is a computably enumerable set of Turing machines. Its members are called* models. *A* universal model *for* C *is a Turing machine* U^C *such that* $U^C(\bar{\imath}p \mid y) = T_i(p \mid y)$ *where* i *is an index over the elements of* C.

Definition 3. *For a given* C *and* U^C *we have* $K^C(x) = \min\{|p| : U^C(p) = x\}$, *called the* model-bounded Kolmogorov complexity.

K^C, unlike K, depends heavily on the choice of enumeration of C. A notation like K_{U^C} or $K^{i,C}$ would express this dependence better, but for the sake of clarity we will use K^C.

We define a model-bounded variant of m as $m^C(x) = \sum_{p:U^C(p)=x} 2^{-|p|}$, which dominates all distributions in C:

Lemma 2. *For any* $T_q \in C$, $m^C(x) \geq c_q p_q(x)$ *for some* c_q *independent of* x.

Proof.

$$m^C(x) = \sum_{i,p:U^C(\bar{\imath}p)=x} 2^{-|\bar{\imath}p|} \geq \sum_{p:U^C(\bar{q}p)=x} 2^{-|\bar{q}|} 2^{-|p|} = 2^{-|\bar{q}|} p_q(x).$$

$\square$

Unlike K and $-\log m$, K^C and $-\log m^C$ do not dominate one another. We can only show that $-\log m^C$ bounds K^C from below ($\sum_{U^C(p)=x} 2^{-|p|} > 2^{-|k^C(x)|}$). In fact, as shown in Theorem 1, $-\log m^C$ and K^C can differ by arbitrary amounts.

Example 1 (Resource-Bounded Kolmogorov Complexity [11, Ch. 7]). *Let* $t(n)$ *be some time-constructible function.[3] Let* T_i^t *be the modification of* $T_i \in \mathscr{C}$ *such that at any point in the computation, it halts immediately if more than* k *cells have been written to on the output tape and the number of steps that have passed is less than* $t(k)$. *In this case, whatever is on the output tape is taken as the output of the computation. If this situation does not occur,* T_i *runs as normal. Let* $U^t(\bar{\imath}p) = T_i^t(p)$. *We call this model class* C^t. *We abbreviate* K^{C^t} *as* K^t.

Since there is no known means of simulating U^t *within* $t(n)$, *we do not know whether* $U^t \in C^t$. *It can be run in* $ct(n) \log t(n)$ *[11, 12], so we do know that* $U^t \in C^{ct \log t}$.

Other model classes include Deterministic Finite Automata, Markov Chains, or the exponential family (suitably discretized). These have all been thoroughly investigated in coding contexts in the field of Minimum Description Length [10].

---

[3] Ie. $t : \mathbb{N} \rightarrow \mathbb{N}$ and t can be computed in $O(t(n))$ [13].

3   Safe Approximation

When a code-length function like K turns out to be incomputable, we may try to find a lower and upper bound, or to find a function which dominates it. Unfortunately, neither of these will help us. Such functions invariably turn out to be incomputable themselves [11, Section 2.3].

To bridge the gap between incomputable and computable functions, we require a softer notion of approximation; one which states that errors of any size may occur, but that the larger errors are so unlikely, that they can be safely ignored:

Definition 4. *Let f and f_a be two functions. We take f_a to be an approximation of f. We call the approximation b-safe (from above) for a distribution (or adversary) p if for all k and some $c > 0$:*

$$p(f_a(x) - f(x) \geq k) \leq cb^{-k}.$$

Since we focus on code-length functions, usually omit "from above". A safe function is b-safe for some $b > 1$. An approximation is safe for a model class C if it is safe for all p_q with $T_q \in C$.

While the definition requires this property to hold for all k, it actually suffices to show that it holds for k above a constant, as we can freely scale c:

Lemma 3. *If $\exists_c \forall_{k:k>k_0} : p(f_a(x) - f(x) \geq k) \leq cb^{-k}$, then f_a is b-safe for f against p.*

Proof. First, we name the k below k_0 for which the ratio between the bound and the probability is the greatest: $k_m = \arg \max_{k \in [0,k_0]} \left[p(f_a(x) - f(x) \geq k)/cb^{-k} \right]$. We also define $b_m = cb^{-k_m}$ and $p_m = p(f_a(x) - f(x) \geq k_m)$. At k_m, we have $p(f_a(x) - f(x) \geq k_m) = p_m = \frac{p_m}{b_m} cb^{-k_m}$. In other words, the bound $c'b^{-k}$ with $c' = \frac{p_m}{b_m} c$ bounds p at k_m, the point where it diverges the most from the old bound. Therefore, it must bound it at all other $k > 0$ as well.   □

Safe approximation, domination and lowerbounding form a hierarchy:

Lemma 4. *Let f_a and f be code-length functions. If f_a is a lower bound on f, it also dominates f. If f_a dominates f, it is also a safe approximation.*

Proof. Domination means that for all x: $f_a(x) - f(x) < c$, if f_a is a lower bound, $c = 0$. If f_a dominates f we have $\forall p, k > c : p(f_a(x) - f(x) \geq k) = 0$.   □

Finally, we show that safe approximation is transitive, so we can chain together proofs of safe approximation; if we have several functions with each safe for the next, we know that the first is also safe for the last.

Lemma 5. *The property of safety is transitive over the space of functions from $\mathbb{B}$ to $\mathbb{B}$ for a fixed adversary.*

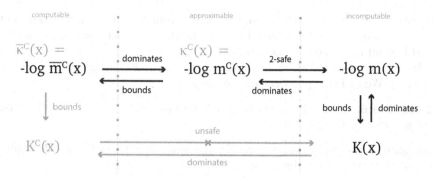

Fig. 1. An overview of how various code-length functions relate to each other in terms of approximation safety. These relations hold under the assumption that the data is generated by a distribution in C and that C is sufficient and complete.

Proof. Let f, g and h be functions such that

$$p(f(x) - g(x) \geq k) \leq c_1 b_1^{-k} \text{ and}$$
$$p(g(x) - h(x) \geq k) \leq c_2 b_2^{-k}.$$

We need to show that $p(f(x) - h(x) \geq k)$ decays exponentially with k. We start with

$$p\left(f(x) - g(x) \geq k \vee g(x) - h(x) \geq k\right) \leq c_1 b_1^{-k} + c_2 b_2^{-k}.$$

$\{x : f(x) - h(x) \geq 2k\}$ is a subset of $\{x : f(x) - g(x) \geq k \vee g(x) - h(x) \geq k\}$, so that the probability of the first set is less than that of the second:

$$p\left(f(x) - h(x) \geq 2k\right) \leq c_1 b_1^{-k} + c_2 b_2^{-k}.$$

Which gives us

$$p\left(f(x) - h(x) \geq 2k\right) \leq cb^{-k} \qquad \text{with } b = \min(b_1, b_2) \text{ and } c = \max(c_1, c_2),$$
$$p\left(f(x) - h(x) \geq k'\right) \leq cb'^{-k'} \qquad \text{with } b' = \sqrt{b}. \qquad \square$$

4 A Safe, Computable Approximation of K

Assuming that our data is produced from a model in C, can we construct a computable function which is safe for K? An obvious first choice is K^C. For it to be computable, we would normally ensure that all programs for all models in C halt. Since the halting programs form a prefix-free set, this is impossible. There is however a property for prefix-free functions that is analogous. We call this *sufficiency*:

Definition 5. *A sufficient model T is a model for which every infinite binary string contains a halting program as a prefix. A sufficient model class contains only sufficient models.*

We can therefore enumerate all inputs for U^C from short to long in series to find $k^C(x)$, so long as C is sufficient. For each input, U^C either halts or attempts to read beyond the length of the input.

In certain cases, we also require that C can represent all $x \in \mathbb{B}$ (ie. $m^C(x)$ is never 0). We call this property *completeness*:

Definition 6. *A model class C is called* complete *if for any x, there is at least one p such that $U^C(p) = x$.*

We can now say, for instance, that K^C is computable for sufficient C. Unfortunately, K^C turns out to be unsafe:

Theorem 1. *There exist model classes C so that $K^C(x)$ is an unsafe approximation for $K(x)$ against some p_q with $T_q \in C$.*

Proof. We first show that K^C is unsafe for $- \log m^C$.

Let C contain a single Turing machine T_q which outputs x for any input of the form $\bar{x}p$ with $|p| = x$ and computes indefinitely for all other inputs.

T_q samples from $p_q(x) = 2^{-|\bar{x}|}$, but it distributes each x's probability mass uniformly over many programs much longer than $|\bar{x}|$.

This gives us $K^C(x) = |\bar{x}| + |p| = |\bar{x}| + x$ and $- \log m^C(x) = |\bar{x}|$, so that $K^C(x) + \log m^C(x) = x$. We get

$$m^C(K^C(x) + \log m^C(x) \geq k) = m^C(x \geq k) =$$
$$\sum_{x:x\geq k} 2^{-|\bar{x}|} \geq \sum_{x:x\geq k} 2^{-2\log x} \geq k^{-2}$$

so that K^C is unsafe for $- \log m^C$.

It remains to show that this implies that K^C is unsafe for K. In Theorem 2, we prove that $- \log m^C$ is safe for K. Assuming that K^C is safe for K (which dominates $- \log m^C$) implies K^C is safe for $- \log m^C$, which gives us a contradiction. $\square$

Note that the use of a model class with a single model is for convenience only. The main requirement for K^C to be unsafe is that the prefix tree of U^C's programs distributes the probability mass for x over many programs of similar length. The greater the difference between K^C and $- \log m^C$, the greater the likelihood that K^C is unsafe.

Our next candidate for a safe approximation of K is $- \log m^C$. This time, we fare better. We first require the following lemma, called the *no-hypercompression theorem* in [10, p103]:

Lemma 6. *Let p_q be a probability distribution. The corresponding code-length function, $- \log p_q$, is a 2-safe approximation for any other code-length function against p_q. For any p_r and $k > 0$: $p_q(- \log p_q(x) + \log p_r(x) \geq k) \leq 2^{-k}$.*

Theorem 2. $- \log m^C(x)$ *is a 2-safe approximation of $K(x)$ against any adversary from C.*

Proof. Let p_q be some adversary in C. We have

$$p_q(-\log m^C(x) - K(x) \geq k)$$
$$\leq cm^C(-\log m^C(x) - K(x) \geq k) \qquad \text{by Lemma 2,}$$
$$\leq c2^{-k} \qquad\qquad\qquad\qquad \text{by Lemma 6.} \qquad \square$$

While we have shown m^C to be safe for K, it may not be computable, even if C is sufficient (since it is an infinite sum). We can, however, define an approximation, which, for sufficient C, is computable and dominates m^C.

Definition 7. *Let the model class D be the union of C and some arbitrary sufficient and complete distribution from $\mathscr{C}$.*

Let $\overline{m}_c^C(x)$ be the function computed by the following algorithm: Dovetail the computation of all programs on $U^D(x)$ in cycles, so that in cycle n, the first n programs are simulated for one further step. After each such step we consider the probability mass s of all programs that have stopped (where each program p contributes $2^{-|p|}$), and the probability mass s_x of all programs that have stopped and produced x. We halt the dovetailing and output s_x if $s_x > 0$ and the following stop condition is met:

$$\frac{1-s}{s_x} \leq 2^c - 1 \,.$$

Note that if C is sufficient so is D, so that s goes to 1 and s_x never decreases. Since all programs halt, the stop condition must be reached. The addition of a complete model is required to ensure that s_x does not remain 0 indefinitely.

Lemma 7. *If C is sufficient, $\overline{m}_c^C(x)$ dominates m^C with a constant multiplicative factor 2^{-c} (ie. their code-lengths differ by at most c bits).*

Proof. We will first show that $\overline{m}_c^C$ dominates m^D. Note that when the computation of $\overline{m}_c^C$ halts, we have $\overline{m}_c^C(x) = s_x$ and $m^D(x) \leq s_x + (1-s)$. This gives us:

$$\frac{m^D(x)}{\overline{m}_c^C(x)} \leq 1 + \frac{1-s}{s_x} \leq 2^c \,.$$

Since $C \subseteq D$, m^D dominates m^C (see Lemma 9 in the appendix) and thus, $\overline{m}_c^C$ dominates m^C. $\qquad \square$

The parameter c in $\overline{m}_c^C$ allows us to tune the algorithm to trade off running time for a smaller constant of domination. We will usually omit it when it is not relevant to the context.

Putting all this together, we have achieved our aim:

Theorem 3. *For a sufficient model class C, $-\log \overline{m}^C$ is a safe, computable approximation of $K(x)$ against any adversary from C*

Proof. We have shown that, under these conditions, $-\log m^C$ safely approximates $-\log m$ which dominates K, and that $-\log \overline{m}^C$ dominates $-\log m^C$. Since domination implies safe approximation (Lemma 4), and safe approximation is transitive (Lemma 5), we have proved the theorem. $\qquad \square$

Figure 1 summarizes this chain of reasoning and other relations between the various code-length functions mentioned.

The negative logarithm of m^C will be our go-to approximation of K, so we will abbreviate it with κ:

Definition 8. $\kappa^C(x) = -\log m^C(x)$ and $\overline{\kappa}^C(x) = -\log \overline{m}^C(x)$.

Finally, if we violate our model assumption we lose the property of safety. For adversaries outside C, we cannot be sure that κ^C is safe:

Theorem 4. *There exist adversaries p_q with $T_q \notin C$ for which neither κ^C nor $\overline{\kappa}^C$ is a safe approximation of K.*

Proof. Consider the following algorithm for sampling from a computable distribution (which we will call p_q):

- Sample $n \in \mathbb{N}$ from some distribution $s(n)$ which decays polynomially.
- Loop over all x of length n return the first x such that $\kappa^C(x) \geq n$.

Note that at least one such x must exist by a counting argument: if all x of length n have $-\log \overline{m}^C(x) < n$ we have a code that assigns 2^n different strings to $2^n - 1$ different codes.

For each x sampled from q, we know that $\overline{\kappa}(x) \geq |x|$ and $K(x) \leq -\log p_q(x) + c_q$. Thus:

$$p_q(\overline{\kappa}^C(x) - K(x) \geq k) \geq p_q(|x| + \log p_q(x) - c_q \geq k)$$

$$= p_q(|x| + \log s(|x|) - c_q \geq k) = \sum\nolimits_{n:n+\log s(n)-c_q \geq k} s(n).$$

Let n_0 be the smallest n for which $2n > n + \log s(n) - c_q$. For all $k > 2n_0$ we have

$$\sum\nolimits_{n:n+\log s(n)-c_q \geq k} s(n) \geq \sum\nolimits_{n:2n \geq k} s(n) \geq s\left(\tfrac{1}{2}k\right). \qquad \square$$

For C^t (as in Example 1), we can sample the p_q constructed in the proof in $O(2^n \cdot t(n))$. Thus, we know that κ^t is safe for K against adversaries from C^t, and we know that it is unsafe against C^{2^t}.

5 Approximating Normalized Information Distance

Definition 9 ([2, 6]). *The normalized information distance between two strings x and y is*

$$NID(x, y) = \frac{\max[K(x \mid y), K(y \mid x)]}{\max[K(x), K(y)]}.$$

The information distance (ID) is the numerator of this function. The NID is neither lower nor upper semicomputable [7]. Here, we investigate whether we can safely approximate either function using κ. We define ID^C and NID^C as the ID and NID functions with K replaced by $\overline{\kappa}^C$. We first show that, even if the adversary only combines functions and distributions in C, ID^C may be an unsafe approximation.

Definition 10. [4]*A function f is a (b-safe) model-bounded one-way function for C if it is injective, and for some $b > 1$, some $c > 0$, all $q \in C$ and all k:*

$$p_q\left(\kappa^C(x) - \kappa^C(x \mid f(x)) \geq k\right) \leq cb^{-k}.$$

Theorem 5. ** *Under the following assumptions:*

- *C contains a model T_0, with $p_0(x) = 2^{-|x|}s(|x|)$, with s a distribution on $\mathbb{N}$ which decays polynomially or slower,*
- *there exists a model-bounded one-way function f for C,*
- *C is normal, ie. for some c and all x: $\kappa^C(x) < |\bar{x}| + c$*

ID^C is an unsafe approximation for ID against an adversary T_q which samples x from p_0 and returns $\bar{x}f(x)$.

If x and y are sampled from C independently, we can prove safety:

Theorem 6. ** *Let T_q be a Turing machine which samples x from p_a, y from p_b and returns $\bar{x}y$. If $T_a, T_b \in C$, $\mathrm{ID}^C(x,y)$ is a safe approximation for $\mathrm{ID}(x,y)$ against any such T_q.*

The proof relies on two facts:

- $\bar{\kappa}^C(x \mid y)$ is safe for $K(x \mid y)$ if x and y are generated this way.
- Maximization is a *safety preserving operation*: if we have two functions f and g with safe approximations f_a and g_a, $\max(f_a(x), g_a(x))$ is a safe approximation of $\max(f(x), g(x))$.

For *normalized* information distance, which is dimensionless, the error k in bits as we have used it so far does not mean much. Instead, we use f/f_a as a measure of approximation error, and we introduce an additional parameter ϵ:

Theorem 7. ** *We can approximate NID with NID^C with the following bound:*

$$p_q\left(\frac{\mathrm{NID}(x,y)}{\mathrm{NID}^C(x,y)} \notin \left(1 - \frac{k}{c}, 1 + \frac{k}{c}\right)\right) \leq c'b^{-k} + 2\epsilon$$

with

$$p_q(\mathrm{ID}^C(x,y) \geq c) \leq \epsilon \text{ and } p_q\left(\max\left[\kappa^C(x), \kappa^C(y)\right] \geq c\right) \leq \epsilon$$

for some $b > 1$ and $c' > 0$, assuming that p_q samples x and y independently from models in C.

---

[4] This is similar to the Kolmogorov one-way function [14, Definition 11].

6 Discussion

We have provided a function $\overline{\kappa}^C(x)$ for a given model class C, which is computable if C is sufficient. Under the assumption that x is produced by a model from C, $\overline{\kappa}^C(x)$ approximates $K(x)$ in a probabilistic sense. We have also shown that $K^C(x)$ is not safe. Finally, we have given some insight into the conditions on C and the adversary, which can affect the safety of NCD as an approximation to NID.

Since, as shown in Example 1, resource-bounded Kolmogorov complexity is a variant of model-bounded Kolmogorov complexity, our results apply to K^t as well: K^t is not necessarily a safe approximation of K, even if the data can be sampled in t and κ^t *is* safe if the data can be sampled in t. Whether K^t is safe ultimately depends on whether a single shortest program dominates among the sum of all programs, as it does in the unbounded case.

For complex model classes, κ^C may still be impractical to compute. In such cases, we may be able to continue the chain of safe approximation proofs. Ideally, we would show that a model which is only locally optimal, found by an iterative method like gradient descent, is still a safe approximation of K. Such proofs would truly close the circuit between the ideal world of Kolmogorov complexity and modern statistical practice.

Acknowledgement. We would like to thank the reviewers for their insightful comments. This publication was supported by the Dutch national program COMMIT, the Netherlands eScience center, the ERDF (European Regional Development Fund) through the COMPETE Programme (Operational Programme for Competitiveness) and by National Funds through the FCT (Fundação para a Ciência e a Tecnologia, the Portuguese Foundation for Science and Technology) within project *FCOMP-01-0124-FEDER-037281*.

References

[1] Gács, P., Tromp, J., Vitányi, P.M.B.: Algorithmic statistics. IEEE Transactions on Information Theory 47(6), 2443–2463 (2001)

[2] Li, M., Chen, X., Li, X., Ma, B., Vitányi, P.M.B.: The similarity metric. IEEE Transactions on Information Theory 50(12), 3250–3264 (2004)

[3] Vitányi, P.M.B.: Meaningful information. IEEE Transactions on Information Theory 52(10), 4617–4626 (2006)

[4] Adriaans, P.: Facticity as the amount of self-descriptive information in a data set. arXiv preprint arXiv:1203.2245 (2012)

[5] Gailly, J., Adler, M.: The GZIP compressor (1991)

[6] Cilibrasi, R., Vitányi, P.M.B.: Clustering by compression. IEEE Transactions on Information Theory 51(4), 1523–1545 (2005)

[7] Terwijn, S.A., Torenvliet, L., Vitányi, P.M.B.: Nonapproximability of the normalized information distance. J. Comput. Syst. Sci. 77(4), 738–742 (2011)

[8] Rissanen, J.: Modeling by shortest data description. Automatica 14(5), 465–471 (1978)

[9] Rissanen, J.: Universal coding, information, prediction, and estimation. IEEE Transactions on Information Theory 30(4), 629–636 (1984)

[10] Grünwald, P.D.: The Minimum Description Length Principle. Adaptive computation and machine learning series. The MIT Press (2007)

[11] Li, M., Vitányi, P.M.B.: An introduction to Kolmogorov complexity and its applications, 2nd edn. Graduate Texts in Computer Science. Springer (1997)

[12] Hennie, F.C., Stearns, R.E.: Two-tape simulation of multitape Turing machines. J. ACM 13(4), 533–546 (1966)

[13] Antunes, L.F.C., Matos, A., Souto, A., Vitányi, P.M.B.: Depth as randomness deficiency. Theory Comput. Syst. 45(4), 724–739 (2009)

[14] Antunes, L.F.C., Matos, A., Pinto, A., Souto, A., Teixeira, A.: One-way functions using algorithmic and classical information theories. Theory Comput. Syst. 52(1), 162–178 (2013)

A   Appendix

A.1   Turing Machines and lsc. Probability Semimeasures (Lemma 1)

Definition 11. *A function $f : \mathbb{B} \to \mathbb{R}$ is* lower semicomputable (lsc.) *iff there exists a total, computable two-argument function $f' : \mathbb{B} \times \mathbb{N} \to \mathbb{Q}$ such that: $\lim_{i \to \infty} f'(x, i) = f(x)$ and for all i, $f'(x, i + 1) \geq f'(x, i)$.*

Lemma 8. *If f is an lsc. probability semimeasure, then there exists a a function $f^*(x, i)$ with the same properties of the function f' from Definition 11, and the additional property that all values returned by f^* have finite binary expansions.*

Proof. Let x_j represent $x \in \mathbb{D}$ truncated at the first j bits of its binary expansion and x^j the remainder. Let $f^*(x, i) = f'(x, i)_i$. Since $f'(x, i) - f^*(x, i)_i$ is a value with $i + 1$ as the highest non-zero bit in its binary expansion, $\lim_{i \to \infty} f^*(x, i) = \lim f'(x, i) = f(x)$.

It remains to show that f^* is nondecreasing in i. Let $x \geq y$. We will show that $x_j \geq y_j$, and thus $x_{j+1} \geq y_j$. If $x = y$ the result follows trivially. Otherwise, we have $x_j = x - x^j > y - x^j = y_j + y^j - x^j \geq y_j - 2^{-j}$. Substituting $x = f'(x, i+1)$ and $y = f'(x, i)$ tells us that $f^*(x, i + 1) \geq f^*(x, i)$ □

Theorem 8. *Any TM, T_q, samples from an lsc. probability semimeasure.*

Proof. We will define a program computing a function $p'_q(x, i)$ to approximate $p_q(x)$: Dovetail the computation of T_q on all inputs $x \in \mathbb{B}$ for i cycles.

Clearly this function is nondecreasing. To show that it goes to $p(x)$ with i, we first note that for a given i_0 there is a j such that, $2^{-j-1} < p_q(x) - p_q(x, i_0) \leq 2^{-j}$. Let $\{p_i\}$ be an ordering of the programs producing x, by increasing length, that have not yet stopped at dovetailing cycle i_0. There is an m such that $\sum_{i=1}^{m} 2^{-|p_i|} \geq 2^{-j-1}$, since $\sum_{i=1}^{\infty} 2^{-|p_i|} > 2^{-j-i}$. Let i_1 be the dovetailing cycle for which the last program below p_{m+1} halts. This gives us $p_q(x) - p_q(x, i_1) \leq 2^{-j-1}$. Thus, by induction, we can choose i to make $p(x) - p'(x, i)$ arbitrarily small.   □

Theorem 9. *Any lsc. probability semimeasure can be sampled by a TM.*

Proof. Let $p(x)$ be an lsc. probability semimeasure and $p^*(x, i)$ as in Lemma 8. We assume—without loss of generality—that $p^*(x, 0) = 0$. Consider the following algorithm:

> **initialize** $s \leftarrow \epsilon,\, r \leftarrow \epsilon$
> **for** $c = 1, 2, \ldots$:
>   **for** $x \in \{b \in \mathbb{B} : |b| \le c\}$
> $d \leftarrow p^*(x, c - i + 1) - p^*(x, c - i)$
> $s \leftarrow s + d$
>   add a random bit to r until it is as long as s
>   **if** $r < s$ **then return** x

The reader may verify that this program dovetails computation of $p^*(x, i)$ for increasing i for all x; the variable s contains the summed probability mass that has been encountered so far. Whenever s is incremented, mentally associate the interval $(s, s + d]$ with outcome x. Since $p^*(x, i)$ goes to $p(x)$ as i increases, the summed length of the intervals associated with x goes to $p(x)$ and s itself goes to $\bar{s} = \sum_x p(x)$. We can therefore sample from p by picking a number r that is uniformly random on $[0, 1]$ and returning the outcome associated with the interval containing r. Since s must have finite length (due to the construction of p^*), we only need to know r up to finite precision to be able to determine which interval it falls in; this allows us to generate r on the fly. The algorithm halts unless r falls in the interval $[\bar{s}, 1]$, which corresponds exactly to the deficiency of p: if p is a semimeasure, we expect the non-halting probability of a TM sampling it to correspond to $1 - \sum_x p(x)$. $\qquad\square$

Theorems 8 and 9 combined prove that the class of distributions sampled by Turing machines equals the lower semicomputable semimeasures (Lemma 1).

A.2   Domination of Model Class Supersets

Lemma 9. *Let C and D be model classes. If $C \subseteq D$, then m^D dominates m^C:*

$$\frac{m^D(x)}{m^C(x)} \ge \alpha$$

for some constant α independent of x.

Proof. We can partition the models of D into those belonging to C and the rest, which we'll call $\overline{C}$. For any given enumeration of D, we get $m^D(x) = \alpha m^C(x) + (1 - \alpha)m^{\overline{C}(x)}$. This gives us:

$$\frac{m^D(x)}{m^C(x)} = \alpha + (1 - \alpha)\frac{m^{\overline{C}(x)}}{m^C(x)} \ge \alpha.$$

$\qquad\square$

A.3 Unsafe Approximation of ID (Theorem 5)

Proof.

$$p_q\left(\mathrm{ID}^C(x,y) - \mathrm{ID}(x,y) \geq k\right) =$$
$$p_0\left(\max\left[\overline{\kappa}^C(x \mid f(x)), \overline{\kappa}^C(f(x) \mid x)\right] - \max\left[K(x \mid f(x)), K(f(x) \mid x))\right] \geq k\right).$$
$$p_q\left(|x| - \mathrm{ID}^C(x,y) \geq 2k\right) \leq p_0\left(|x| - \overline{\kappa}^C(x \mid f(x)) \geq 2k\right)$$
$$\leq p_0\left(|x| - \kappa^C(x) \geq k \vee \kappa^C(x) - \overline{\kappa}^C(x \mid f(x)) \geq k\right)$$
$$\leq p_0\left(|x| - \kappa^C(x) \geq k\right) + p_0\left(\kappa^C(x) - \kappa^C(x \mid f(x)) \geq k\right) \leq 2^{-k} + cb^{-k}.$$

K can invert $f(x)$, so

$$\mathrm{ID}(x,y) = \max\left[K(x \mid f(x)), K(f(x) \mid x)\right] = \max\left[|f^*|, |f^*_{\mathrm{inv}}|\right] < c_f$$

where f^* and f^*_{inv} are the shortest program to compute f on U and the shortest program to compute the inverse of f on U respectively.

$$p_q\left(\mathrm{ID}^C(x,y) - \mathrm{ID}(x,y) \geq k\right) + p_q\left(|x| - \mathrm{ID}^C(x,y) \geq k\right)$$
$$\geq p_q\left(\mathrm{ID}^C(x,y) - \mathrm{ID}(x,y) \geq k \vee |x| - \mathrm{ID}^C(x,y) \geq k\right)$$
$$\geq p_q\left(|x| - \mathrm{ID}(x,y) \geq k\right) \geq p_0\left(|x| - c_f \geq k\right) = \sum_{i \geq k - c_f} s(i).$$

Which gives us:

$$p_q\left(\mathrm{ID}^C(x,y) - \mathrm{ID}(x,y) \geq k\right)$$
$$\geq -p_q(|x| - \mathrm{ID}^C \geq k) + \sum_{i \geq k - |f|} s(i) \geq -cb^{-k} + \sum_{i \geq k - |f|} s(i)$$
$$\geq s(k - |f|) - cb^{-k} \geq c's(k) \qquad \text{for the right } c'. \qquad \square$$

Corollary 1 *Under the assumptions of Theorem 5, $\overline{\kappa}^C(x \mid y)$ is an unsafe approximation for $K(x \mid y)$ against q.*

Proof. Assuming $\overline{\kappa}^C$ is safe, then since max is safety-preserving (Lemma 11), ID^C should be safe for ID. Since it isn't, $\overline{\kappa}^C$ cannot be safe. $\qquad \square$

A.4 Safe Approximation of ID (Theorem 6)

Lemma 10. *If q samples x and y independently from models in C, then $\kappa^C(x \mid y)$ is a 2-safe approximation of $-\log m(x \mid y)$ against q.*

Proof. Let q sample x from p_r and y from p_s.

$$p_q(-\log m^C(x \mid y) + \log m(x \mid y) \geq k) \quad = p_q(m(x \mid y)/m^C(x \mid y) \geq 2^k)$$
$$\leq 2^{-k} E\left[m(x \mid y)/m^C(x \mid y)\right] \quad = 2^{-k} \sum_{x,y} p_s(y) m(x \mid y) \frac{p_r(x)}{m^C(x \mid y)}$$
$$\leq c2^{-k} \sum_{x,y} p_s(y) m(x \mid y) \frac{m^C(x \mid y)}{m^C(x \mid y)} \quad \leq c2^{-k} \sum_{x,y} p_s(y) m(x \mid y) \leq c2^{-k}. \qquad \square$$

Since m and K mutually dominate, $-\log m^C$ is 2-safe for $K(x \mid y)$, as is $\bar{K}(x \mid y)$.

Lemma 11. *If f_a is safe for f against q, and g_a is safe for g against q, then* $\max(f_a, g_a)$ *is safe for* $\max(f, g)$ *against q.*[5]

Proof. We first partition $\mathbb{B}$ into sets A_k and B_k:

$A_k = \{x : f_a(x) - f(x) \geq k \vee g_a - g(x) \geq k\}$ Since both f_a and g_a are safe, we
know that $p_q(A_k)$ will be bounded above by the sum of two inverse exponentials in k, which from a given k_0 is itself bounded by an exponential in k.

$B_k = \{x : f_a(x) - f(x) < k \wedge g_a - g(x) < k\}$ We want to show that B contains
no strings with error over k. If, for a given x the left and right max functions in $\max(f_a, g_a) - \max(f, g)$ select the outcome from matching functions, and the error is below k by definition. Assume then, that a different function is selected on each side. Without loss of generality, we can say that $\max(f_a, g_a) = f_a$ and $\max(f, g) = g$. This gives us: $\max(f_a, g_a) - \max(f, g) = f_a - g \leq f_a - f \leq k$.

We now have $p(B_k) = 0$ and $p(A_k) \leq cb^{-k}$, from which the theorem follows.

$\square$

Corollary 2 ID^C *is a safe approximation of* ID *against sources that sample x and y independently from models in C.*

A.5   Safe Approximation of NID (Theorem 7)

Lemma 12. *Let f and g be two functions, with f_a and g_a their safe approximations against adversary p_q. Let $h(x) = f(x)/g(x)$ and $h_a(x) = f_a(x)/g_a(x)$. Let $c > 1$ and $0 < \epsilon \ll 1$ be constants such that $p_q(f_a(x) \geq c) \leq \epsilon$ and $p_q(g_a(x) \geq c) \leq \epsilon$. We can show that for some $b > 1$ and $c > 0$*

$$p_q\left(\left|\frac{h(x)}{h_a(x)} - 1\right| \geq \frac{k}{c}\right) \leq cb^{-k} + 2\epsilon.$$

Proof. We will first prove the bound from above, using f_a's safety, and then the bound from below using g_a's safety.

$$p_q\left(\frac{h}{h_a} \leq 1 - \frac{k}{c}\right) \leq p_q\left(\frac{h}{h_a} \leq 1 - \frac{k}{c} \,\&\, c < f_a\right) + \epsilon \leq p_q\left(\frac{h}{h_a} \leq 1 - \frac{k}{f_a}\right) + \epsilon$$

$$= p_q\left(\frac{f}{f_a}\frac{g_a}{g} \leq 1 - \frac{k}{f_a}\right) + \epsilon \leq p_q\left(\frac{f}{f_a} \leq 1 - \frac{k}{f_a}\right) + \epsilon$$

$$= p_q\left(\frac{f + k}{f_a} \leq 1\right) + \epsilon = p_q\left(f_a - f \geq k\right) + \epsilon \leq c_f b_f^{-k} + \epsilon.$$

The other bound we prove similarly. Combining the two, we get

$$p_q\left(h/h_a \notin (k/c - 1, k/c + 1)\right) \leq c_f b_f^{-k} + c_g b_g^{-k} + 2\epsilon \leq c'b'^{-k} + 2\epsilon. \quad \square$$

Theorem 7 follows as a corollary.

---

[5] We will call such operations *safety preserving*.

Author Index